Lecture Notes in Mathematics

Volume 2300

This series reports on new developments in all areas of mathematics and their applications - quickly, informally and at a high level. Mathematical texts analysing new developments in modelling and numerical simulation are welcome. The type of material considered for publication includes:

1. Research monographs
2. Lectures on a new field or presentations of a new angle in a classical field
3. Summer schools and intensive courses on topics of current research.

Texts which are out of print but still in demand may also be considered if they fall within these categories. The timeliness of a manuscript is sometimes more important than its form, which may be preliminary or tentative.

Titles from this series are indexed by Scopus, Web of Science, Mathematical Reviews, and zbMATH.

More information about this series at https://link.springer.com/bookseries/304

Takuro Mochizuki

Periodic Monopoles and Difference Modules

Springer

Takuro Mochizuki
Research Institute for Mathematical Sciences
Kyoto University
Kyoto, Japan

ISSN 0075-8434 ISSN 1617-9692 (electronic)
Lecture Notes in Mathematics
ISBN 978-3-030-94499-5 ISBN 978-3-030-94500-8 (eBook)
https://doi.org/10.1007/978-3-030-94500-8

Mathematics Subject Classification: 53C07, 58E15, 14D21, 81T13

This Springer imprint is published by the registered company Springer Nature Switzerland AG
The registered company address is: Gewerbestrasse 11, 6330 Cham, Switzerland

Dedicated to Professor Carlos Simpson on the occasion of his 60th birthday.

Preface

In this monograph, we shall study the relationship between polystable parabolic difference modules and singular monopoles on $\mathbb{R}^3$ with one periodicity. It is a new equivalence between objects in algebraic geometry and in differential geometry.

On one hand, a monopole on $\mathbb{R}^3$ is a vector bundle E with a Hermitian metric h, a unitary connection ∇ and an anti-Hermitian endomorphism ϕ satisfying the Bogomolny equation $F(\nabla) = *\nabla\phi$, where $F(\nabla)$ denotes the curvature of ∇, and $*$ is the Hodge star operation. Recall that the Bogomolny equation is a dimensional reduction of the anti-self duality equation in the gauge theory, and that it is a non-abelian generalization of Maxwell's equation under some constraints. A monopole (E, h, ∇, ϕ) is called periodic if it is invariant under the translation by a vector. Similarly, it is called doubly periodic (resp. triply periodic) if it is invariant under the translation by two (resp. three) linearly independent vectors. More generally, we shall consider a monopole (E, h, ∇, ϕ) which can be singular at a discrete subset S, i.e., (E, h, ∇, ϕ) is defined on $\mathbb{R}^3 \setminus S$. We should impose some asymptotic conditions to the monopoles, called of GCK-type.

On the other hand, for $\varrho \in \mathbb{C}$, a ϱ-difference module means, in this monograph, a finite dimensional $\mathbb{C}(y)$-vector space $\boldsymbol{V}$ equipped with a $\mathbb{C}$-linear automorphism Φ^* such that $\Phi^*(fs) = f(y+\varrho)\Phi^*(s)$ for any $f \in \mathbb{C}(y)$ and $s \in \boldsymbol{V}$. If $\varrho = 0$, it is just a $\mathbb{C}(y)$-vector space with a $\mathbb{C}(y)$-linear automorphism. We shall introduce the notion of *parabolic structure* for ϱ-difference modules in this monograph, as we consider a parabolic structure for a vector bundle on a punctured Riemann surface. A ϱ-difference module $\boldsymbol{V}$ equipped with a parabolic structure is called a parabolic ϱ-difference module, and denoted by $\boldsymbol{V}_*$. We obtain the number $\deg(\boldsymbol{V}_*) \in \mathbb{R}$ called the degree, which is an analogue of the degree of a vector bundle on a compact Riemann surface. The number $\mu(\boldsymbol{V}_*) = \deg(\boldsymbol{V}_*)/\dim_{\mathbb{C}(y)} \boldsymbol{V}$ is called the slope. Any ϱ-difference submodule $\boldsymbol{V}' \subset \boldsymbol{V}$ is naturally enhanced to a parabolic ϱ-difference submodule $\boldsymbol{V}'_* \subset \boldsymbol{V}_*$. In this situation, it is standard in algebraic geometry to define that $\boldsymbol{V}_*$ is stable if the slope of $\boldsymbol{V}'_*$ is strictly smaller than the slope of $\boldsymbol{V}_*$ for any ϱ-difference proper submodule $\boldsymbol{V}'$ of $\boldsymbol{V}$, and that $\boldsymbol{V}_*$ is polystable if it is a direct sum of stable ones $\boldsymbol{V}_{i*}$ with the same slope.

It is our purpose in this monograph to show that there exist equivalences between periodic singular monopoles and polystable parabolic difference modules of degree 0. We also study the triply periodic case [67] and the doubly periodic case [70] in the other papers. I hope that these studies will provide us with one of the starting points of a new interesting investigation for difference modules from the viewpoint of the non-abelian Hodge theory, as in the case of the equivalences for harmonic bundles, flat bundles and Higgs bundles.

I am also partially motivated by the following basic question, which has attracted me for years. Let X_1 be a compact space, and let X_2 be a non-compact space. Then, how we can relate the asymptotic behaviour of good differential geometric objects on $X_1 \times X_2$ with the asymptotic behaviour of good differential geometric objects on X_2 which is obtained as the dimensional reduction? One of the purposes in this monograph is to pursue it in the case $X_1 = S^1$, $X_2 = \mathbb{R}^2$, monopoles on $S^1 \times \mathbb{R}^2$ and harmonic bundles on $\mathbb{R}^2$. Relatedly, it was interesting for me to obtain equivalences between difference modules and mini-holomorphic bundles in the formal level, and that it is useful to compare the asymptotic conditions for monopoles and harmonic bundles.

I originally studied the Nahm transforms between periodic monopoles and harmonic bundles on $\mathbb{P}^1$ [69], as a refinement of [13, 14], and I intended to prove the equivalences between periodic monopoles and parabolic difference modules as a consequence of the Kobayashi-Hitchin correspondence for wild harmonic bundles. However, the progress in the theory of Kobayashi-Hitchin correspondence on a non-compact space with infinite volume [66] allows us to go along a more direct route. Hence, I eventually omitted the Nahm transform from this monograph, but I still include an explanation that a parabolic structure of a filtered λ-flat bundle is transformed to a parabolic structure at finite place of a difference module, which was a little surprising to me.

As mentioned above, it is my most ambitious hope that this monograph will lead us to a new aspect of the non-abelian Hodge theory. More modestly, I hope that this monograph will be useful in the mathematical study of monopoles and the differential geometric study of difference modules. I also expect that several concepts in this monograph will be significant for the further studies of monopoles and difference modules. For example, the notion of parabolic structure seems essential in the algebraic geometric study of the moduli space of difference modules.

Kyoto, Japan Takuro Mochizuki
November 2021

Acknowledgements

It is my pleasure to thank Hiraku Nakajima whose excellent lectures attracted me to the study of monopoles. A part of this study was done during my stay at the University of Melbourne, and I am grateful to Kari Vilonen and Ting Xue for their excellent hospitality and support. I appreciate Carlos Simpson whose works provide the most important foundation with this study. During the study, I realized how deeply influenced I am by the ideas of Nigel Hitchin. This work is under the direct influence of the interesting works of Benoit Charbonneau and Jacques Hurtubise [11], and Sergey Cherkis and Anton Kapustin [13, 14]. I remember that Sergey asked me some questions in 2002 when we met at the Institute for Advanced Study, and I hope that this monograph is a partial answer after almost 20 years. I was inspired by a talk of Hurtubise in a conference held at the Tata Institute of Fundamental Research in 2009. I thank Maxim Kontsevich and Yan Soibelman for their comments and discussions. I am grateful to Claude Sabbah for his kindness and discussions on many occasions. I appreciate Masaki Tsukamoto for asking a question about parabolic structure of doubly periodic instantons, which encouraged me to study instantons and monopoles, not only harmonic bundles. I thank Szilard Szabo who first attracted my attention to Nahm transforms. I appreciate Motohico Mulase for discussions and encouragements. I am grateful to Masaki Yoshino for our discussions, which were particularly useful to keep my interest in monopoles. I am grateful to Indranil Biswas for inviting me several times To the Tata Institute of Fundamental Research, which was quite helpful for my study of monopoles. I appreciate Masa-Hiko Saito and Atsushi Moriwaki for their generous supports. I thank Yoshifumi Tsuchimoto and Akira Ishii for their constant encouragement. I am heartily grateful to the reviewers for their careful and patient readings and for their constructive comments to improve this manuscript. I thank the editorial board of Springer Lecture Notes in Mathematics and Ute McCrory for their helpful comments. Special thanks go to Pierre Deligne, Kenji Fukaya, William Fulton, David Gieseker, Akira Kono, Mikiya Masuda, Tomohide Terasoma and Michael Thaddeus for their supports in my early career.

My interest in "Kobayashi-Hitchin correspondence" was renewed when I made a preparation for a talk in the 16th Oka Symposium, which drove me to this study. I thank the organizers, particularly Junichi Matsuzawa and Ken-ichi Yoshikawa.

I wrote this monograph at the Research Institute for Mathematical Sciences, Kyoto University. I benefited from my visits to the University of Melbourne and Tata Institute of Fundamental Research for my study of monopoles. I thank the institutions for the excellent research environment. It was beneficial in improving this manuscript, to make preparations for the lecture series at Osaka University and Nagoya University, and for the talks at various conferences and workshops, for which I heartily thank the organizers.

I am partially supported by the Grant-in-Aid for Scientific Research (S) (No. 17H06127), the Grant-in-Aid for Scientific Research (S) (No. 16H06335), the Grant-in-Aid for Scientific Research (A) (No. 21H04429), the Grant-in-Aid for Scientific Research (C) (No. 15K04843), and the Grant-in-Aid for Scientific Research (C) (No. 20K03609), Japan Society for the Promotion of Science.

Contents

Chapter 1
Introduction

Abstract First, we shall explain the motivation of our study. In particular, we review some previous results for monopoles and some equivalences for harmonic bundles. Then, we shall closely explain the results of this monograph. Namely, we formulate parabolic difference modules, and how they are induced by periodic monopoles. It is the main result in this monograph that the procedure induces an equivalence between monopoles and parabolic difference modules. Finally, we shall also explain an outline of the study of the asymptotic behaviour of monopoles.

1.1 Background and Motivation

One of the most interesting themes in complex differential geometry is to find an equivalence between objects in differential geometry and in algebraic geometry. An object in differential geometry is defined as a solution of a system of non-linear partial differential equations. Therefore, it is usually difficult to prove even the existence of such an object. In contrast, it is much easier to construct objects in algebraic geometry. Moreover, we can expect to obtain the classification of such objects as an explicit description of the moduli space. It is one reason to study this kind of equivalences. Conversely, we expect to obtain strong consequences for algebraic objects from this type of equivalence. For example, it should imply that the moduli space of the algebraic objects is equipped with many interesting structures such as a hyperkähler structure, and that good properties of algebraic objects such as stability and semisimplicity are preserved by various operations such as the tensor product, the pull back and the push-forward. Therefore, there is a two-way benefit to obtain such an equivalence.

An interesting trend is concerned with a good *metric* of an algebraic vector bundle on the side of differential geometry, and a good algebraic property of the same bundle on the side of algebraic geometry. The most classical is the theorem of Narasimhan-Seshadri [72]; an algebraic vector bundle on a compact Riemann surface of degree 0 has a flat metric if and only if it is stable. Here, a metric of an algebraic vector bundle is called flat if the associated Chern connection is flat, which is an object in differential geometry. The stability condition is a good

T. Mochizuki, *Periodic Monopoles and Difference Modules*, Lecture Notes in Mathematics 2300, https://doi.org/10.1007/978-3-030-94500-8_1

property for an algebraic vector bundle defined in an algebraic way. The higher dimensional generalization is known as "a holomorphic vector bundle on a compact Kähler manifold is equipped with a Hermitian-Einstein metric if and only if it is polystable". It was first pursued by Kobayashi [45–47] and Hitchin [50] (see [35]). The "only if" part was established by Kobayashi [46, 47] and Lübke [55, 56], and the "if" part was established by Donaldson [20, 21], and Uhlenbeck and Yau [89].

Since then, such correspondences have been studied for vector bundles with an additional structure. One of the most fruitful is the equivalence between harmonic bundles on the side of differential geometry, and flat bundles, Higgs bundles and more generally λ-flat bundles ($\lambda \in \mathbb{C}$) on the side of algebraic geometry. It was established by Corlette [15], Donaldson [22], Hitchin [34] and Simpson [79]. It is not only interesting in its own right, but also provides a starting point of an exciting and tremendous research area, so called the non-abelian Hodge theory. (We shall review briefly the equivalence for singular harmonic bundles on a compact Riemann surface in Sect. 1.4.) Nowadays, the non-abelian Hodge theory is so huge, and influential to various other research areas. We just mention that the equivalence for wild harmonic bundles is in particular essentially useful in the proof of the decomposition theorem for semisimple algebraic holonomic D-modules by the projective push-forward, which was an important problem in algebraic analysis (see [64]).

In this monograph, we shall clarify the relationships between singular monopoles on $S^1 \times \mathbb{R}^2$, and polystable parabolic difference modules, which is a new equivalence between differential geometric objects and algebraic geometric objects. As we shall explain (in particular, see Sect. 1.5), it can be regarded as a variant of equivalences between harmonic bundles and λ-flat bundles mentioned above. We also study the triply periodic case [67] and the doubly periodic case [70]. Hopefully, these studies will be one of the starting points of a new investigation for monopoles and difference modules from the viewpoint of the non-abelian Hodge theory.

We should mention that there are many previous studies relating monopoles with algebraic geometric objects in various cases, by using 'scattering maps". The most classical is the work of Donaldson [19] and Hitchin [33] on the equivalence between SU(2)-monopoles on $\mathbb{R}^3$ with L^2-curvature and holomorphic maps from $\mathbb{P}^1$ to $\mathbb{P}^1$. Note that Hurtubise [36] clarified the role of the scattering maps in this equivalence. More recently, inspired by the works of Kapustin and Witten [44] and Norbury [74], Charbonneau and Hurtubise [11] obtained an equivalence between singular monopoles on the product of S^1 and a compact Riemann surface Σ, and bundle pairs on Σ satisfying a stability condition, by using the scattering maps. We shall briefly review the works in Sect. 1.3.

Our study is directly inspired by Charbonneau and Hurtubise [11], and the scattering maps are essentially used, too. However, there are also two main issues which we need to clarify in our particular situation. First, because we study singular monopoles on the non-compact space $S^1 \times \mathbb{R}^2$, it is fundamental to study their asymptotic behaviour around infinity. Note that it is essentially different from the asymptotic behaviour of monopoles on $\mathbb{R}^3$. Roughly speaking, up to the pull back by a ramified covering, a monopole on $S^1 \times \mathbb{R}^2$ is asymptotically close to a direct sum of the tensor product of monopoles of rank one and monopoles induced by wild harmonic bundles. (See Sect. 1.8 for a summary of our study of the asymptotic behaviour of monopoles on $S^1 \times \mathbb{R}^2$.) Second, we would like to emphasize that the equivalences depend on the twistor parameters. It is analogue to the fact that harmonic bundles are equivalent to flat bundles, Higgs bundles and more generally λ-flat bundles ($\lambda \in \mathbb{C}$). We note that even in the case of SU(2)-monopoles on $\mathbb{R}^3$ with L^2-curvature, we obtain an equivalence for each choice of a twistor parameter λ. However, the method proving the equivalence is independent of λ, and hence the dependence on λ has not been attracted. In the case of monopoles on $S^1 \times \mathbb{R}^2$, the corresponding algebraic objects are $2\sqrt{-1}\lambda$-difference modules, i.e., $\mathbb{C}(y)$-modules M equipped with a $\mathbb{C}$-linear automorphism Φ^* such that $\Phi^*(fs) = f(y + 2\sqrt{-1}\lambda)\Phi^*(s)$ for any $f \in \mathbb{C}(y)$ and $s \in M$. In particular, the properties of the algebraic objects are quite different in the cases $\lambda = 0$ and $\lambda \neq 0$, and we need more involved arguments for non-zero λ to prove equivalences. Relatedly, there are many specific issues. For example, we need to formulate precisely the notion of parabolic structure for such difference modules, it is more convenient to generalize "scattering map" to the notion of mini-holomorphic structure, it is useful to clarify the relation between mini-holomorphic bundles and λ-flat bundles, etc. The author hopes that our study on these specific issues will be useful for the further investigations of difference modules and monopoles.

1.2 Monopoles of GCK-Type

For $T > 0$, we set $S^1_T := \mathbb{R}/T\mathbb{Z}$. Let (t, w) denote the standard local coordinate system on $S^1_T \times \mathbb{C}$. We regard $S^1_T \times \mathbb{C}$ as a Riemannian manifold with the metric $dt\,dt + dw\,d\overline{w}$.

A periodic monopole is a monopole (E, h, ∇, ϕ) on $S^1_T \times \mathbb{C}$. Namely, E is a vector bundle on $S^1_T \times \mathbb{C}$ with a Hermitian metric h, a unitary connection ∇ and an anti-Hermitian endomorphism ϕ satisfying the Bogomolny equation

$$F(\nabla) - *\nabla\phi = 0, \tag{1.1}$$

where $F(\nabla)$ denote the curvature of ∇, and $*$ denotes the Hodge star operator. More precisely, we admit that the monopole may have isolated singularities at a finite subset $Z \subset S^1_T \times \mathbb{C}$, i.e., the monopole (E, h, ∇, ϕ) is defined on $(S^1_T \times \mathbb{C}) \setminus Z$. We impose the following conditions on the behaviour of (E, h, ∇, ϕ) around Z and $S^1_T \times \{\infty\}$.

- Each point of Z is Dirac type singularity of (E, h, ∇, ϕ).
- $F(\nabla) \to 0$ and $\phi = O(\log|w|)$ as $|w| \to \infty$.

In this paper, such monopoles are called of GCK-type (generalized Cherkis-Kapustin type).

Cherkis and Kapustin [13, 14] studied such monopoles under some more additional assumptions of genericity on the behaviour around Z and $S^1_T \times \{\infty\}$. In particular, they studied the Nahm transforms between such periodic monopoles and harmonic bundles on $(\mathbb{P}^1, \{0, \infty, p_1, \ldots, p_m\})$. Foscolo studied the deformation theory of such periodic monopoles in [25], and the gluing construction in [26]. More recently, Harland [30] classified SU(2)-monopoles in the case $Z = \emptyset$ in terms of line bundles with parabolic structure on spectral curves. See also [31, 57, 58] on the geometry of the moduli space of some type of periodic monopoles.

In this monograph, we shall study periodic monopoles of GCK-type. The first goal is to clarify their asymptotic behaviour around $S^1_T \times \{\infty\}$. Then, we shall establish that periodic monopoles are equivalent to difference modules.

There are two origins of this study. One is the classifications of monopoles in terms of holomorphic objects. The other is the non-abelian Hodge theory for harmonic bundles on compact Riemann surfaces. Let us briefly recall them in Sects. 1.3 and 1.4.

1.3 Previous Works on Monopoles and Algebraic Objects

There are several interesting and deep studies on the classification of monopoles in terms of algebraic objects.

The most classical and pioneering work is due to Donaldson and Hitchin. Roughly speaking, they obtained an equivalence between SU(2)-monopoles (E, h, ∇, ϕ) on $\mathbb{R}^3$ with finite energy $\int_{\mathbb{R}^3} |F(\nabla)|^2 < \infty$, and based holomorphic maps $\varphi : \mathbb{P}^1 \longrightarrow \mathbb{P}^1$, where φ is called based if $\varphi(0) = \infty$. More precisely, Hitchin [33] established that the Nahm transform induces an equivalence between monopoles and solutions of the Nahm equation, and Donaldson [19] obtained an equivalence between the solutions of the Nahm equation and based holomorphic maps $\varphi : \mathbb{P}^1 \longrightarrow \mathbb{P}^1$. Hurtubise [36] clarified a more direct approach to obtain based holomorphic maps from monopoles. (Atiyah [3] gave an analogue construction for monopoles on hyperbolic spaces.) We shall review a construction of a based holomorphic map from an SU(2)-monopole in Sect. 1.3.1. We remark that the periodic monopoles are excluded by the finite energy condition.

The result was generalized by Hurtubise and Murray [37–39] and Jarvis [42, 43] to the context of a more general compact Lie group G rather than SU(2). Namely, they obtained a classification of G-monopoles in terms of holomorphic maps from $\mathbb{P}^1$ to flag varieties associated to G.

The interest in monopoles was renewed by the work of Kapustin and Witten [44] on the geometric Langlands theory from a physics viewpoint. Inspired by their work, Norbury [74] established that any Hecke transform of a holomorphic bundle on a Riemann surface Σ is represented by a singular monopole on $]0, 1[\times \Sigma$ satisfying the Dirichlet condition at $t = 0$ and the Neumann condition at $t = 1$. Here, for $a < b$, we set $]a, b[:= \{a < t < b\}$. Charbonneau and Hurtubise [11] studied singular monopoles on $S^1 \times \Sigma$, and they established an equivalence between such singular monopoles and holomorphic bundles with a meromorphic automorphism satisfying a stability condition on Σ. Because our study is directly influenced by the work of Charbonneau and Hurtubise, we shall briefly review it in Sect. 1.3.2.

As noted above, we shall briefly recall the constructions of algebraic objects in the correspondences due to Donaldson-Hitchin, and Charbonneau-Hurtubise in Sects. 1.3.1 and 1.3.2, respectively. Though scattering maps are applied in the both cases, there are also different flavors. See also a remark in Sect. 1.3.3.

1.3.1 SU(2)-*Monopoles with Finite Energy on* $\mathbb{R}^3$

We recall an outline of the construction of based holomorphic maps from SU(2)-monopoles with finite energy on $\mathbb{R}^3$, where some of fundamental concepts have already appeared. We follow the explanation in [4, §2 and §16] and [36, 37] though we omit the detail. (See also [42].) Let (E, h, ∇, ϕ) be a monopole with finite energy on $\mathbb{R}^3 = \{(x_1, x_2, x_3) \in \mathbb{R}^3\}$ with the Euclidean metric $\sum dx_i^2$. Following [4], on the basis of the results in [41], we restrict ourselves to the case where the following asymptotic conditions are satisfied.

- $|\phi|_h = 1 - \frac{k}{2r} + O(r^{-2})$ as $r \to \infty$, where $r = \sqrt{\sum x_i^2}$, $|\phi|_h = -\frac{1}{2}\operatorname{Tr}(\phi^2)$ and k denotes a positive integer.
- $\frac{\partial |\phi|_h}{\partial \Omega} = O(r^{-2})$ as $r \to \infty$, where $\frac{\partial}{\partial \Omega}$ denotes the angular derivative.
- $|\nabla\phi|_h = O(r^{-2})$ as $r \to \infty$.

We identify $\mathbb{R}^3$ with $\mathbb{R}_t \times \mathbb{C}_z$ by setting $t = x_1$ and $z = x_2 + \sqrt{-1}x_3$. Note that $dt\,dt + dz\,d\bar{z} = \sum dx_i^2$. We set $E^t := E_{|\{t\}\times\mathbb{C}}$ for $t \in \mathbb{R}$. They are enhanced to holomorphic vector bundles by the differential operator $\nabla_{\bar{z}} = \frac{1}{2}(\nabla_{x_2} + \sqrt{-1}\nabla_{x_3})$. For any $t, t' \in \mathbb{R}$, we have the isomorphism $E^t \simeq E^{t'}$ obtained as the parallel transport with respect to the differential operator $\partial_t := \nabla_{x_1} - \sqrt{-1}\phi$. The isomorphisms are called the *scattering maps*. Because the Bogomolny equation implies the commutativity

$$[\partial_t, \nabla_{\bar{z}}] = 0, \tag{1.2}$$

the scattering maps are holomorphic. This is one of the key facts in the study of the relationship between monopoles and holomorphic or algebraic objects, which goes back to [32].

Remark 1.3.1 In our study of periodic monopoles, it is useful to formulate the integrability condition (1.2) as a mini-complex structure which is an analogue of a complex structure. (See Sect. 2.2 for mini-complex structure.) ■

There exist unitary frames $u_1^{\pm}, u_2^{\pm}$ of E such that the following holds.

- $\nabla(u_1^{\pm}, u_2^{\pm}) = (u_1^{\pm}, u_2^{\pm})(A_2^{\pm}\,dx_2 + A_3^{\pm}\,dx_3)$, i.e., $\nabla_{x_1} u_i^{\pm} = 0$. Moreover, $A_2^{\pm}, A_3^{\pm} \to 0$ as $x_1 \to \pm\infty$.
- Let $\Phi^{\pm}$ be the $\mathfrak{su}(2)$-valued functions determined by $\phi(u_1^{\pm}, u_2^{\pm}) = (u_1^{\pm}, u_2^{\pm})\Phi^{\pm}$. Then, as $x_1 \to \pm\infty$, $\Phi^{\pm}$ converge to a diagonal matrix whose (1, 1)-entry is $\sqrt{-1}$ and (2, 2)-entry is $-\sqrt{-1}$.

Let $(e_1^{\pm}, e_2^{\pm})$ denote the asymptotic gauge of E at $x_1 = \pm\infty$ induced by $(u_1^{\pm}, u_2^{\pm})$. They are well defined up to multiplications of complex numbers with absolute value 1 to $e_i^{\pm}$.

As explained in [32], there exists a frame (v_1, v_2) of E such that (i) $\partial_t v_i = 0$, (ii) $e^t t^{-k/2} v_1 \to e_1^+$ and $e^{-t} t^{k/2} v_2 \to e_2^+$ as $t \to \infty$. It turns out that $\nabla_{\overline{z}} v_1 = 0$. Indeed, by the commutativity (1.2), we obtain $\partial_t(\nabla_{\overline{z}} v_1) = 0$. Hence, we have the expression $\nabla_{\overline{z}} v_1 = \alpha_1(z) v_1 + \beta_1(z) v_2$, which implies

$$\nabla_{\overline{z}}(e^t t^{-k/2} v_1) = \alpha_1(z)(e^t t^{-k/2} v_1) + \beta_1(z) e^{2t} t^{-k} \cdot (e^{-t} t^{k/2} v_2).$$

Because $\nabla_{\overline{z}}(e^t t^{-k/2} v_1) \to 0$ as $t \to \infty$, we obtain $\alpha_1 = \beta_1 = 0$. Similarly, $\nabla_{\overline{z}} v_2 = \alpha_2(z) v_1$ holds for a function $\alpha_2(z)$. For any $t \in \mathbb{R}$, let $L^t \subset E^t$ denote the subbundle generated by $f^t_{+,1} := v_{1|\{t\}\times\mathbb{C}}$. It is characterized by the following property.

- For $(t_1, z_1) \in \mathbb{R} \times \mathbb{C}$ and $s \in L^{t_1}_{+|z_1} \subset E^{t_1}_{|z_1}$, let $\widetilde{s}$ denote the section of $E_{|\mathbb{R}\times\{z_1\}}$ determined by the conditions $\partial_t \widetilde{s} = 0$ and $\widetilde{s}_{|(t_1,z_1)} = s$. Then, $|\widetilde{s}|_h \to 0$ as $t \to \infty$.

Because $\nabla_{\overline{z}} f^t_{+,1} = 0$, L^t_+ is a holomorphic subbundle of E^t, and $f^t_{+,1}$ is a global holomorphic frame of L^t_+. Moreover, v_2 induces a global holomorphic frame $f^t_{+,2}$ of E^t/L^t_+. The scattering map $E^t \simeq E^{t'}$ induces the isomorphisms $L^t_+ \simeq L^{t'}_+$ and $E^t/L^t_+ \simeq E^{t'}/L^{t'}_+$, which preserve the distinguished frames.

By applying a similar consideration to the behaviour as $t \to -\infty$, we obtain a holomorphic subbundle $L^t_- \subset E^t$ characterized by the following condition.

- For $(t_1, z_1) \in \mathbb{R} \times \mathbb{C}$ and $s \in L^{t_1}_{-|z_1} \subset E^{t_1}_{|z_1}$, let $\widetilde{s}$ denote the section of $E_{|\mathbb{R}\times\{z_1\}}$ determined by the conditions $\partial_t \widetilde{s} = 0$ and $\widetilde{s}_{|(t_1,z_1)} = s$. Then, $|\widetilde{s}|_h \to 0$ as $t \to -\infty$.

Moreover, the line bundles L^t_- and E^t/L^t_- are equipped with the distinguished frames $f^t_{-,1}$ and $f^t_{-,2}$, respectively. The subbundles and the distinguished frames are preserved by the scattering maps.

There exists the naturally defined morphism $\gamma : L^t_- \longrightarrow E^t/L^t_+$. By the frames $f^t_{-,1}$ and $f^t_{+,2}$, we regard γ as an entire holomorphic function. It turns out that γ is a non-zero polynomial of degree k. As remarked in [4, §16], there uniquely exists a holomorphic section $\widetilde{f}^t_{+,2}$ of E^t such that (i) $\widetilde{f}^t_{+,2}$ induces $f^t_{+,2}$, (ii) for the expression $f^t_{-,1} = \gamma \widetilde{f}^t_{+,2} + \delta f^t_{+,1}$, δ is a polynomial with $\deg \delta < k$. Thus, we obtain the based holomorphic map $\varphi = \delta/\gamma : \mathbb{P}^1 \longrightarrow \mathbb{P}^1$. Indeed, because of the ambiguity of the choice of an asymptotic gauge (e_1^+, e_2^+), we obtain a family of based holomorphic maps parameterized by S^1. Donaldson and Hitchin proved that this procedure induces an equivalence between SU(2)-monopoles with finite energy and based holomorphic maps.

Remark 1.3.2 The construction also depends on the choice of an $\mathbb{R}$-linear isomorphism $\mathbb{R}^3 \simeq \mathbb{R} \times \mathbb{C}$ such that $\sum dx_i^2 = dt\,dt + dw\,d\overline{w}$. The choice of such an isomorphism corresponds to a twistor parameter. Hence, for each twistor parameter, we obtain a family of based holomorphic maps parameterized by S^1. ■

We may also formulate the associated algebraic object in terms of $\mathbb{C}[z]$-modules, where $\mathbb{C}[z]$ denotes the polynomial ring of the variable z. Let M be a free $\mathbb{C}[z]$-module of rank 2. Let $L_\pm \subset M$ be free $\mathbb{C}[z]$-submodules of rank 1 such that $M/L_\pm$ are also free $\mathbb{C}[z]$-modules of rank 1. Let $g_{\pm,1}$ be frames of $L_\pm$, and let g_2 be a frame of M/L_+. Assume that the induced map $L_- \longrightarrow M/L_+$ is expressed by a non-zero polynomial of degree k.

From the above monopole (E, h, ∇, ϕ), we obtain such $(M, L_\pm, g_{\pm,1}, g_2)$ as follows. Let $\mathcal{O}_{\mathbb{P}^1}(*\infty)$ denote the sheaf of meromorphic functions on $\mathbb{P}^1$ which may allow poles along ∞. We set $\mathcal{E} = \mathcal{O}_{\mathbb{P}^1}(*\infty) f^0_{+,1} \oplus \mathcal{O}_{\mathbb{P}^1}(*\infty) \widetilde{f}^0_{+,2}$. It is equipped with the filtrations $\mathcal{L}_+ = \mathcal{O}_{\mathbb{P}^1}(*\infty) f^0_{+,1} \subset \mathcal{E}$ and $\mathcal{L}_- = \mathcal{O}_{\mathbb{P}^1}(*\infty) f^0_{-,1} \subset \mathcal{E}$. We set $M := H^0(\mathbb{P}^1, \mathcal{E})$, $L_\pm := H^0(\mathbb{P}^1, \mathcal{L}_\pm)$, $g_{\pm,1} := f^0_{\pm,1}$ and $g_2 := f^0_{+,2}$. It is easy to see that the associated based holomorphic map φ is obtained from the associated $(M, L_\pm, g_{\pm,1}, g_2)$, and vice versa.

Remark 1.3.3 In our study of periodic monopoles, we prefer $\mathbb{C}[z]$-modules or $\mathcal{O}$-modules as algebraic objects because we can easily add more structures such as parabolic structures. ■

1.3.2 The Correspondence due to Charbonneau and Hurtubise

We also recall the study of Charbonneau and Hurtubise [11], by omitting the details. Let Σ be a compact Riemann surface equipped with a Kähler metric g_Σ. We set $S^1 = \mathbb{R}/T\mathbb{Z}$ for some $T > 0$, which is equipped with the standard metric $dt\, dt$. Let g denote the induced Riemannian metric of $S^1 \times \Sigma$. Let $Z = \{(t_i, p_i) \mid i = 1, \dots, N\} \subset S^1 \times \Sigma$ be a finite subset. For simplicity, we assume that $t_i \neq t_j$ and $p_i \neq p_j$ if $i \neq j$. We also assume $t_i \neq 0$.

Let (E, h, ∇, ϕ) be a monopole on $(S^1 \times \Sigma) \setminus Z$ such that each point of Z is a Dirac type singularity of (E, h, ∇, ϕ). (We shall review the Dirac type singularity of monopoles in the Euclidean case in Sect. 2.4.3.) We recall the way to obtain an algebraic object in this context. Let $p_\Sigma : S^1 \times \Sigma \longrightarrow \Sigma$ denote the projection. Let $\nabla^{0,1}_{|\Sigma} : E \longrightarrow E \otimes p_\Sigma^{-1} \Omega^{0,1}_\Sigma$ denote the induced differential operator. We set $\partial_t := \nabla_t - \sqrt{-1}\phi$. The Bogomolny equation implies the commutativity $[\partial_t, \nabla^{0,1}_{|\Sigma}] = 0$.

Let $Z_1 \subset S^1$ and $Z_2 \subset \Sigma$ denote the image by the projections of Z. For any $t \in S^1 \setminus Z_1$, we obtain the vector bundle $E^t = E_{|\{t\}\times\Sigma}$ on Σ, which is enhanced to a holomorphic bundle by $\nabla^{0,1}_{|\Sigma}$. As the parallel transport by ∂_t, we obtain the isomorphisms $\Phi_{t',t} : E^t_{|\Sigma\setminus Z_2} \simeq E^{t'}_{|\Sigma\setminus Z_2}$ for $t, t' \in S^1 \setminus Z_1$, called *the scattering maps*. By the commutativity $[\partial_t, \nabla^{0,1}_{|\Sigma}] = 0$, the maps are holomorphic. Under the assumption that each point of Z is a Dirac type singularity of (E, h, ∇, ϕ), it turns out that the scattering maps are at most meromorphic along Z_2. Thus, we obtain a holomorphic vector bundle E^0 with a meromorphic automorphism $\Phi_{T,0} : E^0_{|\Sigma\setminus Z_2} \simeq E^T_{|\Sigma\setminus Z_2} = E^0_{|\Sigma\setminus Z_2}$. Such a pair is called a bundle pair in [11].

Let $\mathcal{E}$ be a holomorphic vector bundle on Σ of rank r equipped with a meromorphic automorphism $\rho : \mathcal{E}_{|\Sigma\setminus Z_2} \simeq \mathcal{E}_{|\Sigma\setminus Z_2}$. For each $p_i \in Z_2$, there exist frames $v_1^{\pm}, \dots, v_r^{\pm}$ of $\mathcal{E}$ around p_i, and a decreasing sequence of integers $k_{i,1}, \dots, k_{i,r}$ such that $\rho(v_j^-) = z^{k_{i,j}} v_j^+$, where z denotes a holomorphic coordinate around p_i such that $z(p_i) = 0$. The tuple $(k_{i,j} \mid j = 1, \dots, r)$ is well defined for each $p_i \in Z_2$. We note that $\sum_{j=1}^r \sum_{i=1}^N k_{i,j} = 0$. Then, Charbonneau and Hurtubise introduced the degree for such a bundle pair $(\mathcal{E}, \rho)$, depending on Z:

$$\deg(\mathcal{E}, \rho) = \deg(\mathcal{E}) - \sum_{i=1}^{N} \frac{t_i}{T} \sum_{j=1}^{r} k_{i,j}.$$

Let $\mathcal{E}' \subset \mathcal{E}$ be a subbundle such that $\rho(\mathcal{E}'_{|\Sigma\setminus Z_2}) = \mathcal{E}'_{|\Sigma\setminus Z_2}$. Then, we obtain the induced meromorphic automorphism ρ' of $\mathcal{E}'$, and $\deg(\mathcal{E}', \rho')$ is defined similarly. We say that $(\mathcal{E}, \rho)$ is stable if

$$\frac{\deg(\mathcal{E}', \rho')}{\operatorname{rank} \mathcal{E}'} < \frac{\deg(\mathcal{E}, \rho)}{\operatorname{rank} \mathcal{E}}$$

for any proper subbundle $\mathcal{E}' \subset \mathcal{E}$ such that $\rho(\mathcal{E}'_{|\Sigma\setminus Z_2}) = \mathcal{E}'_{|\Sigma\setminus Z_2}$. We say that $(\mathcal{E}, \rho)$ is poly-stable if it is a direct sum $(\mathcal{E}, \rho) = \bigoplus(\mathcal{E}_\ell, \rho_\ell)$ such that (i) each $(\mathcal{E}_\ell, \rho_\ell)$ is stable, (ii) $\deg(\mathcal{E}_\ell, \rho_\ell)/\operatorname{rank}(\mathcal{E}_\ell) = \deg(\mathcal{E}, \rho)/\operatorname{rank}(\mathcal{E})$ for any ℓ. Charbonneau and Hurtubise proved that if $(\mathcal{E}, \rho)$ is induced by a monopole (E, ∇, h, ϕ) with Dirac type singularity, then $(\mathcal{E}, \rho)$ is polystable of degree 0. By using the fundamental result of Simpson in [79], they proved that this procedure induces an equivalence between monopoles and polystable bundle pairs of degree 0.

Remark 1.3.4 Let (E, h, ∇, ϕ) be a vector bundle E on $(S^1 \times \Sigma) \setminus Z$ with a Hermitian metric h, a unitary connection ∇ and an anti-Hermitian endomorphism ϕ. As a generalization of the Bogomolny equation, Charbonneau and Hurtubise studied the Hermitian-Einstein-Bogomolny equation

$$F(\nabla) - \sqrt{-1} c \omega_\Sigma \operatorname{id}_E = *\nabla\phi, \tag{1.3}$$

where ω_Σ denotes the Kähler form associated with g_Σ. If $c = 0$, it is the Bogomolny equation. ■

1.3.3 Remark

Though the scattering maps are efficiently used in the both constructions of algebraic objects in Sects. 1.3.1 and 1.3.2, there are also different flavors. Roughly speaking, in Sect. 1.3.1, we obtain a module with filtrations on Y from a monopole on $\mathbb{R} \times Y$, and in Sect. 1.3.2, we obtain a module with an automorphism on Y from a monopole on $S^1 \times Y$.

In this monograph, we construct a difference module with parabolic structure from a monopole depending on the twistor parameter λ. In the case $\lambda = 0$, our construction is exactly an analogue of that in Sect. 1.3.2 in the context of $\Sigma = \mathbb{C}$ though we need to study the additional issues caused by the non-compactness of Σ. In the case $\lambda \neq 0$, our construction is again similar to that in Sect. 1.3.2, but it is more complicated.

We also note that in the case $\lambda \neq 0$ it is also natural to expect to obtain another algebraic object as the associated "Betti" object, and the construction should be in some sense similar to the construction in Sect. 1.3.1 though we shall not discuss it in this monograph. Indeed, in our study of doubly periodic monopoles [70], two kinds of the constructions appear. From a monopole on the product of $\mathbb{R}$ and an elliptic curve, we construct a q^λ-difference module with parabolic structure depending on a twistor parameter λ [70, Theorem 1.5]. If $|q^\lambda| \neq 1$, or equivalently $|\lambda| \neq 1$, by the Riemann-Hilbert correspondence for q^λ-difference modules in the local case due to van der Put and Reversat [90] and Ramis, Sauloy and Zhang [77] and in the global case due to Kontsevich and Soibelman, the monopole also induces a locally free sheaf with some filtrations on the elliptic curve T^λ obtained as the quotient of $\mathbb{C}^*$ by the action of $(q^\lambda)^{\mathbb{Z}}$ [70, Theorem 1.7]. The construction of the q^λ-difference module from the monopole is similar to the construction in Sect. 1.3.2, and the construction of the locally free module with filtrations on T^λ is similar to the construction in Sect. 1.3.1.

1.4 Review of the Kobayashi-Hitchin Correspondences for λ-Flat Bundles

Recall that Simpson established the Kobayashi-Hitchin correspondence for tame harmonic bundles on non-compact curves [80]. It is our main goal to develop an analogue theory in the context of periodic monopoles. Hence, let us begin with a brief review on the theory of harmonic bundles on curves.

1.4.1 Harmonic Bundles and Their Underlying λ-Flat Bundles

Let C be any complex curve. Let $(E, \overline{\partial}_E)$ be a holomorphic vector bundle on C. Let θ be a Higgs field, i.e., a holomorphic section of $\mathrm{End}(E) \otimes \Omega^1_C$. Let h be a Hermitian metric of E. We obtain the Chern connection $\nabla_h = \overline{\partial}_E + \partial_{E,h}$ determined by $\overline{\partial}_E$ and h, and the adjoint $\theta_h^\dagger$ of θ with respect to h. The metric h is called harmonic if the Hitchin equation

$$[\overline{\partial}_E, \partial_{E,h}] + [\theta, \theta_h^\dagger] = 0 \tag{1.4}$$

is satisfied. A Higgs bundle with a harmonic metric is called a harmonic bundle. The Hitchin equation implies that the connection $\overline{\partial}_E + \theta_h^\dagger + \partial_{E,h} + \theta$ is flat. More generally, for any complex number λ, there exists the associated flat λ-connection. Indeed, we set $\mathbb{D}^\lambda := \overline{\partial}_E + \lambda\theta^\dagger + \lambda\partial_E + \theta$. As explained in [82, 83], it is a λ-connection in the sense of Deligne, i.e., it satisfies a twisted Leibniz rule $\mathbb{D}^\lambda(fs) = (\lambda\partial_C + \overline{\partial}_C)f \cdot s + f\mathbb{D}^\lambda(s)$ for any $f \in C^\infty(C, \mathbb{C})$ and $s \in C^\infty(C, E)$. The Hitchin equation means that $\mathbb{D}^\lambda$ is flat, i.e., $\mathbb{D}^\lambda \circ \mathbb{D}^\lambda = 0$. Hence, harmonic bundles have the underlying λ-flat bundles. We may define the concept of harmonic metrics for λ-flat bundles. A λ-flat bundle with a harmonic metric is equivalent to a Higgs bundle with a harmonic metric. See [63], for example.

1.4.2 Kobayashi-Hitchin Correspondences in the Smooth Case

Suppose that C is projective and connected. We set $\deg(F) = \int_C c_1(F)$ for any vector bundle F on C. A λ-flat bundle (V, ∇^λ) is called stable (semistable) if we obtain

$$\deg(V')/\operatorname{rank} V' < (\leq) \deg(V)/\operatorname{rank} V$$

for any λ-flat subbundle $(V', \nabla^\lambda) \subset (V, \nabla^\lambda)$ with $0 < \operatorname{rank}(V') < \operatorname{rank}(V)$. A λ-flat bundle (V, ∇^λ) is called polystable if it is a direct sum of stable λ-flat bundles $\bigoplus(V_i, \nabla^\lambda)$ such that $\deg(V)/\operatorname{rank} V = \deg(V_i)/\operatorname{rank} V_i$ for any i. Note that $\deg(V) = 0$ always holds if $\lambda \neq 0$.

The following is a fundamental theorem in the study of harmonic bundles on projective curves, due to Diederich-Ohsawa [17], Donaldson [22] and Hitchin [34], in the rank 2 case, and Corlette [15] and Simpson [79, 81, 82] in the higher rank case, or even in the higher dimensional case. (See also [63] for the case of general flat λ-connections.)

Theorem 1.4.1 *Suppose that C is projective and connected. A λ-flat bundle (V, ∇^λ) is polystable of degree 0 if and only if (V, ∇^λ) has a harmonic metric. If (V, ∇^λ) has two harmonic metrics h_j $(j = 1, 2)$, there exists a decomposition $(V, \nabla^\lambda) = \bigoplus(V_i, \nabla^\lambda)$ which is orthogonal with respect to both h_1 and h_2, such that $h_{1|V_i} = a_i \cdot h_{2|V_i}$ for some constants $a_i > 0$. In particular, if (V, ∇^λ) is stable, it has a unique harmonic metric up to the multiplication of positive constants.* ∎

Corollary 1.4.2 *For each λ, there exists a natural bijective correspondence between the equivalence classes of polystable λ-flat bundles of degree 0 and the equivalence classes of polystable Higgs bundles of degree 0, through harmonic bundles.* ∎

The Kobayashi-Hitchin correspondence in the theorem provides us with an interesting equivalence between objects in differential geometry and objects in algebraic geometry. It is an origin of the hyperkähler property of the moduli spaces. It is a starting point of the non-abelian Hodge theory of Simpson. (See [82].)

1.4.3 Tame Harmonic Bundles and Regular Filtered λ-Flat Bundles

Simpson studied a generalization of Theorem 1.4.1 to the context of harmonic bundles on quasi-projective curves in [80]. To state his result, let us recall the concepts of tame harmonic bundles and regular filtered λ-flat bundles.

Let Y be a neighbourhood of 0 in $\mathbb{C}$. Let $(E, \overline{\partial}_E, \theta, h)$ be a harmonic bundle on $Y \setminus \{0\}$. We obtain the holomorphic endomorphism f of E such that $\theta = f\,dz/z$. The harmonic bundle is called tame on $(Y, \{0\})$ if the following holds.

- Let $\det(t\,\mathrm{id}_E - f) = t^{\mathrm{rank}\,E} + \sum_{j=0}^{\mathrm{rank}\,E-1} a_j(z)t^j$ be the characteristic polynomial of f. Then, a_j are holomorphic at $z = 0$.

Let us explain the notion of regular filtered λ-flat bundles on $(Y, \{0\})$. Let $\mathcal{O}_Y(*0)$ denote the sheaf of meromorphic functions on Y which may have poles along 0. Recall that a filtered bundle on $(Y, \{0\})$ is a locally free $\mathcal{O}_Y(*0)$-module $\mathcal{V}$ of finite rank with an increasing sequence of locally free $\mathcal{O}_Y$-submodules $\mathcal{P}_a\mathcal{V} \subset \mathcal{V}$ $(a \in \mathbb{R})$ satisfying the following conditions.

- $\mathcal{P}_a\mathcal{V}$ $(a \in \mathbb{R})$ are lattices of $\mathcal{V}$, i.e., $\mathcal{P}_a\mathcal{V}(*\{0\}) = \mathcal{V}$.
- $\mathcal{P}_{a+n}\mathcal{V} = \mathcal{P}_a\mathcal{V}(n\{0\})$ for any $a \in \mathbb{R}$ and $n \in \mathbb{Z}$.
- For any $a \in \mathbb{R}$, there exists $\epsilon > 0$ such that $\mathcal{P}_{a+\epsilon}\mathcal{V} = \mathcal{P}_a\mathcal{V}$.

A regular filtered λ-flat bundle is a filtered bundle $\mathcal{P}_*\mathcal{V}$ with a λ-connection $\nabla^\lambda : \mathcal{V} \longrightarrow \mathcal{V} \otimes \Omega^1$ such that ∇^λ is logarithmic with respect to $\mathcal{P}_*\mathcal{V}$ in the sense of $\nabla^\lambda \mathcal{P}_a\mathcal{V} \subset \mathcal{P}_a\mathcal{V} \otimes \Omega^1(\log\{0\})$ for any $a \in \mathbb{R}$. Note that we obtain the finite dimensional $\mathbb{C}$-vector spaces $\mathrm{Gr}_a^{\mathcal{P}}(\mathcal{V}) := \mathcal{P}_a(\mathcal{V}) / \sum_{b<a} \mathcal{P}_b(\mathcal{V})$.

Let $(E, \overline{\partial}_E, \theta, h)$ be a tame harmonic bundle on $Y \setminus \{0\}$. For any complex number λ, we obtain the holomorphic vector bundle $(E, \overline{\partial}_E + \lambda\theta^\dagger)$ on $Y \setminus \{0\}$. Let E^λ denote the locally free $\mathcal{O}_{Y\setminus\{0\}}$-module obtained as the sheaf of holomorphic sections of $(E, \overline{\partial}_E + \lambda\theta^\dagger)$. It is equipped with a flat λ-connection $\mathbb{D}^\lambda$. For any open subset $U \ni 0$, let $\mathcal{P}^h E^\lambda(U)$ denote the space of holomorphic sections s of E^λ on $U \setminus \{0\}$ such that $|s|_h = O(|z|^{-N})$ for some N, where $|s|_h$ denotes the norm of s with respect to h. Thus, we obtain an $\mathcal{O}_Y(*0)$-module $\mathcal{P}^h E^\lambda$ as a meromorphic prolongation of E^λ. For any $a \in \mathbb{R}$, let $\mathcal{P}_a^h E^\lambda(U)$ denote the space of sections s of E^λ on $U \setminus \{0\}$ such that $|s|_h = O(|z|^{-a-\epsilon})$ for any $\epsilon > 0$. Thus, we obtain an increasing sequence of $\mathcal{O}_Y$-submodules $\mathcal{P}_a^h E^\lambda \subset \mathcal{P}^h E^\lambda$ $(a \in \mathbb{R})$. Simpson proved that $(\mathcal{P}_*^h E^\lambda, \mathbb{D}^\lambda)$ is a regular filtered λ-flat bundle in [80].

Remark 1.4.3 We prefer to consider filtered λ-flat bundle $(\mathcal{P}^h_* E^\lambda, \mathbb{D}^\lambda)$ on $(Y, 0)$ rather than $(E^\lambda, \mathbb{D}^\lambda)$ on $Y \setminus \{0\}$ to keep the information of the behaviour of h around 0. ■

Let C be a smooth connected projective curve with a finite subset $D \subset C$. The concept of tame harmonic bundles on (C, D) is defined in a natural way. Let $\mathcal{O}_C(*D)$ denote the sheaf of meromorphic functions on C which may have poles along D. A filtered bundle on (C, D) is a locally free $\mathcal{O}_C(*D)$-module $\mathcal{V}$ of finite rank with an increasing sequence of $\mathcal{O}_C$-locally free submodules $\mathcal{P}_{\boldsymbol{a}}\mathcal{V}$ $(\boldsymbol{a} = (a_P \mid P \in D) \in \mathbb{R}^D)$ such that the following holds.

- For any $P \in D$, take a small neighbourhood U_P of P in C. Then, $\mathcal{P}_{\boldsymbol{a}}\mathcal{V}_{|U_P}$ depends only on a_P, which we denote by $\mathcal{P}^{(P)}_{a_P}(\mathcal{V}_{|U_P})$.
- $\mathcal{P}^{(P)}_*(\mathcal{V}_{|U_P})$ is a filtered bundle on (U_P, P) in the above sense.

A regular filtered λ-flat bundle is a filtered bundle $\mathcal{P}_*\mathcal{V}$ with a λ-connection ∇^λ such that $\nabla^\lambda \mathcal{P}_{\boldsymbol{a}}\mathcal{V} \subset \mathcal{P}_{\boldsymbol{a}}\mathcal{V} \otimes \Omega^1_C(\log D)$ for any $\boldsymbol{a} \in \mathbb{R}^D$. Moreover, for any regular filtered λ-flat bundle $(\mathcal{P}_*\mathcal{V}, \nabla^\lambda)$ on (C, D), we define

$$\deg(\mathcal{P}_*\mathcal{V}) := \deg(\mathcal{P}_{\boldsymbol{0}}\mathcal{V}) - \sum_{P \in D} \sum_{-1 < b \le 0} b \dim_{\mathbb{C}} \mathrm{Gr}^{\mathcal{P}^{(P)}}_b(\mathcal{V}_{|U_P}).$$

Here, $\boldsymbol{0} = (0, \dots, 0) \in \mathbb{R}^D$. By using the degrees, we can define the stability condition and the polystability condition in the context of regular filtered λ-flat bundles in the natural way.

Any tame harmonic bundle on (C, D) naturally induces a regular filtered λ-flat bundle on (C, D) by the procedure explained above. Simpson established the following theorem in [80], which is the ideal generalization of Theorem 1.4.1 to the context of regular singular case. (See also [61, 63] for the case of λ-connections and the higher dimensional case.)

Theorem 1.4.4 (Simpson) *Suppose that C is projective and connected. The procedure induces a bijection between the equivalence classes of tame harmonic bundles on (C, D) and the equivalence classes of polystable regular filtered λ-flat bundles with degree* 0. ■

1.4.4 Wild Harmonic Bundles and Good Filtered λ-Flat Bundles

Let us give a complement on the generalization of Theorem 1.4.4 to the context of non-regular case, which was studied by Biquard-Boalch [8, 9] and Sabbah [78]. (See also [64, 68].)

Let Y be a neighbourhood of 0 in $\mathbb{C}$. A harmonic bundle $(E, \overline{\partial}_E, \theta, h)$ on $Y \setminus \{0\}$ is called wild on $(Y, 0)$ if the following holds.

- $a_j(z)$ are meromorphic at $z = 0$. (See Sect. 1.4.3 for the construction of a_j from $(E, \overline{\partial}_E, \theta, h)$.)

To explain the concept of good filtered λ-flat bundles, we make several preliminaries. First, we recall the concept of unramifiedly good filtered λ-flat bundle. A filtered bundle $\mathcal{P}_*\mathcal{V}$ with a λ-connection ∇^λ on $(Y, \{0\})$ is called unramifiedly good if there exist a subset $S \subset z^{-1}\mathbb{C}[z^{-1}]$ and the Hukuhara-Levelt-Turrittin decomposition

$$(\mathcal{P}_*\mathcal{V}, \nabla^\lambda)_{|\widehat{0}} = \bigoplus_{\mathfrak{a}\in S} (\mathcal{P}_*\widehat{\mathcal{V}}_\mathfrak{a}, \widehat{\nabla}^\lambda_\mathfrak{a}) \tag{1.5}$$

such that $\widehat{\nabla}^\lambda_\mathfrak{a} - d\mathfrak{a} \cdot \mathrm{id}_{\widehat{\mathcal{V}}_\mathfrak{a}}$ are logarithmic with respect to $\mathcal{P}_*\widehat{\mathcal{V}}_\mathfrak{a}$. Here, for any $\mathcal{O}_Y$-module $\mathcal{F}$, let $\mathcal{F}_{|\widehat{0}}$ denote the formal completion of the stalk $\mathcal{F}_0$ at 0, i.e., $\mathcal{F}_0 \otimes_{\mathcal{O}_{Y,0}} \mathbb{C}[\![z]\!]$.

Second, let us recall the pull back of filtered bundles. Take $q \in \mathbb{Z}_{\geq 1}$. Let $\varphi_q : \mathbb{C} \longrightarrow \mathbb{C}$ be the ramified covering given by $\varphi_q(\zeta) = \zeta^q$. We set $Y_q = \varphi_q^{-1}(Y)$. Let $\mathcal{P}_*\mathcal{V}$ be a filtered bundle on $(Y, \{0\})$. We obtain the locally free $\mathcal{O}_{Y_q}(*\{0\})$-module $\mathcal{V}' := \varphi_q^*\mathcal{V}$. We define

$$\mathcal{P}_a\mathcal{V}' := \sum_{n+qb\leq a} \zeta^{-n}\varphi_q^*\mathcal{P}_b\mathcal{V} \subset \mathcal{V}'.$$

Thus, we obtain a filtered bundle $\mathcal{P}_*\mathcal{V}'$ over $(Y_q, \{0\})$ which is called the pull back of $\mathcal{P}_*\mathcal{V}$, and denoted by $\varphi_q^*(\mathcal{P}_*\mathcal{V})$.

A filtered bundle $\mathcal{P}_*\mathcal{V}$ with a λ-connection ∇^λ is called good if there exists a ramified covering $\varphi_q : Y_q \longrightarrow Y$ as above for an appropriate q such that $\varphi_q^*(\mathcal{P}_*\mathcal{V}, \nabla^\lambda)$ is unramifiedly good.

Let $(E, \overline{\partial}_E, \theta, h)$ be a wild harmonic bundle on $(Y, 0)$. For any λ, we obtain the λ-flat bundle $(E^\lambda, \mathbb{D}^\lambda)$ on $Y \setminus \{0\}$ as above. We obtain the $\mathcal{O}_Y(*\{0\})$-module $\mathcal{P}^h E^\lambda$ and the increasing sequence of $\mathcal{O}_Y$-submodules $\mathcal{P}^h_a E^\lambda$ by the procedure explained in Sect. 1.4.3. As proved in [64], $(\mathcal{P}^h_* E^\lambda, \mathbb{D}^\lambda)$ is a good filtered λ-flat bundle.

Although we explained the concepts of wild harmonic bundles and good filtered λ-flat bundles on a neighbourhood of 0 in $\mathbb{C}$, they are naturally generalized to the context of any complex curve C with a discrete subset D, and any wild harmonic bundle $(E, \overline{\partial}_E, \theta, h)$ on (C, D) induces a good filtered λ-flat bundle $(\mathcal{P}_*\mathcal{E}^\lambda, \mathbb{D}^\lambda)$. The following is essentially proved in [9]. (See also [64, 68].)

Theorem 1.4.5 *Suppose that C is projective and connected. Then, the above procedure induces a bijection between the equivalence classes of wild harmonic bundles on (C, D) and the equivalence classes of polystable good filtered λ-flat bundles with degree 0.* ∎

1.5 Equivariant Instantons and the Underlying Holomorphic Objects

In Sects. 1.5.1–1.5.3, we explain our motivation to study the Kobayashi-Hitchin correspondences for periodic monopoles as analogue of the Kobayashi-Hitchin correspondences for harmonic bundles (see Sect. 1.4). The precise statements are postponed to Sects. 1.6 and 1.7. In Sect. 1.5.4, we recall that monopoles are regarded as harmonic bundles of infinite rank, which is a basic idea behind our study. In particular, it is significant when we study the asymptotic behaviour of periodic monopoles (see Sect. 1.8).

1.5.1 Instantons and the Underlying Holomorphic Bundles

Set $X := \mathbb{R}^4$ with the standard Euclidean metric $\sum_{i=1}^{4} dx_i^2$. An instanton on an open subset $U \subset X$ is a vector bundle E with a Hermitian metric h and a unitary connection ∇ satisfying the anti-self duality (ASD) equation:

$$F(\nabla) + *F(\nabla) = 0. \tag{1.6}$$

We take an $\mathbb{R}$-linear isomorphism $X \simeq \mathbb{C}^2 = \{(z, w)\}$ such that the metric is $dz\,d\overline{z} + dw\,d\overline{w}$, with which we regard X as a Kähler surface. The ASD equation (1.6) holds if and only if (i) the curvature is a $(1, 1)$-form with respect to the complex structure, and (ii) $\Lambda F(\nabla) = 0$. (See [48] for the operator Λ.) The $(0, 1)$-part of ∇ induces a holomorphic structure of E. Hence, from an instanton, we obtain a holomorphic vector bundle $(E, \overline{\partial}_E)$ with a Hermitian metric h satisfying $\Lambda F(h) = 0$, where $F(h)$ denote the curvature of the Chern connection determined by h and $\overline{\partial}_E$.

Recall that such complex structures on X are parameterized by $\mathbb{P}^1$. For each $\lambda \in \mathbb{P}^1 \setminus \{\infty\}$, we obtain the complex coordinate system $(\xi, \eta) = (z + \lambda\overline{w}, w - \lambda\overline{z})$ which induces the complex structure corresponding to λ. We obtain the complex manifold X^λ and the open subset U^λ. An instanton (E, h, ∇) on U induces a holomorphic vector bundle $(E^\lambda, \overline{\partial}_{E^\lambda})$ on U^λ.

Let Γ be a closed subgroup of $\mathbb{R}^4$, which naturally acts on $\mathbb{R}^4$ by the translation. Suppose that U is invariant under the Γ-action. If (E, h, ∇) is Γ-equivariant, then the underlying holomorphic bundles $(E^\lambda, \overline{\partial}_{E^\lambda})$ $(\lambda \in \mathbb{C})$ are also Γ-equivariant.

1.5.2 Instantons and Harmonic Bundles

Recall that the concept of harmonic bundles was discovered by Hitchin [34] as the 2-dimensional reduction of instantons. Namely, in the case $\Gamma = \mathbb{R}^2$, Γ-equivariant instantons on U are equivalent to harmonic bundles on U/Γ. Indeed, we may choose $\Gamma = \mathbb{C}_z \times \{0\} \subset \mathbb{C}_z \times \mathbb{C}_w$. We may regard $U_0 = U/\mathbb{C}_z$ as an open subset in $\mathbb{C}_w$. Let $p : U \longrightarrow U_0$ denote the projection. We obtain the induced vector bundle E_0 on U_0 with a $\mathbb{C}_z$-equivariant isomorphism $p^*E_0 \simeq E$. It is equipped with the induced metric h_0. The $\mathbb{C}_z$-equivariant differential operator $\nabla_{\overline{w}}$ on E naturally induces a holomorphic structure $\partial_{E_0,\overline{w}}$ on E_0. We obtain a holomorphic endomorphism f_0 on E_0 such that $\nabla_{\overline{z}}(p^*(s)) = p^*(f_0 s)$ for any $s \in C^\infty(U_0, E_0)$. Then, the equation $\Lambda F(h) = 0$ is reduced to the Hitchin equation (1.4) for the metric h_0 of the Higgs bundle $(E_0, \overline{\partial}_{E_0}, f_0 dw)$.

Note that $\mathbb{C}_z$-equivariant holomorphic vector bundles on U^λ are equivalent to λ-flat bundles on U_0. We can easily check that the $\mathbb{C}_z$-equivariant holomorphic bundle $(E^\lambda, \overline{\partial}_{E^\lambda})$ on U^λ corresponds to the λ-flat bundle $(E_0^\lambda, \mathbb{D}^\lambda)$ on U_0 associated with $(E, \overline{\partial}_E, \theta, h)$:

$$\begin{array}{ccc}
\begin{pmatrix} \mathbb{C}_z-\text{equivariant} \\ \text{instantons on } U \end{pmatrix} & \longrightarrow & \begin{pmatrix} \mathbb{C}_z-\text{equivariant holomorphic} \\ \text{vector bundles on } U^\lambda \end{pmatrix} \\
=\big\downarrow & & =\big\downarrow \\
\begin{pmatrix} \text{harmonic bundles} \\ \text{on } U_0 \end{pmatrix} & \longrightarrow & \begin{pmatrix} \lambda-\text{flat bundles} \\ \text{on } U_0 \end{pmatrix}.
\end{array}$$

Hence, the Kobayashi-Hitchin correspondence between harmonic bundles and λ-flat bundles in the case $X = \mathbb{P}^1$ can be regarded as a correspondence between $\mathbb{C}_z$-equivariant instantons and $\mathbb{C}_z$-equivariant holomorphic vector bundles on $p^{-1}(\mathbb{C}_w \setminus D)$, under some natural assumptions on the boundary behaviour of the objects.

This kind of problems have been studied in various cases of equivariant instantons, sometimes in relation with the Nahm transforms. For example, see [10, 18, 19, 38, 39, 42, 43, 65, 85], etc. It is useful for the classification of equivariant instantons because holomorphic objects are easier to study. Conversely, it is useful for the study of deeper aspects of such holomorphic objects, for example the hyperkähler property of the moduli spaces. To establish the correspondence, we need a careful study on the asymptotic behaviour of equivariant instantons depending on Γ, and hence there seem to remain many things to be clarified mathematically.

1.5.3 *Instantons and Monopoles*

It is our purpose in this monograph is to study the issue in the case $\Gamma \simeq \mathbb{R} \times \mathbb{Z}$. We recall that monopoles are the 1-dimensional reduction of instantons. Indeed, if $\Gamma \simeq \mathbb{R} \times \mathbb{Z}$, then Γ-equivariant instantons on U are equivalent to monopoles on U/Γ. We can choose the coordinate system (x_1, x_2, x_3, x_4) as $\Gamma = \{(x_1, nT, 0, 0) \mid x_1 \in \mathbb{R}, n \in \mathbb{Z}\}$ for some $T > 0$. We may regard $U_1 = U/\Gamma$ as an open subset of

$(\mathbb{R}/T\mathbb{Z}) \times \mathbb{R}^2_{x_3,x_4}$. Let $p_1 : U \longrightarrow U_1$ denote the projection. If (E, h, ∇) is a Γ-equivariant tuple of a vector bundle E with a Hermitian metric and a unitary connection ∇ on U, we obtain the vector bundle E_1 on U_1 with a Γ-equivariant isomorphism $p_1^*(E_1) \simeq E$. We obtain the induced Hermitian metric h_1 of E_1. There exist the unitary connection ∇_1 and the anti-Hermitian endomorphism ϕ of (E_1, h_1) such that $\nabla = p_1^*(\nabla) + p_1^*(\phi)\, dx_1$. Then, the ASD equation (1.6) holds if and only if (E, h, ∇, ϕ) is a monopole.

We shall introduce the notion of mini-complex structure of a 3-dimensional manifold (see Sect. 2.2 for mini-complex structure). We also introduce the notion of mini-holomorphic bundle on a mini-complex manifold, which is essentially a vector bundle equipped with two commuting differential operators as in Sects. 1.3.1 and 1.3.2. (See Sect. 2.3 for mini-holomorphic bundles.) In our situation, we obtain the mini-complex manifold U_1^λ as U_1 equipped with the mini-complex structure induced by the complex structure of U^λ. (See Sect. 2.7.1.) A Γ-equivariant holomorphic vector bundle on U^λ is equivalent to a mini-holomorphic bundle on U_1^λ. Moreover, the Γ-equivariant holomorphic bundle $(E^\lambda, \overline{\partial}_{E^\lambda})$ underlying (E, h, ∇) is equivalent to the mini-holomorphic bundle $(E_1^\lambda, \overline{\partial}_{E_1^\lambda})$ underlying the monopole $(E_1, h_1, \nabla_1, \phi)$:

$$\begin{array}{ccc}
\begin{pmatrix} \Gamma-\text{equivariant} \\ \text{instantons on } U \end{pmatrix} & \longrightarrow & \begin{pmatrix} \Gamma-\text{equivariant holomorphic} \\ \text{vector bundles on } U^\lambda \end{pmatrix} \\
=\big\downarrow & & =\big\downarrow \\
\left(\text{monopoleson } U_1\right) & \longrightarrow & \begin{pmatrix} \text{mini-holomorphic} \\ \text{bundles on } U_1^\lambda \end{pmatrix}.
\end{array}$$

Very roughly, when U_1 is the complement of a finite subset in $(\mathbb{R}/T\mathbb{Z}) \times \mathbb{R}^2$, our Kobayashi-Hitchin correspondence for periodic monopoles are equivalences between monopoles on U_1 and mini-holomorphic bundles on U_1^λ satisfying some conditions, and the latter objects are translated to difference modules. We shall explain our correspondences for more details in Sects. 1.6 and 1.7.

1.5.4 Instantons and Monopoles as Harmonic Bundles of Infinite Rank

1.5.4.1 Instantons as Harmonic Bundles of Infinite Rank

If $U = \mathbb{C}_z \times U_0 \subset \mathbb{C}_z \times \mathbb{C}_w$, $\mathbb{C}_z$-equivariant harmonic bundles on U are equivalent to harmonic bundles on U_0, as mentioned in Sect. 1.5.2. It also implies that we may regard instantons on U as harmonic bundles of infinite rank on U_0, as we shall explain below. Let (E, ∇, h) be an instanton on U. Let $\mathcal{E}^\infty$ denote the sheaf of C^∞-sections of E on U. Let $C^\infty_{U_0}$ denote the sheaf of C^∞-functions on U_0. Let p :

$U \longrightarrow U_0$ denote the projection, and let $p_!$ denote the proper push-forward functor for sheaves with respect to p. We obtain the $\mathcal{C}^\infty_{U_0}$-module $p_!\mathcal{E}^\infty$ on U_0. We regard $p_!\mathcal{E}^\infty$ as a C^∞-vector bundle on U_0 of infinite rank. The metric h and the integration induces a Hermitian metric $p_!h$ on $p_!\mathcal{E}^\infty$. We may regard $\nabla_{\overline{w}}$ as a holomorphic structure $\partial_{p_!\mathcal{E}^\infty,\overline{w}}$ of $p_!\mathcal{E}^\infty$. We also have the differential operators $\partial_{p_!\mathcal{E}^\infty,w}$ on $p_!\mathcal{E}^\infty$ induced by ∇_w. Then, $\partial_{p_!\mathcal{E}^\infty,\overline{w}}\,d\overline{w} + \partial_{p_!\mathcal{E}^\infty,w}\,dw$ is a connection of $p_!\mathcal{E}^\infty$ which is unitary with respect to $p_!h$. The differential operator $f = \nabla_{\overline{z}}$ on $p_!\mathcal{E}^\infty$ is $\mathcal{C}^\infty_{U_0}$-linear. Because $[\nabla_{\overline{z}}, \nabla_{\overline{w}}] = 0$, we may regard f as a holomorphic endomorphism of $(p_!\mathcal{E}^\infty, \partial_{p_!\mathcal{E}^\infty,\overline{w}})$, and we obtain the Higgs field $\theta = f\,dw$. We also have the $\mathcal{C}^\infty_{U_0}$-linear endomorphism $f^\dagger = -\nabla_{\overline{z}}$. We may regard $f^\dagger$ as the adjoint of f with respect to $p_!h$. We set $\theta^\dagger = f^\dagger\,d\overline{w}$. Let F denote the curvature of the connection $\partial_{p_!\mathcal{E}^\infty,\overline{w}}\,d\overline{w} + \partial_{p_!\mathcal{E}^\infty,w}\,dw$. Then, the equation $[\nabla_{\overline{w}}, \nabla_w] + [\nabla_{\overline{z}}, \nabla_z] = 0$ is exactly the equation $F + [\theta, \theta^\dagger] = 0$. In this sense, $(p_!\mathcal{E}^\infty, \theta, p_!h)$ is a harmonic bundle of infinite rank on U_0.

1.5.4.2 The Underlying λ-Flat Bundles of Infinite Rank

Let us pursue this analogy further. For a complex number λ, if $(\xi, \eta) = (z+\lambda\overline{w}, w-\lambda\overline{z})$, the following holds:

$$(1+|\lambda|^2)\partial_{\overline{\xi}} = \lambda\partial_w + \partial_{\overline{z}}, \quad (1+|\lambda|^2)\partial_{\overline{\eta}} = \partial_{\overline{w}} - \lambda\partial_z.$$

Hence, for any section s of $p_!\mathcal{E}^\infty$ and $g \in \mathcal{C}^\infty_{U_0}$, we have the following equalities:

$$(1+|\lambda|^2)\nabla_{\overline{\xi}}(gs) = g(1+|\lambda|^2)\nabla_{\overline{\xi}}(s) + \lambda\partial_w(g)\cdot s,$$

$$(1+|\lambda|^2)\nabla_{\overline{\eta}}(gs) = g(1+|\lambda|^2)\nabla_{\overline{\eta}}(s) + \partial_{\overline{w}}(g)\cdot s.$$

We consider the differential operator $\nabla^\lambda = (1+|\lambda|^2)\nabla_{\overline{\xi}}\,dw + (1+|\lambda|^2)\nabla_{\overline{\eta}}\,d\overline{w}$ on $p_!\mathcal{E}^\infty$, which is naturally regarded as a λ-connection. The commutativity of $\nabla_{\overline{\xi}}$ and $\nabla_{\overline{\eta}}$ is equivalent to the flatness of the λ-connection ∇^λ, i.e., $\nabla^\lambda \circ \nabla^\lambda = 0$. Moreover, we have $\nabla^\lambda = \partial_{p_!\mathcal{E}^\infty,\overline{w}}\,d\overline{w} + \lambda\theta^\dagger + \lambda\partial_{p_!\mathcal{E}^\infty,w}\,dw + \theta$. Hence, we may regard the induced holomorphic bundle $(E^\lambda, \overline{\partial}_{E^\lambda})$ on U^λ as the λ-flat bundle of infinite rank on U_0 associated with the harmonic bundle $(p_!\mathcal{E}^\infty, \theta, p_!h)$ of infinite rank.

1.5.4.3 Monopoles as Harmonic Bundles of Infinite Rank

Let us consider the case where $\Gamma = \{(x_1+\sqrt{-1}nT, 0) \mid x_1 \in \mathbb{R}, n \in \mathbb{Z}\} \subset \mathbb{C}_z \times \mathbb{C}_w$. Suppose that $U = \mathbb{C}_z \times U_0$ for an open subset $U_0 \subset \mathbb{C}_w$. Set $U_1 = U/\Gamma \subset (\mathbb{R}/T\mathbb{Z}) \times \mathbb{C}_w$. Because $(\mathbb{R}/T\mathbb{Z})$-equivariant monopoles on U_1 are equivalent to $\mathbb{C}_z$-equivariant instantons on U, they are equivalent to harmonic bundles on U_0. It means that we may regard monopoles on U_1 as harmonic bundles on U_0 of infinite rank. We shall explain the more precise relation between monopoles on U_1 and harmonic bundles on U_0 in Sect. 2.6, the relation of the mini-holomorphic bundles on U_1^λ and λ-flat bundles on U_0 in Sect. 2.8.1 (see also Sects. 2.8.3, 3.6, and 4.5), and the compatibility of the relations in Sect. 2.8.2. We shall give a complement on the compatibility in Sect. 2.9.5.

1.6 Difference Modules with Parabolic Structure

As an algebraic objects corresponding to periodic monopoles, we introduce the concept of parabolic structure for difference modules. Let $\mathbb{C}(y)$ denote the field of rational functions of the variable y. Let $\mathbb{C}[y]$ denote the ring of polynomials. For $y_0 \in \mathbb{C}$, $\mathbb{C}[\![y - y_0]\!]$ denote the ring of the formal power series of $y - y_0$, i.e., $\mathbb{C}[\![y - y_0]\!] = \left\{ \sum_{j=0}^{\infty} a_j (y - y_0)^j \,\middle|\, a_j \in \mathbb{C} \right\}$. Let $\mathbb{C}(\!(y - y_0)\!)$ denote the fraction field of $\mathbb{C}[\![y - y_0]\!]$, i.e., $\mathbb{C}(\!(y - y_0)\!)$ is the field of formal Laurent power series of $y - y_0$. Similarly, let $\mathbb{C}[\![y^{-1}]\!]$ denote the ring of the formal power series of y^{-1}, and let $\mathbb{C}(\!(y^{-1})\!)$ denote the ring of the formal Laurent power series of y^{-1}.

1.6.1 *Difference Modules*

Take $\varrho \in \mathbb{C}$. Let Φ^* be the automorphism of the field $\mathbb{C}(y)$ induced by $\Phi^*(y) = y + \varrho$. We prefer to regard it as the pull back of functions by the automorphism Φ of $\mathbb{C}$ or $\mathbb{P}^1$.

Definition 1.6.1 In this monograph, a difference module is a finite dimensional $\mathbb{C}(y)$-vector space $\boldsymbol{V}$ with a $\mathbb{C}$-linear automorphism Φ^* such that

$$\Phi^*(g\,s) = \Phi^*(g) \cdot \Phi^*(s)$$

for any $g(y) \in \mathbb{C}(y)$ and $s \in \boldsymbol{V}$. When we emphasize ϱ, it is called a ϱ-difference module. ■

Remark 1.6.2 We set $\mathcal{A} := \bigoplus_{n\in\mathbb{Z}} \mathbb{C}[y](\Phi^*)^n$. It is naturally a $\mathbb{C}$-vector space. We define the multiplication $\mathcal{A} \times \mathcal{A} \longrightarrow \mathcal{A}$ by $\sum a_n(y)(\Phi^*)^n \cdot \sum b_m(y)(\Phi^*)^m = \sum a_n(y) b_m(y + n\varrho)(\Phi^*)^{m+n}$. Thus, $\mathcal{A}$ is an associative $\mathbb{C}$-algebra. As a difference module, it would be more natural to consider a finitely generated left $\mathcal{A}$-module M such that (i) M is a torsion-free $\mathbb{C}[y]$-module, (ii) $\dim_{\mathbb{C}(y)} \mathbb{C}(y) \otimes_{\mathbb{C}[y]} M < \infty$. The data of such finitely generated $\mathcal{A}$-modules are contained in the parabolic structure at finite place, which will be explained below. Indeed, let $\boldsymbol{V}$ be a difference module in the sense of Definition 1.6.1. For any $\mathbb{C}[y]$-free submodule $V \subset \boldsymbol{V}$ such that $\mathbb{C}(y) \otimes_{\mathbb{C}[y]} V = \boldsymbol{V}$, we obtain the $\mathcal{A}$-module $\mathcal{A} \cdot V \subset \boldsymbol{V}$ which satisfies the above finiteness conditions. ■

Remark 1.6.3 If $\varrho = 0$, the difference module is just a finite dimensional $\mathbb{C}(y)$-vector space with a $\mathbb{C}(y)$-linear automorphism. ■

1.6.2 Parabolic Structure of Difference Modules at Finite Place

Let $\boldsymbol{V}$ be a difference module.

Definition 1.6.4 (Definition 2.10.1) A parabolic structure of $\boldsymbol{V}$ at finite place is a tuple as follows.

- A $\mathbb{C}[y]$-free submodule $V \subset \boldsymbol{V}$ such that $V \otimes_{\mathbb{C}[y]} \mathbb{C}(y) = \boldsymbol{V}$.
- A function $m : \mathbb{C} \longrightarrow \mathbb{Z}_{\geq 0}$ such that $\sum_{x\in\mathbb{C}} m(x) < \infty$. We assume
$$V \otimes_{\mathbb{C}[y]} \mathbb{C}[y]_D = (\Phi^*)^{-1}(V) \otimes_{\mathbb{C}[y]} \mathbb{C}[y]_D,$$
where we set $D := \{x \in \mathbb{C} \mid m(x) > 0\}$, and let $\mathbb{C}[y]_D$ denote the localization of $\mathbb{C}[y]$ with respect to $y - x$ $(x \in D)$.
- A sequence of real numbers $0 \leq \tau_x^{(1)} < \cdots < \tau_x^{(m(x))} < 1$ for each $x \in \mathbb{C}$. If $m(x) = 0$, the sequence is assumed to be empty. The sequence is denoted by $\boldsymbol{\tau}_x$.
- Lattices $L_{x,i} \subset V \otimes_{\mathbb{C}[y]} \mathbb{C}(\!(y - x)\!)$ for $x \in \mathbb{C}$ and $i = 1, \ldots, m(x) - 1$. We formally set $L_{x,0} := V \otimes_{\mathbb{C}[y]} \mathbb{C}[\![y - x]\!]$ and $L_{x,m(x)} := (\Phi^*)^{-1} V \otimes \mathbb{C}[\![y - x]\!]$. The tuple of lattices is denoted by $\boldsymbol{L}_x$. Note that $V \otimes \mathbb{C}[\![y-x]\!] = (\Phi^*)^{-1}(V) \otimes \mathbb{C}[\![y - x]\!]$ if $m(x) = 0$. ■

Remark 1.6.5 We may regard a parabolic structure of finite place of a difference module as a reincarnation of a part of an ordinary parabolic structure of λ-flat bundles. See Sect. 2.10.4. ■

1.6.3 Good Parabolic Structure at ∞

Note that the automorphism Φ^* of $\mathbb{C}(y)$ induces an automorphism of $\mathbb{C}(\!(y^{-1})\!)$, which is also denoted by Φ^*. We set $\widehat{\boldsymbol{V}} := \boldsymbol{V} \otimes_{\mathbb{C}(y)} \mathbb{C}(\!(y^{-1})\!)$, which is a finite dimensional $\mathbb{C}(\!(y^{-1})\!)$-vector space with the induced $\mathbb{C}$-linear automorphism Φ^* such that $\Phi^*(fs) = \Phi^*(f)\Phi^*(s)$ for $f \in \mathbb{C}(\!(y^{-1})\!)$ and $s \in \widehat{\boldsymbol{V}}$. Such $(\widehat{\boldsymbol{V}}, \Phi^*)$ is called a formal difference module.

For any $p \in \mathbb{Z}_{\geq 1}$, we set $S(p) := \left\{ \sum_{j=1}^{p-1} \mathfrak{a}_j y^{-j/p} \,\middle|\, \mathfrak{a}_j \in \mathbb{C} \right\}$. According to the classification of formal difference modules [12, 76, 86], there exist $p \in \mathbb{Z}_{\geq 1}$ and a decomposition of the formal difference module

$$\widehat{\boldsymbol{V}} \otimes_{\mathbb{C}(\!(y^{-1})\!)} \mathbb{C}(\!(y^{-1/p})\!) = \bigoplus_{\omega\in p^{-1}\mathbb{Z}} \bigoplus_{\alpha\in\mathbb{C}^*} \bigoplus_{\mathfrak{a}\in S(p)} \left(\widehat{\boldsymbol{V}}_{p,\omega,\alpha,\mathfrak{a}}, \Phi^*\right), \tag{1.7}$$

such that each $\widehat{V}_{p,\omega,\alpha,\mathfrak{a}}$ has a $\mathbb{C}[\![y^{-1/p}]\!]$-lattice $L_{p,\omega,\alpha,\mathfrak{a}}$ satisfying

$$\big(\alpha^{-1}y^{\omega}\Phi^{*}-(1+\mathfrak{a})\,\mathrm{id}\big)L_{p,\omega,\alpha,\mathfrak{a}}\subset y^{-1}L_{p,\omega,\alpha,\mathfrak{a}}.$$

We note that in the case $\varrho = 0$ we can obtain the decomposition (1.7) from the generalized eigen decomposition of a linear map.

Definition 1.6.6 (Definition 3.3.14, Definition 3.3.16) A good parabolic structure of V at ∞ is a filtered bundle $\mathcal{P}_*\widehat{V}$ over $\widehat{V}$ which is good with respect to Φ^* in the following sense.

- There exists the decomposition $\varphi_p^*(\mathcal{P}_*\widehat{V}) = \bigoplus \mathcal{P}_*\widehat{V}_{p,\omega,\alpha,\mathfrak{a}}$, where $\varphi_p^*(\mathcal{P}_*\widehat{V})$ is the filtered bundle over $\widehat{V}\otimes\mathbb{C}(\!(y^{-1/p})\!)$ obtained as the pull back of $\mathcal{P}_*\widehat{V}$. (See Sect. 1.4.4 or Sect. 3.3.1 for the pull back of filtered bundles via a ramified covering.)
- $\big(\alpha^{-1}y^{\omega}\Phi^{*}-(1+\mathfrak{a})\,\mathrm{id}\big)\mathcal{P}_a\widehat{V}_{p,\omega,\alpha,\mathfrak{a}}\subset y^{-1}\mathcal{P}_a\widehat{V}_{p,\omega,\alpha,\mathfrak{a}}$ for any $a\in\mathbb{R}$ and for any $(\omega,\alpha,\mathfrak{a})$. ■

1.6.4 Parabolic Difference Modules

Definition 1.6.7 A parabolic difference module means a difference module equipped with a parabolic structure at finite place and a good parabolic structure at infinity. ■

Note that in the decomposition (1.7), the numbers

$$r(\omega) := \sum_{\alpha,\mathfrak{a}} \operatorname{rank}\widehat{V}_{p,\omega,\alpha,\mathfrak{a}} \tag{1.8}$$

are independent of p, and well defined for $\omega\in\mathbb{Q}$. Indeed, there exists the slope decomposition $\widehat{V} = \bigoplus_{\omega\in\mathbb{Q}}\mathcal{S}_{\omega}\widehat{V}$ as explained in Sect. 3.2 for which we obtain $\mathcal{S}_{\omega}(\widehat{V})\otimes_{\mathbb{C}(\!(y^{-1})\!)}\mathbb{C}(\!(y_p^{-1})\!) = \bigoplus_{\alpha,\mathfrak{a}}\widehat{V}_{p,\omega,\alpha,\mathfrak{a}}$ for any p as above. Hence, $r(\omega) = \dim_{\mathbb{C}}\mathcal{S}_{\omega}\widehat{V}$. For $\mathcal{P}_*\widehat{V}$ as in Definition 1.6.6, we obtain the decomposition $\mathcal{P}_*\widehat{V} = \bigoplus\mathcal{P}_*\mathcal{S}_{\omega}(\widehat{V})$.

1.6.5 Degree and Stability Condition

Consider a parabolic difference module

$$\boldsymbol{V}_* = (\boldsymbol{V}, (V, m, (\boldsymbol{\tau}_x, \boldsymbol{L}_x)_{x\in\mathbb{C}}), \mathcal{P}_*\widehat{V}).$$

Let $\mathcal{F}_V$ be the $\mathcal{O}_{\mathbb{P}^1}(*\infty)$-module associated with V. We obtain the filtered bundle $\mathcal{P}_*\mathcal{F}_V$ over $\mathcal{F}_V$ induced by $\mathcal{P}_*\widehat{V}$. If $m(x) > 0$, for each $i = 1, \dots, m(x)$, we set

$$\deg(L_{i,x}, L_{i-1,x}) := \dim_{\mathbb{C}}\left(L_{i,x}/(L_{i,x} \cap L_{i-1,x})\right) - \dim_{\mathbb{C}}\left(L_{i-1,x}/(L_{i,x} \cap L_{i-1,x})\right).$$

Then, we define

$$\deg(V_*) := \deg(\mathcal{P}_*\mathcal{F}_V) + \sum_{x \in \mathbb{C}} \sum_{i=1}^{m(x)} \left(1 - \tau_x^{(i)}\right) \deg\left(L_{i,x}, L_{i-1,x}\right) + \sum_{\omega \in \mathbb{Q}} \frac{\omega}{2} r(\omega). \tag{1.9}$$

Here, $r(\omega)$ are defined in (1.8). We also define

$$\mu(V_*) := \deg(V_*) \big/ \dim_{\mathbb{C}(y)} V. \tag{1.10}$$

Let V' be any difference submodule of V, i.e., V' is a $\mathbb{C}(y)$-subspace of V such that $\Phi^*(V') = V'$. By setting $V' := V' \cap V$ and $L'_{i,x} := L_{i,x} \cap (V' \otimes_{\mathbb{C}(y)} \mathbb{C}(\!(y-x)\!))$, we obtain a parabolic structure $\left(V', m, (\tau_x, L'_x)_{x \in \mathbb{C}}\right)$ at finite place of V'. By setting $\widehat{V}' := \mathbb{C}(\!(y^{-1})\!) \otimes_{\mathbb{C}(y)} V'$ and $\mathcal{P}_a\widehat{V}' := \widehat{V}' \cap \mathcal{P}_a\widehat{V}$, we obtain a good parabolic structure $\mathcal{P}_*\widehat{V}'$ at ∞. Thus, we obtain the following induced parabolic difference module from V':

$$V'_* = \left(V', (V', m, (\tau_x, L'_x)_{x \in \mathbb{C}}), \mathcal{P}_*\widehat{V}'\right).$$

Definition 1.6.8 A parabolic difference module V_* is called stable (resp. semistable) if $\mu(V'_*) < \mu(V_*)$ (resp. $\mu(V'_*) \leq \mu(V_*)$) for any difference submodule V' of V with $0 < \dim_{\mathbb{C}(y)} V' < \dim_{\mathbb{C}(y)} V$. It is called polystable if it is semistable and a direct sum of stable ones. ■

Remark 1.6.9 There exists the relation

$$\sum_{x \in \mathbb{C}} \sum_{i=1}^{m(x)} \deg(L_{i,x}, L_{i-1,x}) + \sum_{\omega \in \mathbb{Q}} \omega \cdot r(\omega) = 0. \tag{1.11}$$

Indeed, by the comparison of V and $(\Phi^*)^{-1}(V)$ in V, we obtain

$$\deg(\mathcal{P}_*(\mathcal{F}_{(\Phi^*)^{-1}(V)})) - \deg(\mathcal{P}_*\mathcal{F}_V) = \sum_{x,i} \deg(L_{i,x}, L_{i-1,x}).$$

There exists the isomorphism of $\mathcal{O}_{\mathbb{P}^1}(*\infty)$-modules $\Phi^*(\mathcal{F}_{(\Phi^*)^{-1}(V)}) \simeq \mathcal{F}_V$ induced by Φ^*. For the slope decomposition $\widehat{V} = \bigoplus \mathcal{S}_\omega(\widehat{V})$, it induces $\Phi^*(\mathcal{P}_a\mathcal{S}_\omega(\widehat{V})) \simeq$

$\mathcal{P}_{a-\omega}\mathcal{S}_\omega(\widehat{V})$. Hence, we obtain

$$\deg(\mathcal{P}_*\mathcal{F}_V) = \deg(\mathcal{P}_*\mathcal{F}_{(\Phi^*)^{-1}(V)}) + \sum_{\omega\in\mathbb{Q}} r(\omega)\cdot\omega.$$

Thus, we obtain (1.11). ■

1.6.6 *Easy Examples of Stable Parabolic Difference Modules (1)*

Let us mention some easy examples of parabolic difference modules. Needless to say, we can easily construct many more interesting examples. For simplicity, we consider the case $\varrho = 0$.

We take a non-empty finite subset $S \subset \mathbb{C}$ and a function $\ell : S \longrightarrow \mathbb{Z}$ such that $\sum_{a\in S}|\ell(a)| \neq 0$. We also assume that one of $\ell(a)$ is an odd integer. We set $P(y) := \prod_{a\in S}(y-a)^{\ell(a)} \in \mathbb{C}(y)$. We set $\boldsymbol{V} := \mathbb{C}(y)e_1 \oplus \mathbb{C}(y)e_2$ and $V := \mathbb{C}[y]e_1 \oplus \mathbb{C}[y]e_2$. We define the $\mathbb{C}(y)$-automorphism Φ^* of $\boldsymbol{V}$ by

$$\Phi^*(e_1,e_2) = (e_1,e_2)\cdot\begin{pmatrix} 0 & P(y) \\ 1 & 0 \end{pmatrix}.$$

Let $m : \mathbb{C} \longrightarrow \mathbb{Z}_{\geq 0}$ be given by $m(a) = 1$ $(a \in S)$ and $m(a) = 0$ $(a \notin S)$. We take $0 \leq \tau_a < 1$ for $a \in S$. Thus, we obtain a parabolic structure of $\boldsymbol{V}$ at finite place. Note the following equality for any $a \in S$:

$$\deg(L_{a,1}, L_{a,0}) = -\ell(a).$$

As for good parabolic structures at infinity, there are two cases where $\ell(\infty) := -\sum_{a\in S}\ell(a)$ is even or odd.

1.6.6.1 The Case Where $\ell(\infty)$ Is Even

Suppose $\ell(\infty) = -2k$ for $k \in \mathbb{Z}$. There exists $\tau \in \mathbb{C}((y^{-1}))$ such that $\tau^{2k} = P(y)$. Note that $\tau/y \in \mathbb{C}[\![y^{-1}]\!]$ and that it is invertible. We set $v_1 := \tau^k e_1 + e_2$ and $v_2 := \tau^k e_1 - e_2$. Because $\Phi^*(v_1) = \tau^k v_1$ and $\Phi^*(v_2) = -\tau^k v_2$, we obtain the decomposition $\widehat{\boldsymbol{V}} = \mathbb{C}((y^{-1}))v_1 \oplus \mathbb{C}((y^{-1}))v_2$ compatible with the action of Φ^*. A filtered bundle $\mathcal{P}_*\widehat{\boldsymbol{V}}$ over $\widehat{\boldsymbol{V}}$ is good with respect to Φ^* if and only if it satisfies $\mathcal{P}_*\widehat{\boldsymbol{V}} = \mathcal{P}_*\big(\mathbb{C}((y^{-1}))v_1\big) \oplus \mathcal{P}_*\big(\mathbb{C}((y^{-1}))v_2\big)$. Hence, such good $\mathcal{P}_*\widehat{\boldsymbol{V}}$ is determined by the numbers

$$d_i = \deg^{\mathcal{P}}(v_i) := \min\{c \in \mathbb{R} \mid v_i \in \mathcal{P}_c\widehat{\boldsymbol{V}}\}.$$

We can easily see that $\boldsymbol{V}$ has no non-trivial difference submodule, under the assumption that one of $\ell(a)$ is odd. Hence, $\boldsymbol{V}_* = (\boldsymbol{V}, (V, m, \boldsymbol{\tau}_x, \boldsymbol{L}_x)_{x\in\mathbb{C}}, \mathcal{P}_*\widehat{\boldsymbol{V}})$ is a stable parabolic difference module. It is easy to see $\deg(\mathcal{P}_*\mathcal{F}_V) = k - d_1 - d_2$, and hence

$$\deg(\boldsymbol{V}_*) = -d_1 - d_2 - \sum_{a\in S}(1-\tau_a)\ell(a). \tag{1.12}$$

1.6.6.2 The Case Where $\ell(\infty)$ Is Odd

Let us consider the case where $\ell(\infty)$ is odd. There exists $\tau \in \mathbb{C}((y^{-1/2}))$ such that $\tau^{-2\ell(\infty)} = P(y)$ and $\tau/y^{1/2} \in \mathbb{C}[\![y^{-1/2}]\!]$. We set $v_1 := \tau^{-\ell(\infty)}e_1 + e_2$ and $v_2 := \tau^{-\ell(\infty)}e_1 - e_2$. Because $\Phi^*(v_1) = \tau^{-\ell(\infty)}v_1$ and $\Phi^*(v_2) = -\tau^{-\ell(\infty)}v_2$, we obtain the decomposition $\widehat{\boldsymbol{V}}\otimes\mathbb{C}((y^{-1/2})) = \mathbb{C}((y^{-1/2}))v_1\oplus\mathbb{C}((y^{-1/2}))v_2$ compatible with the action of Φ^*. A filtered bundle $\mathcal{P}_*\widehat{\boldsymbol{V}}$ over $\widehat{\boldsymbol{V}}$ is good with respect to Φ^* if and only if the induced filtered bundle over $\widehat{\boldsymbol{V}}\otimes\mathbb{C}((y^{-1/2}))$ satisfies $\mathcal{P}_*\big(\widehat{\boldsymbol{V}}\otimes\mathbb{C}((y^{-1/2}))\big) = \mathcal{P}_*\big(\mathbb{C}((y^{-1/2}))v_1\big)\oplus\mathcal{P}_*\big(\mathbb{C}((y^{-1/2}))v_2\big)$. Hence, it is determined by the numbers $d_i := \min\big\{c\in\mathbb{R}\;\big|\; v_i\in\mathcal{P}_*(\widehat{\boldsymbol{V}}\otimes\mathbb{C}((y^{-1/2})))\big\}$. As in the previous case, $\boldsymbol{V}_* = (\boldsymbol{V}, (V, m, (\boldsymbol{\tau}_x, \boldsymbol{L}_x)), \mathcal{P}_*\widehat{\boldsymbol{V}})$ is stable.

Because $\mathcal{P}_*(\widehat{\boldsymbol{V}}\otimes\mathbb{C}((y^{-1/2})))$ is preserved by the natural action of the Galois group of $\mathbb{C}((y^{-1/2}))/\mathbb{C}((y^{-1}))$, we obtain $d_1 = d_2 =: d$. It is easy to see $\deg(\mathcal{P}_*\mathcal{F}_V) = -d - \ell(\infty)/2$, and hence

$$\deg(\boldsymbol{V}_*) = -d - \sum_{a\in S}(1-\tau_a)\ell(a). \tag{1.13}$$

1.6.7 *Easy Examples of Stable Parabolic Difference Modules (2)*

Take polynomials $P(y)$ and $Q(y)$ such that the following holds.

- $2\deg(Q)\geq\deg(P)$. If $2\deg(Q)=\deg(P)$, we assume $q_{\deg(Q)}^2 - 4p_{\deg(P)}\neq 0$ for the expressions $Q=\sum_{j=0}^{\deg(Q)} q_j y^j$ and $P=\sum_{j=0}^{\deg(P)} p_j y^j$. We assume that $(\deg(P),\deg(Q))\neq(0,0)$.
- P has simple zeroes.

We set $\boldsymbol{V} := \mathbb{C}(y)e_1\oplus\mathbb{C}(y)e_2$, and we define the $\mathbb{C}(y)$-automorphism Φ^* by

$$\Phi^*(e_1,e_2) = (e_1,e_2)\begin{pmatrix}0 & P\\ -1 & Q\end{pmatrix}.$$

Because P has simple zeroes, it is easy to see that there is no difference submodule $V' \subset V$ such that $\dim_{\mathbb{C}(y)} V' = 1$.

We put $\alpha_1 = 2^{-1}\big(Q + Q(1-4PQ^{-2})^{1/2}\big) \in \mathbb{C}((y^{-1}))$ and $\alpha_2 = \alpha_1^{-1}P$. If $2\deg(Q) > \deg(P)$, we choose $(1-4PQ^{-1})^{1/2} \in \mathbb{C}[\![y^{-1}]\!]$ such that $(1-4PQ^{-1})^{1/2} - 1 \in y^{-1}\mathbb{C}[\![y^{-1}]\!]$. Note that $\alpha_1 \neq \alpha_2$. Then, α_i $(i = 1, 2)$ are the roots of the characteristic polynomial $T^2 - QT + P$ of the automorphism. Note that $y^{-\deg Q}\alpha_1$ and $y^{-\deg P + \deg Q}\alpha_2$ are invertible in $\mathbb{C}[\![y^{-1}]\!]$.

We set $V := \mathbb{C}[y]e_1 \oplus \mathbb{C}[y]e_2$. Set $S := \{a \in \mathbb{C} \mid P(a) = 0\}$. Let $m : \mathbb{C} \longrightarrow \mathbb{Z}_{\geq 0}$ be the function determined by $m(a) = 1$ $(a \in S)$ and $m(a) = 0$ $(a \notin S)$. We take $0 \leq \tau_a < 1$ $(a \in S)$. Then, the tuple $(V, m, (\tau_a)_{a\in S})$ determines a parabolic structure at finite place of V.

Set $v_i := Pe_1 + \alpha_i e_2$, for which $\Phi^*(v_i) = \alpha_i v_i$ holds. Then, a good filtered bundle $\mathcal{P}_*\widehat{V}$ over $\widehat{V}$ with respect to Φ^* is determined by $d_i := \deg^{\mathcal{P}}(v_i)$. We obtain $\deg(\mathcal{P}_*\mathcal{F}_V) := \deg(P) + \deg(Q) - d_1 - d_2$. The degree of $V_* = (V, (V, m, \{\tau_a\}_{a\in S}, \mathcal{P}_*\widehat{V}))$ is as follows:

$$\deg(V_*) =$$

$$\deg(P) + \deg(Q) - d_1 - d_2 - \sum_{a\in S}(1-\tau_a)\ell(a) - \frac{\deg(Q)}{2} - \frac{\deg(P) - \deg(Q)}{2}$$

$$= \frac{1}{2}\deg(P) + \deg(Q) - d_1 - d_2 - \sum_{a\in S}(1-\tau_a). \tag{1.14}$$

For any given $\{\tau_a\}_{a\in S}$, we may choose $(d_1, d_2) \in \mathbb{R}^2$ such that (1.14) vanishes.

Note that the numbers $n := \deg(P)$ and $k := 2^{-1}\big(\deg(Q) - (\deg(P) - \deg(Q)) + n\big) = \deg(Q)$ correspond to the number of singularity and the non-abelian charge in [14, Page 5] in the context of periodic monopoles.

1.7 Kobayashi-Hitchin Correspondences for Periodic Monopoles

Kobayashi-Hitchin correspondences between periodic monopoles and parabolic difference modules are stated as follows.

Theorem 1.7.1 (Corollary 9.1.4) *For any $T > 0$ and $\lambda \in \mathbb{C}$, there exists a natural bijective correspondence between the isomorphism classes of the following objects.*

- *Periodic monopoles of GCK type on $S^1_T \times \mathbb{C}_w$.*
- *Polystable parabolic $(2\sqrt{-1}\lambda T)$-difference modules of degree 0.*

We shall explain an outline to obtain a parabolic $2\sqrt{-1}\lambda T$-difference module from a periodic monopole for $\lambda = 0$ in Sect. 1.7.1, and for general λ in

Sect. 1.7.2. The constructions are given by way of mini-holomorphic bundles on a 3-dimensional manifolds, inspired by the work of Charbonneau-Hurtubise [11], which we briefly recalled in Sect. 1.3.2.

Corollary 1.7.2 *In* Sect. 1.6.6, *if we choose* $(d_1, d_2) \in \mathbb{R}^2$ *(resp.* $d \in \mathbb{R}$*) such that (1.12) (resp. (1.13)) vanishes, we obtain the periodic monopoles of GCK-type on* $S^1_T \times \mathbb{C}$ *for any* $T > 0$*, corresponding to the stable parabolic difference modules. Similarly, in* Sect. 1.6.7, *if we choose* $(d_1, d_2) \in \mathbb{R}^2$ *such that (1.14) vanishes, we obtain the periodic monopoles of GCK-type on* $S^1_T \times \mathbb{C}$ *for any* $T > 0$*, corresponding to the stable parabolic difference modules.* ■

Note that even the existence of periodic monopoles was non-trivial, which was studied by Foscolo in [26].

Remark 1.7.3 A special case of this type of correspondences is independently studied by Elliott and Pestun in [24], where rich studies of the related subjects are described. ■

Remark 1.7.4 In [30], Harland studied a classification of periodic monopoles of rank 2 without any singularity, by using the Nahm transform between harmonic bundles on $\mathbb{C}^*$ and periodic monopoles. It is given in terms of spectral curves with a parabolic line bundle, which corresponds to the classification at $\lambda = 0$ in Theorem 1.7.1. ■

Remark 1.7.5 In [70], we studied similar correspondences between q-difference modules and doubly periodic monopoles, i.e., monopoles on the product of $\mathbb{R}$ and a 2-dimensional torus. In [67], we studied similar correspondences between difference modules on elliptic curves and triply periodic monopoles, i.e., monopoles on a 3-dimensional torus. The triply periodic case is much easier because we do not have to study the asymptotic behaviour of monopoles around infinity. ■

Remark 1.7.6 Kontsevich and Soibelman proposed a non-abelian Hodge theory in the context of their holomorphic Floer theory [49]. Correspondences for monopoles may be regarded as 1-dimensional examples of their theory. ■

1.7.1 *The Correspondence in the Case* $\lambda = 0$

We first explain our construction in the case $\lambda = 0$, which is conceptually and technically easier than the case $\lambda \neq 0$.

1.7.1.1 Mini-complex Structure

Let κ denote the $\mathbb{Z}$-action on $\mathbb{R}_t \times \mathbb{C}_w$ defined by $\kappa_n(t, w) = (t + nT, w)$ $(n \in \mathbb{Z})$. Let $\mathcal{M}$ denote the quotient space of $\mathbb{R}_t \times \mathbb{C}_w$ by the $\mathbb{Z}$-action κ, i.e., $\mathcal{M} = S^1_T \times \mathbb{C}_w$. We note that the complex vector fields ∂_t and $\partial_{\overline{w}}$ are naturally defined on $\mathcal{M}$.

We have the naturally defined mini-complex structure on $\mathbb{R}_t \times \mathbb{C}_w$. (See Sect. 2.2 for mini-complex structure.) When we emphasize to consider this mini-complex structure, we use the notation $\mathcal{M}^0$, i.e., let $\mathcal{M}^0$ denote the 3-dimensional manifold $\mathcal{M}$ equipped with the mini-complex structure induced by (t, w). A C^∞-function f on an open subset $U \subset \mathcal{M}^0$ is called mini-holomorphic if $\partial_t f = \partial_{\overline{w}} f = 0$. Let $\mathcal{O}_{\mathcal{M}^0}$ denote the sheaf of mini-holomorphic functions on $\mathcal{M}^0$.

1.7.1.2 Mini-holomorphic Bundles Associated with Monopoles

Let Z be a finite subset of $\mathcal{M}$. Let (E, h, ∇, ϕ) be a monopole on $\mathcal{M} \setminus Z$. Recall that the Bogomolny equation implies that

$$[\nabla_t - \sqrt{-1}\phi, \nabla_{\overline{w}}] = 0.$$

Hence, by considering the sheaf of local sections s of E such that $(\nabla_t - \sqrt{-1}\phi)s = \nabla_{\overline{w}} s = 0$, we obtain a locally free $\mathcal{O}_{\mathcal{M}^0 \setminus Z}$-module $\mathcal{E}^0$.

1.7.1.3 Dirac Type Singularity

Let $P \in Z$. We take a lift $\widetilde{P} = (t^0, w^0) \in \mathbb{R}_t \times \mathbb{C}_w$ of P. Let $\epsilon > 0$ and $\delta > 0$ be small, and consider $U_{-\epsilon} := \big\{(t^0 - \epsilon, w) \,\big|\, |w| < \delta\big\}$ and $U_\epsilon := \big\{(t^0 + \epsilon, w) \,\big|\, |w| < \delta\big\}$. By taking the restriction of $\mathcal{E}^0$ to $U_{\pm\epsilon}$, we obtain locally free $\mathcal{O}_{U_{\pm\epsilon}}$-modules $\mathcal{E}^0_{|U_{\pm\epsilon}}$. Set $U^*_{\pm\epsilon} := U_{\pm\epsilon} \setminus \{(t^0 \pm \epsilon, w^0)\}$. As in the previous studies recalled in Sects. 1.3.1 and 1.3.2, by the parallel transport with respect to $\nabla_t - \sqrt{-1}\phi$, we obtain the isomorphism $\Pi_P : \mathcal{E}^0_{|U^*_{-\epsilon}} \simeq \mathcal{E}^0_{|U^*_{\epsilon}}$, called the scattering map. If $P \in Z$ is Dirac type singularity of the monopole (E, h, ∇, ϕ), it is easy to see that $\mathcal{E}^0$ is of Dirac type at P in the sense that Π_P is meromorphic. (See Sect. 2.4.3 for Dirac type singularity of monopoles.)

1.7.1.4 Meromorphic Extension and Filtered Extension at Infinity

We have the partial compactification $\mathbb{R}_t \times \mathbb{P}^1_w$ of $\mathbb{R}_t \times \mathbb{C}_w$. The $\mathbb{Z}$-action κ on $\mathbb{R}_t \times \mathbb{C}_w$ naturally extends to a $\mathbb{Z}$-action on $\mathbb{R}_t \times \mathbb{P}^1_w$, which is also denoted by κ. The quotient space $\overline{\mathcal{M}}^0$ is naturally equipped with the mini-complex structure, and it is a compactification of $\mathcal{M}^0$. Let $H^0_\infty := \overline{\mathcal{M}}^0 \setminus \mathcal{M}^0 \simeq S^1_T$. Let $\mathcal{O}_{\overline{\mathcal{M}}^0}(*H^0_\infty)$ denote the sheaf of meromorphic functions on $\overline{\mathcal{M}}^0$ whose poles are contained in H^0_∞. (See Sect. 2.2.5 for meromorphic functions in the context of mini-complex structures.)

Remark 1.7.7 In the study of tame or wild harmonic bundles we prefer to consider meromorphic or filtered Higgs bundles on (C, D) rather than Higgs bundles on $C \setminus$

D. (See Remark 1.4.3.) Similarly, we would like to consider meromorphic objects or filtered objects on $(\overline{\mathcal{M}}^0, H^0_\infty)$ to keep the information of the growth orders with respect to h, rather than the transcendental object on $\mathcal{M}^0$. It is the reason why we consider the compactification $\overline{\mathcal{M}}^0$. ■

Let U be any open subset in $\overline{\mathcal{M}}^0 \setminus Z$. Let $\mathcal{P}^h\mathcal{E}^0(U)$ denote the space of sections of $\mathcal{E}^0$ on $U \setminus H^0_\infty$ satisfying the following condition.

- For any $P \in U \cap H^0_\infty$, there exists a neighbourhood U_P of P in U such that $|s_{|U_P \setminus H^0_\infty}|_h = O\big(|w|^N\big)$ for some N.

Thus, we obtain the $\mathcal{O}_{\overline{\mathcal{M}}^0 \setminus Z}(*H^0_\infty)$-module $\mathcal{P}^h\mathcal{E}^0$. We shall prove the following.

Proposition 1.7.8 (Proposition 7.2.1) *$\mathcal{P}^h\mathcal{E}^0$ is a locally free $\mathcal{O}_{\overline{\mathcal{M}}^0 \setminus Z}(*H^0_\infty)$-module.*

Let $\pi^0 : \overline{\mathcal{M}}^0 \longrightarrow S^1_T$ denote the map induced by $(t, w) \longmapsto t$. We set $\overline{\mathcal{M}}^0\langle t\rangle = (\pi^0)^{-1}(t)$ and $\mathcal{M}^0\langle t\rangle = \mathcal{M}^0 \cap \overline{\mathcal{M}}^0\langle t\rangle$. We also set $Z_t := (\pi^0)^{-1}(t) \cap Z$. By taking the restriction, we obtain the locally free $\mathcal{O}_{\mathbb{C}_w \setminus Z_t}$-module $\mathcal{E}^0_{|\mathcal{M}^0\langle t\rangle \setminus Z}$. As in Sects. 1.4.3 and 1.4.4, we obtain an increasing sequence $\mathcal{P}^h_a\big(\mathcal{E}^0_{|\mathcal{M}^0\langle 0\rangle \setminus Z}\big)$ $(a \in \mathbb{R})$ of $\mathcal{O}_{\mathbb{P}^1 \setminus Z_t}$-modules by considering local sections s satisfying $|s|_h = O(|w|^{a+\epsilon})$ for any $\epsilon > 0$.

Theorem 1.7.9 (Theorem 7.3.1) *The tuple $\big(\mathcal{P}^h_*(\mathcal{E}^0_{|\mathcal{M}^0\langle t\rangle \setminus Z}\big) \,|\, t \in S^1_T\big)$ is a good filtered bundle over $\mathcal{P}^h\mathcal{E}^0$ in the sense of Definitions 3.5.16 and 4.2.3. (See Theorem 7.3.4 for a more detailed description.)*

As we shall see in Proposition 7.3.5, the tuple $\big(\mathcal{P}^h_*(\mathcal{E}^0_{|\mathcal{M}^0\langle t\rangle \setminus Z}\big) \,|\, t \in S^1_T\big)$ determines the behaviour of the metric h around H^0_∞ up to boundedness. We shall also show that the compatibility with such filtrations implies the GCK condition around infinity. (See Sect. 7.4.1.)

Remark 1.7.10 Although we shall often use the abbreviation $\mathcal{P}^h_*\mathcal{E}^0$ to denote the tuple $\big(\mathcal{P}^h_*(\mathcal{E}^0_{|\mathcal{M}^0\langle t\rangle \setminus Z}\big) \,|\, t \in S^1_T\big)$, the filtrations depend on $t \in S^1_T$ as basic examples show. (See Sect. 5.1.) ■

1.7.1.5 Kobayashi-Hitchin Correspondence in the Case λ = 0

Let $\mathcal{V}$ be a locally free $\mathcal{O}_{\overline{\mathcal{M}}^0 \setminus Z}(*H^0_\infty)$-module. Suppose that it is of Dirac type at each point of Z, and that it is equipped with a tuple of filtered bundles $\mathcal{P}_*\mathcal{V} = \big(\mathcal{P}_*(\mathcal{V}_{|\overline{\mathcal{M}}^0\langle t\rangle \setminus Z}) \,|\, t \in S^1_T\big)$ over $\mathcal{V}$ which is good in the sense of Definitions 3.5.16 and 4.2.3. It is denoted by $\mathcal{P}_*\mathcal{V}$, and called a good filtered bundle of Dirac type over $(\overline{\mathcal{M}}^0; Z, H^0_\infty)$.

For any good filtered bundle of Dirac type $\mathcal{P}_*\mathcal{V}$ over $(\overline{\mathcal{M}}^0; Z, H^0_\infty)$, we note that the numbers $\deg\big(\mathcal{P}_*(\mathcal{V}_{|\overline{\mathcal{M}}^0\langle t\rangle\setminus Z})\big)$ are well defined for $t \in S^1_T \setminus \pi(Z)$, which induces affine functions on the connected components of $S^1_T \setminus \pi(Z)$. We define

$$\deg(\mathcal{P}_*\mathcal{V}) := \frac{1}{T}\int_0^T \deg\big(\mathcal{P}_*(\mathcal{V}_{|\overline{\mathcal{M}}^0\langle t\rangle\setminus Z})\big)\, dt. \tag{1.15}$$

We define the stability condition for good filtered bundles of Dirac type over $(\overline{\mathcal{M}}^0; Z, H^0_\infty)$ by using the degree as in the standard way.

The following theorem is a natural analogue of Theorem 1.4.5 in the case $\lambda = 0$, in the context of periodic monopoles, for which we apply the Kobayashi-Hitchin correspondence for analytic stable bundles studied in [66].

Theorem 1.7.11 (Theorem 9.1.2) *The construction from (E, h, ∇, ϕ) to $\mathcal{P}_*\mathcal{E}^0$ induces a bijective correspondence between the equivalence classes of monopoles of GCK-type on $\mathcal{M}\setminus Z$ and the equivalence classes of polystable good filtered bundles of Dirac type with degree 0 on $(\overline{\mathcal{M}}^0; Z, H^0_\infty)$.*

Remark 1.7.12 Let Σ be a compact Riemann surface, and let $D \subset \Sigma$ denote a finite subset. Let $g_{\Sigma\setminus D}$ be a Kähler metric of $\Sigma \setminus D$. Suppose that for each $P \in D$ there exists a neighbourhood Σ_P such that $\Sigma_P \setminus \{P\}$ with $g_{\Sigma\setminus D}$ is isomorphic to $\{|z| > R\}$ with the metric $dz\, d\overline{z}$. Let Z denotes a finite subset of $S^1_T \times (\Sigma \setminus D)$. Theorem 1.7.22 can be generalized to correspondences for monopoles on $(S^1_T \times (\Sigma \setminus D)) \setminus Z$ with the metric $dt\, dt + g_{\Sigma\setminus D}$ satisfying similar conditions around $(S^1_T \times D) \cup Z$. ■

1.7.1.6 $\mathcal{O}_{\overline{\mathcal{M}}^0\setminus Z}(*H^0_\infty)$-Modules and $\mathbb{C}(w)$-Modules with an Automorphism

Let $\mathcal{V}$ be a locally free $\mathcal{O}_{\overline{\mathcal{M}}^0\setminus Z}(*H^0_\infty)$-module of Dirac type at Z. Let D denote the image of Z by the projection $p_w : \mathcal{M}^0 = S^1_T \times \mathbb{C}_w \longrightarrow \mathbb{C}_w$. Take a sufficiently small positive number ϵ. By the scattering map along the loop $-\epsilon + s$ $(0 \le s \le T)$, we obtain the automorphism ρ of the $\mathcal{O}_{\mathbb{P}^1}(*(D \cup \{\infty\}))$-module $\mathcal{V}_{|\overline{\mathcal{M}}^0\langle -\epsilon\rangle}$.

We set $V := H^0\big(\mathbb{P}^1, \mathcal{V}_{|\overline{\mathcal{M}}^0\langle -\epsilon\rangle}\big)$, and $\boldsymbol{V} := \mathbb{C}(w) \otimes_{\mathbb{C}[w]} V$. It is equipped with a $\mathbb{C}(w)$-linear automorphism ρ, i.e., and $(\boldsymbol{V}, \rho)$ is a 0-difference module.

For any $x \in \mathbb{C}$, we set $m(x) := \big|p_w^{-1}(x) \cap Z\big|$. If $m(x) > 0$, we obtain $0 \le t_x^{(1)} < t_x^{(2)} < \cdots < t_x^{m(x)} < T$ defined by $p_w^{-1}(x) \cap Z = \{(t_x^{(i)}, x)\}$. We set $\tau_x^{(i)} := t_x^{(i)}/T$. For $i = 1, \ldots, m(x) - 1$, by choosing $t_x^{(i)} < s_i < t_x^{(i+1)}$, we set $L_{x,i}$ as $\mathcal{V}_{(s_i,x)} \otimes \mathbb{C}[\![w - x]\!]$, where $\mathcal{V}_{(s_i,x)}$ denotes the stalk of $\mathcal{V}$ at (s_i, x). Thus, we obtain a parabolic structure at finite place $(V, \{\boldsymbol{\tau}_x, \boldsymbol{L}_x\})$ of $\boldsymbol{V}$.

Let $(\mathcal{P}_*(\mathcal{V}_{|\overline{\mathcal{M}}^0\langle t\rangle}) \,|\, t \in S^1_T)$ be a good filtered bundle over $\mathcal{V}$. By the definition of good filtered bundles over $\mathcal{V}$ (Definition 3.5.16), $\mathcal{P}_*\mathcal{V}_{|\overline{\mathcal{M}}^0\langle 0\rangle}$ induces a good parabolic structure of $\boldsymbol{V}$ at ∞.

By this correspondence, the degree (1.15) for $\mathcal{P}_*\mathcal{V}$ is translated to the degree (1.9) for the parabolic difference module $\boldsymbol{V}_* = \big(V, V, m, (\boldsymbol{\tau}_x, \boldsymbol{L}_x)_{x\in\mathbb{C}}, \mathcal{P}_*\widehat{V}\big)$. (See Lemma 4.2.8.) Hence, the stability condition for $\mathcal{P}_*\mathcal{V}$ is equivalent to the stability condition for $\boldsymbol{V}_*$. Thus, we obtain the following proposition.

Proposition 1.7.13 (Proposition 4.2.16, Proposition 4.2.17) *The above construction induces an equivalence between (stable, polystable) good filtered bundle of Dirac type of degree* 0 *and (stable, polystable) parabolic* 0*-difference modules of degree* 0. ■

Theorem 1.7.1 in the case $\lambda = 0$ follows from Theorem 1.7.11 and Proposition 1.7.13.

1.7.2 The Correspondences in the General Case

Let us outline the construction for general λ in a way as parallel to the case $\lambda = 0$ as possible.

1.7.2.1 Preliminary Consideration

In Sect. 1.7.1, we explained the construction of a 0-difference module from a monopole of GCK-type on $\mathcal{M} \setminus Z$, where one of the keys is the integrability $[\nabla_t - \sqrt{-1}\phi, \nabla_{\overline{w}}] = 0$ as in the previous works in Sect. 1.3. If we choose another $\mathbb{R}$-linear isomorphism

$$\mathbb{R}_t \times \mathbb{C}_w \simeq \mathbb{R}_{t_0} \times \mathbb{C}_{\beta_0} \tag{1.16}$$

such that $dt\,dt + dw\,d\overline{w} = dt_0\,dt_0 + d\beta_0\,d\overline{\beta}_0$, we also obtain $[\nabla_{t_0} - \sqrt{-1}\phi, \nabla_{\overline{\beta}_0}] = 0$. It is natural to ask what objects we could obtain by using the mini-complex structure induced by (t_0, β_0) instead of (t, w). (See Sect. 2.2 for mini-complex structure.)

1.7.2.2 Mini-complex Structure Corresponding to the Twistor Parameter λ

We can easily observe that isomorphisms (1.16) satisfying $dt\,dt + dw\,d\overline{w} = dt_0\,dt_0 + d\beta_0\,d\overline{\beta}_0$ are naturally parameterized by $S^2 = \mathbb{P}^1$. Conceptually, such a decomposition is induced by a complex structure underlying the hyperkähler manifold $\mathbb{R}^4$ with the standard Euclidean metric $\sum_{i=1}^4 dx_i^2$. We choose a complex structure $(z, w) = (x_1+\sqrt{-1}x_2, x_3+\sqrt{-1}x_4)$ for which we have $dz\,d\overline{z}+dw\,d\overline{w} = \sum dx_i^2$. We can regard the mini-complex manifold $\mathbb{R}_t \times \mathbb{C}_w$ as the quotient of $\mathbb{C}^2$ by the $\mathbb{R}_s$-action defined as $s \bullet (z, w) = (z + s, w)$. (See Sect. 2.7.1.) In other words, we may regard that the coordinate system (t, w) on $\mathbb{R}_t \times \mathbb{C}_w$ is induced by $(t, w) = (\operatorname{Im} z, w)$. For the twistor parameter $\lambda \in \mathbb{C}$, we have the corresponding complex structure of $(\mathbb{R}^4, \sum dx_i^2)$, for which a complex coordinate system is given by $(\xi, \eta) = (z + \lambda\overline{w}, w - \lambda\overline{z})$. It induces a mini-complex structure on $\mathbb{R}_t \times \mathbb{C}_w$, which is different from the mini-complex structure induced by (t, w) unless $\lambda = 0$. To emphasize the mini-complex structure depending on λ, we use the notation $(\mathbb{R}_t \times \mathbb{C}_w)^\lambda$.

Note that the $\mathbb{R}_s$-action is expressed as $s \bullet (\xi, \eta) = (\xi + s, \eta - \lambda s)$ in terms of (ξ, η). Let (α_0, β_0) be the complex coordinate system defined as

$$(\alpha_0, \beta_0) = \frac{1}{1 + |\lambda|^2}(\xi - \overline{\lambda}\eta, \eta + \lambda\xi).$$

Because the $\mathbb{R}_s$-action is expressed as $s \bullet (\alpha_0, \beta_0) = (\alpha_0 + s, \beta_0)$, we obtain the induced mini-complex coordinate system $(t_0, \beta_0) = (\operatorname{Im}\alpha_0, \beta_0)$ on $(\mathbb{R}_t \times \mathbb{C}_w)^\lambda$, i.e.,

$$t_0 = \frac{1 - |\lambda|^2}{1 + |\lambda|^2}t + \frac{2}{1 + |\lambda|^2}\operatorname{Im}(\lambda\overline{w}), \qquad \beta_0 = \frac{1}{1 + |\lambda|^2}\left(w + 2\sqrt{-1}\lambda t + \lambda^2\overline{w}\right). \tag{1.17}$$

We note $dt\,dt + dw\,d\overline{w} = dt_0\,dt_0 + d\beta_0\,d\overline{\beta}_0$. The $\mathbb{Z}$-action κ is described as follows in terms of (t_0, β_0):

$$\kappa_n(t_0, \beta_0) = (t_0, \beta_0) + nT \cdot \left(\frac{1 - |\lambda|^2}{1 + |\lambda|^2}, \frac{2\sqrt{-1}\lambda}{1 + |\lambda|^2}\right). \tag{1.18}$$

Because the complex vector fields ∂_{t_0} and $\partial_{\overline{\beta}_0}$ are κ-invariant, we obtain the induced complex vector fields on $\mathcal{M}$, which are also denoted by ∂_{t_0} and $\partial_{\overline{\beta}_0}$.

Let $\mathcal{M}^\lambda$ denote the 3-dimensional manifold obtained as the $(\mathbb{R}_t \times \mathbb{C}_w)^\lambda$ by κ. Though $\mathcal{M}^\lambda = \mathcal{M} = \mathcal{M}^{\lambda'}$ as C^∞-manifolds, the mini-complex structures of $\mathcal{M}^\lambda$ and $\mathcal{M}^{\lambda'}$ are different if $\lambda \neq \lambda'$. A C^∞-function f on an open subset $U \subset \mathcal{M}^\lambda$ is called mini-holomorphic if $\partial_{t_0} f = \partial_{\overline{\beta}_0} f = 0$. Let $\mathcal{O}_{\mathcal{M}^\lambda}$ denote the sheaf of mini-holomorphic functions on $\mathcal{M}^\lambda$.

Remark 1.7.14 Once the twistor parameter λ is fixed, the mini-complex coordinate system (t_0, β_0) is uniquely determined by $dt_0\,dt_0 + d\beta_0\,d\overline{\beta}_0 = dt\,dt + dw\,d\overline{w}$ up to the multiplication of a complex number $e^{\sqrt{-1}\varphi}$ ($\varphi \in \mathbb{R}$) to β_0. ■

1.7.2.3 Another Coordinate System and the Compactification of $\mathcal{M}^{\lambda}$

As recalled in Sect. 1.3, it has been standard to use efficiently the mini-complex coordinate system like (t_0, β_0) in the study of monopoles, which induces the orthogonal decomposition of the 3-dimensional Euclidean space. However, in our study of periodic monopoles, we also use another convenient coordinate system (t_1, β_1) given as follows:

$$(t_1, \beta_1) = \big(t_0 + \mathrm{Im}(\overline{\lambda}\beta_0), (1+|\lambda|^2)\beta_0\big) = \big(t + \mathrm{Im}(\lambda\overline{w}), w + 2\lambda\sqrt{-1}t + \lambda^2\overline{w}\big). \tag{1.19}$$

We note that a C^{∞}-function f on $U \subset \mathcal{M}^{\lambda}$ is mini-holomorphic if and only if $\partial_{t_1} f = \partial_{\overline{\beta}_1} f = 0$. The $\mathbb{Z}$-action κ is described as

$$\kappa_n(t_1, \beta_1) = (t_1, \beta_1) + n \cdot (T, 2\sqrt{-1}\lambda T). \tag{1.20}$$

We have the partial compactification $(\mathbb{R}_t \times \mathbb{C}_w)^{\lambda} = \mathbb{R}_{t_1} \times \mathbb{C}_{\beta_1} \subset \mathbb{R}_{t_1} \times \mathbb{P}^1_{\beta_1}$. By the same formula, we define the $\mathbb{Z}$-action κ on $\mathbb{R}_{t_1} \times \mathbb{P}^1_{\beta_1}$. The quotient space $\overline{\mathcal{M}}^{\lambda}$ is naturally equipped with the induced mini-complex structure, and it is a compactification of $\mathcal{M}^{\lambda}$. Let $H^{\lambda}_{\infty} := \overline{\mathcal{M}}^{\lambda} \setminus \mathcal{M}^{\lambda} \simeq S^1_T$. Let $\mathcal{O}_{\overline{\mathcal{M}}^{\lambda}}(*H^{\lambda}_{\infty})$ denote the sheaf of meromorphic functions on $\overline{\mathcal{M}}^{\lambda}$ whose poles are contained in H^{λ}_{∞}.

Remark 1.7.15 As in the case of $\lambda = 0$ or the case of harmonic bundles (see Remark 1.7.7), we would like to consider meromorphic objects or filtered objects on $(\overline{\mathcal{M}}^{\lambda}, H^{\lambda}_{\infty})$ to keep the information of the growth orders with respect to h, rather than the transcendental object on $\mathcal{M}^{\lambda}$. It is the reason why we consider the compactification $\overline{\mathcal{M}}^{\lambda}$. ■

Remark 1.7.16 When we regard the coordinate change from (t_1, β_1) to (t_0, β_0) as a diffeomorphism $\mathbb{R}_{t_1} \times \mathbb{C}_{\beta_1} \simeq \mathbb{R}_{t_0} \times \mathbb{C}_{\beta_0}$, it does not extend to a diffeomorphism $\mathbb{R}_{t_1} \times \mathbb{P}^1_{\beta_1} \simeq \mathbb{R}_{t_0} \times \mathbb{P}^1_{\beta_0}$. Similarly, the diffeomorphism $\mathbb{R}_{t_1} \times \mathbb{C}_{\beta_1} \simeq \mathbb{R}_t \times \mathbb{C}_w$ does not extend to a diffeomorphism $\mathbb{R}_{t_1} \times \mathbb{P}^1_{\beta_1} \simeq \mathbb{R}_t \times \mathbb{P}^1_w$. Note that we may regard $\mathbb{R}_{t_1} \times \mathbb{P}^1_{\beta_1}$ as the quotient of $\mathbb{C}_{\xi} \times \mathbb{P}^1_{\eta}$ by the naturally induced $\mathbb{R}_s$-action. ■

There are two reasons to use (t_1, β_1) instead of (t_0, β_0). It is one reason that for the coordinate system (t_0, β_0) the $\mathbb{Z}$-action in the $\mathbb{R}_{t_0}$-direction is trivial if $|\lambda| = 1$. (Compare (1.18) with (1.20).) See Remark 1.7.20 for another reason. Clearly, we have $(t_0, \beta_0) = (t_1, \beta_1)$ in the case $\lambda = 0$.

Remark 1.7.17 As explained above, we obtain (t_i, β_i) $(i = 0, 1)$ from (ξ, η). If $\lambda \neq 0$, there exists another pair of mini-complex coordinate systems $(t_i^{\dagger}, \beta_i^{\dagger})$ $(i = 0, 1)$ obtained from $(\xi^{\dagger}, \eta^{\dagger}) = (\lambda^{-1}(w - \lambda\overline{z}), \lambda^{-1}(z + \lambda\overline{w})) = (\lambda^{-1}\eta, \lambda^{-1}\xi)$ which we may use to obtain difference modules from monopoles. See Remark 2.7.3. ■

1.7.2.4 Mini-holomorphic Bundles Associated with Monopoles

Let Z be a finite subset of $\mathcal{M}$. Let (E, h, ∇, ϕ) be a monopole on $\mathcal{M}\setminus Z$. Because the Bogomolny equation implies that $[\nabla_{t_0} - \sqrt{-1}\phi, \nabla_{\overline{\beta}_0}] = 0$, we obtain the locally free $\mathcal{O}_{\mathcal{M}^\lambda\setminus Z}$-module $\mathcal{E}^\lambda$ as the sheaf of local sections s of E such that $(\nabla_{t_0} - \sqrt{-1}\phi)s = \nabla_{\overline{\beta}_0} s = 0$.

If $P \in Z$ is a Dirac type singularity of (E, h, ∇, ϕ), then $\mathcal{E}^\lambda$ is of Dirac type at P as in Sect. 1.7.1.3.

1.7.2.5 Meromorphic Extension and Filtered Extension at Infinity

Let U be any open subset in $\overline{\mathcal{M}}^\lambda \setminus Z$. As in Sect. 1.7.1.4, let $\mathcal{P}^h\mathcal{E}^\lambda(U)$ denote the space of sections of $\mathcal{E}^\lambda$ on $U \setminus H^\lambda_\infty$ satisfying the following.

- For any $P \in U \cap H^\lambda_\infty$, there exists a neighbourhood U_P of P in U such that $|s_{|U_P\setminus H^\lambda_\infty}|_h = O\big(|w|^N\big)$ for some N.

Thus, we obtain the $\mathcal{O}_{\overline{\mathcal{M}}^\lambda\setminus Z}(*H^\lambda_\infty)$-module $\mathcal{P}^h\mathcal{E}^\lambda$. We shall prove the following.

Proposition 1.7.18 (Proposition 7.2.1) *$\mathcal{P}^h\mathcal{E}^\lambda$ is a locally free $\mathcal{O}_{\overline{\mathcal{M}}^\lambda\setminus Z}(*H^\lambda_\infty)$-module.*

Let $\pi^\lambda : \overline{\mathcal{M}}^\lambda \longrightarrow S^1_T$ denote the map induced by $(t_1, \beta_1) \longmapsto t_1$. We set $\overline{\mathcal{M}}^\lambda\langle t_1\rangle = (\pi^\lambda)^{-1}(t_1)$ and $\mathcal{M}^\lambda\langle t_1\rangle = \mathcal{M}^\lambda \cap \overline{\mathcal{M}}^\lambda\langle t_1\rangle$ for $t_1 \in S^1_T$. By taking the restriction, we obtain the locally free $\mathcal{O}_{\mathcal{M}^\lambda\langle t_1\rangle\setminus Z}(*\infty)$-module $\mathcal{P}^h(\mathcal{E}^\lambda_{|\mathcal{M}^\lambda\langle t_1\rangle\setminus Z})$. As in Sects. 1.4.3 and 1.4.4, we obtain an increasing sequence $\mathcal{P}^h_a\big(\mathcal{E}^\lambda_{|\mathcal{M}^\lambda\langle t_1\rangle\setminus Z}\big)$ $(a \in \mathbb{R})$ of $\mathcal{O}_{\overline{\mathcal{M}}^\lambda\langle t_1\rangle\setminus Z}$-modules by considering local sections s satisfying $|s|_h = O(|w|^{a+\epsilon})$ for any $\epsilon > 0$.

Theorem 1.7.19 (Theorem 7.3.1) *The tuple $\big(\mathcal{P}^h_*(\mathcal{E}^\lambda_{|\mathcal{M}^\lambda\langle t_1\rangle\setminus Z}) \,\big|\, t_1 \in S^1_T\big)$ is a good filtered bundle over $\mathcal{P}^h\mathcal{E}^\lambda$ in the sense of Definitions* 3.5.16 *and* 4.2.3. *(See Theorem* 7.3.4 *for a more detailed description.)*

As we shall see in Proposition 7.3.5, the tuple $\big(\mathcal{P}^h_*(\mathcal{E}^\lambda_{|\mathcal{M}^\lambda\langle t_1\rangle\setminus Z}) \,\big|\, t_1 \in S^1_T\big)$ determines the behaviour of the metric h around H^λ_∞ up to boundedness. We shall also show that the compatibility with such filtrations implies the GCK condition around infinity. (See Sect. 7.4.1.)

We shall often use the abbreviation $\mathcal{P}^h_*\mathcal{E}^\lambda$ to denote $\big(\mathcal{P}^h_*(\mathcal{E}^\lambda_{|\mathcal{M}^\lambda\langle t_1\rangle\setminus Z}) \,\big|\, t_1 \in S^1_T\big)$, but we remark that the filtrations depend on $t_1 \in S^1_T$ as basic examples show (see Sect. 5.1).

Remark 1.7.20 In the proof of Proposition 1.7.18 and Theorem 1.7.19, one of the key facts is that the holomorphic vector bundle with a Hermitian metric $(\mathcal{E}^\lambda, h)_{|\mathcal{M}^\lambda\langle t_1\rangle\setminus Z}$ is acceptable (Lemma 7.2.2). (See Sect. 2.11.2 for the acceptability condition.) Hence, we may apply the general result for acceptable bundles in [80] to extend $(\mathcal{E}^\lambda, h)_{|\mathcal{M}^\lambda\langle t_1\rangle\setminus Z}$ across $\infty \in \overline{\mathcal{M}}^\lambda\langle t_1\rangle \simeq \mathbb{P}^1_{\beta_1}$ as a filtered bundle. See Sect. 2.9.7 for a more detailed explanation.

Let $\pi_0^\lambda : \mathcal{M}^\lambda \longrightarrow S^1_T$ denote the projection induced by $(t_0, \beta_0) \longmapsto t_0$. We note that $(\mathcal{E}^\lambda, h)_{|(\pi_0^\lambda)^{-1}(t_0)\setminus Z}$ is not acceptable, in general.

We may use (t_1, β_1) not only (t_0, β_0) thanks to the notion of mini-complex structure instead of a pair of differential operators $\nabla_{t_0} - \sqrt{-1}\phi$ and $\nabla_{\overline{\beta}_0}$ satisfying the integrability condition $[\nabla_{t_0} - \sqrt{-1}\phi, \nabla_{\overline{\beta}_0}]$. This is a merit to consider mini-complex structure. ■

Remark 1.7.21 To define the concept of good filtered bundle in Theorem 1.7.19, we need to know the classification of locally free $\mathcal{O}_{\widehat{H}^\lambda_\infty}(*H^\lambda_\infty)$-modules of finite rank, where $\widehat{H}^\lambda_\infty$ is the formal space obtained as the completion of $\overline{\mathcal{M}}^\lambda$ along H^λ_∞. Because such modules are naturally equivalent to formal difference modules (see Sect. 3.4.3), we may apply the classical results on the classification of formal difference modules mentioned in Sect. 1.6.3. For our purpose, it is also convenient to use another equivalence of formal difference modules of level ≤ 1 and formal differential modules whose Poincaré rank is strictly smaller than 1, which will be explained in Sect. 3.6. ■

1.7.2.6 Kobayashi-Hitchin Correspondence of Periodic Monopoles of GCK Type

Let $\mathcal{V}$ be a locally free $\mathcal{O}_{\overline{\mathcal{M}}^\lambda\setminus Z}(*H^\lambda_\infty)$-module. Suppose that it is of Dirac type at each point of Z, and that it is equipped with a tuple of filtered bundles $\mathcal{P}_*\mathcal{V} = \big(\mathcal{P}_*(\mathcal{V}_{|\overline{\mathcal{M}}^\lambda\langle t_1\rangle}) \,|\, t_1 \in S^1_T\big)$ over $\mathcal{V}$ which is good in the sense of Definitions 3.5.16 and 4.2.3. We denote such object by $\mathcal{P}_*\mathcal{V}$, and call a good filtered bundle of Dirac type over $(\overline{\mathcal{M}}^\lambda; Z, H^\lambda_\infty)$.

For any good filtered bundle of Dirac type $\mathcal{P}_*\mathcal{V}$ over $(\overline{\mathcal{M}}^\lambda; Z, H^\lambda_\infty)$, we note that the numbers $\deg\big(\mathcal{P}_*(\mathcal{V}_{|\overline{\mathcal{M}}^\lambda\langle t_1\rangle})\big)$ are well defined for $t_1 \in S^1_T\setminus\pi^\lambda(Z)$, which induces affine functions on the connected components of $S^1_T \setminus \pi^\lambda(Z)$. We define

$$\deg(\mathcal{P}_*\mathcal{V}) := \frac{1}{T}\int_0^T \deg\big(\mathcal{P}_*(\mathcal{V}_{|\overline{\mathcal{M}}^\lambda\langle t_1\rangle})\big)\,dt_1. \tag{1.21}$$

We define the stability condition for good filtered bundles of Dirac type over $(\overline{\mathcal{M}}^\lambda; Z, H^\lambda_\infty)$ by using the degree as in the standard way.

The following theorem is a generalization of Theorem 1.7.11 to the case of general λ, and it is a natural analogue of Theorem 1.4.5 in the context of periodic monopoles. We again apply the Kobayashi-Hitchin correspondence for analytic stable bundles studied in [66].

Theorem 1.7.22 (Theorem 9.1.2) *The construction from* (E, h, ∇, ϕ) *to* $\mathcal{P}_*\mathcal{E}^\lambda$ *induces a bijective correspondence between the equivalence classes of monopoles of GCK-type on* $\mathcal{M} \setminus Z$ *and the equivalence classes of polystable good filtered bundles of Dirac type with degree 0 on* $(\overline{\mathcal{M}}^\lambda; Z, H_\infty^\lambda)$.

The following is an analogue of Corlette-Simpson correspondence between flat bundles and Higgs bundles, which is an immediate consequence of Theorem 1.7.22.

Corollary 1.7.23 (Corollary 9.1.3) *For any* $\lambda \in \mathbb{C}$, *there exists the natural bijective correspondence of the following objects through periodic monopoles of GCK-type:*

- *Polystable good filtered bundles of Dirac type with degree 0 on* $(\overline{\mathcal{M}}^0; Z, H_\infty^0)$.
- *Polystable good filtered bundles of Dirac type with degree 0 on* $(\overline{\mathcal{M}}^\lambda; Z, H_\infty^\lambda)$. ■

1.7.2.7 Difference Modules and $\mathcal{O}_{\overline{\mathcal{M}}^\lambda \setminus Z}(*H_\infty^\lambda)$-Modules

Let $\mathcal{V}$ be a locally free $\mathcal{O}_{\overline{\mathcal{M}}^\lambda \setminus Z}(*H_\infty^\lambda)$-module of Dirac type at Z. Let $p_1 : \mathbb{R}_{t_1} \times \mathbb{P}^1_{\beta_1} \longrightarrow \mathbb{R}_{t_1}$ and $p_2 : \mathbb{R}_{t_1} \times \mathbb{P}^1_{\beta_1} \longrightarrow \mathbb{P}^1_{\beta_1}$ denote the projections. Let $\widetilde{Z} \subset \mathbb{R}_{t_1} \times \mathbb{P}^1_{\beta_1}$ be the pull back of Z by the projection $\varpi^\lambda : \mathbb{R}_{t_1} \times \mathbb{P}^1_{\beta_1} \longrightarrow \overline{\mathcal{M}}^\lambda$. We put

$$D := p_2\Big(\widetilde{Z} \cap p_1^{-1}\big(\{0 \leq t_1 < T\}\big)\Big).$$

Take a sufficiently small positive number ϵ. By the scattering map, we obtain the following isomorphism of $\mathcal{O}_{\mathbb{P}^1}(*(D \cup \{\infty\}))$-modules:

$$(\varpi^\lambda)^*(\mathcal{V})_{|p_1^{-1}(-\epsilon)}(*D) \simeq (\varpi^\lambda)^*(\mathcal{V})_{|p_1^{-1}(T-\epsilon)}(*D). \tag{1.22}$$

We also have the following natural isomorphism

$$\Phi^*(\varpi^\lambda)^*(\mathcal{V})_{|p_1^{-1}(T-\epsilon)} \simeq (\varpi^\lambda)^*(\mathcal{V})_{|p_1^{-1}(-\epsilon)}. \tag{1.23}$$

We set $V := H^0\big(\mathbb{P}^1, (\varpi^\lambda)^*(\mathcal{V})_{|p_1^{-1}(-\epsilon)}\big)$. Let $\mathbb{C}[\beta_1]_D$ denote the localization of $\mathbb{C}[\beta_1]$ by $\prod_{x \in D}(\beta_1 - x)$. By (1.22) and (1.23), we obtain the $\mathbb{C}$-linear isomorphism

$$\Phi^* : V \otimes_{\mathbb{C}[\beta_1]} \mathbb{C}[\beta_1]_D \longrightarrow V \otimes_{\mathbb{C}[\beta_1]} \mathbb{C}[\beta_1]_{\Phi^{-1}(D)}. \tag{1.24}$$

We set $\boldsymbol{V} := \mathbb{C}(\beta_1) \otimes V$. It is equipped with a $\mathbb{C}$-linear automorphism Φ^* induced by (1.24), and $(\boldsymbol{V}, \Phi^*)$ is a difference module.

For any $x \in \mathbb{C}$, we set $m(x) := \left|p_2^{-1}(x) \cap \widetilde{Z}\right|$. If $m(x) > 0$, we obtain $0 \leq t_{1,x}^{(1)} < t_{1,x}^{(2)} < \cdots < t_{1,x}^{m(x)} < T$ as $p_2^{-1}(x) \cap \widetilde{Z}$. We set $\tau_x^{(i)} := t_{1,x}^{(i)}/T$. For $i = 1, \dots, m(x)-1$, by choosing $t_{1,x}^{(i)} < s_i < t_{1,x}^{(i+1)}$, we set $L_{x,i}$ as $\mathcal{V}_{(s_i,x)} \otimes \mathbb{C}[\![\beta_1 - x]\!]$, where $\mathcal{V}_{(s_i,x)}$ denotes the stalk of $\mathcal{V}$ at (s_i, x). Thus, we obtain a parabolic structure at finite place $(V, \{\boldsymbol{\tau}_x, \boldsymbol{L}_x\})$ of $\boldsymbol{V}$.

Let $\left(\mathcal{P}_*(\mathcal{V}_{|\overline{\mathcal{M}}^{\lambda}\langle t_1\rangle \setminus Z}) \,|\, t_1 \in S_T^1\right)$ be a good filtered bundle over $\mathcal{V}$. We obtain the filtered bundle $\mathcal{P}_*\left((\varpi^{\lambda})^{-1}(\mathcal{V})_{|p_1^{-1}(0)}\right)$ over $(\varpi^{\lambda})^{-1}(\mathcal{V})_{|p_1^{-1}(0)}$ induced by the filtered bundle $\mathcal{P}_*(\mathcal{V}_{|\overline{\mathcal{M}}^{\lambda}\langle 0\rangle})$. By the definition of good filtered bundles over $\mathcal{V}$ (Definition 3.5.16), it induces a good parabolic structure of $\boldsymbol{V}$ at ∞.

By this correspondence, the degree in (1.21) for $\mathcal{P}_*\mathcal{V}$ is translated to the degree (1.9) for the parabolic difference module $\boldsymbol{V}_* = \left(\boldsymbol{V}, V, m, (\boldsymbol{\tau}_x, \boldsymbol{L}_x)_{x\in\mathbb{C}}, \mathcal{P}_*\widehat{V}\right)$. (See Lemma 4.2.8.) Hence, the stability condition for $\mathcal{P}_*\mathcal{V}$ is equivalent to the stability condition for $\boldsymbol{V}_*$. Thus, we obtain the following proposition.

Proposition 1.7.24 (Proposition 4.2.16, Proposition 4.2.17) *The above construction induces an equivalence between (stable, polystable) good filtered bundle of Dirac type of degree* 0 *and (stable, polystable) difference modules of degree* 0. ■

Theorem 1.7.1 for general λ follows from Theorem 1.7.22 and Proposition 1.7.24.

1.8 Asymptotic Behaviour of Periodic Monopoles of GCK-Type

In the theory of wild harmonic bundles $(E, \overline{\partial}_E, \theta, h)$, the first task is to study the asymptotic behaviour of h and θ around the singularity, called Simpson's main estimate. See [80, Theorem 1] for the tame case, and [64, §7.2] for the wild case. The estimate is fundamental not only for the study of the filtered extension $(\mathcal{P}_*^h E, \theta)$ of the Higgs bundle $(E, \overline{\partial}_E, \theta)$ but also for the study of the filtered extensions $(\mathcal{P}_*^h E^{\lambda}, \mathbb{D}^{\lambda})$ $(\lambda \in \mathbb{C})$ of the λ-flat bundles $(E^{\lambda}, \mathbb{D}^{\lambda})$ underlying the harmonic bundle.

We pursue a similar route in our study of periodic monopoles by following the idea that periodic monopoles are regarded as harmonic bundles of infinite rank. (See Sect. 1.5.4.) Indeed, for the results in Sects. 1.7.1.4 and 1.7.1.5 and Sects. 1.7.2.5 and 1.7.2.6, it is fundamental to understand the asymptotic behaviour of monopoles (E, h, ∇, ϕ) on $\mathcal{B}^*(R) = S_T^1 \times \{w \in \mathbb{C} \,|\, |w| > R\}$ such that $F(\nabla) \to 0$ and $|\phi|_h = O(\log|w|)$ as $|w| \to \infty$, which we call GCK-condition. We shall briefly describe the results in this section.

1.8.1 Setting

It is more convenient to study monopoles on a ramified covering. Namely, we set $U^*_{w,q}(R_1) := \{w_q \in \mathbb{C} \mid |w_q^q| > R_1\}$ for $q \in \mathbb{Z}_{\geq 1}$ and for $R_1 > 0$. There exists the natural map $U^*_{w,q}(R_1) \longrightarrow \mathbb{C}_w$ defined by $w_q \longmapsto w_q^q$. We consider monopoles (E, h, ∇, ϕ) on $\mathcal{B}^*_q(R_1) = S^1_T \times U^*_{w,q}(R_1)$ with respect to the Riemannian metric $dt\, dt + dw\, d\overline{w} = dt\, dt + q^2|w_q|^{2(q-1)} dw_q\, d\overline{w}_q$. In the rest of this section, we impose the GCK-condition, i.e., $F(\nabla) \to 0$ and $|\phi|_h = O(\log|w_q|)$ as $|w_q| \to \infty$.

For each $\lambda \in \mathbb{C}$, $\mathcal{B}^*_q(R_1)$ has the mini-complex structure induced by the covering map $\mathcal{B}^*_q(R_1) \longrightarrow \mathcal{B}^*(R)$ and the mini-complex structure of $\mathcal{B}^*(R)$ as an open subset of $\mathcal{M}^\lambda$. (See Sect. 1.7.2.2 for $\mathcal{M}^\lambda$.) When we emphasize the mini-complex structure depending on λ, we use the notation $\mathcal{B}^{\lambda *}_q(R_1)$. Let $(E^\lambda, \overline{\partial}_{E^\lambda})$ denote the mini-holomorphic vector bundle on $\mathcal{B}^{\lambda *}_q(R_1)$ underlying the monopole (E, h, ∇, ϕ). (See Sect. 2.3 for mini-holomorphic bundles.)

As in the case of harmonic bundles, we begin with the analysis of $(E^0, \overline{\partial}_{E^0})$ with h, and we prove Theorem 1.8.2 below which is an analogue of the asymptotic orthogonality for wild harmonic bundles contained in Simpson's main estimate (see [64, Theorem 8.2.1]). It is fundamental even for the study of $(E^\lambda, \overline{\partial}_{E^\lambda}, h)$ as in the case of harmonic bundles. To show it, we outline our proof of Proposition 1.7.18 and Theorem 1.7.19.

1.8.2 Decomposition of Mini-holomorphic Bundles

By considering the monodromy along $S^1_T \times \{w_q\}$, we obtain the automorphism $M(w_q)$ of $E^0_{|\{0\}\times U^*_{w,q}(R_1)}$. The eigenvalues of $M(w_q)$ ($w_q \in U^*_{w,q}$) determine a complex curve $\mathcal{S}p(E^0)$ in $U^*_{w,q}(R_1) \times \mathbb{C}^*$, called the spectral curve of the periodic monopole. (See [13, 14].) Set $U_{w,q}(R_1) := U^*_{w,q}(R_1) \cup \{\infty\}$ in $\mathbb{P}^1_{w_q}$. Under the GCK-condition, the closure $\overline{\mathcal{S}p(E^0)}$ of the spectral curve $\mathcal{S}p(E^0)$ in $U_{w,q}(R_1) \times \mathbb{P}^1$ is also a complex analytic curve. (See Proposition 6.3.13.) Hence, after taking a ramified covering, we may assume that there exists the following decomposition:

$$\overline{\mathcal{S}p(E^0)} = \coprod_{i \in \Lambda} S_i.$$

Here, each S_i is a graph of a meromorphic function g_i on $(U_{w,q}, \infty)$. There exist $(\ell_i, \alpha_i) \in \mathbb{Z} \times \mathbb{C}^*$ such that $g_i \sim \alpha_i w_q^{-\ell_i}$. We obtain

$$\overline{\mathcal{S}p(E^0)} = \coprod_{(\ell,\alpha) \in \mathbb{Z} \times \mathbb{C}^*} \Big(\coprod_{\substack{i \in \Lambda \\ (\ell_i, \alpha_i) = (\ell, \alpha)}} S_i \Big).$$

We obtain the corresponding decomposition of mini-holomorphic bundles:

$$(E^0, \overline{\partial}_{E^0}) = \bigoplus_{(\ell,\alpha)} (E_{\ell,\alpha}, \overline{\partial}_{E_{\ell,\alpha}}). \tag{1.25}$$

1.8.3 The Induced Higgs Bundles

1.8.3.1 Preliminary (1)

Let V be a locally free $\mathcal{O}_{U^*_{w,q}(R_1)}$-module with an endomorphism g. Let $\Psi_q : \mathcal{B}^*_q(R_1) \longrightarrow U^*_{w,q}(R_1)$ denote the projection. Let $\widetilde{V}$ be the C^∞-bundle on $\mathcal{B}^*_q(R_1)$ obtained as the pull back of V by Ψ_q. We obtain the naturally defined operator $\partial_{\widetilde{V},\overline{w}}$ on $\widetilde{V}$ determined by $\partial_{\widetilde{V},\overline{w}}(f\Psi_q^{-1}(s)) = \partial_{\overline{w}}(f)\Psi_q^{-1}(s) + f\Psi_q^{-1}(\partial_{V,\overline{w}}s)$ for any $f \in C^\infty(\mathcal{B}^*_q(R_1))$ and C^∞ section s of V on $U^*_{w,q}(R_1)$. We also have the operator $\partial_{\widetilde{V},t}$ on $\widetilde{V}$ determined by the following condition for f and s as above:

$$\partial_{\widetilde{V},t}(f\Psi_q^{-1}(s)) = \partial_t(f)\Psi_q^{-1}(s) - 2\sqrt{-1}f\Psi_q^{-1}(gs).$$

Thus, we obtain a mini-holomorphic structure $\overline{\partial}_{\widetilde{V}}$ on $\widetilde{V}$ on $\mathcal{B}^{0*}_q(R_1)$. We set $\Psi^*_q(V,g) := (\widetilde{V}, \overline{\partial}_{\widetilde{V}})$.

Remark 1.8.1 The monodromy along $S^1_T \times \{w_q\}$ is $\exp(2\sqrt{-1}Tg(w_q))$. ■

1.8.3.2 Preliminary (2)

For each $(\ell,\alpha) \in \mathbb{Z} \times \mathbb{C}^*$, there exists a basic example of monopole

$$(\mathbb{L}^*_q(\ell,\alpha), h_{\mathbb{L},q,\ell,\alpha}, \nabla_{\mathbb{L},q,\ell,\alpha}, \phi_{\mathbb{L},q,\ell,\alpha}) \tag{1.26}$$

on $\mathcal{B}^*_q(R_1)$, for which the monodromy along $S^1_T \times \{w_q\}$ with respect to $\nabla_t - \sqrt{-1}\phi$ is given as the multiplication of $\alpha w_q^{-\ell}$. (See Sect. 5.2.4.) Let $\mathbb{L}^{0*}_q(\ell,\alpha)$ denote the underlying mini-holomorphic bundle on $\mathcal{B}^{0*}_q(R_1)$.

1.8.3.3 The Induced Higgs Bundles

It is not difficult to see that, for each $(E_{\ell,\alpha}, \overline{\partial}_{E_{\ell,\alpha}})$ in (1.25), there exist a holomorphic vector bundle with an endomorphism $(V_{\ell,\alpha}, \overline{\partial}_{V_{\ell,\alpha}}, g_{\ell,\alpha})$ on $U^*_{w,q}(R_1)$ and an isomorphism

$$(E_{\ell,\alpha}, \overline{\partial}_{E_{\ell,\alpha}}) \simeq \mathbb{L}^{0*}_q(\ell,\alpha) \otimes \Psi^*_q(V_{\ell,\alpha}, \overline{\partial}_{V_{\ell,\alpha}}, g_{\ell,\alpha}).$$

We may choose $g_{\ell,\alpha}$ such that the eigenvalues of $g_{\ell,\alpha|w_q}$ goes to 0 as $|w_q| \to \infty$. By setting $\theta_{\ell,\alpha} := g_{\ell,\alpha}\, dw$, we obtain Higgs bundles $(V_{\ell,\alpha}, \overline{\partial}_{V_{\ell,\alpha}}, \theta_{\ell,\alpha})$.

1.8.4 Asymptotic Orthogonality

Let $h_{\ell,\alpha}$ be the restriction of h to $E_{\ell,\alpha}$. We obtain the metric $h_{\ell,\alpha} \otimes h^{-1}_{\mathbb{L},q,\ell,\alpha}$ of $\Psi_q^*(V_{\ell,\alpha}, \overline{\partial}_{V_{\ell,\alpha}}, g_{\ell,\alpha})$. There exists the Fourier expansion of $h_{\ell,\alpha} \otimes h^{-1}_{\mathbb{L},q,\ell,\alpha}$ along the fibers $S^1_T \times \{w_q\}$, and it turns out that the invariant part induces a metric $h_{V,\ell,\alpha}$ of $V_{\ell,\alpha}$. We obtain the following metric of E:

$$h^\sharp := \bigoplus_{(\ell,\alpha)} h_{\mathbb{L},q,\ell,\alpha} \otimes \Psi_q^{-1}\big(h_{V,\ell,\alpha}\big).$$

Let s be the automorphism of E determined by $h = h^\sharp \cdot s$. The following theorem implies that the difference of h and $h^\sharp$ decays rapidly.

Theorem 1.8.2 (Theorem 6.3.4, Proposition 6.3.8) *For any* $m \in \mathbb{Z}_{\geq 0}$, *there exist positive constants* $C_i(m)$ $(i = 1, 2)$ *such that*

$$\Big|\nabla_{\kappa_1} \circ \cdots \circ \nabla_{\kappa_m}(s - \mathrm{id}_E)\Big|_h \leq C_1(m) \exp\big(-C_2(m)|w_q^q|\big)$$

for any $(\kappa_1, \ldots, \kappa_m) \in \{t, w, \overline{w}\}^m$.

As a consequence of Theorem 1.8.2, $(V_{\ell,\alpha}, \overline{\partial}_{V_{\ell,\alpha}}, \theta_{\ell,\alpha}, h_{V,\ell,\alpha})$ asymptotically satisfies the Hitchin equation. Namely, let $F(h_{V,\ell,\alpha})$ denote the curvature of the Chern connection determined by $\overline{\partial}_{V_{\ell,\alpha}}$ and $h_{V,\ell,\alpha}$, and let $\theta^\dagger_{\ell,\alpha}$ be the adjoint of $\theta_{\ell,\alpha}$ with respect to $h_{V,\ell,\alpha}$. Then, we obtain the following decay for some $\epsilon > 0$:

$$F(h_{V,\ell,\alpha}) + \big[\theta_{\ell,\alpha}, \theta^\dagger_{\ell,\alpha}\big] = O\Big(\exp(-\epsilon|w_q^q|)\Big).$$

We also obtain similar estimates for any derivatives of the left hand side. In this sense, $(V_{\ell,\alpha}, \overline{\partial}_{V_{\ell,\alpha}}, \theta_{\ell,\alpha}, h_{V,\ell,\alpha})$ is an asymptotic harmonic bundle.

1.8.5 Curvature Decay

Many of the estimates for harmonic bundles can be generalized to estimates for asymptotic harmonic bundles. (See Sect. 10.1.1 and [65, §5.5].) It allows us to obtain estimates for periodic monopoles of GCK-type. For example, we obtain the estimate for the curvature $F(\nabla)$ of the monopole (E, h, ∇, ϕ) of GCK-type.

Corollary 1.8.3 (Corollary 6.3.12) *For the expression*

$$F(\nabla) = F(\nabla)_{w,\overline{w}} dw\, d\overline{w} + F(\nabla)_{w,t} dw\, dt + F(\nabla)_{\overline{w},t} d\overline{w}\, dt,$$

the following estimates hold:

$$\left|F(\nabla)_{w\overline{w}}\right|_h = O\Big(|w_q^q|^{-2}(\log|w_q|)^{-2}\Big),$$

$$\left|F(\nabla)_{wt}\right|_h = O(|w_q^q|^{-1}), \quad \left|F(\nabla)_{\overline{w}t}\right|_h = O(|w_q^q|^{-1}).$$

The estimates in Corollary 1.8.3 are useful in the study of $(E^\lambda, \overline{\partial}_{E^\lambda})$ for any λ. Let (t_1, β_1) be the mini-complex coordinate system as in Sect. 1.7.2.3. Note that the complex vector fields $\partial_{\overline{\beta}_1}$ are defined on $\mathcal{B}_q^*(R_1)$. We obtain the differential operator $\partial_{E^\lambda,\overline{\beta}_1}$ on E^λ induced by $\partial_{\overline{\beta}_1}$ and the mini-holomorphic structure $\overline{\partial}_{E^\lambda}$. We also obtain $\partial_{E^\lambda,h,\beta_1}$ induced by h and $\partial_{E^\lambda,\overline{\beta}_1}$ in the standard way. According to Proposition 2.9.5, there exists the following relation:

$$\left[\partial_{E^\lambda,h,\beta_1}, \partial_{E^\lambda,\overline{\beta}_1}\right] = \frac{1}{1+|\lambda|^2} F(\nabla)_{w,\overline{w}}.$$

We also note that $\beta_1 = (1+|\lambda|^2)w + 2\lambda\sqrt{-1}t_1$. Hence, we obtain the following estimate as $|\beta_1| \to \infty$ where t_1 varies in a compact set:

$$\left|\left[\partial_{E^\lambda,h,\beta_1}, \partial_{E^\lambda,\overline{\beta}_1}\right]\right|_h = O\Big(|\beta_1|^{-2}(\log|\beta_1|)^{-2}\Big). \tag{1.27}$$

This means the acceptability mentioned in Remark 1.7.20, which is useful in the proof of Proposition 1.7.18 and Theorem 1.7.19.

1.8.6 The Filtered Extension in the Case $\lambda = 0$

Let us explain an outline of the proof of Proposition 1.7.8 and Theorem 1.7.9 in the case $\lambda = 0$, which are easier than the claims for general λ (Proposition 1.7.18 and Theorem 1.7.19). We discuss it in the ramified case under the setting in Sect. 1.8.1.

Let $U_{w,q}$ be a neighbourhood of ∞ in $\mathbb{P}^1_{w_q}$, and $U^*_{w,q} = U_{w,q} \setminus \{\infty\}$. We assume that $U^*_{w,q} \subset U^*_{w,q}(R_1)$. We have the natural partial compactification $\mathcal{B}_q^0 = S^1_T \times U_{w,q}$ of the mini-complex manifold $\mathcal{B}_q^{0*} = S^1_T \times U^*_{w,q}$. Let $H^0_{\infty,q} := \mathcal{B}_q^0 \setminus \mathcal{B}_q^{0*} = S^1_T \times \{\infty\}$. Let $\pi_q^0 : \mathcal{B}_q^0 \longrightarrow S^1_T$ denote the projection induced by $(t, w_q) \longmapsto t$. For $t \in S^1_T$, we set $\mathcal{B}_q^0\langle t\rangle = (\pi_q^0)^{-1}(t)$ and $\mathcal{B}_q^{0*}\langle t\rangle = \mathcal{B}^{0*} \cap \mathcal{B}_q^0\langle t\rangle$.

We define the $\mathcal{O}_{\mathcal{B}_q^0}(*H^0_{\infty,q})$-module $\mathcal{P}^h E^0$ by the procedure in Sect. 1.7.1.4 from E^0 with h. It is easy to prove that $\mathcal{P}^h E^0$ is a locally free $\mathcal{O}_{\mathcal{B}_q^0}(*H^0_{\infty,q})$-module by the acceptability (1.27) as in Proposition 1.7.8. Moreover, we obtain a tuple of filtered bundles $\{\mathcal{P}^h_*(E^0_{|\mathcal{B}^{0*}_q\langle t\rangle}) \mid t \in S^1_T\}$. To obtain Theorem 1.7.9, we need to prove that the tuple is good.

By using the estimates for the asymptotic harmonic bundles, we obtain the filtered bundle $\mathcal{P}_* V_{\ell,\alpha}$ on $(U_{w,q}, \infty)$ from the holomorphic bundle $(V_{\ell,\alpha}, \overline{\partial}_{V_{\ell,\alpha}})$ with the metric $h_{V,\ell,\alpha}$. (See Sect. 10.1.) Note that the projection $\Psi_q : \mathcal{B}^{0*}_q \longrightarrow U^*_{w,q}$ given by $(t, w_q) \longmapsto w_q$ naturally extends to $\Psi^0_q : \mathcal{B}^0_q \longrightarrow U_{w,q}$. We can naturally generalize the construction in Sect. 1.8.3.1 to the construction of filtered bundle on $(\mathcal{B}^0_q, H^0_{\infty,q})$ from a filtered bundle with endomorphism on $(U_{w,q}, \infty)$. Hence, we obtain a filtered extension $(\Psi^0_q)^*(\mathcal{P}_* V_{\ell,\alpha}, g_{\ell,\alpha})$ of $\Psi^*_q(V_{\ell,\alpha}, g_{\ell,\alpha})$, which is good (see Proposition 4.5.2 which goes back to Proposition 3.6.15). It is easy to see that $\Psi^*_q(\mathcal{P}_* V_{\ell,\alpha}, g_{\ell,\alpha})$ is equal to the filtered extension of $\Psi^*_q(V_{\ell,\alpha}, g_{\ell,\alpha})$ with respect to $\Psi^{-1}_q(h_{V,\ell,\alpha})$.

By explicit computations in Sect. 5, we obtain that $\bigl\{\mathcal{P}_*\bigl(\mathbb{L}^{0*}_q(\ell,\alpha)_{|\mathcal{B}^{0*}_q\langle t\rangle}\bigr) \bigm| t \in S^1_T\bigr\}$ is good. Because

$$\mathcal{P}^h_*(E^0_{|\mathcal{B}^{0*}_q\langle t\rangle}) = \mathcal{P}_*\bigl(\mathbb{L}^{0*}_q(\ell,\alpha)_{|\mathcal{B}^{0*}_q\langle t\rangle}\bigr) \otimes \Psi^*_q(\mathcal{P}_* V_{\ell,\alpha}, g_{\ell,\alpha})_{|\mathcal{B}^{0*}_q\langle t\rangle} \quad (t \in S^1_T)$$

we obtain that $\{\mathcal{P}^h_*(E^0_{|\mathcal{B}^{0*}_q\langle t\rangle}) \mid t \in S^1_T\}$ is good. This is an outline of the proof of Theorem 1.7.9 in the case $\lambda = 0$.

1.8.7 The Filtered Extension for General λ

1.8.7.1 Ramified Covering Space

We continue to use the notation in Sect. 1.8.6. There exists the map $\overline{\mathcal{M}}^\lambda \longrightarrow \mathbb{P}^1_w$ induced by $(t_1, \beta_1) \longmapsto (1+|\lambda|^2)^{-1}(\beta_1 - 2\sqrt{-1}t_1)$. Let $\mathcal{B}^\lambda_q$ denote the fiber product of $\overline{\mathcal{M}}^\lambda$ and $U_{w,q}$ over $\mathbb{P}^1_w$. Let $\Psi^\lambda_q : \mathcal{B}^\lambda_q \longrightarrow U_{w,q}$ denote the projection. We set $H^\lambda_{\infty,q} := (\Psi^\lambda_q)^{-1}(\infty)$ and $\mathcal{B}^{\lambda*}_q := (\Psi^\lambda_q)^{-1}(U^*_{w,q})$. Because the induced map $\mathcal{B}^{\lambda*}_q \longrightarrow \mathcal{M}^\lambda$ is a local diffeomorphism, $\mathcal{B}^{\lambda*}_q$ inherits the locally Euclidean metric and the mini-complex structure. Note that $\mathcal{B}^{\lambda*}_q = \mathcal{B}^{0*}_q$ as Riemannian manifolds because $\mathcal{M}^\lambda = \mathcal{M}^0$ as Riemannian manifolds. We can observe that the mini-complex structure of $\mathcal{B}^{\lambda*}_q$ uniquely extends to a mini-complex structure of $\mathcal{B}^\lambda_q$. (See Sect. 4.3.2.) We also have the naturally defined map $\pi^\lambda_q : \mathcal{B}^\lambda_q \longrightarrow S^1_T$ which is induced by the projection $(t_1, \beta_1) \longmapsto t_1$ where (t_1, β_1) denotes the mini-complex coordinate system in Sect. 1.7.2.3. We set $\mathcal{B}^\lambda_q\langle t_1\rangle = (\pi^\lambda_q)^{-1}(t_1)$ and $\mathcal{B}^{\lambda*}_q\langle t_1\rangle = \mathcal{B}^\lambda_q\langle t_1\rangle \cap \mathcal{B}^{\lambda*}_q$.

We obtain the mini-holomorphic bundle $(E^\lambda, \overline{\partial}_{E^\lambda})$ on $\mathcal{B}^{\lambda*}_q$ as in Sect. 1.8.1. We obtain the $\mathcal{O}_{\mathcal{B}^\lambda_q}(*H^\lambda_{\infty,q})$-module $\mathcal{P}^h(E^\lambda)$ from $(E^\lambda, \overline{\partial}_{E^\lambda}, h)$ by the procedure in

Sect. 1.7.2.5. We can prove that it is a locally free $\mathcal{O}_{\mathcal{B}_q^\lambda}(*H^\lambda_{\infty,q})$-module by the acceptability (1.27), as in Proposition 1.7.18. Moreover, we obtain the tuple of filtered bundles $\{\mathcal{P}^h_*(E^\lambda_{|\mathcal{B}_q^{\lambda*}\langle t_1\rangle}) \mid t_1 \in S^1_T\}$. To obtain Theorem 1.7.19, we need to prove that the tuple is good.

1.8.7.2 Approximation

As an analogue of the Chern connection for holomorphic vector bundles with a Hermitian metric, we construct $\partial^\sharp_{E^0}$ and $\phi^\sharp$ from $(E^0, \overline{\partial}_{E^0})$ with the metric $h^\sharp$. By Theorem 1.8.2, we obtain $\nabla - (\overline{\partial}_{E^0} + \partial^\sharp_{E^0}) = O\big(\exp(-\epsilon|w_q^q|)\big)$ and $\phi - \phi^\sharp = O\big(\exp(-\epsilon|w_q^q|)\big)$ for some $\epsilon > 0$. We construct the differential operator $\overline{\partial}^\sharp_{E^\lambda}$ of E by using the connection $\overline{\partial}_{E^0} + \partial^\sharp_{E^0}$ and $\phi^\sharp$ as in Sect. 7.3.4. Then, $\overline{\partial}^\sharp_{E^\lambda}$ is asymptotically a mini-holomorphic structure, i.e., $\overline{\partial}^\sharp_{E^\lambda} \circ \overline{\partial}^\sharp_{E^\lambda} = O\big(\exp(-\epsilon|w_q|^q)\big)$. We also have $\overline{\partial}_{E^\lambda} - \overline{\partial}^\sharp_{E^\lambda} = O\big(\exp(-\epsilon|w_q|^q)\big)$. Let $C^\infty_{\mathcal{B}_q^\lambda}$ denote the sheaf of C^∞-functions on $\mathcal{B}_q^\lambda$. It turns out that $\overline{\partial}^\sharp_{E^\lambda}$ induces a C^∞-differential operator on $C^\infty_{\mathcal{B}_q^\lambda} \otimes_{\mathcal{O}_{\mathcal{B}_q^\lambda}} \mathcal{P}^h(E^\lambda)$, and $\overline{\partial}^\sharp_{E^\lambda}$ and $\overline{\partial}_{E^\lambda}$ induce the same operator on the formal completion of $\mathcal{P}^h(E^\lambda)$ along $H^\lambda_{\infty,q}$.

1.8.7.3 Formal Completion of Asymptotic Harmonic Bundles at Infinity

Let $(V_{\ell,\alpha}, \overline{\partial}_{V_{\ell,\alpha}}, \theta_{\ell,\alpha}, h_{V,\ell,\alpha})$ be the asymptotic harmonic bundle in Sect. 1.8.3. Let $\overline{\partial}_{V_{\ell,\alpha}} + \partial_{V_{\ell,\alpha}}$ denote the Chern connection associated with $h_{V,\ell,\alpha}$. We obtain the holomorphic vector bundle $V^\lambda_{\ell,\alpha} = (V_{\ell,\alpha}, \overline{\partial}_{V_{\ell,\alpha}} + \lambda\theta^\dagger_{\ell,\alpha})$. It turns out that $(V^\lambda_{\ell,\alpha}, h_{V,\ell,\alpha})$ is acceptable, and hence we obtain the filtered extension $\mathcal{P}_* V^\lambda_{\ell,\alpha}$ on $(U_{w,q}, \infty)$. We have the λ-connection $\mathbb{D}^\lambda_{\ell,\alpha} = \overline{\partial}_{V_{\ell,\alpha}} + \lambda\theta^\dagger_{\ell,\alpha} + \lambda\partial_{V_{\ell,\alpha}} + \theta_{\ell,\alpha}$ of $V_{\ell,\alpha}$. Though it is not necessarily flat, we have $\mathbb{D}^\lambda_{\ell,\alpha} \circ \mathbb{D}^\lambda_{\ell,\alpha} = O\big(\exp(-\epsilon|w_q^q|)\big)$. It turns out that as the formal completion at infinity, we obtain a formal good filtered λ-flat bundle $(\mathcal{P}_* \widehat{V}^\lambda_{\ell,\alpha}, \widehat{\mathbb{D}}^\lambda_{\ell,\alpha})$ as in the case of wild harmonic bundles. (See Sects. 10.1.3 and 10.1.4.)

1.8.7.4 The Formal Structure of $\mathcal{P}^h E^\lambda$ at Infinity

We can consider the formal space $\widehat{H}^\lambda_{\infty,q}$ as the formal completion of $\mathcal{B}_q^\lambda$ along $H^\lambda_{\infty,q}$. (See Sect. 3.4.) We set $H^\lambda_{\infty,q}\langle t_1\rangle = \mathcal{B}_q^\lambda \cap H^\lambda_{\infty,q}$, and let $\widehat{H}^\lambda_{\infty,q}\langle t_1\rangle$ denote the formal completion of $\mathcal{B}_q^\lambda\langle t_1\rangle$ along $H^\lambda_{\infty,q}\langle t_1\rangle$.

We obtain the $\mathcal{O}_{\widehat{H}^{\lambda}_{\infty,q}}(*H^{\lambda}_{\infty,q})$-module $\mathcal{P}(E^{\lambda})_{|\widehat{H}^{\lambda}_{\infty,q}}$ induced by $\mathcal{P}(E^{\lambda})$. We also have the tuple of filtered bundles $\mathcal{P}_*(E^{\lambda})_{|\widehat{H}^{\lambda}_{\infty,q}\langle t_1\rangle}$ $(t_1 \in S^1_T)$.

We have the mini-holomorphic bundles $\mathbb{L}^{\lambda *}_q(\ell,\alpha)$ on $\mathcal{B}^{\lambda *}_q$ underlying the monopoles (1.26). As their filtered extension, we obtain $\mathcal{O}_{\mathcal{B}^{\lambda}_q}(*H^{\lambda}_{\infty,q})$-modules $\mathbb{L}^{\lambda}_q(\ell,\alpha)$ and the tuples of filtered bundles $\{\mathcal{P}_*\mathbb{L}^{\lambda}_q(\ell,\alpha)_{|\mathcal{B}^{\lambda}_q\langle t_1\rangle} \mid t_1 \in S^1_T\}$. As the formal completion along $H^{\lambda}_{\infty,q}$, we obtain $\mathcal{O}_{\widehat{H}^{\lambda}_{\infty,q}}(*H^{\lambda}_{\infty,q})$-modules $\mathbb{L}^{\lambda}_q(\ell,\alpha)_{|\widehat{H}^{\lambda}_{\infty,q}}$ and the good tuple of the filtered bundles $\{\mathcal{P}_*\mathbb{L}^{\lambda}_q(\ell,\alpha)_{|\widehat{H}^{\lambda}_{\infty,q}\langle t_1\rangle} \mid t_1 \in S^1_T\}$.

As we shall explain in Sect. 3.6, from the formal good filtered λ-flat bundles $(\mathcal{P}_*\widehat{V}_{\ell,\alpha}, \widehat{\mathbb{D}}^{\lambda}_{\ell,\alpha})$, good filtered $\mathcal{O}_{\widehat{H}^{\lambda}_{\infty,q}}(*H^{\lambda}_{\infty,q})$-modules $(\Psi^{\lambda}_q)^*(\mathcal{P}_*\widehat{V}_{\ell,\alpha}, \widehat{\mathbb{D}}^{\lambda}_{\ell,\alpha})$ are obtained. (See Proposition 3.6.15.)

By the approximation in Sect. 1.8.7.3, and by the compatibility as in Proposition 2.8.3 and Lemma 4.5.12, there exists an isomorphism of $\mathcal{O}_{\widehat{H}^{\lambda}_{\infty,q}}(*H^{\lambda}_{\infty,q})$-modules

$$\mathcal{P}(E^{\lambda})_{|\widehat{H}^{\lambda}_{\infty,q}} \simeq \bigoplus \mathbb{L}^{\lambda}_q(\ell,\alpha)_{|\widehat{H}^{\lambda}_{\infty,q}} \otimes (\Psi^{\lambda}_q)^*(\widehat{V}_{\ell,\alpha}, \widehat{\mathbb{D}}^{\lambda}_{\ell,\alpha}),$$

which induces an isomorphism of the tuples of filtered bundles

$$\mathcal{P}(E^{\lambda})_{|\widehat{H}^{\lambda}_{\infty,q}\langle t_1\rangle} \simeq \bigoplus \mathcal{P}_*\mathbb{L}^{\lambda}_q(\ell,\alpha)_{|\widehat{H}^{\lambda}_{\infty,q}\langle t_1\rangle} \otimes (\Psi^{\lambda}_q)^*(\mathcal{P}_*\widehat{V}_{\ell,\alpha}, \widehat{\mathbb{D}}^{\lambda}_{\ell,\alpha})_{|\widehat{H}^{\lambda}_{\infty,q}\langle t_1\rangle}.$$

Thus, we obtain that the tuple $\{\mathcal{P}_*(E^{\lambda})_{|\widehat{H}^{\lambda}_{\infty,q}\langle t_1\rangle} \mid t_1 \in S^1_T\}$ is good. This is an outline of our proof of Theorem 1.7.19.

Chapter 2
Preliminaries

Abstract We introduce some basic notions such as mini-holomorphic bundles which are useful in our study of monopoles. We also recall the dimensional reductions to relate instantons, monopoles, harmonic bundles and their underlying holomorphic objects.

2.1 Outline of This Chapter

In Sect. 2.2, we introduce the notion of mini-complex structure of three-dimensional manifolds. In Sect. 2.3, we introduce the notion of mini-holomorphic vector bundles on mini-complex three-dimensional manifolds. The notions of mini-complex structure and mini-holomorphic bundle have been implicitly used in the study of monopoles. In Sect. 2.4, we recall the notion of monopole as mini-holomorphic bundles equipped with a Hermitian metric satisfying (2.3). We also recall the notion of Dirac type singularity of monopoles [51], and a characterization which easily follows from [71].

In Sect. 2.5, we recall the dimensional reduction of instantons to monopoles. We also explain the underlying dimensional reduction of holomorphic bundles to mini-holomorphic bundles. In Sect. 2.6, we explain the dimensional reduction of monopoles to harmonic bundles. We also explain the underlying dimensional reduction of mini-holomorphic bundles to Higgs bundles in Sect. 2.6.2.

In Sect. 2.7, we study the twistor family of mini-complex structures on $\mathcal{M} = (\mathbb{R}/T\mathbb{Z}) \times \mathbb{C}$ for $T > 0$. For $\lambda \in \mathbb{C}$, we obtain the mini-complex manifold $\mathcal{M}^\lambda$ as $\mathcal{M}$ equipped with a mini-complex structure corresponding to λ. We introduce two convenient mini-complex local coordinate systems (t_0, β_0) and (t_1, β_1) for $\mathcal{M}^\lambda$. We introduce a compactification $\overline{\mathcal{M}}^\lambda$ of $\mathcal{M}^\lambda$ by using the coordinate system (t_1, β_1). Set $H^\lambda_\infty := \overline{\mathcal{M}}^\lambda \setminus \mathcal{M}^\lambda$.

In Sect. 2.8, we study the dimensional reduction from $\mathcal{O}_{\mathcal{M}^\lambda}$-modules to λ-flat bundles on $\mathbb{C}_w$ for general λ, and its compatibility with the dimensional reduction from monopoles to harmonic bundles. The case of $\lambda = 0$ is already essentially explained in Sect. 2.6. The case of $\lambda \neq 0$ is also useful for our study. We also explain

T. Mochizuki, *Periodic Monopoles and Difference Modules*, Lecture Notes in Mathematics 2300, https://doi.org/10.1007/978-3-030-94500-8_2

the dimensional reduction from $\mathcal{O}_{\overline{\mathcal{M}}^\lambda}$-modules to an $\mathcal{O}_{\mathbb{P}^1_w}$-modules equipped with a meromorphic λ-connection.

In Sect. 2.9, we shall introduce a section $G(h)$ of $\mathrm{End}(E)$ for a mini-holomorphic bundle $(E, \overline{\partial}_E)$ with a Hermitian metric h on an open subset of $\mathcal{M}^\lambda$. It is an analogue of the contraction of the curvature of the Chern connection for a holomorphic vector bundle with a Hermitian metric on a Kähler manifold. In Sects. 2.9.2–2.9.4, we explain some general formulas for $G(h)$ which are analogues of standard and useful formulas in the study of Hermitian-Einstein metrics [79]. We give a complement to Sect. 2.8 in Sect. 2.9.5, which is mainly a preliminary for the proof of Proposition 7.4.3. In Sect. 2.9.6, we recall the dimensional reduction of instantons with respect to the Hopf fibration due to Kronheimer, and prove a formula which will be useful in the proof of Proposition 9.2.2.

In Sect. 2.10, we explain an equivalence between difference modules with parabolic structure at finite place and locally free $\mathcal{O}_{\mathcal{M}^\lambda \setminus Z}(*H^\infty_\lambda)$-modules for finite subsets Z.

In Sect. 2.11, we review filtered bundles on punctured curves, and we recall the general construction of a filtered bundle from a holomorphic bundle with a Hermitian metric satisfying the acceptability condition.

2.2 Mini-Complex Structures on 3-Manifolds

2.2.1 Mini-Holomorphic Functions on $\mathbb{R} \times \mathbb{C}$

Let t and w be the standard coordinates of $\mathbb{R}$ and $\mathbb{C}$, respectively. The orientation of $\mathbb{R} \times \mathbb{C}$ is given as the product of the orientation of $\mathbb{R}$ and $\mathbb{C}$. Let U be an open subset in $\mathbb{R} \times \mathbb{C}$. A C^∞-function f on U is called mini-holomorphic if $\partial_t f = 0$ and $\partial_{\overline{w}} f = 0$.

Let U_i $(i = 1, 2)$ be open subsets in $\mathbb{R} \times \mathbb{C}$. Let $F : U_1 \longrightarrow U_2$ be a diffeomorphism. It is called mini-holomorphic if (i) F preserves the orientations, (ii) $F^*(f)$ is mini-holomorphic for any mini-holomorphic function f on U_2.

Lemma 2.2.1 *Let* $F = (F_t(t, w), F_w(t, w)) : U_1 \longrightarrow U_2$ *be a diffeomorphism. Then,* F *is mini-holomorphic if and only if we have* $\partial_t F_t > 0$, $\partial_t F_w = 0$ *and* $\partial_{\overline{w}} F_w = 0$.

Proof Suppose that F is mini-holomorphic. Because $F^*(w) = F_w$ is mini-holomorphic, we obtain $\partial_t F_w = \partial_{\overline{w}} F_w = 0$. Because F is orientation preserving, we obtain $\partial_t F_t \cdot |\partial_w F_w|^2 > 0$, which implies $\partial_t F_t > 0$. The converse is also easily proved. ■

2.2.2 Mini-Complex Structure on Three-Dimensional Manifolds

Let us define the notion of mini-complex structure for Three-dimensional manifolds as in the case of smooth structure. (For example, see [53]). Let M be an oriented three-dimensional C^∞-manifold. A mini-complex atlas on M is a family of open subsets U_λ ($\lambda \in \Lambda$) with an orientation-preserving embedding $\varphi_\lambda : U_\lambda \longrightarrow \mathbb{R} \times \mathbb{C}$ satisfying the following conditions.

- $M = \bigcup_{\lambda \in \Lambda} U_\lambda$.
- The coordinate change $\varphi_\lambda(U_\lambda \cap U_\mu) \longrightarrow \varphi_\mu(U_\lambda \cap U_\mu)$ is mini-holomorphic.

Such $(U_\lambda, \varphi_\lambda)$ is called a mini-complex chart. We shall often use (t, w) instead of φ_λ. Two mini-complex atlas $\{(U_\lambda, \varphi_\lambda) \mid \lambda \in \Lambda\}$ and $\{(V_\gamma, \psi_\gamma) \mid \gamma \in \Gamma\}$ are defined to be equivalent if the union is also a mini-complex atlas. There exists the partial order on the family of mini-complex atlases defined by inclusions. Each equivalence class of mini-complex atlases has a unique maximal mini-complex atlas. A mini-complex structure on M is defined to be a maximal mini-complex atlas on M.

When M is equipped with a mini-complex structure, a C^∞-function f on M is called mini-holomorphic if its restriction to any mini-complex chart is mini-holomorphic. Let $\mathcal{O}_M$ denote the sheaf of mini-holomorphic functions.

2.2.3 Tangent Bundles

Suppose that M is equipped with a mini-complex structure. Let $(U; t, w)$ be a mini-complex chart. The real vector field ∂_t determines an oriented subbundle $T_S U$ of TU. The quotient bundle $T_Q U$ is equipped with the complex structure J, where J is an automorphism of $T_Q U$ such that $J^2 = -1$. Because the mini-complex coordinate change preserves the subbundles, we obtain a globally defined subbundle $T_S M$ of TM of rank 1. We also obtain the quotient bundle $T_Q M$, which is equipped with the complex structure.

Let us consider $T^{\mathbb{C}} M := TM \otimes_{\mathbb{R}} \mathbb{C}$. There exists the decomposition of complex vector bundles $T_Q^{\mathbb{C}} M = T_Q^{1,0} M \oplus T_Q^{0,1} M$, where $T_Q^{1,0} M$ is the $\sqrt{-1}$-eigen bundle of J, and $T_Q^{0,1} M$ is the $-\sqrt{-1}$-eigen bundle of J. We obtain the exact sequence of complex vector bundles on M:

$$0 \longrightarrow T_S^{\mathbb{C}} M \longrightarrow T^{\mathbb{C}} M \xrightarrow{a_1} T_Q^{1,0} M \oplus T_Q^{0,1} M \longrightarrow 0.$$

Let $\Theta_M^{1,0}$ and $\Theta_M^{0,1}$ denote the inverse image of $T_Q^{1,0} M$ and $T_Q^{0,1} M$ by a_1, respectively. Note that $\Theta_M^{1,0} \cap \Theta_M^{0,1} = T_S^{\mathbb{C}} M$. The complex conjugate on $\mathbb{C}$ induces the complex conjugate on $TM \otimes \mathbb{C}$ for which we obtain $\overline{\Theta_M^{1,0}} = \Theta_M^{0,1}$. It also induces the complex conjugate on $T_Q^{1,0}M \oplus T_Q^{0,1}M$ for which $\overline{T_Q^{1,0}M} = T_Q^{0,1}M$ holds.

Lemma 2.2.2 *Let $F : M_1 \longrightarrow M_2$ be a diffeomorphism of oriented three-dimensional manifolds. Suppose that M_i are equipped with mini-complex structure, and that the following holds:*

- *F is orientation preserving.*
- *$dF(T_S M_1) = T_S M_2$.*
- *The induced isomorphism $T_Q M_1 \simeq T_Q M_2$ is $\mathbb{C}$-linear.*

Then, F preserves the mini-complex structures.

Proof It is enough to study the claim locally around any point of M_i. Let U_i $(i = 1, 2)$ be open subsets of $\mathbb{R} \times \mathbb{C}$ which are naturally equipped with mini-complex structures. Let $F : U_1 \longrightarrow U_2$ be an orientation preserving diffeomorphism such that (i) $dF(T_S U_1) = T_S U_2$, (ii) the induced isomorphism $T_Q U_1 \simeq T_Q U_2$ is $\mathbb{C}$-linear. We have the expression $F = (F_t(t, w), F_w(t, w))$. By the condition (i), we obtain $\partial_t F_w(t, w) = 0$. By the condition (ii), we obtain $\partial_{\overline{w}} F_w(t, w) = 0$. Hence, F is mini-holomorphic. ■

2.2.4 Cotangent Bundles

Let $\Omega_Q^{1,0}$ and $\Omega_Q^{0,1}$ denote the complex dual bundles of $T_Q^{1,0}M$ and $T_Q^{0,1}M$, respectively. Let $\Omega_M^{1,0}$ and $\Omega_M^{0,1}$ denote the complex dual bundle of $\Theta_M^{1,0}$ and $\Theta_M^{0,1}$. We obtain the following exact sequences:

$$0 \longrightarrow \Omega_Q^{1,0} \oplus \Omega_Q^{0,1} \longrightarrow T^*M \otimes \mathbb{C} \longrightarrow (T_S^{\mathbb{C}}M)^{\vee} \longrightarrow 0,$$

$$0 \longrightarrow \Omega_Q^{1,0} \longrightarrow \Omega_M^{1,0} \longrightarrow (T_S^{\mathbb{C}}M)^{\vee} \longrightarrow 0,$$

$$0 \longrightarrow \Omega_Q^{0,1} \longrightarrow \Omega_M^{0,1} \longrightarrow (T_S^{\mathbb{C}}M)^{\vee} \longrightarrow 0,$$

$$0 \longrightarrow \Omega_Q^{1,0} \longrightarrow T^*M \otimes \mathbb{C} \longrightarrow \Omega_M^{0,1} \longrightarrow 0,$$

$$0 \longrightarrow \Omega_Q^{0,1} \longrightarrow T^*M \otimes \mathbb{C} \longrightarrow \Omega_M^{1,0} \longrightarrow 0.$$

Note $\operatorname{rank}_{\mathbb{C}} \Omega_M^{0,1} = \operatorname{rank}_{\mathbb{C}} \Omega_M^{1,0} = 2$. We set $\Omega_M^{0,2} := \bigwedge^2 \Omega_M^{0,1}$ and $\Omega_M^{2,0} := \bigwedge^2 \Omega_M^{1,0}$.

For any C^∞-vector bundle E on M, let $C^\infty(M, E)$ denote the space of C^∞-sections of E.

The exterior derivative $d : C^\infty(M, \mathbb{C}) \longrightarrow C^\infty(M, T^*M \otimes \mathbb{C})$ induces the following differential operators:

$$\overline{\partial}_M : C^\infty(M, \mathbb{C}) \longrightarrow C^\infty(M, \Omega_M^{0,1}), \qquad \partial_M : C^\infty(M, \mathbb{C}) \longrightarrow C^\infty(M, \Omega_M^{1,0}).$$

We also obtain the following operators induced by $d : C^\infty(M, T^*M \otimes \mathbb{C}) \longrightarrow C^\infty(M, \bigwedge^2 T^*M \otimes \mathbb{C})$:

$$\overline{\partial}_M : C^\infty(M, \Omega_M^{0,1}) \longrightarrow C^\infty(M, \Omega_M^{0,2}),$$

$$\partial_M : C^\infty(M, \Omega_M^{1,0}) \longrightarrow C^\infty(M, \Omega_M^{2,0}).$$

We have $\overline{\partial}_M \circ \overline{\partial}_M = 0$ and $\partial_M \circ \partial_M = 0$. It is easy to see that a C^∞-function f on M is mini-holomorphic if and only if $\overline{\partial}_M f = 0$.

2.2.5 *Meromorphic Functions*

Let M be a three-dimensional manifold with a mini-complex structure. Let $H \subset M$ be a one-dimensional submanifold such that the tangent bundle TH is contained in $(T_S M)_{|H}$. Let U be an open subset of M. A holomorphic function f on $U \setminus H$ is called meromorphic of at most order N along H if the following holds.

- Let P be any point of $H \cap U$. We take a mini-complex coordinate neighbourhood (U_P, t, w) around P. Note that H is described as $\{w = w_0\}$. Then, $(w - w_0)^N f_{|U_P}$ gives a mini-holomorphic function on U_P.

Let $\mathcal{O}_M(NH)$ denote the sheaf of meromorphic functions of order N along H. We obtain the sheaf $\mathcal{O}_M(*H) := \varinjlim_N \mathcal{O}_M(NH)$. A local section of $\mathcal{O}_M(*H)$ is called a meromorphic function on M whose poles are contained in H.

2.3 Mini-Holomorphic Bundles

2.3.1 *Mini-Holomorphic Bundles*

Let M be a three-dimensional manifold with a mini-complex structure. Let E be a C^∞-bundle on M of finite rank. Let us consider a differential operator

$$\overline{\partial}_E : C^\infty(M, E) \longrightarrow C^\infty(M, E \otimes \Omega_M^{0,1})$$

satisfying $\overline{\partial}_E(fs) = \overline{\partial}_M(f)s + f\overline{\partial}_E(s)$ for any $f \in C^\infty(M, \mathbb{C})$ and $s \in C^\infty(M, E)$. As in the ordinary case of vector bundles on complex manifolds, it induces $\overline{\partial}_E : C^\infty(M, E \otimes \Omega_M^{0,1}) \longrightarrow C^\infty(M, E \otimes \Omega_M^{0,2})$. Such a differential operator $\overline{\partial}_E$ is called a mini-holomorphic structure of E if $\overline{\partial}_E \circ \overline{\partial}_E = 0$.

In terms of a mini-complex coordinate system (t, w), a mini-holomorphic structure is equivalent to a pair of differential operators $\partial_{E,t}$ and $\partial_{E,\overline{w}}$ on $C^\infty(E)$ such that $\partial_{E,t}(fs) = \partial_t(f)s + f\partial_{E,t}(s)$ and $\partial_{E,\overline{w}}(fs) = \partial_{\overline{w}}(f)s + f\partial_{E,\overline{w}}(s)$ for $f \in C^\infty(M, \mathbb{C})$ and $s \in C^\infty(M, E)$, satisfying the commutativity condition $[\partial_{E,t}, \partial_{E,\overline{w}}] = 0$. Indeed, the operators $\partial_{E,\overline{w}}$ and $\partial_{E,t}$ are induced by the inner product of $\partial_{\overline{w}}$ and ∂_t with $\overline{\partial}_E(s)$ $(s \in C^\infty(U, E))$, respectively.

A local section s of E is called mini-holomorphic if $\overline{\partial}_E(s) = 0$. By considering the sheaf of mini-holomorphic sections of E, we obtain a locally free $\mathcal{O}_U$-module, which is often denoted by the same notation E. The following is standard.

Lemma 2.3.1 *The above procedure induces an equivalence of mini-holomorphic bundles of finite rank and locally free $\mathcal{O}_M$-modules of finite rank.* ∎

2.3.2 Metrics and the Induced Operators

Let $(E, \overline{\partial}_E)$ be a mini-holomorphic bundle on M. Let h be a C^∞-metric of a mini-holomorphic bundle E. As in the case of complex differential geometry (see [48]), we obtain the induced differential operator $\partial_{E,h} : C^\infty(M, E) \longrightarrow C^\infty(M, E \otimes \Omega_M^{1,0})$ such that $\partial_{E,h}(fs) = \partial_M(f)s + f\partial_{E,h}(s)$ for $f \in C^\infty(M, \mathbb{C})$ and $s \in C^\infty(M, E)$, determined by the condition $\overline{\partial}_M h(u, v) = h(\overline{\partial}_E u, v) + h(u, \partial_{E,h} v)$ for any $u, v \in C^\infty(M, E)$. We have $\partial_{E,h} \circ \partial_{E,h} = 0$.

2.3.3 Splittings

Let $T_Q M \longrightarrow TM$ be a splitting of the exact sequence $0 \longrightarrow T_S M \longrightarrow TM \longrightarrow T_Q M \longrightarrow 0$. We obtain the induced splitting $(T_S^{\mathbb{C}} M)^\vee \longrightarrow (T^{\mathbb{C}} M)^\vee$, and the decomposition:

$$(T^{\mathbb{C}} M)^\vee = (T_S^{\mathbb{C}} M)^\vee \oplus \Omega_Q^{1,0} \oplus \Omega_Q^{0,1}.$$

We also obtain the decompositions:

$$\Omega_M^{1,0} = \Omega_Q^{1,0} \oplus (T_S^{\mathbb{C}} M)^\vee, \qquad \Omega_M^{0,1} = \Omega_Q^{0,1} \oplus (T_S^{\mathbb{C}} M)^\vee. \tag{2.1}$$

Let $(E, \overline{\partial}_E)$ be a mini-holomorphic bundle with a metric h. We obtain the decompositions $\overline{\partial}_E = \overline{\partial}_E^Q + \overline{\partial}_E^S$ and $\partial_{E,h} = \partial_{E,h}^Q + \partial_{E,h}^S$ according to (2.1).

We obtain the unitary connection ∇_h and the anti-self-adjoint section ϕ_h of $\mathrm{End}(E) \otimes (T_S M \otimes \mathbb{C})^\vee$:

$$\nabla_h := \overline{\partial}_E^Q + \partial_{E,h}^Q + \frac{1}{2}(\overline{\partial}_E^S + \partial_{E,h}^S), \qquad \phi_h = \frac{\sqrt{-1}}{2}(\overline{\partial}_E^S - \partial_{E,h}^S).$$

They are called the Chern connection and the Higgs field. Note that the construction of the Chern connection and the Higgs field depend on the choice of the splitting $T_Q M \longrightarrow TM$.

Remark 2.3.2 If M is equipped with a Riemannian metric, the orthogonal complement of $T_S M$ in TM is naturally isomorphic to $T_Q M$. Hence, we obtain a splitting, which we shall use without mention. Moreover, by the Riemannian metric, $T_S M$ is identified with the product bundle $\mathbb{R} \times M$. Hence, we regard ϕ_h as an anti-Hermitian section of $\mathrm{End}(E)$. ■

2.3.4 Scattering Maps

Let $(E, \overline{\partial}_E)$ be a mini-holomorphic bundle on M. Let $\gamma : [0, 1] \longrightarrow M$ be a path such that $T\gamma(\partial_s)$ is contained in $T_S M$. Then, the mini-holomorphic structure of E induces a connection of $\gamma^* E$. Hence, we obtain the isomorphism $E_{\gamma(0)} \longrightarrow E_{\gamma(1)}$ as the parallel transport of the connection, called the scattering map [11].

Let U be an open subset of M with a mini-complex coordinate φ such that $\varphi(U) \supset]-2\epsilon, 2\epsilon[\times U_w$, where U_w is an open subset in $\mathbb{C}$. Here, for $a < b$, we set $]a, b[:= \{t \in \mathbb{R} \mid a < t < b\}$. We regard $[-\epsilon, \epsilon] \times U_w$ as a subset of M. By the mini-holomorphic structure, $(E^t, \overline{\partial}_{E^t}) := (E, \overline{\partial}_E)_{|\{t\} \times U_w}$ is naturally a holomorphic vector bundle on U_w for any $t \in [-\epsilon, \epsilon]$. By considering the scattering map along the path $\gamma_w : [0, 1] \longrightarrow \big(t_1 + s(t_2 - t_1), w\big)$ $(w \in U_w)$, we obtain the isomorphism $(E^{t_1}, \overline{\partial}_{E^{t_1}}) \simeq (E^{t_2}, \overline{\partial}_{E^{t_2}})$ because of the commutativity $[\partial_{E,t}, \partial_{E,\overline{w}}] = 0$.

2.3.5 Dirac Type Singularity of Mini-Holomorphic Bundles

Let U be a neighbourhood of $(0, 0)$ in $\mathbb{R} \times \mathbb{C}$. We take $\epsilon > 0$ and $\delta > 0$ such that $]-2\epsilon, 2\epsilon[\times \{|w| < \delta\}$ is contained in U. Set $U_w := \{|w| < \delta\}$ and $U_w^* := U_w \setminus \{0\}$. Let (E, ∂_E) be a mini-holomorphic bundle on $U \setminus \{(0, 0)\}$. For any $t \in \mathbb{R}$ with $0 < |t| < 2\epsilon$, we obtain the holomorphic vector bundle $(E^t, \overline{\partial}_{E^t})$ on U_w as the restriction of $(E, \overline{\partial}_E)$. We have the scattering map $\Psi : E^{-\epsilon}_{|U_w^*} \simeq E^{\epsilon}_{|U_w^*}$. Recall that $(0, 0)$ is called Dirac type singularity of $(E, \overline{\partial}_E)$ if Ψ is meromorphic at 0. (See [71].) In that case, we say that $(E, \overline{\partial}_E)$ is of Dirac type on $(U, (0, 0))$.

Let M be a three-dimensional manifold with a mini-complex manifold. Let Z be a discrete subset in M. Let $(E, \overline{\partial}_E)$ be a mini-holomorphic bundle on $M \setminus Z$. We say that $(E, \overline{\partial}_E)$ is of Dirac type on (M, Z) if the following holds.

- For any $P \in Z$, we take a mini-complex chart (M_P, φ_P) such that (i) $\varphi_P(P) = (0, 0)$, (ii) the induced mini-holomorphic bundle on $\varphi_P(M_P \setminus \{P\})$ is of Dirac type on $(\varphi_P(M_P), (0, 0))$.

We can easily see that the condition is independent of the choice of a mini-complex chart (M_P, φ_P).

2.3.6 Kronheimer Resolution of Dirac Type Singularity

Let U be a neighbourhood of $(0, 0)$ in $\mathbb{R} \times \mathbb{C}$. Let $(E, \overline{\partial}_E)$ be a mini-holomorphic bundle on $U^* := U \setminus \{(0, 0)\}$. Let $\varphi : \mathbb{C}^2 \longrightarrow \mathbb{R} \times \mathbb{C}$ be the map given by

$$\varphi(u_1, u_2) = (|u_1|^2 - |u_2|^2, 2u_1u_2). \tag{2.2}$$

We set $\widetilde{U} := \varphi^{-1}(U)$ and $\widetilde{U}^* := \widetilde{U} \setminus \{(0, 0)\}$. The induced map $\widetilde{U}^* \longrightarrow U^*$ is also denoted by φ. We set $\widetilde{E} := \varphi^{-1}(E)$. We obtain the holomorphic structure $\overline{\partial}_{\widetilde{E}}$ on $\widetilde{E}$ determined by the following condition:

$$\partial_{\widetilde{E},\overline{u}_1}\varphi^{-1}(s) = u_1\varphi^{-1}(\partial_{E,t}s) + 2\overline{u}_2\varphi^{-1}(\partial_{E,\overline{w}}s),$$

$$\partial_{\widetilde{E},\overline{u}_2}\varphi^{-1}(s) = -u_2\varphi^{-1}(\partial_{E,t}s) + 2\overline{u}_1\varphi^{-1}(\partial_{E,\overline{w}}s).$$

Here, s denotes a C^∞-section of E. (See [11, 71].)

Lemma 2.3.3 ([11, 71]) *$(0, 0)$ is Dirac type singularity of $(E, \overline{\partial}_E)$ if and only if $(\widetilde{E}, \overline{\partial}_{\widetilde{E}})$ extends to a holomorphic vector bundle $(\widetilde{E}_0, \overline{\partial}_{\widetilde{E}_0})$ on $\widetilde{U}$.* ■

We call such $(\widetilde{E}_0, \overline{\partial}_{\widetilde{E}_0})$ the Kronheimer resolution of $(E, \overline{\partial}_E)$ at $(0, 0)$.

Remark 2.3.4 Note that there exists a natural morphism $\varphi^{-1}(\mathcal{O}_{U^*}) \longrightarrow \mathcal{O}_{\widetilde{U}^*}$. Let $\mathcal{E}$ denote the $\mathcal{O}_{U^*}$-module obtained as the sheaf of mini-holomorphic sections of E. It is easy to see that the sheaf of holomorphic sections of $(\widetilde{E}, \overline{\partial}_{\widetilde{E}})$ is naturally isomorphic to $\mathcal{O}_{\widetilde{U}^*} \otimes_{\varphi^{-1}(\mathcal{O}_U)} \varphi^{-1}(\mathcal{E})$. ■

2.3.7 Precise Description of Dirac Type Singularities

Let us describe mini-holomorphic bundles of Dirac type more precisely. For simplicity, we assume that $U =]-2\epsilon, 2\epsilon[\times U_w$, where U_w is as above. We put $U^* := U \setminus \{(0, 0)\}$. We set $H_{>0} :=]0, 2\epsilon[\times\{0\}$.

Lemma 2.3.5 *For any mini-holomorphic bundle $\mathcal{E}$ of rank r, there exist a tuple of integers $\ell_1 \leq \ell_2 \leq \cdots \leq \ell_r$ and an isomorphism $\mathcal{E} \simeq \bigoplus_{i=1}^r \mathcal{O}_{U^*}(\ell_i H_{>0})$. The tuple of the integers $(\ell_1, \ldots, \ell_r)$ is called the weight of $\mathcal{E}$ at the Dirac type singularity* $(0, 0)$.

Proof Let $\mathcal{E}_{(\pm\epsilon,0)}$ denote the stalks of $\mathcal{E}$ at $(\pm\epsilon, 0)$, which are free $\mathcal{O}_{\mathbb{C},0}$-modules. We have the isomorphism $\mathcal{E}_{(-\epsilon,0)}(*0) \simeq \mathcal{E}_{(\epsilon,0)}(*0)$. It is a standard fact that there exists a frame $v_1, \ldots, v_r$ of $\mathcal{E}_{(-\epsilon,0)}$ such that $\mathcal{E}_{(\epsilon,0)}(*0) = \bigoplus_{i=1}^r \mathcal{O}_{\mathbb{C},0} w^{-\ell_i} v_i$. Then, the claim follows. ■

Lemma 2.3.6 *Let $\mathcal{E}$ be as in Lemma 2.3.5. We set $S_\epsilon^2 = \{(t, w) \mid t^2 + |w|^2 = \epsilon^2\} \subset U \setminus \{(0, 0)\}$ with the orientation as the boundary of $\{(t, w) \mid t^2 + |w|^2 \leq \epsilon^2\}$. Then, we have $\int_{S_\epsilon^2} c_1(\mathcal{E}) = \sum_{i=1}^r \ell_i$.*

Proof It is enough to check the claim in the case $\operatorname{rank}\mathcal{E} = 1$. There exists a frame v of $\mathcal{E}_{(-\epsilon,0)}$ such that $w^{-\ell}v$ is a frame of $\mathcal{E}_{(\epsilon,0)}$ under the isomorphism $\mathcal{E}_{(-\epsilon,0)}(*0) \simeq \mathcal{E}_{(\epsilon,0)}$. In the case $\mathcal{E}' = \mathcal{O}_U \cdot u$, we clearly have $\int_{S_\epsilon^2} c_1(\mathcal{E}') = 0$. There exists a complex structure of S_ϵ^2 such that (i) it is compatible with the orientation, (ii) w induces a holomorphic coordinate around $(\epsilon, 0)$. Then, by the correspondence $v \mapsto u$, we can identify $\mathcal{E}_{|S_\epsilon^2}$ with $\mathcal{E}'_{|S_\epsilon^2}\big(\ell(\epsilon, 0)\big)$. Hence, we obtain $\int_{S_\epsilon^2} c_1(\mathcal{E}) = \ell$. ■

We set $H_{<0} :=]-2\epsilon, 0[\times\{0\}$. Note that for any integers ℓ_i $(i = 1, 2)$ there exists the isomorphism

$$\mathcal{O}_{U^*}((\ell_1 + \ell_2)H_{>0}) \simeq \mathcal{O}_{U^*}(\ell_1 H_{>0} - \ell_2 H_{<0})$$

induced by the multiplication of w^{ℓ_2}.

Let $j : U^* \longrightarrow U$ denote the inclusion. We set $H_{\geq 0} := [0, 2\epsilon[\times\{0\}$. We obtain

$$j_*\mathcal{O}_{U^*}(\ell H_{>0}) \simeq \begin{cases} \mathcal{O}_U(\ell H_{\geq 0}) & (\ell < 0), \\ \mathcal{O}_U(\ell H_{>0}) & (\ell \geq 0). \end{cases}$$

Let ι_0 denote the inclusion $\{0\} \times U_w \longrightarrow U$. We obtain

$$\iota_0^{-1}\big(j_*\mathcal{O}_{U^*}(\ell H_{>0})\big) \simeq \begin{cases} \mathcal{O}_{U_w}(\ell\{0\}) & (\ell < 0), \\ \mathcal{O}_{U_w} & (\ell \geq 0). \end{cases}$$

2.3.8 *Subbundles and Quotient Bundles*

Let U be as in Sect. 2.3.7. Let $\mathcal{E}$ be a mini-holomorphic bundle of Dirac type on $(U, 0)$.

Lemma 2.3.7

- *Let $\mathcal{E}'$ be an $\mathcal{O}_{U^*}$-submodule of $\mathcal{E}$ which is also locally free. Then,* $(0, 0)$ *is Dirac type singularity of $\mathcal{E}'$.*
- *Let $\mathcal{E}''$ be a quotient $\mathcal{O}_{U^*}$-module of $\mathcal{E}$ which is also locally free. Then,* $(0, 0)$ *is Dirac type singularity of $\mathcal{E}''$.*

Proof It is easy to see that the scattering maps of $\mathcal{E}'$ and $\mathcal{E}''$ are meromorphic at 0. Hence, $(0, 0)$ is Dirac type singularity. ■

Remark 2.3.8 Let $0 \longrightarrow \mathcal{E}' \longrightarrow \mathcal{E} \longrightarrow \mathcal{E}'' \longrightarrow 0$ be an exact sequence of mini-holomorphic bundles of Dirac type on U^*. We have the Kronheimer resolutions $\widetilde{\mathcal{E}}'$, $\widetilde{\mathcal{E}}$ and $\widetilde{\mathcal{E}}''$ at $(0, 0)$, as in Sect. 2.3.6. Note that $0 \longrightarrow \widetilde{\mathcal{E}}' \longrightarrow \widetilde{\mathcal{E}} \longrightarrow \widetilde{\mathcal{E}}'' \longrightarrow 0$ is not necessarily exact at $(0, 0)$. ■

2.3.9 Basic Functoriality

Let $(E_i, \overline{\partial}_{E_i})$ $(i = 1, 2)$ be a three-dimensional mini-complex manifolds. Let $(E_1, \overline{\partial}_{E_1}) \oplus (E_2, \overline{\partial}_{E_2})$ denote the mini-holomorphic bundle $(E_1 \oplus E_2, \overline{\partial}_{E_1 \oplus E_2})$, where the operator $\overline{\partial}_{E_1 \oplus E_2} : C^\infty(M, E_1 \oplus E_2) \longrightarrow C^\infty(M, (E_1 \oplus E_2) \otimes \Omega_M^{0,1})$ is determined by $\overline{\partial}_{E_1 \oplus E_2}(s_1 \oplus s_2) = \overline{\partial}_{E_1}(s_1) \oplus \overline{\partial}_{E_2}(s_2)$ for $s_i \in C^\infty(M, E_i)$. Let $(E_1, \overline{\partial}_{E_1}) \otimes (E_2, \overline{\partial}_{E_2})$ denote the mini-holomorphic bundle $(E_1 \otimes E_2, \overline{\partial}_{E_1 \otimes E_2})$, where the operator $\overline{\partial}_{E_1 \otimes E_2} : C^\infty(M, E_1 \otimes E_2) \longrightarrow C^\infty(M, (E_1 \otimes E_2) \otimes \Omega_M^{0,1})$ satisfies

$$\overline{\partial}_{E_1 \otimes E_2}(s_1 \otimes s_2) = \overline{\partial}_{E_1}(s_1) \otimes s_2 + s_1 \otimes \overline{\partial}_{E_2}(s_2)$$

for $s_i \in C^\infty(M, E_i)$. Let $\mathrm{Hom}\big((E_1, \overline{\partial}_{E_1}), (E_2, \overline{\partial}_{E_2})\big)$ denote the mini-holomorphic bundle

$$(\mathrm{Hom}(E_1, E_2), \overline{\partial}_{\mathrm{Hom}(E_1, E_2)}),$$

where the operator

$$\overline{\partial}_{\mathrm{Hom}(E_1, E_2)} : C^\infty(M, \mathrm{Hom}(E_1, E_2)) \longrightarrow C^\infty(M, \mathrm{Hom}(E_1, E_2) \otimes \Omega_M^{0,1})$$

satisfies

$$\overline{\partial}_{E_2}(f(s_1)) = \overline{\partial}_{\mathrm{Hom}(E_1, E_2)}(f)(s_1) + f(\overline{\partial}_{E_1} s_1)$$

for $f \in C^\infty(M, \mathrm{Hom}(E_1, E_2))$, and $s_1 \in C^\infty(M, E_1)$.

For a mini-holomorphic bundle $(E, \overline{\partial}_E)$ on M, the dual $(E, \overline{\partial}_E)^\vee$ is defined as $\mathrm{Hom}((E, \overline{\partial}_E), (\mathbb{C} \times M, \overline{\partial}_M))$.

2.4 Monopoles

2.4.1 Monopoles and Mini-Holomorphic Bundles

Let M be an oriented three-dimensional manifold with a Riemannian metric g_M.

Definition 2.4.1 Let E be a complex vector bundle with a Hermitian metric h, a unitary connection ∇ and an anti-self-adjoint endomorphism ϕ. The tuple (E, h, ∇, ϕ) is called a monopole on M if the Bogomolny equation $F(\nabla) - *\nabla\phi = 0$, where $F(\nabla)$ denotes the curvature of ∇, and $*$ denotes the Hodge star operator with respect to g_M. ■

Suppose that M is equipped with a mini-complex structure such that the orthogonal decomposition $TM = T_S M \oplus (T_S M)^{\perp}$ induces a splitting of $TM \longrightarrow T_Q M$, and that the multiplication of $\sqrt{-1}$ on $T_Q M$ is an isometry with respect to the induced metric of $T_Q M$. We identify $T_S M$ with the product bundle $\mathbb{R} \times M$.

Let $(E, \overline{\partial}_E)$ be a mini-holomorphic bundle on M. For any Hermitian metric h of E, the Chern connection ∇_h and the Higgs field ϕ_h are associated.

Definition 2.4.2 We say that $(E, \overline{\partial}_E, h)$ is a monopole if (E, h, ∇_h, ϕ_h) is a monopole. In the case, we say that $(E, \overline{\partial}_E)$ is the mini-holomorphic bundle underlying the monopole. ■

2.4.2 Euclidean Monopoles

In this paper, we study monopoles on spaces which are locally isomorphic to $\mathbb{R} \times \mathbb{C}$ with the natural metric and the natural mini-complex structure. Let us look at the condition more explicitly in terms of local coordinate systems.

Let U be an open subset of $\mathbb{R} \times \mathbb{C}$. It is equipped with the metric $dt\,dt + dw\,d\overline{w}$. It is also equipped with the mini-complex structure. Let $(E, \overline{\partial}_E)$ be a mini-holomorphic bundle on U. Let h be a Hermitian metric of E. We obtain the differential operator $\partial_{E,h,w} : E \longrightarrow E$ determined by the condition

$$\partial_{\overline{w}} h(u, v) = h(\partial_{E,\overline{w}} u, v) + h(u, \partial_{E,h,w} v).$$

We also obtain the differential operator $\partial'_{E,h,t} : E \longrightarrow E$ determined by the condition

$$\partial_t h(u, v) = h(\partial_{E,t} u, v) + h(u, \partial'_{E,h,t} v).$$

We set

$$\nabla_{h,x} := \partial_{E,\overline{w}} + \partial_{E,w}, \quad \nabla_{h,y} := -\sqrt{-1}\big(\partial_{E,\overline{w}} - \partial_{E,w}\big),$$

$$\nabla_{h,t} = \frac{1}{2}\big(\partial_{E,t} + \partial'_{E,h,t}\big), \quad \phi_h = \frac{\sqrt{-1}}{2}\big(\partial_{E,t} - \partial'_{E,h,t}\big).$$

Thus, from the metric h, we obtain the unitary connection $\nabla_h(s) = \nabla_{h,x}(s)\,dx + \nabla_{h,y}(s)\,dy + \nabla_{h,y}(s)\,dt$, and the anti-self-adjoint section ϕ_h of $\mathrm{End}(E)$. Note that the mini-holomorphic structure is recovered by

$$\partial_{E,\overline{w}} = \frac{1}{2}(\nabla_{h,x} + \sqrt{-1}\nabla_{h,y}), \qquad \partial_{E,t} = \nabla_{h,t} - \sqrt{-1}\phi_h.$$

We also write $\partial_{E,\overline{w}}$ and $\partial_{E,h,w}$ as $\nabla_{h,\overline{w}}$ and $\nabla_{h,w}$. Let $F(h)$ denote the curvature of ∇_h, which is expressed as $F(h) = F(h)_{\overline{w},t} d\overline{w}\,dt + F(h)_{w,t} dw\,dt + F(h)_{\overline{w},w} d\overline{w}\,dw$.

Lemma 2.4.3 *$F(h)_{\overline{w},t} - \sqrt{-1}\nabla_{h,\overline{w}}\phi_h = 0$ and $F(h)_{w,t} + \sqrt{-1}\nabla_{h,w}\phi_h = 0$ hold.*

Proof We have the commutativity $[\partial_{E,\overline{w}}, \partial_{E,t}] = 0$, which implies

$$\big[\nabla_{h,\overline{w}}, \nabla_{h,t} - \sqrt{-1}\phi_h\big] = F(h)_{\overline{w},t} - \sqrt{-1}\nabla_{h,\overline{w}}\phi_h = 0.$$

As the adjoint, we obtain $F(h)_{w,t} + \sqrt{-1}\nabla_{h,w}\phi_h = 0$. ■

Corollary 2.4.4 *A mini-holomorphic bundle $(E, \overline{\partial}_E)$ with a Hermitian metric h is a monopole, i.e., the associated tuple (E, h, ∇_h, ϕ_h) is a monopole if and only if the following equation is satisfied.*

$$\frac{\sqrt{-1}}{2}\nabla_{h,t}\phi_h + F(h)_{\overline{w},w} = 0. \tag{2.3}$$

Proof Because $*(dw) = -\sqrt{-1}dw\,dt$, $*(d\overline{w}) = \sqrt{-1}d\overline{w}\,dt$, and $*(dt) = -\frac{\sqrt{-1}}{2}d\overline{w}\,dw$, the claim follows from Lemma 2.4.3. ■

Conversely, let E be a C^∞-vector bundle on U with a Hermitian metric h, a unitary connection ∇ and an anti-Hermitian endomorphism ϕ. We set $\partial_{E,\overline{w}} := \nabla_{\overline{w}}$ and $\partial_{E,t} := \nabla_t - \sqrt{-1}\phi$. We obtain the differential operator $\overline{\partial}_E : C^\infty(U, E) \longrightarrow C^\infty(U, E \otimes \Omega_U^{0,1})$ by $\overline{\partial}_E(s) = \partial_{E,\overline{w}}(s)\,d\overline{w} + \partial_{E,t}(s)\,dt$ for any $s \in C^\infty(U, E)$. In general, $(E, \overline{\partial}_E)$ is not necessarily mini-holomorphic. If (E, h, ∇, ϕ) is a monopole, the Bogomolny equation implies the commutativity $\big[\partial_{E,\overline{w}}, \partial_{E,t}\big] = 0$, and hence $(E, \overline{\partial}_E)$ is a mini-holomorphic bundle, which is called the underlying mini-holomorphic bundle of the monopole (E, h, ∇, ϕ). Note that (E, h, ∇, ϕ) is recovered from $(E, \overline{\partial}_E)$ with h.

Remark 2.4.5 Let $(E, \overline{\partial}_E)$ be a mini-holomorphic bundle with a Hermitian metric h. Let ∇_h and ϕ_h be the Chern connection and the Higgs field. We also have the operator $\partial'_{E,h,t} = \nabla_{h,t} + \sqrt{-1}\phi_h$. By the construction, we have

$$[\nabla_{h,\overline{w}}, \partial'_{E,h,t}] = 2[\nabla_{h,\overline{w}}, \nabla_{h,t}], \quad [\nabla_{h,w}, \partial_{E,t}] = 2[\nabla_{h,w}, \nabla_{h,t}].$$

We also have

$$\nabla_{h,\overline{w}}\phi_h = -\frac{\sqrt{-1}}{2}[\nabla_{h,\overline{w}}, \partial'_{E,h,t}], \quad \nabla_{h,w}\phi_h = \frac{\sqrt{-1}}{2}[\nabla_{h,w}, \partial_{E,t}],$$

$$\nabla_{h,t}(\phi_h) = -\frac{\sqrt{-1}}{2}[\partial_{E,t}, \partial'_{E,h,t}].$$

■

2.4.3 Dirac Type Singularity

We recall the notion of Dirac type singularity of monopoles. It is originally introduced by Kronheimer [51], and it was later generalized by Pauly [75] to the context of general three-dimensional Riemannian manifolds, and by Charbonneau and Hurtubise [11] in the higher rank case.

Let $\varphi : \mathbb{C}^2 \longrightarrow \mathbb{R} \times \mathbb{C}$ be the map (2.2). Let U be a neighbourhood of $(0, 0)$ in $\mathbb{R} \times \mathbb{C}$. We set $U^* := U \setminus \{(0, 0)\}$.

Let (E, h, ∇, ϕ) be a monopole on U^*. We set $\xi := -u_1 d\overline{u}_1 + \overline{u}_1 du_1 - \overline{u}_2 du_2 + u_2 d\overline{u}_2$. We put $(\widetilde{E}, \widetilde{h}) := \varphi^{-1}(E, h)$. We obtain the unitary connection $\widetilde{\nabla} := \varphi^*(\nabla) + \sqrt{-1}\varphi^*(\phi) \otimes \xi$. As proved in [51], $(\widetilde{E}, \widetilde{h}, \widetilde{\nabla})$ is an instanton on $\varphi^{-1}(U^*)$. Recall that $(0, 0)$ is called Dirac type singularity of (E, h, ∇, ϕ) if $(\widetilde{E}, \widetilde{h}, \widetilde{\nabla})$ extends to an instanton on $\varphi^{-1}(U)$. It particularly implies that the underlying mini-holomorphic bundle $(E, \overline{\partial}_E)$ of (E, h, ∇, ϕ) is of Dirac type on $(U, (0, 0))$.

We have the following simple characterization of Dirac type singularity

Theorem 2.4.6 ([71, Theorem 1]) $(0, 0)$ *is a Dirac type singularity of* (E, h, ∇, ϕ) *if and only if* $|\phi|_h = O\big((t^2 + |w|^2)^{-1/2}\big)$. ■

We also have the following characterization which easily follows from [71, Theorem 2].

Proposition 2.4.7 *Let* $(E, \overline{\partial}_E, h)$ *be a monopole on* U^*. *It has a Dirac type singularity at* $(0, 0)$ *if and only if the following holds.*

- $(E, \overline{\partial}_E)$ *is a mini-holomorphic bundle of Dirac type on* $(U, (0, 0))$. *In particular, we obtain the Kronheimer resolution* $(\widetilde{E}_0, \overline{\partial}_{\widetilde{E}_0})$ *on* $\varphi^{-1}(U)$. *(See* Sect. 2.3.6.*)*

- *Let h_1 be a C^∞-metric of E such that $\varphi^{-1}(h_1)$ induces a C^∞-metric on $\widetilde{E}_0$. Then, there exist positive constants $C_1 > 1$ and N_1 such that $C_1^{-1}(|t|^2+|w|^2)^{-N_1}\cdot h_1 \leq h \leq C_1(|t|^2+|w|^2)^{N_1}\cdot h_1$.* ■

2.4.3.1 Dirac Monopoles (Examples)

The fundamental examples are Dirac monopoles. Let us recall the description from [71, §5.2] with a minor correction. We set $U = \mathbb{R}\times\mathbb{C}$, $U^* = U \setminus \{(0,0)\}$, and $A_\pm := U \setminus \{(t,0) \mid \pm t \leq 0\}$. We have $U^* = A_+ \cup A_-$. We regard U as a mini-complex manifold in a natural way. For any integer m, let $L^{(m)}$ be the mini-holomorphic line bundle on U^* equipped with mini-holomorphic frames $\sigma_\pm^{(m)}$ of $L^{(m)}_{|A_\pm}$ such that $\sigma_-^{(m)} = (w/2)^m\sigma_+^{(m)}$ on $A_+\cap A_-$. Let $h^{(m)}$ be the Hermitian metric of $L^{(m)}$ determined by

$$h^{(m)}(\sigma_+^{(m)},\sigma_+^{(m)}) = 2^m(t+R)^{-m},$$

$$h^{(m)}(\sigma_-^{(m)},\sigma_-^{(m)}) = 2^{-m}(-t+R)^{-m},$$

where $R=\sqrt{t^2+|w|^2}$. We obtain the Chern connection $\nabla^{(m)}$ and the Higgs field $\phi^{(m)}$ associated with $(L^{(m)},\overline{\partial}_{L^{(m)}},h^{(m)})$.

We obtain the holomorphic line bundle $\widetilde{L}^{(m)}$ on $\mathbb{C}^2\setminus\{(0,0)\} = \varphi^{-1}(U^*)$ as in Sect. 2.3.6. It is equipped with a holomorphic frame $e^{(m)}$ such that $e^{(m)} = u_1^m\varphi^{-1}(\sigma_+^{(m)})$ on $\varphi^{-1}(A_+)$, and $e^{(m)} = u_2^{-m}\varphi^{-1}(\sigma_-^{(m)})$ on $\varphi^{-1}(A_-)$. Let $\widetilde{h}_0^{(m)}$ be the Hermitian metric of $\widetilde{L}^{(m)}$ defined as $\widetilde{h}_0^{(m)}(e^{(m)},e^{(m)}) = 1$. It is easy to check that $\widetilde{h}_0^{(m)}$ is the pull back of $h^{(m)}$. Because $(\widetilde{L}^{(m)},\widetilde{h}^{(m)})$ is an instanton, we obtain that $(L^{(m)},h^{(m)},\nabla^{(m)},\phi^{(m)})$ is a monopole.

Let us compute $\nabla^{(m)}$ and $\phi^{(m)}$ explicitly. $\partial_{L^{(m)},\overline{w}}\sigma_+^{(m)} = \partial_{L(m),t}\sigma_+^{(m)} = 0$. We have

$$\partial_{L^{(m)},w}\sigma_+^{(m)} = \sigma_+^{(m)}\partial_w\log h^{(m)}(\sigma_+^{(m)},\sigma_+^{(m)}) = \sigma_+^{(m)}\frac{-m\overline{w}}{2R(t+R)}.$$

We also have

$$\partial'_{L^{(m)},t}\sigma_+^{(m)} = \sigma_+^{(m)}\partial_t\log h^{(m)}(\sigma_+^{(m)},\sigma_+^{(m)}) = \sigma_+^{(m)}\frac{-m}{R}.$$

Therefore, we obtain

$$\nabla^{(m)}\sigma_+^{(m)} = \sigma_+^{(m)}\Big(-\frac{m\overline{w}}{2R(t+R)}\,dw - \frac{m}{2R}\,dt\Big),\qquad \phi^{(m)} = \frac{\sqrt{-1}m}{2R}.$$

Remark 2.4.8 In [71, §5.2], $\phi^{(m)} = \frac{\sqrt{-1}m}{R}$ should be corrected to $\phi^{(m)} = \frac{\sqrt{-1}m}{2R}$. ■

2.4.4 Basic Functoriality

Let M be a three-dimensional mini-complex manifold with a Riemannian metric. Assume that it is locally isomorphic to $\mathbb{R} \times \mathbb{C}$ with the canonical mini-complex structure and the Euclidean metric. Let $(E_i, h_i, \nabla_i, \phi_i)$ $(i = 1, 2)$ be monopoles on M. The underlying mini-holomorphic bundle is denoted by $(E_i, \overline{\partial}_{E_i})$.

The vector bundle $E_1 \oplus E_2$ is equipped with the naturally induced Hermitian metrics $h_{E_1 \oplus E_2}$, the unitary connection $\nabla_{E_1 \oplus E_2}$, and the Higgs field $\phi_{E_1 \oplus E_2} = \phi_{E_1} \oplus \phi_{E_2}$. The tuple

$$(E_1, h_1, \nabla_1, \phi_1) \oplus (E_2, h_2, \nabla_2, \phi_2) = (E_1 \oplus E_2, h_{E_1 \oplus E_2}, \nabla_{E_1 \oplus E_2}, \phi_{E_1 \oplus E_2})$$

is a monopole. The underlying mini-holomorphic bundle is naturally isomorphic to $(E_1, \overline{\partial}_{E_1}) \oplus (E_2, \overline{\partial}_{E_2})$.

The vector bundle $E_1 \otimes E_2$ is equipped with the naturally induced Hermitian metrics $h_{E_1 \otimes E_2}$, the unitary connection $\nabla_{E_1 \otimes E_2}$, and the Higgs field $\phi_{E_1 \otimes E_2} = \phi_{E_1} \otimes \mathrm{id}_{E_2} + \mathrm{id}_{E_1} \otimes \phi_{E_2}$. The tuple

$$(E_1, h_1, \nabla_1, \phi_1) \otimes (E_2, h_2, \nabla_2, \phi_2) = (E_1 \otimes E_2, h_{E_1 \otimes E_2}, \nabla_{E_1 \otimes E_2}, \phi_{E_1 \otimes E_2})$$

is a monopole. The underlying mini-holomorphic bundle is naturally isomorphic to $(E_1, \overline{\partial}_{E_1}) \otimes (E_2, \overline{\partial}_{E_2})$.

The vector bundle $\mathrm{Hom}(E_1, E_2)$ is equipped with the naturally induced Hermitian metrics $h_{\mathrm{Hom}(E_1,E_2)}$, the unitary connection $\nabla_{\mathrm{Hom}(E_1,E_2)}$, and the Higgs field $\phi_{\mathrm{Hom}(E_1,E_2)}$ given as $\phi_{\mathrm{Hom}(E_1,E_2)}(s) = \phi_2 \circ s - s \circ \phi_1$. The tuple

$$\begin{aligned}
&\mathrm{Hom}\Big((E_1, h_1, \nabla_1, \phi_1), (E_2, h_2, \nabla_2, \phi_2)\Big) \\
&\qquad = (\mathrm{Hom}(E_1, E_2), h_{\mathrm{Hom}(E_1,E_2)}, \nabla_{\mathrm{Hom}(E_1,E_2)}, \phi_{\mathrm{Hom}(E_1,E_2)}) \qquad (2.4)
\end{aligned}$$

is a monopole. The underlying mini-holomorphic bundle is naturally isomorphic to $\mathrm{Hom}\big((E_1, \overline{\partial}_{E_1}), (E_2, \overline{\partial}_{E_2})\big)$.

The product line bundle $\mathbb{C} \times M$ on M is equipped with the metric h_0 defined by $h_0(1, 1) = 1$, and the trivial unitary connection ∇_0 and the Higgs field 0, and the tuple $(\mathbb{C} \times M, h_0, \nabla_0, 0)$ is a monopole. For any monopole (E, h, ∇, ϕ) on M, we obtain the induced monopole $(E, h, \nabla, \phi)^{\vee} = \mathrm{Hom}\Big((E, h, \nabla, \phi), (\mathbb{C} \times M, h_0, \nabla_0, 0)\Big)$ as the dual. The underlying mini-holomorphic bundle of $(E, h, \nabla, \phi)^{\vee}$ is naturally isomorphic to the dual of the mini-holomorphic bundle underlying (E, h, ∇, ϕ).

2.5 Dimensional Reduction from 4*D* to 3*D*

2.5.1 *Instantons Induced by Monopoles*

Monopoles are regarded as the one-dimensional reduction of instantons. Namely, monopoles are equivalent to $\mathbb{R}$-equivariant instantons. Let us recall the construction explicitly in terms of coordinate systems.

Let U be an open subset of $\mathbb{R} \times \mathbb{C}$. We use real and complex coordinates t and $w = x + \sqrt{-1}y$ for $\mathbb{R}$ and $\mathbb{C}$, respectively. Let $p : \mathbb{C} \times \mathbb{C} \longrightarrow \mathbb{R} \times \mathbb{C}$ be the map given by $(s + \sqrt{-1}t, w) \longmapsto (t, w)$. We use the standard Euclidean metrics $dt\,dt + dw\,d\overline{w}$ on $\mathbb{R} \times \mathbb{C}$, and $dz\,d\overline{z} + dw\,d\overline{w}$ on $\mathbb{C} \times \mathbb{C}$.

Let E be a C^∞-bundle on U with a Hermitian metric h, a unitary connection ∇_E and an anti-self-adjoint endomorphism ϕ of E. We set $(\widetilde{E}, \widetilde{h}) := p^*(E, h)$ and $\nabla_{\widetilde{E}} := p^*\nabla_E + p^*\phi\,ds$ on $p^{-1}(U)$.

Proposition 2.5.1 (Hitchin) *(E, h, ∇_E, ϕ) is a monopole if and only if $(\widetilde{E}, \widetilde{h}, \nabla_{\widetilde{E}})$ is an instanton.*

Proof Let $*_3$ denote the Hodge star operator of $\mathbb{R}^3$ with the orientation $dt \wedge dx \wedge dy$. For example, we have $*_3(dx \wedge dy) = dt$ and $*_3(dx) = dy \wedge dt$. Note that $*_3 \circ *_3 = \mathrm{id}$. Let $*_4$ denote the Hodge star operator on $\mathbb{R}^4$ with the orientation $ds \wedge dt \wedge dx \wedge dy$. For example, we have $*_4(dx \wedge dy) = ds \wedge dt$ and $*_4(dx \wedge ds) = -dy \wedge dt$.

By the construction of $\nabla_{\widetilde{E}}$, we obtain $F(\nabla_{\widetilde{E}}) = p^*F(\nabla_E) + p^*(\nabla_E\phi)\,ds$, which implies

$$*_4F(\nabla_{\widetilde{E}}) = -p^*(*_3F(\nabla_E))\,ds - p^*(*_3\nabla_E\phi).$$

Hence, we obtain

$$*_4F(\nabla_{\widetilde{E}}) + F(\nabla_{\widetilde{E}}) = p^*\big(F(\nabla_E) - *_3\nabla_E\phi\big) + p^*\big(\nabla_E\phi - *_3F(\nabla_E)\big)\,ds.$$

Therefore, $*_4F(\nabla_{\widetilde{E}}) + F(\nabla_{\widetilde{E}}) = 0$ holds if and only if $F(\nabla_E) = *_3\nabla_E\phi$ holds. ■

Remark 2.5.2 Let us consider an $\mathbb{R}$-action on $p^{-1}(U)$ by $T \cdot (z, w) = (z + T, w)$. Then, the proposition means that $\mathbb{R}$-equivariant instantons on $p^{-1}(U)$ correspond to monopoles on U. ■

The following lemma is clear by the construction.

Lemma 2.5.3 *Let f be any section of E. Then, $(\nabla_{\widetilde{E},s} + \sqrt{-1}\nabla_{\widetilde{E},t})p^*f = 0$ holds if and only if $(\nabla_{E,t} - \sqrt{-1}\phi)f = 0$ holds. In other words, $\mathbb{R}$-invariant $\widetilde{\nabla}_{\overline{z}}$-holomorphic sections on $p^{-1}(U)$ correspond to $(\nabla_{E,t} - \sqrt{-1}\phi)$-flat sections on U.* ■

The following lemma follows from the estimate for instantons due to Uhlenbeck [87].

Lemma 2.5.4 *Let (E, h, ∇_E, ϕ) be a monopole on U. Suppose $\left|\nabla_E \phi\right|_h \leq \epsilon$, or equivalently $\left|F(\nabla_E)\right|_h \leq \epsilon$ for a positive number ϵ. Let K be any relatively compact subset of U. If ϵ is sufficiently small, the higher derivatives of $\nabla_E \phi$ and $F(\nabla_E)$ are dominated by $C_K \epsilon$ on K for some $C_K > 0$, where C_K depends on the order of derivatives and K, but independent of (E, h, ∇_E, ϕ).* ■

We write some formulas. We clearly have $\nabla_{\widetilde{E},\overline{w}}(p^{-1}(u)) = p^{-1}(\nabla_{E,\overline{w}} u)$ and $\nabla_{\widetilde{E},w}(p^{-1}(u)) = p^{-1}(\nabla_{E,w} u)$ for any $u \in C^\infty(U, E)$. As for the derivative in the z-direction, we have

$$\nabla_{\widetilde{E},\overline{z}}(p^{-1}(u)) = \frac{\sqrt{-1}}{2} p^{-1}\big((\nabla_t - \sqrt{-1}\phi)u\big),$$

$$\nabla_{\widetilde{E},z}(p^{-1}(u)) = -\frac{\sqrt{-1}}{2} p^{-1}\big((\nabla_t + \sqrt{-1}\phi)u\big).$$

For the real coordinate system (s, t) induced by $z = s + \sqrt{-1}t$, we obtain $\nabla_{\widetilde{E},s} p^{-1}(u) = p^{-1}(\phi u)$ and $\nabla_{\widetilde{E},t} p^{-1}(u) = p^{-1}(\nabla_t u)$.

2.5.2 *Holomorphic Bundles and Mini-Holomorphic Bundles*

We have the corresponding procedure to construct a holomorphic bundle from a mini-holomorphic bundle. Let E be a C^∞-vector bundle on U with a mini-holomorphic structure $\overline{\partial}_E : C^\infty(U, E) \longrightarrow C^\infty(U, E \otimes \Omega^{0,1})$. We have the operators $\partial_{E,\overline{w}}$ and $\partial_{E,t}$ on $C^\infty(U, E)$ satisfying $[\partial_{E,\overline{w}}, \partial_{E,t}] = 0$. We set $\widetilde{E} := p^{-1}(E)$. We have the holomorphic structure $\overline{\partial}_{\widetilde{E}}$ satisfying the following formula for any C^∞-section v of E:

$$\overline{\partial}_{\widetilde{E}}\big(p^{-1}(v)\big) = d\overline{w} \wedge p^{-1}\big(\partial_{\overline{w}} v\big) + d\overline{z} \wedge \frac{\sqrt{-1}}{2} p^{-1}\big(\partial_t v\big). \tag{2.5}$$

Let h be a Hermitian metric of E. It induces a Hermitian metric $\widetilde{h}$ of $\widetilde{E}$. We can easily check that $(E, \overline{\partial}_E, h)$ is a monopole on U with respect to the Euclidean metric $dt\,dt + dw\,d\overline{w}$ if and only if $(\widetilde{E}, \overline{\partial}_{\widetilde{E}}, \widetilde{h})$ is an instanton on $p^{-1}(U)$ with respect to the Euclidean metric $dz\,d\overline{z} + dw\,d\overline{w}$. This is compatible with the construction in Sect. 2.4.1.

2.6 Dimensional Reduction from 3*D* to 2*D*

2.6.1 *Monopoles Induced by Harmonic Bundles*

Let us recall that the concept of harmonic bundle was discovered by Hitchin [34] as the two-dimensional reduction of instantons. Because monopoles are regraded as the one-dimensional reduction of instantons, harmonic bundles are the one-dimensional reduction of monopoles. We recall the construction in an explicit way.

Let U be a Riemann surface equipped with a holomorphic 1-form φ which is nowhere vanishing on U. Let $(V, \overline{\partial}_V, h_V, \theta_V)$ be a harmonic bundle on U. We obtain the holomorphic endomorphism f of V determined by $\theta_V = f\varphi$. Let Y be a real one-dimensional manifold equipped with a closed real 1-form τ which is nowhere vanishing on Y. The product $Y \times U$ is naturally equipped with a mini-holomorphic structure and the metric $\tau\tau + \varphi\overline{\varphi}$.

Let $p_2 : Y \times U \longrightarrow U$ be the projection. We set

$$(E, h_E) := p_2^*(V, h_V), \quad \nabla_E = p_2^*\nabla_V - \sqrt{-1}p_2^*(f + f^\dagger)\tau, \quad \phi_E = p_2^*(f - f^\dagger). \tag{2.6}$$

Here, $p_2^*\nabla_V$ denotes the connection of E induced as the pull back of ∇_V. Then, $(E, h_E, \nabla_E, \phi_E)$ is a monopole on $Y \times U$ with the metric $\tau\tau + \varphi\overline{\varphi}$. We set $\mathrm{Hit}_2^3(V, \overline{\partial}_V, \theta_V, h_V) := (E, h_E, \nabla_E, \phi_E)$.

Let $\mathfrak{v}$ be the vector field of Y such that $\langle \tau, \mathfrak{v} \rangle = 1$. It naturally defines a vector field on $Y \times U$, which is also denoted by $\mathfrak{v}$. Let $L_{\mathfrak{v}}$ be the differential operator of E defined by $L_{\mathfrak{v}}(g\, p_2^* v) = \mathfrak{v}(g) p_2^* v$ for any $g \in C^\infty(Y \times U)$ and any C^∞-section v of V. Then, we have

$$\nabla_{E,\mathfrak{v}} - \sqrt{-1}\phi_E = L_{\mathfrak{v}} - \sqrt{-1}p_2^*(f + f^\dagger) - \sqrt{-1}p_2^*(f - f^\dagger) = L_{\mathfrak{v}} - 2\sqrt{-1}p_2^* f.$$

2.6.2 *Mini-Holomorphic Bundles Induced by Holomorphic Bundles with a Higgs Field*

We have the corresponding procedure to construct a mini-holomorphic bundle on $Y \times U$ from a holomorphic bundle V with an endomorphism f on U. We use the natural splitting $\Omega^{0,1}_{Y\times U} = (T^{\mathbb{C}}Y)^\vee \oplus \Omega^{0,1}_U$.

Let $(V, \overline{\partial}_V, \theta_V)$ be a Higgs bundle on U. We have the expression $\theta_V = f\varphi$. We set

$$E := p_2^{-1}(V), \quad \overline{\partial}_E := p_2^*(\overline{\partial}_V) - 2\sqrt{-1}p_2^* f\, \tau.$$

Here, $p_2^*(\overline{\partial}_V)$ is the mini-holomorphic structure of E obtained as the pull back of $\overline{\partial}_V$. We obtain a mini-holomorphic bundle $(E, \overline{\partial}_E)$ on $Y \times U$. We denote it by $p_2^*(V, \overline{\partial}_V, \theta_V)$ or $p_2^*(V, \overline{\partial}_V, f)$.

If V is equipped with a Hermitian metric h_V, we have the induced metric $h = p_2^*(h_V)$ on $p_2^*(V, \overline{\partial}_V, \theta_V)$. The Chern connection and the Higgs fields are induced by the formula (2.6). If h_V is a harmonic metric of the Higgs bundle $(V, \overline{\partial}_V, \theta_V)$, then h satisfies the Bogomolny equation for $p_2^*(V, \overline{\partial}_V, \theta_V)$ with respect to the metric $\tau\overline{\tau} + \varphi\overline{\varphi}$. This is compatible with the construction in Sect. 2.6.1.

Remark 2.6.1 The above procedure describes the dimensional reduction of the underlying objects of monopoles and harmonic bundles at the twistor parameter 0. There exists a similar procedure for each twistor parameter λ which we shall describe in Sect. 2.8. ∎

2.6.3 Mini-Holomorphic Sections and Monodromy

Let us consider $Y_1 = \mathbb{R}$ equipped with the nowhere vanishing closed 1-form $\tau_1 = dt$, where t is the standard coordinate of $\mathbb{R}$. Let $p_{2,1} : Y_1 \times U \longrightarrow U$ denote the projection. For any Higgs bundle $(V, \overline{\partial}_V, \theta_V)$ on U, we obtain the mini-holomorphic bundle $p_{2,1}^*(V, \overline{\partial}_V, \theta_V)$ on $Y_1 \times U$ as explained in Sect. 2.6.2. The following lemma is clear by the construction.

Lemma 2.6.2 *For any holomorphic section v of V on U,*

$$\exp\big(2\sqrt{-1}tp_{2,1}^*(f)\big)p_{2,1}^*(v)$$

is a mini-holomorphic section of $p_{2,1}^*(V, \overline{\partial}_V, \theta_V)$. ∎

Let $Y_2 = \mathbb{R}/T\mathbb{Z}$ equipped with $\tau_2 = dt$. Let $p_{2,2} : Y_2 \times U \longrightarrow U$ denote the projection. For any Higgs bundle $(V, \overline{\partial}_V, \theta_V)$ on U, we obtain the mini-holomorphic bundle $p_{2,2}^*(V, \overline{\partial}_V, \theta_V)$ on $Y_2 \times U$.

For $P \in U$, let γ_P denote the loop $[0, T] \longrightarrow Y \times U$ given by $s \longmapsto (s, P)$. We identify $V_{|P}$ and $E_{|(0,P)}$. By the mini-holomorphic structure, we obtain the parallel transport along γ_P which induces the monodromy automorphism $M(\gamma_P)$ of $V_{|P}$. We obtain the following lemma from the previous one.

Lemma 2.6.3 $M(\gamma_P) = \exp(2\sqrt{-1}Tf_{|P})$. ∎

2.6.4 Appendix: Monopoles as Harmonic Bundles of Infinite Rank

Let $Y \times U$ with $\tau\overline{\tau}+\varphi\overline{\varphi}$ be as in Sect. 2.6.1. Let E be a C^{∞}-vector bundle on $Y \times U$ equipped with a Hermitian metric h, a unitary connection ∇ and an anti-Hermitian endomorphism ϕ. Let $\mathcal{E}^{\infty}$ denote the sheaf of C^{∞}-sections of E. Let $\mathcal{C}_U^{\infty}$ denote the sheaf of C^{∞}-functions on U. We obtain the $\mathcal{C}_U^{\infty}$-module $p_{2!}\mathcal{E}^{\infty}$, where $p_{2!}$ denote the proper push-forward with respect to p_2. It is equipped with the Hermitian metric $p_{2!}h$ defined as

$$p_{2!}h(s_1,s_2)=\int_Y h(s_1,s_2)\tau$$

for sections s_i of $p_{2!}\mathcal{E}^{\infty}$, which we can naturally regard as C^{∞}-sections of E whose supports are proper over U. By using the $\Omega^{0,1}(U)$-part $\overline{\partial}_E^Q$ of ∇ in the decomposition $T^*(Y\times U)\otimes\mathbb{C}=T^*(Y)\otimes\mathbb{C}\oplus\Omega^{1,0}(U)\oplus\Omega^{0,1}(U)$, we obtain the holomorphic structure $\overline{\partial}_{p_{2!}\mathcal{E}^{\infty}}$ of $p_{2!}\mathcal{E}^{\infty}$ by $\overline{\partial}_{p_{2!}\mathcal{E}^{\infty}}(s)=\overline{\partial}_E^Q(s)$. Similarly, by using the $\Omega^{1,0}(U)$-part ∂_E^Q of ∇, we obtain the differential operator $\partial_{p_{2!}\mathcal{E}^{\infty}}$ of $p_{2!}\mathcal{E}^{\infty}$ by $\partial_{p_{2!}\mathcal{E}^{\infty}}(s)=\partial_E^Q(s)$. It is easy to check that $\overline{\partial}_{p_{2!}\mathcal{E}^{\infty}}+\partial_{p_{2!}\mathcal{E}^{\infty}}$ is unitary with respect to $p_{2!}h$. We define the $\mathcal{C}_U^{\infty}$-endomorphisms f and $f^{\dagger}$ of $p_{2!}\mathcal{E}^{\infty}$ by

$$f(s)=\frac{\sqrt{-1}}{2}(\nabla_{\mathfrak{v}}-\sqrt{-1}\phi)s,\quad f^{\dagger}(s)=\frac{\sqrt{-1}}{2}(\nabla_{\mathfrak{v}}+\sqrt{-1}\phi)s.$$

We set $\theta=f\varphi$ and $\theta^{\dagger}=f^{\dagger}\overline{\varphi}$, which are mutually adjoint with respect to $p_{2!}h$. It is easy to see that (E,h,∇,ϕ) is a monopole if and only if $(p_{2!}\mathcal{E}^{\infty},\overline{\partial}_{p_{2!}\mathcal{E}^{\infty}},\theta,p_{2!}h)$ is a harmonic bundle of infinite rank in the sense that $\overline{\partial}_{p_{2!}\mathcal{E}^{\infty}}(f)=0$ and $[\overline{\partial}_{p_{2!}\mathcal{E}^{\infty}},\partial_{p_{2!}\mathcal{E}^{\infty}}]+[\theta,\theta^{\dagger}]=0$ are satisfied.

Let $(E_1,\overline{\partial}_{E_1})$ be a mini-holomorphic bundle on $Y\times U$ with the natural mini-complex structure. Similarly, let $\mathcal{E}_1^{\infty}$ denote the sheaf of C^{∞}-sections of E_1, and we obtain the $\mathcal{C}_U^{\infty}$-module $p_{2!}\mathcal{E}_1^{\infty}$. It is equipped with the holomorphic structure $\overline{\partial}_{p_{2!}\mathcal{E}_1^{\infty}}$ induced by the $\Omega^{0,1}(U)$-component of $\overline{\partial}_{E_1}$. Let $\partial_{E_1,\mathfrak{v}}$ be the differential operator of E_1 induced by $\overline{\partial}_{E_1}$ and $\mathfrak{v}$. We obtain the endomorphism f_1 of $p_{2!}\mathcal{E}_1^{\infty}$ by $f_1(s)=\partial_{E_1,\mathfrak{v}}(s)$, which is holomorphic with respect to $\overline{\partial}_{p_{2!}\mathcal{E}_1^{\infty}}$. We obtain the Higgs field $\theta_1=f_1\varphi$. Thus, we obtain a Higgs bundle $(p_{2!}\mathcal{E}_1^{\infty},\overline{\partial}_{p_{2!}\mathcal{E}_1^{\infty}},\theta_1)$ of infinite rank on U. Clearly, if the mini-holomorphic bundle $(E_1,\overline{\partial}_{E_1})$ underlies a monopole, the Higgs bundle $(p_{2!}\mathcal{E}_1^{\infty},\overline{\partial}_{p_{2!}\mathcal{E}_1^{\infty}},\theta_1)$ underlies the corresponding harmonic bundle of infinite rank.

2.7 Twistor Families of Mini-Complex Structures on $\mathbb{R} \times \mathbb{C}$ and $(\mathbb{R}/T\mathbb{Z}) \times \mathbb{C}$

2.7.1 *Preliminary*

Let X be a two-dimensional complex vector space. Let $L \subset X$ be an oriented one-dimensional real vector space. Let $M := X/L$ be the quotient three-dimensional real vector space. Note that M is equipped with a naturally induced mini-complex structure. Indeed, there exists a $\mathbb{C}$-linear isomorphism $\varphi : X \simeq \mathbb{C}^2 = \{(z, w)\}$ such that $\varphi(L) = \{(s, 0) \mid s \in \mathbb{R}\}$ in a way compatible with the orientations. Let t be the imaginary part of z. Then, by the mini-complex coordinate system (t, w), we obtain the isomorphism $M \simeq \mathbb{R} \times \mathbb{C}$ which induces a mini-complex structure on M. We can check the following claim by a direct computation.

Lemma 2.7.1 *The mini-complex structure is independent of a choice of φ.* ■

Remark 2.7.2 Let $L_{\mathbb{C}} \subset X$ denote the one-dimensional complex vector space generated by L. There exists the natural exact sequence of $\mathbb{R}$-vector spaces

$$0 \longrightarrow L_{\mathbb{C}}/L \longrightarrow M \longrightarrow X/L_{\mathbb{C}} \longrightarrow 0. \tag{2.7}$$

Once we fix isomorphisms $L_{\mathbb{C}}/L \simeq \mathbb{R}$ and $X/L_{\mathbb{C}} \simeq \mathbb{C}$, a choice of linear mini-complex coordinate system (t, w) as above induces a splitting of the exact sequence (2.7). ■

2.7.2 *Spaces*

Let X be a two-dimensional $\mathbb{C}$-vector space. We take a $\mathbb{C}$-linear isomorphism $X \simeq \mathbb{C}^2 = \{(z, w)\}$, and consider the hyperkähler metric $g_X := dz\, d\overline{z} + dw\, d\overline{w}$. We consider the $\mathbb{R}$-subspace $L = \{(s, 0) \mid s \in \mathbb{R}\} \subset X$ with the natural orientation. We set $M := X/L$. Let $t := \mathrm{Im}(z)$. We obtain the mini-complex coordinate system (t, w) on M.

We take $T > 0$. We consider the $\mathbb{Z}$-action κ on M given by $\kappa_n(t, w) = (t + Tn, w)$ $(n \in \mathbb{Z})$. The quotient space is denoted by $\mathcal{M}$.

2.7.3 *Twistor Family of Complex Structures*

For any complex number λ, we have the complex structure on X whose twistor parameter is λ. We use the notation X^{λ} to denote the complex manifold X equipped with the complex structure corresponding to λ. We define the complex coordinate system (ξ, η) on X^{λ} as follows:

$$\xi = z + \lambda \overline{w}, \quad \eta = w - \lambda \overline{z}. \tag{2.8}$$

The inverse transform is described as follows:

$$z = \frac{1}{1+|\lambda|^2}(\xi - \lambda \overline{\eta}), \quad w = \frac{1}{1+|\lambda|^2}(\eta + \lambda \overline{\xi}).$$

The metric $dz\, d\overline{z} + dw\, d\overline{w}$ is equal to $(1+|\lambda|^2)^{-2}(d\xi\, d\overline{\xi} + d\eta\, d\overline{\eta})$. We obtain the following relations of complex vector fields:

$$\partial_{\overline{\xi}} = \frac{1}{1+|\lambda|^2}(\partial_{\overline{z}} + \lambda \partial_w), \quad \partial_{\overline{\eta}} = \frac{1}{1+|\lambda|^2}(\partial_{\overline{w}} - \lambda \partial_z),$$

$$\partial_{\xi} = \frac{1}{1+|\lambda|^2}(\partial_z + \overline{\lambda} \partial_{\overline{w}}), \quad \partial_{\eta} = \frac{1}{1+|\lambda|^2}(\partial_w - \overline{\lambda} \partial_{\overline{z}}).$$

The $\mathbb{R}$-subspace L and the $\mathbb{Z}$-action κ are described as follows in terms of the coordinate system (ξ, η):

$$L = \big\{ s(1, -\lambda) \,\big|\, s \in \mathbb{R} \big\},$$

$$\kappa_n(\xi, \eta) = (\xi + \sqrt{-1}Tn, \eta + \lambda\sqrt{-1}Tn) = (\xi, \eta) + nT \cdot (\sqrt{-1}, \lambda\sqrt{-1}).$$

Remark 2.7.3 Recall that the space of the twistor parameters is $\mathbb{P}^1$. Let $\mathbb{P}^1 = \mathbb{C}_{\lambda} \cup \mathbb{C}_{\mu}$ be the covering where λ and μ are related by $\lambda\mu = 1$. Let $X^{\dagger\mu}$ denote the complex manifold X equipped with the complex structure corresponding to μ. Let $M^{\dagger\mu}$ denote the mini-complex manifold obtained as $X^{\dagger\mu}/L$. We have $X^{\dagger\mu} = X^{\lambda}$ and $M^{\dagger\mu} = M^{\lambda}$ if $\lambda\mu = 1$. At $\mu = 0$, i.e., $\lambda = \infty$, the complex structure of $X^{\dagger 0}$ is induced by $(z^{\dagger}, w^{\dagger}) = (-\overline{z}, \overline{w})$. We note that $L = \{(s, 0)\}$ with respect to $(z^{\dagger}, w^{\dagger})$, but that the orientations of L induced by z and $z^{\dagger} = -\overline{z}$ are mutually reversed. The complex structure of $X^{\dagger\mu}$ is induced by

$$(\xi^{\dagger}, \eta^{\dagger}) = (z^{\dagger} + \mu\overline{w}^{\dagger}, w^{\dagger} - \mu\overline{z}^{\dagger}) = (\lambda^{-1}(w - \lambda\overline{z}), \lambda^{-1}(z + \lambda\overline{w})) = (\lambda^{-1}\eta, \lambda^{-1}\xi).$$

We shall explain how to obtain difference modules from monopoles behind which the complex coordinate system (ξ, η) is implicitly used. For $\lambda = \mu^{-1} \neq 0$, we may also obtain another difference module from the monopole similarly by using $(\xi^{\dagger}, \eta^{\dagger})$. It is analogue to the situation that a harmonic bundle $(E, \overline{\partial}_E, \theta, h)$ on a complex manifold Y induces a λ-flat bundle $(E, \overline{\partial}_E + \lambda\theta^{\dagger} + \lambda\partial_E + \theta)$ on Y and a μ-flat bundle $(E, \partial_E + \mu\theta + \mu\overline{\partial}_E + \theta^{\dagger})$ on $Y^{\dagger}$ which is the conjugate of Y. The analogy can be enhanced to the more precise correspondence by the Nahm transforms between wild harmonic bundles on $\mathbb{P}^1$ and periodic monopoles, which will be explained elsewhere. ■

2.7.4 Family of Mini-Complex Structures

Corresponding to the complex structure of X^λ, we obtain the mini-complex structures on M and $\mathcal{M}$. (See Sect. 2.7.1.) The three-dimensional manifolds with mini-complex structure are denoted by M^λ and $\mathcal{M}^\lambda$.

In the case $\lambda = 0$, we use the mini-complex coordinate system $(t, w) = (\mathrm{Im}(z), w)$ on M^0. It induces local mini-complex coordinate systems on $\mathcal{M}^0$.

We shall introduce two mini-complex coordinate systems (t_i, β_i) $(i = 0, 1)$ on M^λ, which induce mini-complex local coordinate systems on $\mathcal{M}^\lambda$. For that purpose, we shall introduce two complex coordinate systems (α_i, β_i) $(i = 0, 1)$ on X^λ from which (t_i, β_i) are induced.

Remark 2.7.4 The coordinate systems (t_i, β_i) depend on λ, but we omit to denote the dependence to simplify the description. ■

2.7.5 The Mini-Complex Coordinate System (t_0, β_0)

Let (α_0, β_0) be the complex coordinate system of X^λ given by the following relation:

$$(\xi, \eta) = \alpha_0(1, -\lambda) + \beta_0(\overline{\lambda}, 1) = (\alpha_0 + \overline{\lambda}\beta_0, \beta_0 - \lambda\alpha_0).$$

The inverse transform is described as follows:

$$\alpha_0 = \frac{\xi - \overline{\lambda}\eta}{1 + |\lambda|^2}, \qquad \beta_0 = \frac{\eta + \lambda\xi}{1 + |\lambda|^2}.$$

It is easy to check $d\alpha_0\, d\overline{\alpha}_0 + d\beta_0\, d\overline{\beta}_0 = dz\, d\overline{z} + dw\, d\overline{w}$.

The $\mathbb{R}$-subspace L and the $\mathbb{Z}$-action κ are described as follows in terms of (α_0, β_0):

$$L = \{(s, 0) \mid s \in \mathbb{R}\}, \tag{2.9}$$

$$\kappa_n(\alpha_0, \beta_0) = (\alpha_0, \beta_0) + nT \cdot \left(\frac{1 - |\lambda|^2}{1 + |\lambda|^2}\sqrt{-1}, \frac{2\lambda\sqrt{-1}}{1 + |\lambda|^2}\right). \tag{2.10}$$

We set $t_0 := \mathrm{Im}(\alpha_0)$. Because L is described as (2.9), (t_0, β_0) is a mini-complex coordinate system of M^λ. The induced $\mathbb{Z}$-action κ on M^λ is described as follows in terms of (t_0, β_0):

$$\kappa_n(t_0, \beta_0) = (t_0, \beta_0) + nT \cdot \left(\frac{1 - |\lambda|^2}{1 + |\lambda|^2}, \frac{2\lambda\sqrt{-1}}{1 + |\lambda|^2}\right).$$

Clearly, κ_n is given along the integral curve of the real vector field:

$$\partial_t = \frac{1-|\lambda|^2}{1+|\lambda|^2}\partial_{t_0} + \frac{2\lambda}{1+|\lambda|^2}\sqrt{-1}\partial_{\beta_0} - \frac{2\overline{\lambda}}{1+|\lambda|^2}\sqrt{-1}\partial_{\overline{\beta}_0}. \tag{2.11}$$

Remark 2.7.5 If (t_0', β_0') is another mini-complex coordinate system of M^λ such that $dt_0'\,dt_0' + d\beta_0'\,d\overline{\beta}_0'$, we obtain $(t_0', \beta_0') = (t_0, a\beta_0)$, where a is a complex number such that $|a| = 1$. ■

Remark 2.7.6 For the twistor parameter $\mu \in \mathbb{C}_\mu \subset \mathbb{P}^1$, we obtain $(\alpha_0^\dagger, \beta_0^\dagger)$ from $(\xi^\dagger, \mu^\dagger)$ as above:

$$(\alpha_0^\dagger, \beta_0^\dagger) = \frac{1}{1+|\lambda|^2}(\xi^\dagger - \overline{\mu}\eta^\dagger, \eta^\dagger + \mu\xi^\dagger)$$

Because the orientations of L induced by z and $z^\dagger = -\overline{z}$ are mutually reversed, the induced mini-complex coordinate system of $M^{\dagger\mu}$ is $(-\operatorname{Im}(\alpha_0^\dagger), \beta_0^\dagger)$. If $\mu = \lambda^{-1} \neq 0$, we have $(\alpha_0^\dagger, \beta_0^\dagger) = \big(-\alpha_0, (\lambda^{-1}\overline{\lambda})\beta_0\big)$, and hence $(t_0^\dagger, \beta_0^\dagger) = (-\operatorname{Im}(\alpha_0^\dagger), \beta_0^\dagger) = \big(t_0, (\lambda^{-1}\overline{\lambda})\beta_0\big)$. ■

2.7.6 The Mini-Complex Coordinate System (t_1, β_1)

Let (α_1, β_1) be the complex coordinate system of X^λ determined by the relation $(\xi, \eta) = \alpha_1(1, -\lambda) + \beta_1(0, 1)$. The transformations are described as follows:

$$\begin{cases} \xi = \alpha_1, \\ \eta = \beta_1 - \lambda\alpha_1, \end{cases} \qquad \begin{cases} \alpha_1 = \xi, \\ \beta_1 = \eta + \lambda\xi. \end{cases}$$

The $\mathbb{R}$-subspace L and the $\mathbb{Z}$-action κ are described as follows in terms of (α_1, β_1):

$$L = \{(s, 0) \mid s \in \mathbb{R}\}, \tag{2.12}$$

$$\kappa_n(\alpha_1, \beta_1) = (\alpha_1, \beta_1) + nT(\sqrt{-1}, 2\lambda\sqrt{-1}). \tag{2.13}$$

We set $t_1 := \operatorname{Im}\alpha_1$. Because L is described as (2.12), we obtain the mini-complex coordinate system (t_1, β_1) on M^λ. The induced $\mathbb{Z}$-action κ on M^λ is described as follows:

$$\kappa_n(t_1, \beta_1) = (t_1, \beta_1) + nT(1, 2\lambda\sqrt{-1}).$$

Remark 2.7.7 For the twistor parameter $\mu \in \mathbb{C}_\mu \subset \mathbb{P}^1$, we obtain the complex coordinate system $(\alpha_1^\dagger, \beta_1^\dagger) = (\xi^\dagger, \eta^\dagger + \mu\xi^\dagger)$ of $X^{\dagger\mu}$, and the mini-complex coordinate system $(t_1^\dagger, \beta_1^\dagger) = (-\mathrm{Im}(\alpha_1)^\dagger, \beta_1^\dagger)$ of $M^{\dagger\mu}$. If $\mu = \lambda^{-1}$, we have $(\alpha_1^\dagger, \beta_1^\dagger) = (\lambda^{-1}\eta, \lambda^{-2}(\eta + \lambda\xi))$, and hence $(t_1^\dagger, \beta_1^\dagger) = (t_1 - \mathrm{Im}(\lambda^{-1}\beta_1), \lambda^{-2}\beta_1)$. We obtain the splittings of $M^\lambda = M^{\dagger\mu}$ induced by (t_1, β_1) and $(t_1^\dagger, \beta_1^\dagger)$ as in Remark 2.7.2, and they are different. Therefore, we obtain two difference modules depending on the choice of (ξ, η) or $(\xi^\dagger, \eta^\dagger)$ as mentioned in Remark 2.7.3. There are essentially only these two choices. See Sect. 2.9.7. ■

Remark 2.7.8 If $\lambda = 0$, we have $(t_0, \beta_0) = (t_1, \beta_1) = (t, w)$. ■

2.7.7 Coordinate Change

We have the following relation:

$$\begin{cases} \alpha_1 = \alpha_0 + \overline{\lambda}\beta_0 \\ \beta_1 = (1 + |\lambda|^2)\beta_0, \end{cases} \qquad \begin{cases} \alpha_0 = \alpha_1 - (1 + |\lambda|^2)\overline{\lambda}\beta_1 \\ \beta_0 = (1 + |\lambda|^2)^{-1}\beta_1. \end{cases}$$

Hence, we have the following relation:

$$\begin{cases} t_1 = t_0 + \mathrm{Im}(\overline{\lambda}\beta_0) \\ \beta_1 = (1 + |\lambda|^2)\beta_0, \end{cases} \qquad \begin{cases} t_0 = t_1 - (1 + |\lambda|^2)^{-1}\,\mathrm{Im}(\overline{\lambda}\beta_1) \\ \beta_0 = (1 + |\lambda|^2)^{-1}\beta_1. \end{cases}$$

We obtain the following relation of vector fields:

$$\partial_{t_1} = \partial_{t_0}, \quad \partial_{\overline{\beta}_1} = \frac{\lambda}{1 + |\lambda|^2}\frac{1}{2\sqrt{-1}}\partial_{t_0} + \frac{1}{1 + |\lambda|^2}\partial_{\overline{\beta}_0}. \tag{2.14}$$

2.7.8 Compactification

Set $\overline{M}^\lambda := \mathbb{R}_{t_1} \times \mathbb{P}^1_{\beta_1}$, which is equipped with the natural mini-complex structure. We have the $\mathbb{Z}$-action κ on $\overline{M}^\lambda$ given by $\kappa_n(t_1, \beta_1) = (t_1 + Tn, \beta_1 + 2\sqrt{-1}\lambda Tn)$. The quotient space is denoted by $\overline{\mathcal{M}}^\lambda$. It is a compactification of $\mathcal{M}^\lambda$, and equipped with the naturally induced mini-complex structure. We set $H_\infty^{\lambda\,\mathrm{cov}} := \overline{M}^\lambda \setminus M^\lambda$ and $H_\infty^\lambda := \overline{\mathcal{M}}^\lambda \setminus \mathcal{M}^\lambda$. Let $\varpi^\lambda : M^\lambda \longrightarrow \overline{\mathcal{M}}^\lambda$ denote the projection. For $\lambda_1 \neq \lambda_2$, we have $\overline{\mathcal{M}}^{\lambda_1} \neq \overline{\mathcal{M}}^{\lambda_2}$ though $\mathcal{M}^{\lambda_1} = \mathcal{M}^{\lambda_2}$ as C^∞-manifolds.

Remark 2.7.9 If $\lambda = \mu^{-1} \neq 0$, we may obtain another natural compactification induced by $(t_1^\dagger, \beta_1^\dagger)$. The compactifications depend on the splittings induced by (t_1, β_1) and $(t_1^\dagger, \beta_1^\dagger)$ (see also Remark 2.7.2). ■

2.7.9 *Mini-Holomorphic Bundles Associated with Monopoles*

Let $\varphi : U \longrightarrow \mathcal{M}$ be a local diffeomorphism. We regard U as the Riemannian manifold whose metric is obtained as the pull back of $dt\,dt + dw\,d\overline{w}$. For $\lambda \in \mathbb{C}$, let U^{λ} denote the three-dimensional manifold with the mini-complex structure obtained as the pull back of the mini-complex structure of $\mathcal{M}^{\lambda}$.

For any monopole (E, h, ∇, ϕ) on U, we obtain the mini-holomorphic bundle on U^{λ} underlying (E, h, ∇, ϕ) for each λ (see Sect. 2.4.2), which we denote by $(E^{\lambda}, \overline{\partial}_{E^{\lambda}})$.

2.7.9.1 Compatibility with the Dimensional Reduction from 4D to 3D

Let U be as above. We set $\widetilde{U} := \mathbb{R}_s \times U$ on which we consider the Riemannian metric $ds\,ds + dt\,dt + dw\,d\overline{w}$. Let $\widetilde{U}^0$ denote the complex manifold whose complex structure is given by local coordinate systems $(z, w) = (s + \sqrt{-1}t, w)$. For any $\lambda \in \mathbb{C}$, let $\widetilde{U}^{\lambda}$ denote the complex manifold whose complex structure is given by the local complex coordinate systems (ξ, η) as in Sect. 2.7.3.

Let (E, h, ∇, ϕ) be a monopole on U. We obtain the induced instanton $(\widetilde{E}, \widetilde{h}, \widetilde{\nabla})$ on $\widetilde{U}$ as in Sect. 2.5.1. Let $(\widetilde{E}^{\lambda}, \overline{\partial}_{\widetilde{E}^{\lambda}})$ denote the holomorphic bundle on $\widetilde{U}^{\lambda}$ underlying the instanton. The following lemma is clear by the constructions, or can be checked by a direct computation.

Lemma 2.7.10 *$(\widetilde{E}^{\lambda}, \overline{\partial}_{\widetilde{E}^{\lambda}})$ is equal to the holomorphic bundle induced by the mini-holomorphic bundle $(E^{\lambda}, \overline{\partial}_{E^{\lambda}})$ as in* Sect. 2.5.2. ■

2.8 $\mathcal{O}_{\mathcal{M}^{\lambda}}$-Modules and λ-Connections

2.8.1 *Dimensional Reduction from $\mathcal{O}_{\mathcal{M}^{\lambda}}$-Modules to λ-Flat Bundles*

2.8.1.1 Setting

Let $\Psi : \mathcal{M} = S^1_T \times \mathbb{C}_w \longrightarrow \mathbb{C}_w$ denote the projection. Let $U_w \longrightarrow \mathbb{C}_w$ be a local diffeomorphism. Let U denote the fiber product of U_w and $\mathcal{M}$ over $\mathbb{C}_w$. The induced morphism $U \longrightarrow U_w$ is also denoted by Ψ. There exists a natural isomorphism $U \simeq S^1_T \times U_w$ which is equipped with the naturally defined S^1_T-action. We obtain the naturally induced local diffeomorphism $U \longrightarrow \mathcal{M}$. We regard U as a Riemannian manifold whose metric is obtained as the pull back of $dt\,dt + dw\,d\overline{w}$. For $\lambda \in \mathbb{C}$, let U^{λ} denote the three-dimensional manifold equipped with the mini-complex structure obtained as the pull back of the mini-complex structure of $\mathcal{M}^{\lambda}$.

We shall explain that λ-flat bundles on U_w are the dimensional reduction of mini-holomorphic bundles on U^λ (Corollary 2.8.2), as the special case of a more general equivalence including the non-integrable case (Lemma 2.8.1).

2.8.1.2 Some Vector Fields and Forms

We have the local complex coordinates on U_w induced by w, which induce the differential forms dw and $d\overline{w}$, and the vector fields ∂_w and $\partial_{\overline{w}}$ globally defined on U_w. We have the local mini-complex coordinate systems on U^λ induced by (t_1, β_1), and we obtain the globally defined differential forms dt_1, $d\beta_1$ and $d\overline{\beta}_1$, and the globally defined complex vector fields ∂_{t_1}, ∂_{β_1} and $\partial_{\overline{\beta}_1}$ on U. Note that Ψ is described as $\Psi(t_1, \beta_1) = (1+|\lambda|^2)^{-1}(\beta_1 - 2\sqrt{-1}\lambda t_1)$ in terms of the coordinate systems (t_1, β_1) and w. The tangent map of Ψ is described as follows:

$$T\Psi(\partial_{t_1}) = \frac{-2\sqrt{-1}\lambda}{1+|\lambda|^2}\partial_w + \frac{2\sqrt{-1}\cdot\overline{\lambda}}{1+|\lambda|^2}\partial_{\overline{w}}, \qquad T\Psi(\partial_{\overline{\beta}_1}) = \frac{1}{1+|\lambda|^2}\partial_{\overline{w}}. \tag{2.15}$$

Similarly, we have the local mini-complex coordinate systems on U^λ induced by (t_0, β_0), the globally defined differential forms dt_0, $d\beta_0$ and $d\overline{\beta}_0$, and the globally defined complex vector fields ∂_{t_0}, ∂_{β_0} and $\partial_{\overline{\beta}_0}$ on U. We also have the local mini-complex coordinate systems on U^0 induced by (t, w), the globally defined differential forms dt, dw and $d\overline{w}$, and the globally defined complex vector fields ∂_t, ∂_w and $\partial_{\overline{w}}$ on U.

2.8.1.3 A General Equivalence

Let V be a C^∞-vector bundle on U_w. Recall that a λ-connection of V in the C^∞-sense is a linear differential operator $\mathbb{D}^\lambda : C^\infty(U_w, V) \longrightarrow C^\infty(U_w, V \otimes \Omega^1_{U_w})$ such that $\mathbb{D}^\lambda(gs) = (\lambda\partial + \overline{\partial})g \cdot s + f\mathbb{D}^\lambda(s)$ for any $g \in C^\infty(U_w)$ and $s \in C^\infty(U_w, V)$. (See [63, §2.2].) It is called flat if $\mathbb{D}^\lambda \circ \mathbb{D}^\lambda = 0$.

Let $\partial_{V,\overline{w}}$ (resp. $\mathbb{D}^\lambda_w$) denote the differential operator of V induced by $\partial_{\overline{w}}$ (resp. ∂_w) and $\mathbb{D}^\lambda$. The flatness of $\mathbb{D}^\lambda$ is equivalent to $[\partial_{V,\overline{w}}, \mathbb{D}^\lambda_w] = 0$.

We set $\widetilde{V} = \Psi^{-1}(V)$. We say that a linear differential operator

$$\overline{\partial}_{\widetilde{V}} : C^\infty(U^\lambda, \widetilde{V}) \longrightarrow C^\infty(U^\lambda, \widetilde{V} \otimes \Omega^{0,1}_{U^\lambda})$$

satisfies a mini-complex Leibniz rule if $\overline{\partial}_{\widetilde{V}}(fu) = \overline{\partial}_{U^\lambda}(f)u + f\overline{\partial}_{\widetilde{V}}(u)$ for any $f \in C^\infty(U)$ and $u \in C^\infty(U, \widetilde{V})$.

Let $\mathbb{D}^\lambda$ be a λ-connection of V in the C^∞-sense. Noting the relation (2.15), we obtain the linear differential operators $\partial_{\widetilde{V},\overline{\beta}_1}$ and $\partial_{\widetilde{V},t_1}$ of $\widetilde{V}$ such that the following holds for any $s \in C^\infty(U_w, V)$ and $f \in C^\infty(U)$:

$$\partial_{\widetilde{V},\overline{\beta}_1}(f\Psi^{-1}(s)) = \partial_{\overline{\beta}_1} f \cdot \Psi^{-1}(s) + \frac{1}{1+|\lambda|^2} f\Psi^{-1}(\partial_{V,\overline{w}} s). \quad (2.16)$$

$$\partial_{\widetilde{V},t_1}(f\Psi^{-1}(s)) = \partial_{t_1} f \cdot \Psi^{-1}(s) + \frac{1}{1+|\lambda|^2} f\Psi^{-1} \times \big(-2\sqrt{-1}\mathbb{D}^\lambda_w s + 2\sqrt{-1}\cdot\overline{\lambda}\partial_{V,\overline{w}} s\big). \quad (2.17)$$

We define $\overline{\partial}_{\widetilde{V}} : C^\infty(U, \widetilde{V}) \longrightarrow C^\infty(U, \widetilde{V} \otimes \Omega^{0,1}_{U^\lambda})$ by $\overline{\partial}_{\widetilde{V}}(u) = \partial_{\widetilde{V},\overline{\beta}_1}(u)d\overline{\beta}_1 + \partial_{\widetilde{V},t_1}(u)\,dt_1$ for any $u \in C^\infty(U, \widetilde{V})$. By the relation (2.15), $\overline{\partial}_{\widetilde{V}}$ satisfies the mini-complex Leibniz rule. Note that $(\widetilde{V}, \overline{\partial}_{\widetilde{V}})$ is S^1_T-equivariant. By the construction, the following holds:

$$\big[\partial_{\widetilde{V},\overline{\beta}_1}, \partial_{\widetilde{V},t_1}\big] = \frac{-2\sqrt{-1}}{(1+|\lambda|^2)^2}\Psi^{-1}\big([\partial_{\overline{w}}, \mathbb{D}^\lambda_w]\big). \quad (2.18)$$

Lemma 2.8.1 *The above procedure induces an equivalence between the following objects.*

- *Vector bundles V on U_w equipped with a λ-connection in the C^∞-sense.*
- *S^1_T-equivariant vector bundles $\widetilde{V}$ on U equipped with an S^1_T-equivariant linear differential operator $\overline{\partial}_{\widetilde{V}} : C^\infty(U^\lambda, \widetilde{V}) \longrightarrow C^\infty(U^\lambda, \widetilde{V} \otimes \Omega^{0,1}_{U^\lambda})$ satisfying the mini-complex Leibniz rule.*

Proof Let us indicate the inverse construction. Let $(\widetilde{V}, \overline{\partial}_{\widetilde{V}})$ be as in the statement of Lemma 2.8.1. Let $\partial_{\widetilde{V},\overline{\beta}_1}$ (resp. $\partial_{\widetilde{V},t_1}$) denote the S^1_T-equivariant differential operator on $\widetilde{V}$ induced by $\overline{\partial}_{\widetilde{V}}$ and $\partial_{\overline{\beta}_1}$ (resp. ∂_{t_1}). There exists a C^∞-bundle V on U_w with an S^1_T-equivariant isomorphism $\Psi^{-1}(V) \simeq \widetilde{V}$. We obtain the differential operators $\partial_{V,\overline{w}}$ and $\mathbb{D}^\lambda_w$ of V by the following condition for any $s \in C^\infty(U_w, V)$:

$$\Psi^{-1}(\partial_{V,\overline{w}}(s)) = (1+|\lambda|^2)\partial_{\widetilde{V},\overline{\beta}_1}\Psi^{-1}(s),$$

$$\Psi^{-1}(\mathbb{D}^\lambda_w s) = \frac{\sqrt{-1}}{2}(1+|\lambda|^2)\partial_{\widetilde{V},t_1}\Psi^{-1}(s) + (1+|\lambda|^2)\overline{\lambda}\partial_{\widetilde{V},\overline{\beta}_1}\Psi^{-1}(s).$$

Then, by the relation (2.15), we have $\partial_{V,\overline{w}}(gs) = \partial_{\overline{w}}(g)s + g\partial_{V,\overline{w}}(s)$ and $\mathbb{D}^\lambda_w(gs) = \lambda\partial_w(g)s + g\mathbb{D}^\lambda_w(s)$ for any $g \in C^\infty(U_w)$ and $s \in C^\infty(U_w, V)$. Hence, $\partial_{V,\overline{w}}$ and $\mathbb{D}^\lambda_w$ induces a λ-connection of V. Two constructions are mutually inverse. ■

2.8.1.4 Mini-Holomorphic Bundles and Flat λ-Connections

As a consequence of Lemma 2.8.1 and (2.18), we obtain the following.

Corollary 2.8.2 *The construction in* Sect. 2.8.1.3 *induces an equivalence between λ-flat bundles on U_w and S^1_T-equivariant mini-holomorphic vector bundles on U.* ■

2.8.1.5 λ-Flat Bundles of Infinite Rank

Let E be a C^∞-bundle on U^λ equipped with a differential operator $\overline{\partial}_E$: $C^\infty(U^\lambda, E) \longrightarrow C^\infty(U^\lambda, E \otimes \Omega^{0,1}_{U^\lambda})$ satisfying the mini-complex Leibniz rule. Let E_{C^∞} denote the sheaf of C^∞-sections of E. Let $\mathcal{C}^\infty_{U_w}$ denote the sheaf of C^∞-functions on U_w. We obtain the $\mathcal{C}^\infty_{U_w}$-module $V_1 := \Psi_!(E_{C^\infty})$, where $\Psi_!$ denotes the proper push-forward with respect to Ψ. We define $\partial_{V_1,\overline{w}}$ and $\mathbb{D}^\lambda_{V_1,w}$ on the sheaf V_1 by

$$\partial_{V_1,\overline{w}}(u) = (1 + |\lambda|^2)\partial_{E,\overline{\beta}_1}(u),$$

$$\mathbb{D}^\lambda_{V_1,w}(u) = \frac{\sqrt{-1}}{2}(1 + |\lambda|^2)\partial_{E,t_1}(u) + \overline{\lambda}(1 + |\lambda|^2)\partial_{E,\overline{\beta}_1}(u)$$

for sections u of V_1. We can easily observe that $\partial_{V_1,\overline{w}}$ and $\mathbb{D}^\lambda_{V_1,w}$ determines a λ-connection $\mathbb{D}^\lambda_{V_1}$ of V_1 in the C^∞-sense. If $(E, \overline{\partial}_E)$ is a mini-holomorphic bundle, then $\mathbb{D}^\lambda_{V_1}$ is a flat λ-connection.

2.8.1.6 Remark

This analogy between mini-holomorphic bundles on U^λ and λ-flat bundles on U_w is one of the motivations of this study, as explained in Sect. 1.5. (See also Sect. 2.9.5.2.) It is also useful when we study the property of the mini-holomorphic bundles underlying monopoles induced by harmonic bundles. (See Corollary 2.8.4 and the computations in Sects. 5.2 and 5.3 below.) We shall generalize this analogy to the case where U_w is a ramified covering of a neighbourhood of $\infty \in \mathbb{P}^1$ (Corollary 2.8.8 and Corollary 4.5.10). We study the formal and ramified version in Sect. 3.6.3. Note that in the formal version we can remove the S^1_T-equivariance condition (see Proposition 3.6.8). We obtain the filtered version (Proposition 3.6.15 and Proposition 4.5.2) as a refinement. It is also useful to study the non-integrable case (Lemma 2.8.1, Proposition 2.8.3, Lemma 2.8.7 and Lemma 4.5.9).

We recall that in the study of harmonic bundles we understand the asymptotic behaviour of harmonic bundles through the filtered extension of the underlying λ-flat bundles, pioneered by Simpson [80], and further studied in [62, 64]. Similarly, it is our viewpoint to understand the asymptotic behaviour of monopoles through the filtered extension of the underlying mini-holomorphic bundles, studied in Sect. 7. Moreover, we would like to apply the results for filtered extension of λ-flat bundles underlying harmonic bundles to study the filtered extension of the mini-holomorphic bundles. That is the main reason to pursue this analogy. (See Sect. 1.8.7 for an outline of the argument.)

2.8.2 Comparison of Some Induced Operators

Let U_w and U be as in Sect. 2.8.1.1. Let V be a C^∞-bundle on U_w with a Hermitian metric h. Let $\overline{\partial}_V$ be a holomorphic structure of V, and let $\theta \in C^\infty(U_w, \mathrm{End}(V) \otimes \Omega^{1,0})$ which does not necessarily satisfy $\overline{\partial}_V\theta = 0$. Let h be a Hermitian metric of V. We obtain the Chern connection $\overline{\partial}_V + \partial_V$, and the adjoint $\theta^\dagger \in C^\infty(U_w, \mathrm{End}(V) \otimes \Omega^{0,1})$ of θ. Let $\partial_{V,\overline{w}}$ (resp. $\partial_{V,w}$) denote the differential operator of V induced by $\overline{\partial}_V$ and $\partial_{\overline{w}}$ (resp. ∂_V and ∂_w). Let f and $f^\dagger$ be the endomorphisms of V determined by $\theta = f\,dw$ and $\theta^\dagger = f^\dagger d\overline{w}$, respectively. We set $\widetilde{V} = \Psi^{-1}(V)$ on U with the metric $\widetilde{h} = \Psi^{-1}(h)$.

On one hand, we obtain the unitary connection ∇ and an anti-Hermitian endomorphism ϕ of $\widetilde{V}$ satisfying the following condition for $s \in C^\infty(U_w, V)$ as in (2.6):

$$\nabla_{\overline{w}}(\Psi^{-1}(s)) = \Psi^{-1}(\partial_{V,\overline{w}}s), \quad \nabla_w(\Psi^{-1}(s)) = \Psi^{-1}(\partial_{V,w}s),$$

$$\nabla_t(\Psi^{-1}(s)) = \Psi^{-1}\big(-\sqrt{-1}(f + f^\dagger)(s)\big), \quad \phi(\Psi^{-1}(s)) = \Psi^{-1}\big((f - f^\dagger)s\big).$$

From $\widetilde{V}$ with ∇ and ϕ, we obtain a linear differential operator $\overline{\partial}^{(1)}_{\widetilde{V}} : C^\infty(U^\lambda, \widetilde{V}) \longrightarrow C^\infty(U^\lambda, \widetilde{V} \otimes \Omega^{0,1}_{U^\lambda})$ as in Sect. 2.4.2 which satisfies the mini-complex Leibniz rule.

On the other hand, we obtain the λ-connection $\mathbb{D}^\lambda = \overline{\partial}_V + \lambda\theta^\dagger + \lambda\partial_V + \theta$ of V in the C^∞-sense. From $(V, \mathbb{D}^\lambda)$, we obtain $\overline{\partial}^{(2)}_{\widetilde{V}} : C^\infty(U^\lambda, \widetilde{V}) \longrightarrow C^\infty(U^\lambda, \widetilde{V} \otimes \Omega^{0,1}_{U^\lambda})$ as in Sect. 2.8.1.

Proposition 2.8.3 *We have* $\overline{\partial}^{(1)}_{\widetilde{V}} = \overline{\partial}^{(2)}_{\widetilde{V}}$.

Proof Note that

$$\nabla_{t_0} = \frac{1 - |\lambda|^2}{1 + |\lambda|^2}\nabla_t - \frac{2\lambda\sqrt{-1}}{1 + |\lambda|^2}\nabla_w + \frac{2\overline{\lambda}\sqrt{-1}}{1 + |\lambda|^2}\nabla_{\overline{w}},$$

$$\nabla_{\overline{\beta}_0} = \frac{\sqrt{-1}\lambda}{1 + |\lambda|^2}\nabla_t + \frac{\lambda^2}{1 + |\lambda|^2}\nabla_w + \frac{1}{1 + |\lambda|^2}\nabla_{\overline{w}}.$$

We also obtain the following:

$$\partial^{(1)}_{\widetilde{V},\overline{\beta}_1} = \frac{\lambda}{1+|\lambda|^2}\frac{1}{2\sqrt{-1}}\partial^{(1)}_{\widetilde{V},t_0} + \frac{1}{1+|\lambda|^2}\partial^{(1)}_{\widetilde{V},\overline{\beta}_0}$$

$$= \frac{1}{1+|\lambda|^2}\nabla_{\overline{\beta}_0} - \frac{\lambda}{1+|\lambda|^2}\frac{1}{2}(\phi+\sqrt{-1}\nabla_{t_0}), \tag{2.19}$$

$$\partial^{(1)}_{\widetilde{V},t_1} = \partial^{(1)}_{\widetilde{V},t_0} = \nabla_{t_0} - \sqrt{-1}\phi. \tag{2.20}$$

Hence, we obtain the following:

$$\partial^{(1)}_{\widetilde{V},\overline{\beta}_1} = \frac{1}{1+|\lambda|^2}\Big(\frac{\lambda\sqrt{-1}}{2}\nabla_t + \nabla_{\overline{w}} - \frac{\lambda}{2}\phi\Big), \tag{2.21}$$

$$\partial^{(1)}_{\widetilde{V},t_1} = \frac{1-|\lambda|^2}{1+|\lambda|^2}\nabla_t - \frac{2\lambda\sqrt{-1}}{1+|\lambda|^2}\nabla_w + \frac{2\overline{\lambda}\sqrt{-1}}{1+|\lambda|^2}\nabla_{\overline{w}} - \sqrt{-1}\phi. \tag{2.22}$$

By the construction of ∇ and ϕ, we obtain the following:

$$\partial^{(1)}_{\widetilde{V},\overline{\beta}_1}\Psi^{-1}(s) = \frac{1}{1+|\lambda|^2}\Psi^{-1}\big((\partial_{V,\overline{w}} + \lambda f^{\dagger})s\big),$$

$$\partial^{(1)}_{\widetilde{V},t_1}\Psi^{-1}(s) = \Psi^{-1}\Big(\frac{2\overline{\lambda}\sqrt{-1}}{1+|\lambda|^2}(\partial_{V,\overline{w}} + \lambda f^{\dagger})s - \frac{2\sqrt{-1}}{1+|\lambda|^2}(\lambda\partial_{V,w} + f)s\Big).$$

This is equal to $\partial^{(2)}_{\widetilde{V}}$ constructed from $\mathbb{D}^\lambda = \overline{\partial}_V + \lambda\theta^{\dagger} + \lambda\partial_V + \theta$ in Sect. 2.8.1. ■

2.8.2.1 Comparison of Mini-Holomorphic Bundles Induced by Harmonic Bundles

Suppose that $(V, \overline{\partial}_V, \theta, h)$ is a harmonic bundle on U_w. On one hand, $(\widetilde{V}, \widetilde{h}, \nabla, \phi)$ is a monopole on U associated with $(V, \overline{\partial}_V, \theta, h)$ as in Sect. 2.6.1, and $(\widetilde{V}, \overline{\partial}^{(1)}_{\widetilde{V}})$ is the mini-holomorphic bundle on U^λ underlying $(\widetilde{V}, \widetilde{h}, \nabla, \phi)$. On the other hand, $\mathbb{D}^\lambda$ is a flat λ-connection associated with $(V, \overline{\partial}_V, \theta, h)$, and $(\widetilde{V}, \overline{\partial}^{(2)}_{\widetilde{V}})$ is the mini-holomorphic bundle associated with $(V, \mathbb{D}^\lambda)$ in Corollary 2.8.2. We obtain the following from Proposition 2.8.3.

Corollary 2.8.4 *The associated mini-holomorphic bundles $(\widetilde{V}, \overline{\partial}^{(1)}_{\widetilde{V}})$ and $(\widetilde{V}, \overline{\partial}^{(2)}_{\widetilde{V}})$ on U^λ are the same.* ■

2.8.3 $\mathcal{O}_{\overline{\mathcal{M}}^{\lambda}}$-Modules and λ-Connections

2.8.3.1 Setting

Let $\Psi^{\lambda} : \overline{\mathcal{M}}^{\lambda} \longrightarrow \mathbb{P}^1_w$ be the map induced by $\Psi^{\lambda}(t_1, \beta_1) = (1+|\lambda|^2)^{-1}(\beta_1 - 2\sqrt{-1}\lambda t_1)$. Note that the restriction of Ψ^{λ} to $\mathcal{M}^{\lambda}$ is equal to Ψ.

Let U_w be any open subset of $\mathbb{P}^1_w$ such that $\infty \in U_w$. We set $U^{\lambda} = (\Psi^{\lambda})^{-1}(U_w) \subset \overline{\mathcal{M}}^{\lambda}$, which is equipped with the mini-complex structure as an open subset of $\overline{\mathcal{M}}^{\lambda}$. We shall generalize Lemma 2.8.1 and Corollary 2.8.2 to this context.

Remark 2.8.5 We shall later generalize the construction to the case where $U_w \longrightarrow \mathbb{P}^1$ is a ramified covering around ∞. (See Sect. 4.5.5.) ■

2.8.3.2 A General Equivalence

Let $(V, \overline{\partial}_V)$ be a holomorphic vector bundle on U_w. Let $\partial_{V,\overline{w}}$ denote the differential operator of V induced by $\overline{\partial}_V$ and $\partial_{\overline{w}}$. Let $\mathbb{D}^{\lambda}_w$ be a linear differential operator of V satisfying $\mathbb{D}^{\lambda}_w(fs) = \lambda\partial_w(f)\cdot s + f\mathbb{D}^{\lambda}_w(s)$ for any $f \in C^{\infty}(U_w)$ and $s \in C^{\infty}(U_w, V)$. Note that $\mathbb{D}^{\lambda}_w$ is equivalent to a linear differential operator $\mathbb{D}^{\lambda\,(1,0)} : C^{\infty}(U_w, V) \longrightarrow C^{\infty}(U_w, V \otimes \Omega^{1,0}_{U_w}(2\infty))$ such that $\mathbb{D}^{\lambda\,(1,0)}(fs) = \lambda\partial_{U_w}(f)s + f\mathbb{D}^{\lambda\,(1,0)}(s)$ for any $f \in C^{\infty}(U_w)$ and $s \in C^{\infty}(U_w, V)$. The restriction of $\overline{\partial}_V + \mathbb{D}^{\lambda(1,0)}$ to $U_w \setminus \{\infty\}$ is a λ-connection of $V_{|U_w\setminus\{\infty\}}$ in the C^{∞}-sense.

We obtain a vector bundle $\widetilde{V} := (\Psi^{\lambda})^{-1}(V)$ on U^{λ}. We set $U^{\lambda *} := U^{\lambda} \setminus H^{\lambda}_{\infty}$ and $\widetilde{V}^* := \widetilde{V}_{|U^{\lambda *}}$. We have already constructed the S^1_T-equivariant differential operator $\overline{\partial}_{\widetilde{V}^*} : C^{\infty}(U^{\lambda *}, \widetilde{V}^*) \longrightarrow C^{\infty}(U^{\lambda *}, \widetilde{V}^* \otimes \Omega^{0,1}_{U^{\lambda *}})$ satisfying $\overline{\partial}_{\widetilde{V}^*}(fu) = \overline{\partial}_{U^{\lambda *}}(f)\cdot u + f\overline{\partial}_{\widetilde{V}^*}(u)$ for any $f \in C^{\infty}(U^{\lambda *})$ and $u \in C^{\infty}(U^{\lambda *}, \widetilde{V}^*)$.

Lemma 2.8.6 *The differential operator $\overline{\partial}_{\widetilde{V}^*}$ uniquely extends to a linear differential operator $\overline{\partial}_{\widetilde{V}} : C^{\infty}(U^{\lambda}, \widetilde{V}) \longrightarrow C^{\infty}(U^{\lambda}, \widetilde{V} \otimes \Omega^{0,1}_{U^{\lambda}})$ satisfying the mini-complex Leibniz rule.*

Proof We may assume that U_w is an open disc around ∞. Let $\boldsymbol{v}$ be a holomorphic frame of $(V, \overline{\partial}_V)$ on U_w. We have $\partial_{V,\overline{w}}\boldsymbol{v} = 0$. Let A be the matrix valued C^{∞}-function on U_w determined by $\mathbb{D}^{\lambda}_w\boldsymbol{v} = \boldsymbol{v}A$. The pull back $\widetilde{\boldsymbol{v}} := (\Psi^{\lambda})^{-1}\boldsymbol{v}$ is a C^{∞}-frame of $\widetilde{V}$ on U^{λ}.

For any $P \in H^{\lambda}_{\infty}$, we choose $\widetilde{P} \in H^{\lambda\,\mathrm{cov}}_{\infty}$ such that $\varpi^{\lambda}(\widetilde{P}) = P$. Then, $(t_1, \tau_1) = (t_1, \beta_1^{-1})$ around $\widetilde{P}$ induces a local mini-complex coordinate system on a neighbourhood U^{λ}_1 of P. On $U^{\lambda}_1 \setminus H^{\lambda}_{\infty}$, the operators $\partial_{\widetilde{V}^*,t_1}$ and $\partial_{\widetilde{V}^*,\overline{\tau}_1}$ are described as follows with respect to the frame $\widetilde{\boldsymbol{v}}$:

$$\partial_{\widetilde{V}^*,t_1}\widetilde{\boldsymbol{v}} = \widetilde{\boldsymbol{v}}\frac{-2\sqrt{-1}}{1+|\lambda|^2}(\Psi^{\lambda})^*(A), \qquad \partial_{\widetilde{V}^*,\overline{\tau}_1}\widetilde{\boldsymbol{v}} = -\overline{\beta}_1^2\partial_{\widetilde{V}^*,\overline{\beta}_1}\widetilde{\boldsymbol{v}} = 0$$

Note that A is C^∞ with respect to w^{-1} and $\overline{w}^{-1}$, and that $(\Psi^\lambda)^*(w^{-1}) = (1 + |\lambda|^2)\tau_1(1 - 2\sqrt{-1}\lambda t_1\tau_1)^{-1}$ is C^∞ around P. Hence, $\partial_{\widetilde{V}^*,t_1}$ and $\partial_{\widetilde{V}^*,\overline{\tau}_1}$ uniquely extend to operators $\partial_{\widetilde{V},t_1}$ and $\partial_{\widetilde{V},\overline{\tau}_1}$ on $C^\infty(U^\lambda, \widetilde{V})$. Then, the claim of the lemma is clear. ■

Lemma 2.8.7 *The above construction induces an equivalence between the following objects.*

- *Holomorphic vector bundles $(V, \overline{\partial}_V)$ on U_w equipped with a linear differential operator $\mathbb{D}^\lambda_w$ on $C^\infty(U_w, V)$ such that $\mathbb{D}^\lambda_w(fs) = \lambda\partial_w(f)s + f\mathbb{D}^\lambda_w(s)$ for any $f \in C^\infty(U_w)$ and $s \in C^\infty(U_w, V)$.*
- *S^1_T-equivariant vector bundles $\widetilde{V}$ on U^λ equipped with an S^1_T-equivariant differential linear operator $\overline{\partial}_{\widetilde{V}} : C^\infty(U^\lambda, \widetilde{V}) \longrightarrow C^\infty(U^\lambda, \widetilde{V} \otimes \Omega^{0,1}_{U^\lambda})$ satisfying the mini-complex Leibniz rule.*

Proof We indicate the inverse construction. Let $(\widetilde{V}, \overline{\partial}_{\widetilde{V}})$ be as in the statement of Lemma 2.8.7. There exists a C^∞-vector bundle V on U_w equipped with an S^1_T-equivariant isomorphism $\widetilde{V} \simeq (\Psi^\lambda)^{-1}(V)$. We set $V^* := V_{|U_w \setminus \{\infty\}}$. We obtain a λ-connection $\mathbb{D}^\lambda_{V^*}$ corresponding to $\overline{\partial}_{\widetilde{V}|U^{\lambda *}}$.

There exists the map $\pi^\lambda : U^\lambda \longrightarrow S^1_T$ induced by $(t_1, \beta_1) \longmapsto t_1$. Fix $t_1^0 \in S^1_T$. The fiber $C := (\pi^\lambda)^{-1}(t_1^0)$ is naturally equipped with a complex structure, and $(\widetilde{V}, \overline{\partial}_{\widetilde{V}})_{|C}$ is naturally a holomorphic vector bundle on C. The induced morphism $\Psi^\lambda_C : C \longrightarrow U_w$ is holomorphic, and the image is an open neighbourhood of ∞. We have the isomorphism of C^∞-vector bundles $\widetilde{V}_{|C} \simeq (\Psi^\lambda)^{-1}(V)_{|C}$. Note that $(\Psi^\lambda)^{-1}(V)_{|C \setminus H^\lambda_\infty}$ is equipped with the holomorphic structure induced by $\overline{\partial}_{V^*}$, and that $\widetilde{V}_{|C \setminus H^\lambda_\infty} \simeq (\Psi^\lambda)^{-1}(V)_{|C \setminus H^\lambda_\infty}$ is holomorphic. It implies that $\overline{\partial}_{V^*}$ uniquely extends to a holomorphic structure $\overline{\partial}_V$ of V.

Let $\boldsymbol{v}$ be a holomorphic frame of $(V, \overline{\partial}_V)$ on U_w. We obtain a frame $\widetilde{\boldsymbol{v}} = (\Psi^\lambda)^{-1}(\boldsymbol{v})$ of $\widetilde{V}$. There exists a matrix-valued C^∞-function A on U_w determined by

$$\partial_{\widetilde{V},t_1}\widetilde{\boldsymbol{v}} = \widetilde{\boldsymbol{v}} \cdot \frac{-2\sqrt{-1}}{1 + |\lambda|^2}(\Psi^\lambda)^*(A).$$

Let $\mathbb{D}^\lambda_{V^*,w}$ denote the differential operator of V^* induced by $\mathbb{D}^\lambda_{V^*}$ and ∂_w. By the construction of $\mathbb{D}^\lambda_{V^*}$, we have $\mathbb{D}^\lambda_{V^*,w}\boldsymbol{v}_{|U_w \setminus \{\infty\}} = \boldsymbol{v}_{|U_w \setminus \{\infty\}} \cdot A$. It implies that $\mathbb{D}^\lambda_{V^*,w}$ uniquely extends to a differential operator $\mathbb{D}^\lambda_w$ of V.

In this way, we obtain $(V, \overline{\partial}_V, \mathbb{D}^\lambda_w)$ from $(\widetilde{V}, \overline{\partial}_{\widetilde{V}})$. The two constructions are mutually inverse. ■

2.8.3.3 Mini-Holomorphic Bundles and Meromorphic Flat λ-Connections

The following corollary is an immediate consequence of Lemma 2.8.7 and (2.18).

Corollary 2.8.8 *The construction in* Sect. 2.8.3.2 *induces an equivalence of the following objects.*

- *Locally free $\mathcal{O}_{U_w}$-modules $\mathcal{V}$ equipped with a meromorphic λ-connection $\mathbb{D}^\lambda : \mathcal{V} \longrightarrow \mathcal{V} \otimes \Omega^1_{U_w}(2\infty)$.*
- *S^1_T-equivariant locally free $\mathcal{O}_{U^\lambda}$-modules $\widetilde{\mathcal{V}}$.* ■

2.8.3.4 Another Description of the Construction

Let us explain another more direct description of the equivalence in Corollary 2.8.8.

For any open subset $\mathcal{U} \subset \overline{\mathcal{M}}^\lambda$, let $\mathcal{K}_{\overline{\mathcal{M}}^\lambda}(\mathcal{U})$ denote the space of C^∞-functions f on $\mathcal{U}$ such that $\partial_{\overline{\beta}_1}\big(f_{|\mathcal{U}\setminus H^\lambda_\infty}\big) = 0$. For $P \in \mathcal{U} \cap H^\lambda_\infty$, a choice $\widetilde{P} \in (\varpi^\lambda)^{-1}(P)$ induces a local mini-complex coordinate system $(t_1, \tau_1) = (t_1, \beta_1^{-1})$ around P. For any $f \in \mathcal{K}_{\overline{\mathcal{M}}^\lambda}(\mathcal{U})$, we obtain $\partial_{\overline{\tau}_1} f = 0$. Because $[\partial_{t_1}, \partial_{\overline{\beta}_1}] = 0$, we obtain the naturally defined action of ∂_{t_1} on $\mathcal{K}_{\overline{\mathcal{M}}^\lambda}(\mathcal{U})$, and $f \in \mathcal{K}_{\overline{\mathcal{M}}^\lambda}(\mathcal{U})$ is mini-holomorphic if and only if $\partial_{t_1} f = 0$.

Thus, we obtain a sheaf $\mathcal{K}_{\overline{\mathcal{M}}^\lambda}$. We have the natural action ∂_{t_1} on $\mathcal{K}_{\overline{\mathcal{M}}^\lambda}$, and the kernel is $\mathcal{O}_{\overline{\mathcal{M}}^\lambda}$. For any open subset $\mathcal{U} \subset \overline{\mathcal{M}}^\lambda$, let $\mathcal{K}_\mathcal{U}$ denote the restriction of $\mathcal{K}_{\overline{\mathcal{M}}^\lambda}$ to $\mathcal{U}$.

Let U_w be an open subset of $\mathbb{P}^1_w$, and $U^\lambda = (\Psi^\lambda)^{-1}(U_w) \subset \overline{\mathcal{M}}^\lambda$. We have the naturally defined monomorphism of sheaves $(\Psi^\lambda)^{-1}(\mathcal{O}_{U_w}) \longrightarrow \mathcal{K}_{U^\lambda}$.

Let $\mathcal{V}$ be a locally free $\mathcal{O}_{U_w}$-module equipped with a meromorphic λ-connection $\mathbb{D}^\lambda : \mathcal{V} \longrightarrow \mathcal{V} \otimes \Omega^1_{U_w}(2\infty)$. From $\mathcal{O}_{U_w}$-module $\mathcal{V}$, we obtain the locally free $\mathcal{K}_{U^\lambda}$-module

$$\widetilde{\mathcal{V}}^\infty := \mathcal{K}_{U^\lambda} \otimes_{(\Psi^\lambda)^{-1}(\mathcal{O}_{U_w})} (\Psi^\lambda)^{-1}(\mathcal{V}).$$

We obtain the differential operator $\partial_{\widetilde{\mathcal{V}}^\infty, t_1}$ on $\widetilde{\mathcal{V}}^\infty$ determined by the following condition for local sections f and s of $\mathcal{K}_U$ and $\mathcal{V}$ as follows:

$$\partial_{\widetilde{\mathcal{V}}^\infty, t_1}\big(f(\Psi^\lambda)^{-1}(s)\big) = \partial_{t_1}(f)(\Psi^\lambda)^{-1}(s) + \frac{1}{1+|\lambda|^2} f \cdot (\Psi^\lambda)^{-1}(-2\sqrt{-1}\mathbb{D}^\lambda_w s).$$

Note that $\widetilde{\mathcal{V}}^\infty$ is equal to the sheaf of C^∞-sections u of $\widetilde{V}$ such that $\partial_{\widetilde{V},\overline{\beta}_1} u = 0$. Hence, the kernel of $\partial_{\widetilde{\mathcal{V}}^\infty, t_1}$ is naturally isomorphic to the sheaf of mini-holomorphic sections of $\widetilde{V}$.

Remark 2.8.9 Although we use the map $\Psi^\lambda(t_1, \beta_1) = (1+|\lambda|^2)^{-1}(\beta_1 - 2\sqrt{-1}\lambda t_1)$, it is more natural to consider $(t_1, \beta_1) \longmapsto \beta_1 - 2\sqrt{-1}\lambda t_1$. We adopt Ψ^λ for the consistency with the dimensional reduction from monopoles to harmonic bundles in Sect. 2.6.1. See Remark 3.6.7. ■

2.9 Curvatures of Mini-Holomorphic Bundles with Metric on $\mathcal{M}^\lambda$

2.9.1 *Contraction of Curvature and Analytic Degree*

Let $U \longrightarrow \mathcal{M}^\lambda$ be a local diffeomorphism. We obtain the induced metric and the induced mini-complex structure on U. We also obtain the complex vector fields ∂_{β_i}, $\partial_{\overline{\beta}_i}$, and ∂_{t_i} $(i = 0, 1)$ on U.

Let $(E, \overline{\partial}_E)$ be a mini-holomorphic bundle on U. We have the operators $\partial_{E,\overline{\beta}_0}$ and ∂_{E,t_0}. Let h be a Hermitian metric of E, which induces the Higgs field ϕ_h and the Chern connection ∇_h. Let $F(h)$ denote the curvature of ∇_h. We have the expression $F(h) = F(h)_{\beta_0\overline{\beta}_0} d\beta_0\, d\overline{\beta}_0 + F(h)_{\beta_0,t_0} d\beta_0\, dt_0 + F(h)_{\overline{\beta}_0,t_0} d\overline{\beta}_0\, dt_0$. We set

$$G(h) := 2F(h)_{\beta_0\overline{\beta}_0} - \sqrt{-1}\nabla_{h,t_0}\phi_h. \tag{2.23}$$

Note that the Bogomolny equation is equivalent to $G(h) = 0$ (see Corollary 2.4.4).

Definition 2.9.1 Suppose that $\operatorname{Tr} G(h)$ is described as a sum $g_1 + g_2$, where g_1 is L^1 on U and g_2 is non-positive everywhere. Then, we set $\deg(E, \overline{\partial}_E, h) := \int_U \operatorname{Tr} G(h)\, \mathrm{dvol}_U$, where dvol_U is the volume form induced by the Riemannian metric. Note that $\deg(E, \overline{\partial}_E, h) \in \mathbb{R} \cup \{-\infty\}$. ■

Remark 2.9.2 Let $\mathbb{R}_{s_0} \times U \longrightarrow \mathbb{R}_{s_0} \times \mathcal{M}^\lambda$ be the induced local diffeomorphism. We obtain the complex structure on $\mathbb{R}_{s_0} \times U$ such that $(\alpha_0, \beta_0) = (s_0 + \sqrt{-1}t_0, \beta_0)$ induces local complex coordinate systems. Let $(\widetilde{E}, \overline{\partial}_{\widetilde{E}})$ be the holomorphic vector bundle with the metric $\widetilde{h}$ induced by $(E, \overline{\partial}_E)$ with h as in Sect. 2.5.2. Let $\widetilde{F}$ denote the curvature of the Chern connection associated with $(\widetilde{E}, \overline{\partial}_{\widetilde{E}}, \widetilde{h})$. We have the expression $\widetilde{F} = \widetilde{F}_{\alpha_0,\overline{\alpha}_0} d\alpha_0\, d\overline{\alpha}_0 + \widetilde{F}_{\alpha_0,\overline{\beta}_0} d\alpha_0\, d\overline{\beta}_0 + \widetilde{F}_{\beta_0,\overline{\alpha}_0} d\beta_0\, d\overline{\alpha}_0 + \widetilde{F}_{\beta_0,\overline{\beta}_0} d\beta_0\, d\overline{\beta}_0$. Then, $\sqrt{-1}\Lambda\widetilde{F} = 2(\widetilde{F}_{\alpha_0,\overline{\alpha}_0} + \widetilde{F}_{\beta_0,\overline{\beta}_0})$ is equal to the pull back of $G(h)$ by the projection $\mathbb{R}_{s_0} \times U \longrightarrow U$, where Λ is the adjoint of the multiplication of the Kähler form of $\mathbb{R}_{s_0} \times U$. (See [48].) Hence, $\deg(E, \overline{\partial}_E, h)$ is an analogue of the analytic degree in [79]. ■

Remark 2.9.3 In [67] and a previous version of this monograph, we set "$G(h) = F(h)_{\beta_0\overline{\beta}_0} - \frac{\sqrt{-1}}{2}\nabla_{h,t_0}\phi_h$", which is the half of (2.23). There is no essential difference though constants in some formulas are changed. ■

2.9.2 *Chern-Weil Formula*

The standard Chern-Weil formula [79, 80] is translated as follows, which is implicitly contained in [11].

Lemma 2.9.4 *Let V be a mini-holomorphic subbundle of E. Let h_V be the induced metric of V. Let p_V denote the orthogonal projection of E onto V, which we regard as an endomorphism of E in a natural way. Then, we obtain the following formula:*

$$\operatorname{Tr} G(h_V) = \operatorname{Tr}\big(G(h)p_V\big) - 2\big|\partial_{E,\overline{\beta}_0} p_V\big|^2 - \frac{1}{2}\big|\partial_{E,t_0} p_V\big|^2. \tag{2.24}$$

Proof We use the notation in Remark 2.9.2. We obtain the induced holomorphic subbundle $\widetilde{V}$ of $\widetilde{E}$ with the metric $\widetilde{h}_{\widetilde{V}}$ induced by V. We obtain the Chern connection $\nabla_{\widetilde{h}} = \overline{\partial}_{\widetilde{E}} + \partial_{\widetilde{E},h}$. Let $p_{\widetilde{V}}$ be the projection of $\widetilde{E}$ onto $\widetilde{V}$, which is the pull back of p_V. We regard $p_{\widetilde{V}}$ as the endomorphism of $\widetilde{E}$ in a natural way. Let $\iota_{\widetilde{V}}$ and $\iota_{\widetilde{V}^\perp}$ denote the inclusions of $\widetilde{V}$ and $\widetilde{V}^\perp$ into $\widetilde{E}$, respectively. We set $A := p_{\widetilde{V}} \circ \overline{\partial}_{\widetilde{E}} \circ \iota_{\widetilde{V}^\perp}$. Because $p_{\widetilde{V}} \circ F(\widetilde{h}) \circ \iota_{\widetilde{V}} = F(\widetilde{h}_{\widetilde{V}}) - A \circ A^\dagger$, we obtain

$$\sqrt{-1}\operatorname{Tr} \Lambda F(\widetilde{h}_{\widetilde{V}}) = \sqrt{-1}\operatorname{Tr}(p_{\widetilde{V}} \circ \Lambda F(\widetilde{h}) \circ \iota_{\widetilde{V}}) + \sqrt{-1}\Lambda \operatorname{Tr}(A \circ A^\dagger).$$

Note that $\sqrt{-1}\Lambda F(\widetilde{h})$ and $\sqrt{-1}\Lambda F(\widetilde{h}_{\widetilde{V}})$ are the pull back of $G(h)$ and $G(h_V)$, respectively. Because of (2.5), $\sqrt{-1}\Lambda \operatorname{Tr}(A \circ A^\dagger)$ is the pull back of $-2\big|\partial_{E,\overline{\beta}_0} p_V\big|^2 - \frac{1}{2}\big|\partial_{E,t_0} p_V\big|^2$. Then, we obtain the claim of the lemma. ∎

As a direct consequence of the lemma, if $|G(h)|_h$ is L^1 on U, then $\deg(V, h_V)$ makes sense in $\mathbb{R} \cup \{-\infty\}$ for any mini-holomorphic subbundle V of E.

2.9.3 Another Description of $G(h)$

We have the differential operator $\partial_{E,\overline{\beta}_1}$ on $C^\infty(U, E)$, which is given by the inner product of $\partial_{\overline{\beta}_1}$ and $\overline{\partial}_E$. We have the differential operator ∂_{E,h,β_1} on $C^\infty(U, E)$ determined by the condition $\partial_{\overline{\beta}_1} h(u, v) = h(\partial_{E,\overline{\beta}_1} u, v) + h(u, \partial_{E,h,\beta_1} v)$.

Proposition 2.9.5 *The following formula holds:*

$$\begin{aligned} G(h) &= 2(1+|\lambda|^2)^2\big[\partial_{E,h,\beta_1}, \partial_{E,\overline{\beta}_1}\big] \\ &\quad - \sqrt{-1}\Big((1-|\lambda|^2)\nabla_{h,t_0}\phi + 2\lambda\sqrt{-1}\nabla_{h,\beta_0}\phi - 2\overline{\lambda}\sqrt{-1}\nabla_{h,\overline{\beta}_0}\phi\Big) \\ &= 2(1+|\lambda|^2)^2\big[\partial_{E,h,\beta_1}, \partial_{E,\overline{\beta}_1}\big] - \sqrt{-1}(1+|\lambda|^2)\nabla_{h,t}\phi. \end{aligned} \tag{2.25}$$

Here, $\nabla_{h,t}$ denote the inner product of ∇_h and ∂_t. (See (2.11).*)*

Proof To simplify the description, we omit to denote the dependence on h, i.e., ∂_{E,h,β_1} is denoted by ∂_{E,β_1}, for example. By the relation (2.14), we obtain the following:

$$\partial_{E,\overline{\beta}_1} = \frac{\lambda}{1+|\lambda|^2}\frac{1}{2\sqrt{-1}}\partial_{E,t_0} + \frac{1}{1+|\lambda|^2}\partial_{E,\overline{\beta}_0}$$
$$= \frac{1}{1+|\lambda|^2}\nabla_{\overline{\beta}_0} - \frac{\lambda}{1+|\lambda|^2}\frac{1}{2}(\phi+\sqrt{-1}\nabla_{t_0}). \tag{2.26}$$

We obtain the following:

$$\partial_{E,\beta_1} = \frac{1}{1+|\lambda|^2}\nabla_{\beta_0} - \frac{\overline{\lambda}}{1+|\lambda|^2}\frac{1}{2}(\phi-\sqrt{-1}\nabla_{t_0}).$$

We have the following:

$$[\partial_{E,\beta_1},\partial_{E,\overline{\beta}_1}] = \Big(\frac{1}{1+|\lambda|^2}\Big)^2\Big([\nabla_{\beta_0},\nabla_{\overline{\beta}_0}] - \frac{\lambda}{2}(\nabla_{\beta_0}\phi+\sqrt{-1}[\nabla_{\beta_0},\nabla_{t_0}])$$
$$+\frac{\overline{\lambda}}{2}(\nabla_{\overline{\beta}_0}\phi-\sqrt{-1}[\nabla_{\overline{\beta}_0},\nabla_{t_0}]) + \frac{|\lambda|^2}{4}[\phi-\sqrt{-1}\nabla_{t_0},\phi+\sqrt{-1}\nabla_{t_0}]\Big). \tag{2.27}$$

By Lemma 2.4.3, we obtain

$$-\frac{\lambda}{2}\nabla_{\beta_0}\phi - \frac{\lambda}{2}\sqrt{-1}[\nabla_{\beta_0},\nabla_{t_0}] = -\lambda\nabla_{\beta_0}\phi,$$
$$\frac{\overline{\lambda}}{2}\nabla_{\overline{\beta}_0}\phi - \frac{\sqrt{-1}}{2}\overline{\lambda}[\nabla_{\overline{\beta}_0},\nabla_{t_0}] = \overline{\lambda}\nabla_{\overline{\beta}_0}\phi.$$

We also have

$$\frac{|\lambda|^2}{4}[\phi-\sqrt{-1}\nabla_{t_0},\phi+\sqrt{-1}\nabla_{t_0}] = -\frac{|\lambda|^2}{2}\sqrt{-1}\nabla_{t_0}\phi.$$

Hence, we obtain the following:

$$[\partial_{E,\beta_1},\partial_{E,\overline{\beta}_1}] = \Big(\frac{1}{1+|\lambda|^2}\Big)^2\Big(F_{\beta_0,\overline{\beta}_0} - \lambda\nabla_{\beta_0}\phi + \overline{\lambda}\nabla_{\overline{\beta}_0}\phi - \frac{|\lambda|^2}{2}\sqrt{-1}\nabla_{t_0}\psi\Big)$$
$$= \Big(\frac{1}{1+|\lambda|^2}\Big)^2\Big(\frac{1}{2}G(h) - \lambda\nabla_{\beta_0}\phi + \overline{\lambda}\nabla_{\overline{\beta}_0}\phi + \frac{1-|\lambda|^2}{2}\sqrt{-1}\nabla_{t_0}\phi\Big). \tag{2.28}$$

Then, we obtain (2.25). ■

We state some consequences.

Corollary 2.9.6 *Suppose that $U := \mathcal{M}^{\lambda} \setminus Z$, where Z is a finite subset of $\mathcal{M}^{\lambda}$. Suppose that $\mathrm{Tr}\left[\partial_{E,h,\beta_1}, \partial_{E,\overline{\beta}_1}\right]$ and $\mathrm{Tr}\,\nabla_{h,t}\phi_h$ are L^1 on $\mathcal{M}^{\lambda} \setminus Z$. Then, we obtain*

$$\deg(E, \overline{\partial}_E, h) = \int_0^T dt_1 \int_{\mathbb{C}_{\beta_1}} \mathrm{Tr}\left(\left[\partial_{E,h,\beta_1}, \partial_{E,\overline{\beta}_1}\right]\right)\sqrt{-1}d\beta_1\, d\overline{\beta}_1. \tag{2.29}$$

Proof The following formula holds:

$$\mathrm{dvol} = \frac{\sqrt{-1}}{2} d\beta_0 d\overline{\beta}_0\, dt_0 = \frac{\sqrt{-1}}{2}(1+|\lambda|^2)^{-2} d\beta_1\, d\overline{\beta}_1\, dt_1 = \frac{\sqrt{-1}}{2} dw\, d\overline{w}\, dt.$$

By using Fubini theorem, we obtain the following:

$$\int_{\mathcal{M}^{\lambda}\setminus Z} \mathrm{Tr}\left(\nabla_t \phi\right) \mathrm{dvol} = \frac{\sqrt{-1}}{2} \int_{\mathbb{C}_w} dw\, d\overline{w} \int_{\mathbb{R}/T\mathbb{Z}} \partial_t\, \mathrm{Tr}(\phi_h)\, dt = 0.$$

Hence, by Proposition 2.9.5, we obtain the following.

$$\begin{aligned}\int_{\mathcal{M}^{\lambda}\setminus Z} \mathrm{Tr}\, G(h)\, \mathrm{dvol} &= \int_{\mathcal{M}^{\lambda}} \mathrm{Tr}\left([\widetilde{\nabla}_{\beta_1}, \widetilde{\nabla}_{\overline{\beta}_1}]\right)\sqrt{-1}d\beta_1\, d\overline{\beta}_1\, dt_1 \\ &= \int_0^T dt_1 \int_{\mathbb{C}_{\eta_1}} \mathrm{Tr}\left([\widetilde{\nabla}_{\beta_1}, \widetilde{\nabla}_{\overline{\beta}_1}]\right)\sqrt{-1}d\beta_1\, d\overline{\beta}_1.\end{aligned} \tag{2.30}$$

Thus, we obtain (2.29). ■

Note that for almost all $0 \le t_1 \le T$, the analytic degree of $(E, \overline{\partial}_E, h)_{|\{t_1\}\times\mathbb{C}_{\beta_1}}$ is defined as the integration of the first Chern form (see [79]):

$$\deg\left((E, \overline{\partial}_E, h)_{|\{t_1\}\times\mathbb{C}_{\beta_1}}\right) = \frac{\sqrt{-1}}{2\pi} \int_{\mathbb{C}_{\beta_1}} \mathrm{Tr}\left([\partial_{E,h,\beta_1}, \partial_{E,\overline{\beta}_1}]\, d\beta_1\, d\overline{\beta}_1\right). \tag{2.31}$$

Then, (2.29) is rewritten as follows:

$$\deg(E, \overline{\partial}_E, h) = \int_0^T 2\pi \deg\left((E, \overline{\partial}_E, h)_{|\{t_1\}\times\mathbb{C}_{\beta_1}}\right) dt_1. \tag{2.32}$$

We also obtain the following useful formula to relate the curvatures of the mini-holomorphic bundles on $\mathcal{M}^{\lambda}$ and $\mathcal{M}^0$, underlying a monopole.

Corollary 2.9.7 *Suppose that $(E, \overline{\partial}_E, h)$ is a monopole on U, i.e., $G(h) = 0$. Then, we obtain*

$$\left[\partial_{E,h,\beta_1}, \partial_{E,\overline{\beta}_1}\right] = \frac{\sqrt{-1}}{2} \frac{1}{1+|\lambda|^2} \nabla_{h,t}\phi = \frac{1}{1+|\lambda|^2} F_{w,\overline{w}}.$$

■

2.9.4 Change of Metrics

Let h_1 be another Hermitian metric of E. Let s be the automorphism of E determined by $h_1 = h \cdot s$, which is self-adjoint with respect to both h and h_1. The following is a variant of [79, Lemma 3.1].

Lemma 2.9.8 *The following holds:*

$$G(h_1) = G(h) - 2\partial_{E,\overline{\beta}_0}\big(s^{-1}\partial_{E,h,\beta_0}s\big) - \frac{1}{2}\Big[\nabla_{h,t_0} - \sqrt{-1}\phi_h,\ s^{-1}\big[\nabla_{h,t_0} + \sqrt{-1}\phi_h, s\big]\Big]. \tag{2.33}$$

Proof Because $\partial_{E,h_1} = \partial_{E,h} + s^{-1}\partial_{E,h}s$, the following holds:

$$F(h_1)_{\beta_0,\overline{\beta}_0} = \big[\partial_{E,h_1,\beta_0}, \partial_{E,\overline{\beta}_0}\big] = F(h)_{\beta_0,\overline{\beta}_0} - \partial_{E,\overline{\beta}_0}(s^{-1}\partial_{E,h,\beta_0}s).$$

We also obtain the following:

$$\nabla_{h_1,t_0} = \nabla_{h,t_0} + \frac{1}{2}s^{-1}[\nabla_{h,t_0} + \sqrt{-1}\phi_h, s],$$

$$\phi_{h_1} = \phi_h - \frac{\sqrt{-1}}{2}s^{-1}[\nabla_{h,t_0} + \sqrt{-1}\phi_h, s].$$

We obtain the following:

$$-\sqrt{-1}\nabla_{h_1,t_0}\phi_{h_1} = -\sqrt{-1}\nabla_{h,t}\phi_h - \frac{1}{2}\Big[\nabla_{h,t_0} - \sqrt{-1}\phi_h,\ s^{-1}\big[\nabla_{h,t_0} + \sqrt{-1}\phi_h, s\big]\Big].$$

Hence, we obtain (2.33). ■

We obtain the following direct consequence, which is also a variant of [79, Lemma 3.1].

Corollary 2.9.9 *The following equality holds:*

$$-\Big(\partial_{\overline{\beta}_0}\partial_{\beta_0} + \frac{1}{4}\partial_{t_0}^2\Big)\operatorname{Tr} s = \frac{1}{2}\operatorname{Tr}\Big(s\big(G(h_1) - G(h)\big)\Big) - \big|s^{-1/2}\partial_{E,h,\beta_0}s\big|_h^2 - \frac{1}{4}\big|s^{-1/2}\partial'_{E,h,t_0}s\big|_h^2. \tag{2.34}$$

We also have the following inequality:

$$-\Big(\partial_{\overline{\beta}_0}\partial_{\beta_0} + \frac{1}{4}\partial_{t_0}^2\Big)\log\big(\operatorname{Tr}(s)\big) \le \frac{1}{2}\Big(\big|G(h)\big|_h + \big|G(h_1)\big|_{h_1}\Big).$$

■

Corollary 2.9.10 *If* rank $E = 1$, *we have* $G(h_1) - G(h) = 2^{-1}\Delta \log s$ *on* U. *Here,* Δ *denote the Laplacian of the Riemannian manifold* $\mathcal{M}^\lambda$. ■

2.9.5 Relation with λ-Connections

This subsection is a complement to Sect. 2.8. We use the notation there.

Let us recall the condition for harmonic bundles given in terms of λ-connections [63, §2.2]. Let $(V, \mathbb{D}^\lambda)$ be a λ-flat bundle on U_w with a Hermitian metric h_V. Let $\mathbb{D}^\lambda = d''_V + d'_V$ denote the decomposition into the $(0, 1)$-part and the $(1, 0)$-part. We have the $(1, 0)$-operator δ' and the $(0, 1)$-operator δ'' determined by the conditions

$$\overline{\partial} h_V(u, v) = h_V(d''_V u, v) + h_V(u, \delta' v), \quad \lambda\partial h_V(u, v) = h_V(d'_V u, v) + h_V(u, \delta'' v)$$

for $u, v \in C^\infty(U_w, V)$. We have the $(0, 1)$-operator $\overline{\partial}_V$, the $(1, 0)$-operator ∂_V, the section $\theta \in C^\infty(U_w, V \otimes \Omega^{1,0})$ and the section $\theta^\dagger \in C^\infty(U_w, V \otimes \Omega^{0,1})$ determined by

$$d''_V = \overline{\partial}_V + \lambda\theta^\dagger, \quad d'_V = \lambda\partial_V + \theta, \quad \delta' = \partial_V - \overline{\lambda}\theta, \quad \delta'' = \overline{\lambda}\partial_V - \theta^\dagger.$$

Then, $(V, \mathbb{D}^\lambda, h)$ is called a harmonic bundle if $(V, \overline{\partial}_V, \theta, h)$ is a harmonic bundle. If $\lambda \neq 0$, it is equivalent to $\overline{\partial}_V\theta = 0$. We set $\mathbb{D}^{\lambda\star} := \delta' - \delta''$. Then, $(V, \mathbb{D}^\lambda, h_V)$ is a harmonic bundle if and only if $[\mathbb{D}^\lambda, \mathbb{D}^{\lambda\star}] = 0$.

Let $d''_{V,\overline{w}}$ denote the inner product of d''_V and $\partial_{\overline{w}}$. We use the notation $d'_{V,w}$, $\delta''_{\overline{w}}$, etc., in similar meanings. Note that $d'_w - \lambda\delta'_w$ and $\delta''_{\overline{w}} - \overline{\lambda} d'_{\overline{w}}$ are endomorphisms of V, which follow from $d' - \lambda\delta' = (1 + |\lambda|^2)\theta$ and $\delta'' - \overline{\lambda} d' = -(1 + |\lambda|^2)\theta^\dagger$.

As explained in Sect. 2.8, we obtain a mini-holomorphic bundle $(\widetilde{V}, \overline{\partial}_{\widetilde{V}})$ on $U^\lambda = \Psi^{-1}(U_w) \subset \mathcal{M}^\lambda$ from $(V, \mathbb{D}^\lambda)$. We also obtain the Hermitian metric $\widetilde{h} := (\Psi^\lambda)^{-1}(h_V)$. Let $\nabla_{\widetilde{V}}$ and ϕ denote the Chern connection and the Higgs field of $(\widetilde{V}, \overline{\partial}_{\widetilde{V}}, \widetilde{h})$. Let $\partial'_{\widetilde{V},t_0}$ be the differential operator induced by $\partial_{\widetilde{V},t_0}$ and h as in Sect. 2.4.2.

Lemma 2.9.11 *We obtain the following formulas:*

$$[\nabla_{\widetilde{V},\overline{\beta}_0}, \nabla_{\widetilde{V},\beta_0}] = \frac{1}{(1 + |\lambda|^2)^2}(\Psi^\lambda)^{-1}\Big([d''_{\overline{w}}, \delta'_w] + \overline{\lambda}[d''_{\overline{w}}, \delta''_{\overline{w}}] + \lambda[d'_w, \delta'_w] + |\lambda|^2[d'_w, \delta''_{\overline{w}}]\Big),$$

$$[\nabla_{\widetilde{V},\overline{\beta}_0}, \partial'_{\widetilde{V},t_0}] = \frac{2\sqrt{-1}}{(1 + |\lambda|^2)^2}(\Psi^\lambda)^{-1}\Big([d''_{\overline{w}}, \delta''_{\overline{w}}] - \lambda[d''_{\overline{w}}, \delta'_w] + \lambda[d'_w, \delta''_{\overline{w}}] - \lambda^2[d'_w, \delta'_w]\Big),$$

$$[\nabla_{\widetilde{V},\beta_0}, \partial_{\widetilde{V},t_0}] = \frac{-2\sqrt{-1}}{(1 + |\lambda|^2)^2}(\Psi^\lambda)^{-1}\Big([\delta'_w, d'_w] - \overline{\lambda}[\delta'_w, d''_{\overline{w}}] + \overline{\lambda}[\delta''_{\overline{w}}, d'_w] - \overline{\lambda}^2[\delta''_{\overline{w}}, d''_{\overline{w}}]\Big),$$

$$[\partial_{\widetilde{V},t_0}, \partial'_{\widetilde{V},t_0}] = \frac{4}{(1 + |\lambda|^2)^2}(\Psi^\lambda)^{-1}\Big([d'_w, \delta''_{\overline{w}}] - \lambda[d'_w, \delta'_w] - \overline{\lambda}[d''_{\overline{w}}, \delta''_{\overline{w}}] + |\lambda|^2[d''_{\overline{w}}, \delta'_w]\Big).$$

We also have the following formula:

$$\phi = \frac{1}{1+|\lambda|^2}(\Psi^\lambda)^{-1}\big((d'_w - \lambda\delta'_w) + (\delta''_{\overline{w}} - \overline{\lambda} d'_{\overline{w}})\big).$$

Proof By the construction, the following holds for any $s \in C^\infty(U_w, V)$:

$$\partial_{\widetilde{V},t_1}(\Psi^\lambda)^{-1}(s) = \frac{-2\sqrt{-1}}{1+|\lambda|^2}(\Psi^\lambda)^{-1}\Big(\big(d'_w - \overline{\lambda} d''_{\overline{w}}\big)s\Big).$$

We denote it as

$$\partial_{\widetilde{V},t_1} = \frac{-2\sqrt{-1}}{1+|\lambda|^2}(\Psi^\lambda)^{-1}\Big(d'_w - \overline{\lambda} d''_{\overline{w}}\Big).$$

Similarly, we have

$$\begin{cases} \partial'_{\widetilde{V},t_1} = \frac{2\sqrt{-1}}{1+|\lambda|^2}(\Psi^\lambda)^{-1}\Big(\delta''_{\overline{w}} - \lambda\delta'_w\Big) \\ \partial_{\widetilde{V},\overline{\beta}_1} = \frac{1}{1+|\lambda|^2}(\Psi^\lambda)^{-1}\big(d''_{\overline{w}}\big) \\ \partial_{\widetilde{V},\beta_1} = \frac{1}{1+|\lambda|^2}(\Psi^\lambda)^{-1}(\delta'_w). \end{cases} \tag{2.35}$$

Note that $\partial_{\widetilde{V},t_0} = \partial_{\widetilde{V},t_1}$ and $\partial'_{\widetilde{V},t_0} = \partial'_{\widetilde{V},t_1}$. We obtain

$$\begin{aligned} \nabla_{\widetilde{V},\overline{\beta}_0} &= (1+|\lambda|^2)\partial_{\widetilde{V}\overline{\beta}_1} - \frac{\lambda}{2\sqrt{-1}}\partial_{\widetilde{V},t_1} \\ &= \frac{1}{1+|\lambda|^2}(\Psi^\lambda)^{-1}(d''_{\overline{w}}) + \frac{\lambda}{1+|\lambda|^2}(\Psi^\lambda)^{-1}(d'_w). \end{aligned} \tag{2.36}$$

Similarly, we obtain

$$\nabla_{\widetilde{V},\beta_0} = \frac{1}{1+|\lambda|^2}(\Psi^\lambda)^{-1}(\delta'_w) + \frac{\overline{\lambda}}{1+|\lambda|^2}(\Psi^\lambda)^{-1}(\delta''_{\overline{w}}).$$

Then, we obtain the desired equalities by direct computations. ■

Lemma 2.9.12 *We have the following formula:*

$$G(\widetilde{h}) = \frac{1}{(1+|\lambda|^2)}(\Psi^\lambda)^{-1}\Big(\sqrt{-1}\Lambda_{U_w}[\mathbb{D}^\lambda, \mathbb{D}^{\lambda\star}]\Big). \tag{2.37}$$

Here, $\Lambda_{U_w} : \Omega^{1,1}_{U_w} \longrightarrow \Omega^{0,0}_{U_w}$ *is determined by* $\Lambda_{U_w}(dw\, d\overline{w}) = -2\sqrt{-1}$.

Proof By definition, we have

$$-\sqrt{-1}\nabla_{\widetilde{V},t_0}\phi = -\frac{1}{2}\big[\partial_{\widetilde{V},t_0}, \partial'_{\widetilde{V},t_0}\big].$$

By a direct computation, we obtain

$$\begin{aligned}
&2\big[\nabla_{\widetilde{V},\beta_0}, \nabla_{\widetilde{V},\overline{\beta}_0}\big] - \sqrt{-1}\nabla_{t_0}\phi \\
&= \frac{2}{(1+|\lambda|^2)^2}(\Psi^\lambda)^{-1}\Big(-\big[d''_{\overline{w}} + \lambda d'_w, \delta'_w + \overline{\lambda}\delta''_{\overline{w}}\big] - \big[d'_w - \overline{\lambda} d''_{\overline{w}}, \delta''_{\overline{w}} - \lambda\delta'_w\big]\Big) \\
&= \frac{2}{1+|\lambda|^2}(\Psi^\lambda)^{-1}\Big(\big[\delta'_w, d''_{\overline{w}}\big] + \big[\delta''_{\overline{w}}, d'_w\big]\Big). \qquad (2.38)
\end{aligned}$$

We also have

$$\sqrt{-1}\Lambda_{U_w}\big[\mathbb{D}^\lambda, \mathbb{D}^{\lambda\star}\big] = \sqrt{-1}\Lambda_{U_w}\Big([d'', \delta'] - [d', \delta'']\Big) = 2\Big(\big[\delta'_w, d''_{\overline{w}}\big] + \big[\delta''_{\overline{w}}, d'_w\big]\Big).$$

Hence, we obtain (2.37). ■

Corollary 2.9.13 *$(\widetilde{V}, \overline{\partial}_{\widetilde{V}}, \widetilde{h})$ is a monopole if and only if $(V, \mathbb{D}^\lambda, h)$ is a harmonic bundle* ■

Lemma 2.9.14 *Let $\partial_{\widetilde{V},\widetilde{h},\beta_1}$ be the operator induced by $\partial_{\widetilde{V},\overline{\beta}_1}$ and $\widetilde{h}$ as in*Sect. 2.9.3. *Then, we obtain the following equality:*

$$\big[\partial_{\widetilde{V};\overline{\beta}_1}, \partial_{\widetilde{V},\widetilde{h},\beta_1}\big] = \frac{1}{(1+|\lambda|^2)^2}(\Psi^\lambda)^{-1}\Big([d''_{\overline{w}}, \delta'_{h,w}]\Big).$$

Proof It follows from (2.35). ■

2.9.5.1 λ-Flat Bundles of Infinite Rank with a Harmonic Metric

Let $(E, \overline{\partial}_E)$ be a mini-holomorphic bundle on U^λ. We obtain the Chern connection ∇ and the Higgs field ϕ. Let E_{C^∞} denote the sheaf of C^∞-sections of E. We obtain the $C^\infty_{U_w}$-module $V_1 = \Psi_!(E_{C^\infty})$ equipped with the induced λ-connection $\mathbb{D}^\lambda_{V_1}$ as explained in Sect. 2.8.1.5.

Let h be a Hermitian metric of E. It induces a Hermitian metric $h_1 = \Psi_!(h)$ of V_1. We obtain the operators

$$\delta'_w = (1+|\lambda|^2)\partial_{E,h,\beta_1}, \quad \delta''_{\overline{w}} = \frac{-\sqrt{-1}}{2}(1+|\lambda|^2)\partial'_{E,h,t_0} + \lambda(1+|\lambda|^2)\partial_{E,h,\beta_1}$$

on V_1. We set $\delta'(s) = \delta'_w(s)\,dw$ and $\delta''(s) = \delta''_{\overline{w}}(s)\,d\overline{w}$ for any local section u of V_1. We obtain $\mathbb{D}^{\lambda\star}_{V_1} = \delta' - \delta''$ on V_1. We have the decomposition $\mathbb{D}^\lambda_{V_1} = d'' + d'$

into the (0, 1)-part and the (1, 0)-part. By the construction, we have $\overline{\partial} h_1(u_1, u_2) = h_1(d''u_1, u_2) + h_1(u_1, \delta' u_2)$ and $\partial h_1(u_1, u_2) = h_1(d'u_1, u_2) + h_1(u_1, \delta'' u_2)$ for local sections u_i of V_1. By similar computations, we obtain that $\sqrt{-1}\Lambda_{U_w}[\mathbb{D}^\lambda_{V_1}, \mathbb{D}^{\lambda\star}_{V_2}]$ is the multiplication of $(1+|\lambda|^2)G(h)$. Hence, $[\mathbb{D}^\lambda_{V_1}, \mathbb{D}^{\lambda\star}_{V_1}] = 0$ if and only if $(E, \overline{\partial}_E, h)$ is a monopole. In this sense, we may regard a monopole on U^λ as a λ-flat bundle with a harmonic metric of infinite rank.

2.9.5.2 Remark

The analogy between monopoles on U^λ and λ-flat bundles with a harmonic metric on U_w (Corollary 2.9.13 and Sect. 2.9.5.1) is one of the motivation of this study, as explained in Sect. 1.5. It is also useful for the construction of a Hermitian metric h_1 of a given mini-holomorphic bundle such that $G(h_1)$ is small in Proposition 7.4.3.

In the study of harmonic bundles, we study the asymptotic behaviour of a harmonic metric h of $(V, \mathbb{D}^\lambda)$ by constructing a Hermitian metric h_0 of V such that $\mathcal{P}^h_* V = \mathcal{P}^{h_0}_* V$, and that $[\mathbb{D}^\lambda_{h_0}, \mathbb{D}^{\lambda\star}_{h_0}]$ is small. We can explicitly construct such h_0, and we can prove that h and h_0 are mutually bounded. This kind of argument was pioneered by Simpson [80], and further applied in [62, 64]. Such a Hermitian metric is also useful in the proof of Kobayashi-Hitchin correspondence, i.e., the existence of globally defined harmonic metrics. By adopting a similar strategy, in Proposition 7.4.3 below, we shall construct a Hermitian metric h_1 of a mini-holomorphic bundle such that $G(h_1)$ is small. For the construction of such h_1, we use a metric h_0 for λ-flat bundles as above through Lemma 2.9.11, Lemma 2.9.12 and Lemma 2.9.14.

2.9.6 *Dimensional Reduction of Kronheimer*

Let us recall the dimensional reduction of Kronheimer [51]. Let $\varphi : \mathbb{C}^2 \longrightarrow \mathbb{R} \times \mathbb{C}$ be the map (2.2). Let U be a neighbourhood of (0, 0) in $\mathbb{R}\times\mathbb{C}$. We set $\widetilde{U} := \varphi^{-1}(U)$. We put $U^* := U \setminus \{(0, 0)\}$ and $\widetilde{U}^* := \widetilde{U} \setminus \{(0, 0)\}$.

Let $(E, \overline{\partial}_E)$ be a mini-holomorphic bundle on U^* with a Hermitian metric h_E. We obtain the Chern connection ∇ and the Higgs field ϕ. Let F be the curvature of ∇.

We put $\widetilde{E} := \varphi^{-1}(E)$ on $\widetilde{U}^*$. It is equipped with the unitary connection $\widetilde{\nabla} := \varphi^*(\nabla) + \sqrt{-1}\varphi^*(\phi) \otimes \xi$, where

$$\xi = -u_1\, d\overline{u}_1 + \overline{u}_1\, du_1 - \overline{u}_2\, du_2 + u_2 d\overline{u}_2.$$

The curvature $\widetilde{F}$ is equal to $\varphi^*(F) + \sqrt{-1}\varphi^*(\nabla\phi) \wedge \xi$.

Lemma 2.9.15 *We have the following equality of currents on U:*

$$\varphi_*\Big(\operatorname{Tr}\big(\Lambda\widetilde{F}\big)\operatorname{dvol}_{\widetilde{U}}\Big)=\pi\operatorname{Tr}G(h)\operatorname{dvol}_U\,. \tag{2.39}$$

Here, φ_* denote the push-forward of currents by the proper map φ.

Proof Note $(|u_1|^2+|u_2|^2)^2=\varphi^*(|w|^2+|t|^2)$. We have $\varphi^*dw=2u_1du_2+u_2du_1$, $\varphi^*d\overline{w}=2\overline{u}_1d\overline{u}_2+\overline{u}_2d\overline{u}_1$ and $\varphi^*dt=\overline{u}_1du_1+u_1d\overline{u}_1-\overline{u}_2du_2-u_2d\overline{u}_2$. The forms φ^*dw, $\varphi^*d\overline{w}$, φ^*dt and ξ are orthogonal. Moreover, we have $|\varphi^*dw|^2=|\varphi^*d\overline{w}|^2=8(|u_1|^2+|u_2|^2)$, $|\varphi^*dt|^2=4(|u_1|^2+|u_2|^2)$ and $|\xi|^2=4(|u_1|^2+|u_2|^2)$. Hence, we have

$$\big|\varphi^*(dw\,d\overline{w}\,dt)\,\xi\big|=8(|u_1|^2+|u_2|^2)^2\big|du_1\,d\overline{u}_1\,du_2\,d\overline{u}_2\big|.$$

Let $\psi_{(u_1,u_2)}:\mathbb{R}/2\pi\longrightarrow\mathbb{C}^2$ be given by $\psi_{(u_1,u_2)}(e^{\sqrt{-1}\theta})=\big(u_1e^{\sqrt{-1}\theta},u_2e^{-\sqrt{-1}\theta}\big)$. We have

$$\psi^*_{(u_1,u_2)}\xi=2\sqrt{-1}(|u_1|^2+|u_2|^2)d\theta.$$

Hence, we obtain

$$\begin{aligned}\varphi_*\big(|du_1\,d\overline{u}_1\,du_2\,d\overline{u}_2|\big)&=\frac{1}{8(|w|^2+|t|^2)}\big|dw\,d\overline{w}\,dt\big|\cdot 4\pi(|w|^2+|t|^2)^{1/2}\\&=\frac{\pi}{2(|w|^2+|t|^2)^{1/2}}\big|dw\,d\overline{w}\,dt\big|.\end{aligned}\tag{2.40}$$

We have $|dw\,d\overline{w}\,dt|=2\operatorname{dvol}_U$ and $|du_1\,d\overline{u}_1\,du_2\,d\overline{u}_2|=4\operatorname{dvol}_{\widetilde{U}}$. Hence, we obtain the following:

$$\varphi_*\operatorname{dvol}_{\widetilde{U}}=\frac{\pi}{4(|w|^2+|t|^2)^{1/2}}\operatorname{dvol}_U\,.$$

By direct computations, we have

$$\widetilde{F}_{u_1,\overline{u}_1}=4u_1u_2\varphi^*F_{wt}-4\overline{u}_1\overline{u}_2\varphi^*F_{\overline{w}t}+4|u_2|^2F_{w\overline{w}}-2|u_1|^2\sqrt{-1}\varphi^*\nabla_t\phi,$$

$$\widetilde{F}_{u_2\overline{u}_2}=-4u_1u_2\varphi^*F_{wt}+4\overline{u}_1\overline{u}_2\varphi^*F_{\overline{w}t}+4|u_1|^2F_{w\overline{w}}-2|u_2|^2\sqrt{-1}\varphi^*\nabla_t\phi.$$

Hence, we obtain

$$\begin{aligned}\widetilde{F}_{u_1\overline{u}_1}+\widetilde{F}_{u_2\overline{u}_2}&=4(|u_1|^2+|u_2|^2)\varphi^*\big(F_{w\overline{w}}-\frac{\sqrt{-1}}{2}\nabla_t\phi\big)\\&=\varphi^*\Big(2(|w|^2+|t|^2)^{1/2}G(h)\Big).\end{aligned}\tag{2.41}$$

Thus, we obtain (2.39). ■

2.9.7 Appendix: Ambiguity of the Choice of a Splitting

We explain one of the main reasons why we use (t_1, β_1) (or $(t_1^\dagger, \beta_1^\dagger)$). Let Z be a finite subset of $\mathcal{M}$. Let (E, ∇, h, ϕ) be a monopole on $\mathcal{M} \setminus Z$. Let $F(\nabla) = F(\nabla)_{w\overline{w}} dw\, d\overline{w} + F(\nabla)_{tw} dt\, dw + F(\nabla)_{t\overline{w}}\, dt\, d\overline{w}$ denote the curvature of ∇. Suppose the following conditions.

Condition 2.9.16

- $|F(\nabla)_{w\overline{w}}|_h = O\big(|w|^{-2}(\log|w|)^{-2}\big)$ *as* $|w| \to \infty$. *It implies* $|\nabla_t \phi|_h = O\big(|w|^{-2}(\log|w|)^{-2}\big)$ *as* $|w| \to \infty$.
- *There exist positive numbers* $R, C, \epsilon > 0$, *a finite subset* $S \subset \mathbb{Q}$ *and an orthogonal decomposition*

$$E = \bigoplus_{\omega\in S} E_\omega$$

on $S^1_T \times \{|w| > R\}$ *such that* $F(\nabla)_{tw} - \bigoplus_{\omega\in S} \omega w^{-1}\, \mathrm{id}_{E_\omega} = O\big(|w|^{-1-\epsilon}\big)$ *as* $|w| \to \infty$. *Note that it implies that* $F(\nabla)_{t\overline{w}} + \bigoplus_{\omega\in S} \omega \overline{w}^{-1}\, \mathrm{id}_{E_\omega} = O\big(|w|^{-1-\epsilon}\big)$ *as* $|w| \to \infty$. *It also implies that* $\nabla_w\phi - \bigoplus(\sqrt{-1}\omega) w^{-1}\, \mathrm{id}_{E_\omega} = O\big(|w|^{-1-\epsilon}\big)$ *and* $\nabla_{\overline{w}}\phi - \bigoplus(\sqrt{-1}\omega)\overline{w}^{-1}\, \mathrm{id}_{E_\omega} = O\big(|w|^{-1-\epsilon}\big)$ *as* $|w| \to \infty$.
- *We assume that* $S \neq \{0\}$.

Let $p_1 : M \longrightarrow \mathcal{M}$ denote the projection. We set $Z_1 := p_1^{-1}(Z)$. We set $(E_1, h_1, \nabla_1, \phi_1) := p_1^{-1}(E, h, \nabla, \phi)$ on $M \setminus Z_1$. Let $\lambda \neq 0$. Let $(E_1^\lambda, \overline{\partial}_{E_1^\lambda})$ denote the holomorphic vector bundle on $M^\lambda \setminus Z_1$ underlying $(E_1, h_1, \nabla_1, \phi_1)$.

Let (t_2, β_2) be a mini-complex coordinate system of M^λ. Let $F^{t_2}_{\beta_2,\overline{\beta}_2} d\beta_2\, d\overline{\beta}_2$ denote the curvature of the Chern connection of the holomorphic vector bundle $(E_1^\lambda, \overline{\partial}_{E_1^\lambda})_{|\{t_2\}\times\mathbb{C}_{\beta_2}}$ with the metric induced by h_1. Suppose the following conditions

Condition 2.9.17

- $|F^{t_2}_{\beta_2,\overline{\beta}_2}|_h = O\big(|\beta_2|^{-2}(\log|\beta_2|)^{-2}\big)$ *as* $|\beta_2| \to \infty$. *It is equivalent to* $|F^{t_2}_{\beta_2,\overline{\beta}_2}|_h = O\big(|w|^{-2}(\log|w|)^{-2}\big)$ *as* $|w| \to \infty$.

Proposition 2.9.18 *Either one of the following holds.*

- *There exist a positive number* c_1 *and a non-zero complex number* c_2 *such that* $(t_2, \beta_2) = c_1(t_1, c_2\beta_1)$.
- *There exist a positive number* c_1 *a non-zero complex number* c_2 *such that* $(t_2, \beta_2) = c_1(t_1^\dagger, c_2\beta_1^\dagger) = c_1(t_1 - \mathrm{Im}(\lambda^{-1}\beta_1), c_2\lambda^{-2}\beta_1)$.

Proof After changing (t_2, β_2) to $c_1(t_2, \beta_2)$ for some $c_1 > 0$, we may assume that there uniquely exists a linear complex coordinate system (α_2, β_2) of X^λ such that (i) the $\mathbb{R}$-action is described as $s \bullet (\alpha_2, \beta_2) = (\alpha_2 + s, \beta_2)$, (ii) (α_2, β_2) induces (t_2, β_2) on M^λ. We have the instanton $(\widetilde{E}, \widetilde{\nabla}, \widetilde{h})$ on $X \setminus \widetilde{Z}_1$, corresponding to the monopole $(E_1, h_1, \nabla_1, \phi_1)$, where $\widetilde{Z}$ denote the pull back of Z by the projection $X^\lambda \longrightarrow M^\lambda$. Let $(\widetilde{E}^\lambda, \overline{\partial}_{\widetilde{E}^\lambda})$ be the holomorphic vector bundle on $X^\lambda \setminus \widetilde{Z}_1$ underlying $(\widetilde{E}, \widetilde{h}, \widetilde{\nabla})$. Note that $(\widetilde{E}^\lambda, \overline{\partial}_{\widetilde{E}^\lambda})_{|\{\sqrt{-1}t_2\}\times\mathbb{C}_{\beta_2}} \simeq (E_1^\lambda, \overline{\partial}_{E_1^\lambda})_{|\{t_2\}\times\mathbb{C}_{\beta_2}}$. Hence, according to Condition 2.9.17, for the expression

$$F(\widetilde{\nabla}) = F(\widetilde{\nabla})_{\alpha_2\overline{\alpha}_2} d\alpha_2 d\overline{\alpha}_2 + F(\widetilde{\nabla})_{\alpha_2\overline{\beta}_2} d\alpha_2 d\overline{\beta}_2 + F(\widetilde{\nabla})_{\beta_2\overline{\alpha}_2} d\beta_2 d\overline{\alpha}_2 + F(\widetilde{\nabla})_{\beta_2\overline{\beta}_2} d\beta_2 d\overline{\beta}_2, \qquad (2.42)$$

we have $F(\widetilde{\nabla})_{\beta_2\overline{\beta}_2|\{\sqrt{-1}t_2\}\times\mathbb{C}_{\beta_2}} = O\big(|w|^{-2}(\log|w|)^{-2}\big)$ as $|\beta_2| \to \infty$. Note that for the expression

$$F(\widetilde{\nabla}) = F(\widetilde{\nabla})_{z\overline{z}} dz\, d\overline{z} + F(\widetilde{\nabla})_{z\overline{w}_2} dz\, d\overline{w} + F(\widetilde{\nabla})_{w\overline{z}} dw\, dz + F(\widetilde{\nabla})_{w\overline{w}} dw\, d\overline{w},$$

Condition 2.9.16 implies that $|F(\widetilde{\nabla})_{z\overline{z}}|_{\widetilde{h}} = |F(\widetilde{\nabla})_{w\overline{w}}|_{\widetilde{h}} = O\big(|w|^{-2}(\log|w|)^{-2}\big)$. Moreover, there exists a decomposition $\widetilde{E} = \bigoplus \widetilde{E}_\omega$ such that

$$\big|F(\widetilde{\nabla})_{z\overline{w}} + \bigoplus C(\omega\overline{w}^{-1})\, \mathrm{id}_{E_\omega}\big|_{\widetilde{h}} = O\big(|w|^{-1-\epsilon}\big).$$

Because β_2 is $\mathbb{R}$-invariant, there exists a non-zero complex number a such that $\beta_2 = a(\eta + \lambda\xi) = a\beta_1$. There exists a complex number b such that $\alpha_2 = \xi + b(\eta + \lambda\xi)$. We obtain

$$z = \frac{1}{1+|\lambda|^2}\big(\alpha_2 - a^{-1}b\beta_2 + |\lambda|^2\overline{\alpha}_2 - \lambda\overline{a}^{-1}\overline{\beta}_2(1+\overline{\lambda b})\big),$$

$$w = \frac{1}{1+|\lambda|^2}\big(-\lambda\alpha_2 + a^{-1}\beta_2(1+\lambda b) + \lambda\overline{\alpha}_2 - \lambda\overline{a}^{-1}\overline{b\beta_2}\big).$$

Hence, the restriction of $dz\, d\overline{w}$ to $\{\sqrt{-1}t_2\} \times \mathbb{C}_{\beta_2}$ is equal to

$$b(1+\overline{\lambda b})\frac{-1}{(1+|\lambda|^2)}|a|^2 d\beta_2\, d\overline{\beta}_2.$$

Similarly, the restriction of $dw\, d\overline{z}$ to $\{\sqrt{-1}t_2\} \times \mathbb{C}_{\beta_2}$ is equal to

$$\overline{b}(1+\lambda b)\frac{-1}{(1+|\lambda|^2)}|a|^2 d\beta_2\, d\overline{\beta}_2.$$

We obtain $b(1+\overline{\lambda b}) = 0$, i.e., $b = 0$ or $b = -\lambda^{-1}$. If $b = 0$, we obtain $\alpha_2 = \xi = \alpha_1$, and hence $t_2 = t_1$. If $b = -\lambda^{-1}$, we obtain $\alpha_2 = -\lambda^{-1}\eta = -\alpha_1^\dagger$, and hence $t_2 = t_1^\dagger$. Thus, we are done. ■

2.10 Difference Modules and $\mathcal{O}_{\overline{\mathcal{M}}^{\lambda}\setminus Z}(*H^{\lambda}_{\infty})$-Modules

2.10.1 Difference Modules with Parabolic Structure at Finite Place

Let $\varrho \in \mathbb{C}$. Let Φ^* be an automorphism of the rational function field $\mathbb{C}(y)$ induced by $\Phi^*(y) = y + \varrho$. Let $\boldsymbol{V}$ be a difference module as in Definition 1.6.1, i.e., $\boldsymbol{V}$ is a finite dimensional $\mathbb{C}(y)$-module equipped with a $\mathbb{C}$-linear automorphism Φ^* such that $\Phi^*(fs) = \Phi^*(f) \cdot \Phi^*(s)$ for $f \in \mathbb{C}(y)$ and $s \in \boldsymbol{V}$.

Let V be a lattice of $\boldsymbol{V}$, i.e., a $\mathbb{C}[y]$-free module $V \subset \boldsymbol{V}$ such that $\mathbb{C}(y) \otimes_{\mathbb{C}[y]} V = \boldsymbol{V}$. We set $V' := V + (\Phi^*)^{-1}(V)$. We obtain the associated free $\mathcal{O}_{\mathbb{P}^1}(*\infty)$-module $\mathcal{F}_V$. We also obtain the associated free $\mathcal{O}_{\mathbb{P}^1}(*\infty)$-modules $\mathcal{F}_{(\Phi^*)^{-1}V}$ and $\mathcal{F}_{V'}$. There exists the natural isomorphism $\mathcal{F}_V \simeq \Phi^*\mathcal{F}_{(\Phi^*)^{-1}(V)}$. We may regard $\mathcal{F}_V$ and $\mathcal{F}_{(\Phi^*)^{-1}(V)}$ as $\mathcal{O}_{\mathbb{P}^1}(*\infty)$-submodules of $\mathcal{F}_{V'}$, and $\mathcal{F}_{V'} = \mathcal{F}_V + \mathcal{F}_{(\Phi^*)^{-1}(V)}$ holds.

Definition 2.10.1 A parabolic structure of $\boldsymbol{V}$ at finite place is a tuple as follows.

- A $\mathbb{C}[y]$-free submodule $V \subset \boldsymbol{V}$ such that $\mathbb{C}(y) \otimes_{\mathbb{C}[y]} V = \boldsymbol{V}$.
- A function $m : \mathbb{C} \longrightarrow \mathbb{Z}_{\geq 0}$ such that $\sum_{x\in\mathbb{C}} m(x) < \infty$. We assume that $\mathcal{F}_V(*D) = \mathcal{F}_{(\Phi^*)^{-1}(V)}(*D)$, where $D := \{x \in \mathbb{C} \mid m(x) > 0\}$.
- A sequence of real numbers $0 \leq \tau_x^{(1)} < \cdots < \tau_x^{(m(x))} < T$ for each $x \in \mathbb{C}$. If $m(x) = 0$, the sequence is assumed to be empty. The sequence is denoted by $\boldsymbol{\tau}_x$.
- Lattices $L_{x,i} \subset V \otimes \mathbb{C}((y-x))$ for $x \in \mathbb{C}$ and $i = 1, \ldots, m(x) - 1$. We formally set $L_{x,0} := V \otimes \mathbb{C}[\![y-x]\!]$ and $L_{x,m(x)} := (\Phi^*)^{-1}V \otimes \mathbb{C}[\![y-x]\!]$. The tuple of lattices is denoted by $\boldsymbol{L}_x$.

Here, $\mathbb{C}[\![y-x]\!]$ denotes the ring of formal power series with the variable $y-x$, and $\mathbb{C}((y-x))$ denotes the ring of formal Laurent power series with the variable $y-x$. ■

Remark 2.10.2 The notion of parabolic structure of a difference module at finite place is apparently different from the ordinary notion of parabolic structure in the context of harmonic bundles. However, as we shall explain in Sect. 2.10.4, a parabolic structure of $2\sqrt{-1}\lambda$-difference module at finite place is a reincarnation of an ordinary parabolic structure of the regular part of λ-flat bundle at $\{0, \infty\}$ through the Mellin transform or the algebraic Nahm transform. ■

2.10.2 Construction of Difference Modules from $\mathcal{O}_{\overline{\mathcal{M}}^{\lambda}\setminus Z}(*H^{\lambda}_{\infty})$-Modules

Let Z be a finite subset in $\mathcal{M}^{\lambda}$. Let $\mathcal{E}$ be a locally free $\mathcal{O}_{\overline{\mathcal{M}}^{\lambda}\setminus Z}(*H^{\lambda}_{\infty})$-module of Dirac type. Let us observe that we obtain the associated difference module with parabolic structure at finite place with $\varrho = 2\sqrt{-1}\lambda T$.

Let $\varpi^{\lambda} : \overline{M}^{\lambda} \longrightarrow \overline{\mathcal{M}}^{\lambda}$ denote the projection. Set $Z^{\mathrm{cov}} := (\varpi^{\lambda})^{-1}(Z)$. Let $p_1 : \overline{M}^{\lambda} \longrightarrow \mathbb{R}_{t_1}$ and $p_2 : \overline{M}^{\lambda} \longrightarrow \mathbb{P}^1_{\beta_1}$ be the projections. We obtain the $\mathcal{O}_{\overline{M}^{\lambda}\setminus Z}(*H^{\lambda\,\mathrm{cov}}_{\infty})$-module $\mathcal{E}^{\mathrm{cov}} := (\varpi^{\lambda})^{*}\mathcal{E}$. Let $j : \overline{M}^{\lambda} \setminus Z \longrightarrow \overline{M}^{\lambda}$ denote the inclusion. We obtain the $\mathcal{O}_{\overline{M}^{\lambda}}(*H^{\lambda\,\mathrm{cov}}_{\infty})$-module $j_*\mathcal{E}^{\mathrm{cov}}$. For any $a \in \mathbb{R}$, let $\iota_a : p_1^{-1}(a) \longrightarrow \overline{M}^{\lambda}$. We obtain the locally free $\mathcal{O}_{\mathbb{P}^1}(*\infty)$-modules $\iota_a^{-1} j_*\mathcal{E}^{\mathrm{cov}}$ ($a \in \mathbb{R}$).

Because $j_*\mathcal{E}^{\mathrm{cov}}$ is naturally $\mathbb{Z}$-equivariant, we have the following isomorphism for any $a \in \mathbb{R}$ and $n \in \mathbb{Z}$:

$$(\Phi^n)^*\iota^{-1}_{a+Tn}\big(j_*(\mathcal{E}^{\mathrm{cov}})\big) \simeq \iota_a^{-1}\big(j_*(\mathcal{E}^{\mathrm{cov}})\big). \tag{2.43}$$

Here, $\Phi : \mathbb{P}^1 \longrightarrow \mathbb{P}^1$ is given by $\Phi(\beta_1) = \beta_1 + 2\sqrt{-1}\lambda T$.

We set $D_{a,b} := p_2\Big(Z^{\mathrm{cov}} \cap \big(\{a \le t_1 \le b\} \times \mathbb{P}^1_{\beta_1}\big)\Big)$ for any $a \le b$. By the scattering map, we obtain the isomorphism

$$\iota_a^{-1}\big(j_*(\mathcal{E}^{\mathrm{cov}})\big)(*D_{a,b}) \simeq \iota_b^{-1}\big(j_*(\mathcal{E}^{\mathrm{cov}})\big)(*D_{a,b}). \tag{2.44}$$

For any $a \in \mathbb{R}$, we set

$$V_a := H^0\Big(\mathbb{P}^1, \iota_a^{-1}\big(j_*(\mathcal{E}^{\mathrm{cov}})\big)\Big).$$

For any subset $S \subset \mathbb{C}$, let $\mathbb{C}[\beta_1]_S$ denote the localization of $\mathbb{C}[\beta_1]$ with respect to $\beta_1 - x$ ($x \in S$). By (2.44), we obtain the isomorphism of $\mathbb{C}[\beta_1]_{D_{a,b}}$-modules:

$$V_a \otimes \mathbb{C}[\beta_1]_{D_{a,b}} \simeq V_b \otimes \mathbb{C}[\beta_1]_{D_{a,b}}.$$

Hence, we obtain the isomorphism of $\mathbb{C}(\beta_1)$-modules for any $a \le b$.

$$V_a \otimes \mathbb{C}(\beta_1) \simeq V_b \otimes \mathbb{C}(\beta_1). \tag{2.45}$$

From (2.43), we have the $\mathbb{C}$-linear isomorphism

$$(\Phi^*)^n : V_{a+nT} \simeq V_a$$

such that $(\Phi^*)^n(\beta_1^\ell s) = (\beta_1 + 2\sqrt{-1}\lambda nT)^\ell (\Phi^*)^n(s)$. We obtain the $\mathbb{C}$-linear isomorphism

$$(\Phi^*)^n : V_{a+nT} \otimes \mathbb{C}(\beta_1) \simeq V_a \otimes \mathbb{C}(\beta_1) \tag{2.46}$$

such that $(\Phi^*)^n(g(\beta_1)s) = g(\beta_1 + 2\sqrt{-1}\lambda nT) \cdot (\Phi^*)^n(s)$ for any $g(\beta_1) \in \mathbb{C}(\beta_1)$.

We take a small $\epsilon > 0$ such that $Z^{\mathrm{cov}} \cap \big(\{-\epsilon \le t_1 < 0\} \times \mathbb{P}^1_{\beta_1}\big) = \emptyset$. Set $V(\mathcal{E}) := V_{-\epsilon} \otimes \mathbb{C}(\beta_1)$ which is a finite dimensional $\mathbb{C}(\beta_1)$-vector space. It is identified with $V_b \otimes \mathbb{C}(\beta_1)$ for any $b \in \mathbb{R}$ by (2.45). It is equipped with $\mathbb{C}$-linear automorphism Φ^* by (2.46).

Set $V(\mathcal{E}) := V_{-\epsilon}$ for which we obtain $\mathbb{C}(\beta_1) \otimes V(\mathcal{E}) = V(\mathcal{E})$. Let $m_Z : \mathbb{C} \longrightarrow \mathbb{Z}_{\ge 0}$ be the function such that

$$m_Z(x) = \left| p_2^{-1}(x) \cap p_1^{-1}(\{0 \le t_1 < T\}) \cap Z^{\mathrm{cov}} \right|. \tag{2.47}$$

For each $x \in \mathbb{C}$, the sequence $0 \le t^{(1)}_{1,Z,x} < \cdots < t^{(m_Z(x))}_{1,Z,x} < T$ determined by

$$\{(t^{(i)}_{1,Z,x}, x) \mid i = 1, \ldots, m_Z(x)\} = p_2^{-1}(x) \cap p_1^{-1}(\{0 \le t_1 < T\}) \cap Z^{\mathrm{cov}}. \tag{2.48}$$

We set

$$\tau^{(i)}_{Z,x} := t^{(i)}_{1,Z,x}/T. \tag{2.49}$$

The tuple $\tau^{(i)}_{Z,x}$ $(i = 1, \ldots, m_Z(x))$ is denoted by $\boldsymbol{\tau}_{Z,x}$. For $x \in \mathbb{C}$ and for $i = 1, \ldots, m_Z(x) - 1$, we choose $t^{(i)}_{1,Z,x} < b(x,i) < t^{(i+1)}_{1,Z,x}$, and we set

$$L_{Z,x,i}(\mathcal{E}) := V_{b(x,i)} \otimes \mathbb{C}[\![\beta_1 - x]\!],$$

which is independent of the choice of $b(x,i)$. The tuple $L_{Z,x,i}$ $(i = 1, \ldots, m_Z(x) - 1)$ is denoted by $\boldsymbol{L}_{Z,x}$. Thus, we obtain a parabolic structure at finite place

$$\big(V(\mathcal{E}), m_Z, (\boldsymbol{\tau}_{Z,x}, \boldsymbol{L}_{Z,x}(\mathcal{E}))_{x \in \mathbb{C}}\big)$$

of $V(\mathcal{E})$.

2.10.3 Construction of $\mathcal{O}_{\overline{\mathcal{M}}^\lambda\setminus Z}(*H^\lambda)$-Modules from Difference Modules

Let V be a $(2\sqrt{-1}\lambda T)$-difference module with a parabolic structure at finite place $(V, m, (\boldsymbol{\tau}_x, \boldsymbol{L}_x)_{x \in \mathbb{C}})$. We set $Z_1 := \coprod_{x \in \mathbb{C}} \big\{(T\tau^{(i)}_x, x) \,\big|\, \tau^{(i)}_x \in \boldsymbol{\tau}_x\big\} \subset \mathbb{R}_{t_1} \times \mathbb{C}_{\beta_1}$.

We set $Z := \varpi^{\lambda}(Z_1) \subset \mathcal{M}^{\lambda}$. Let us construct an $\mathcal{O}_{\overline{\mathcal{M}}^{\lambda}\setminus Z}(*H^{\lambda})$-module $\mathcal{E}(\boldsymbol{V}, V, m, (\boldsymbol{\tau}_x, \boldsymbol{L}_x)_{x\in\mathbb{C}})$.

We set $U :=]-\epsilon, T-\epsilon/2[\times\mathbb{P}^1_{\beta_1}$ and $H^{\lambda}_{\infty,\epsilon} :=]-\epsilon, T-\epsilon/2[\times\{\infty\}$. We set $D := \{x \in \mathbb{C} \mid m(x) > 0\}$. For $x \in D$, we set $t_x^{(i)} = T\tau_x^{(i)}$ $(i = 1, \dots, m(x))$, $t_x^{(0)} = -\epsilon/2$ and $t_x^{(m(x)+1)} = T - \epsilon$. The following conditions determine a unique $\mathcal{O}_{U\setminus Z_1}(*H^{\lambda}_{\infty,\epsilon})$-module $\mathcal{E}_1$:

- $\mathcal{E}_1(*p_2^{-1}(D))$ is isomorphic to the pull back of $\mathcal{F}_V(*D)$ by the projection $U \longrightarrow \mathbb{P}^1_{\beta_1}$.
- Take $t_x^{(i)} < a_i < t_x^{(i+1)}$ for $i = 0, \dots, m(x)$. Take a small neighbourhood U_x of x in $\mathbb{C}$. Let $\iota_{\{a_i\}\times U_x}$ denote the inclusion of $\{a_i\} \times U_x$ to $U \setminus Z_1$. Then, $\iota^*_{\{a_i\}\times U_x}\mathcal{E}_1 \subset \mathcal{F}_V(*D)_{|U_x}$ is equal to the $\mathcal{O}_{U_x}$-submodule determined by $L_{x,i}$.

Let $\kappa_1 :]-\epsilon, -\epsilon/2[\times\mathbb{P}^1 \longrightarrow]T-\epsilon, T-\epsilon/2[\times\mathbb{P}^1$ be the isomorphism defined by $\kappa_1(t_1, \beta_1) = (t_1 + T, \beta_1 + 2\sqrt{-1}\lambda T)$. By the construction, there exists the isomorphism

$$\kappa_1^*\Bigl(\mathcal{E}_{1|]T-\epsilon,T-\epsilon/2[\times\mathbb{P}^1}\Bigr) \simeq \mathcal{E}_{1|]-\epsilon,-\epsilon/2[\times\mathbb{P}^1}.$$

Hence, we obtain a locally free $\mathcal{O}_{\overline{\mathcal{M}}^{\lambda}\setminus Z}(*H^{\lambda}_{\infty})$-module $\mathcal{E}(\boldsymbol{V}, V, m, (\tau_x, \boldsymbol{L}_x)_{x\in\mathbb{C}})$ of Dirac type. By the construction, the following is clear.

Lemma 2.10.3 *The constructions in* Sect. 2.10.2 *and* Sect. 2.10.3 *are mutually inverse up to canonical isomorphisms.* ■

Remark 2.10.4 The relation between difference modules and $\mathcal{O}_{\overline{\mathcal{M}}^{\lambda}\setminus Z}(*H^{\lambda}_{\infty})$-modules of Dirac type is analogue to the relation between $\mathcal{O}_C(*D)$-modules and filtered bundles over (C, D) for a compact Riemann surface C with a finite subset $D \subset C$. ■

2.10.4 Appendix: Mellin Transform and Parabolic Structures at Finite Place

Let us explain that the notion of parabolic structure at finite place of difference modules is related with the usual notion of parabolic structure of filtered λ-flat bundles through the algebraic Nahm transform, which is an analogue of Mellin transform.

2.10.4.1 Mellin Transform

Let λ be any complex number. Let M be a $\mathbb{C}[u,u^{-1}]$-module equipped with a $\mathbb{C}$-linear endomorphism $\mathbb{D}^{\lambda}_{u}: M \longrightarrow M$ such that $\mathbb{D}^{\lambda}_{u}(fs) = f\mathbb{D}^{\lambda}_{u}(s) + \lambda\partial_u(f)s$.

We define the automorphism Φ^* of $\mathbb{C}$-algebra $\mathbb{C}[\beta_1]$ by $\Phi^*(\beta_1) = \beta_1 + 2\sqrt{-1}\lambda$. We define $\mathbb{C}[\beta_1]$-action on M by $\beta_1 s = -2\sqrt{-1}u\mathbb{D}^{\lambda}_{u}(s)$ for any $s \in M$. We define the $\mathbb{C}$-linear automorphism Φ^*_M of M by $\Phi^*(s) = us$ for any $s \in M$. Because $\Phi^*_M(\beta_1 s) = \Phi^*(\beta_1)\Phi^*_M(s)$, we obtain $2\sqrt{-1}\lambda$-difference module, which is called the Mellin transform.

We have the natural geometrization of the transformation of the Mellin transform. Let p_i $(i = 1, 2)$ denote the projections of $\mathbb{P}^1_u \times \mathbb{C}_{\beta_1}$ onto the i-th component. Let $\mathcal{V}$ be an algebraic quasi-coherent $\mathcal{O}_{\mathbb{P}^1}(*\{0,\infty\})$-module with a λ-connection $\mathbb{D}^{\lambda}$. We obtain the following complex $\mathcal{C}^{\bullet}(\mathcal{V},\mathbb{D}^{\lambda})$ on $\mathbb{P}^1_u \times \mathbb{C}_{\beta_1}$:

$$p_1^*\mathcal{V} \xrightarrow{u\mathbb{D}^{\lambda}_{u} - \frac{\sqrt{-1}}{2}\beta_1} p_1^*\mathcal{V},$$

where the first term sits in the degree 0. We have $R^i p_{2*}(\mathcal{C}^{\bullet}(\mathcal{V},\mathbb{D}^{\lambda})) = 0$ unless $i = 1$. We obtain the algebraic $\mathcal{O}_{\mathbb{C}_{\beta_1}}$-module $\mathfrak{M}(\mathcal{V},\mathbb{D}^{\lambda}) := R^1 p_{2*}(\mathcal{C}^{\bullet}(\mathcal{V},\mathbb{D}^{\lambda}))$.

Let $f_u : \mathcal{V} \longrightarrow \mathcal{V}$ be the automorphism defined as the multiplication of u. It induces the isomorphism $(\mathcal{V},\mathbb{D}^{\lambda} + \lambda du/u) \simeq (\mathcal{V},\mathbb{D}^{\lambda})$.

Let $\Phi : \mathbb{C}_{\beta_1} \longrightarrow \mathbb{C}_{\beta_1}$ be defined by $\Phi(\beta_1) = \beta_1 + 2\sqrt{-1}\lambda$. We obtain the following morphisms:

$$(\mathrm{id}\times\Phi)^*\mathcal{C}^{\bullet}(\mathcal{V},\mathbb{D}^{\lambda}) \simeq \mathcal{C}^{\bullet}(\mathcal{V},\mathbb{D}^{\lambda} + \lambda du/u) \xrightarrow{p_1^* f_u} \mathcal{C}^{\bullet}(\mathcal{V},\mathbb{D}^{\lambda}).$$

Hence, we obtain the isomorphism $\Phi^*\mathfrak{M}(\mathcal{V},\mathbb{D}^{\lambda}) \simeq \mathfrak{M}(\mathcal{V},\mathbb{D}^{\lambda})$.

If $M = H^0(\mathbb{P}^1,\mathcal{V})$, then $H^0(\mathbb{P}^1,\mathfrak{M}(\mathcal{V},\mathbb{D}^{\lambda}))$ is the Mellin transform.

Remark 2.10.5 The stationary phase formula for Mellin transform has been studied in [27] and [29]. ■

2.10.4.2 Algebraic Nahm Transform for Filtered λ-Flat Bundles (Special Case)

Let $D \subset \mathbb{C}^*$ be a finite subset. We set $\overline{D} = D \cup \{0,\infty\}$. Let $\mathcal{V}$ be a locally free $\mathcal{O}_{\mathbb{P}^1}(*\overline{D})$-module of finite rank equipped with a λ-connection $\mathbb{D}^{\lambda}$. Let $(\mathcal{P}_*\mathcal{V},\mathbb{D}^{\lambda})$ be a good filtered λ-flat bundle on $(\mathbb{P}^1,\overline{D})$. For simplicity, we assume the following.

Condition 2.10.6

- *D is non-empty. Moreover, at each point P of D, $(\mathcal{P}_*\mathcal{V},\mathbb{D}^{\lambda})$ does not have the regular part, i.e., in the Hukuhara-Levelt-Turrittin decomposition* (1.5) *of the* *pull back by an appropriate ramified covering of a neighborhood of P, we have* $\mathcal{P}_*\widehat{\mathcal{V}}_0 = 0$.

- $\mathcal{P}_*\mathcal{V}$ *is regular at* 0 *and* ∞, *i.e.,* $\mathbb{D}^\lambda$ *is logarithmic with respect to* $\mathcal{P}_{\boldsymbol{a}}(\mathcal{V})$ *for any* $\boldsymbol{a} \in \mathbb{R}^{\overline{D}}$ *at* 0 *and* ∞. ■

We shall construct an $\mathcal{O}_{\overline{\mathcal{M}}^\lambda\setminus Z}(*H^\lambda)$-module from $(\mathcal{P}_*\mathcal{V}, \mathbb{D}^\lambda)$, where Z denotes a finite subset explained below. We assume $T = 1$ in the construction of $\mathcal{M}^\lambda$.

Let $\mathcal{V}_0$ denote the stalk of $\mathcal{V}$ at 0. We have the filtration $\mathcal{P}_a(\mathcal{V}_0)$ $(a \in \mathbb{R})$. We set $\mathrm{Gr}^{\mathcal{P}}_a(\mathcal{V}_0) = \mathcal{P}_a(\mathcal{V}_0)/\mathcal{P}_{<a}(\mathcal{V}_0)$. Because $\mathbb{D}^\lambda$ is logarithmic at 0, we obtain the endomorphism $\mathrm{Res}_0(\mathbb{D}^\lambda)$ of $\mathrm{Gr}^{\mathcal{P}}_a(\mathcal{V}_0)$ as the residue of $\mathbb{D}^\lambda$. Let $\mathbb{E}_\alpha \mathrm{Gr}^{\mathcal{P}}_a(\mathcal{V}_0)$ denote the generalized eigen space corresponding to $\alpha \in \mathbb{C}$. Let $\mathcal{KMS}(\mathcal{P}_*\mathcal{V}, 0)$ denote the set of $(a, \alpha) \in \mathbb{R} \times \mathbb{C}$ such that $\mathbb{E}_\alpha \mathrm{Gr}^{\mathcal{P}}_a(\mathcal{V}_0) \neq 0$. Similarly, let $\mathcal{V}_\infty$ denote the stalk of $\mathcal{V}$ at ∞, which is equipped with the filtration $\mathcal{P}_*(\mathcal{V}_\infty)$ such that $\mathbb{D}^\lambda$ is logarithmic. We obtain $\mathrm{Gr}^{\mathcal{P}}_a(\mathcal{V}_\infty) := \mathcal{P}_a(\mathcal{V}_\infty)/\mathcal{P}_{<a}(\mathcal{V}_\infty)$ which is equipped with the endomorphism $\mathrm{Res}_\infty(\mathbb{D}^\lambda)$ obtained as the residue of $\mathbb{D}^\lambda$. Let $\mathbb{E}_\alpha \mathrm{Gr}^{\mathcal{P}}_a(\mathcal{V}_\infty)$ denote the generalized eigen space corresponding to $\alpha \in \mathbb{C}$. We obtain the set $\mathcal{KMS}(\mathcal{P}_*\mathcal{V}, \infty)$ of $(a, \alpha) \in \mathbb{R} \times \mathbb{C}$ such that $\mathbb{E}_\alpha \mathrm{Gr}^{\mathcal{P}}_a(\mathcal{V}_\infty) \neq 0$.

Let $\widetilde{Z} \subset \mathbb{R} \times \mathbb{C}$ denote the union of the following sets

$$\Big\{(t_1, \beta_1) \in \mathbb{R} \times \mathbb{C} \,\Big|\, \Big(t_1, \frac{\sqrt{-1}}{2}\beta_1\Big) \in \mathcal{KMS}(\mathcal{P}_*\mathcal{V}, 0)\Big\}$$

$$\Big\{(t_1, \beta_1) \in \mathbb{R} \times \mathbb{C} \,\Big|\, \Big(-t_1, -\frac{\sqrt{-1}}{2}\beta_1\Big) \in \mathcal{KMS}(\mathcal{P}_*\mathcal{V}, \infty)\Big\}.$$

Note that $\widetilde{Z}$ is preserved by the $\mathbb{Z}$-action κ on $\mathbb{R} \times \mathbb{C}$ defined as $\kappa_n(t_1, \beta_1) = (t_1 + n, \beta_1 + 2\sqrt{-1}\lambda n)$. Let $Z \subset \mathcal{M}^\lambda$ denote the quotient of $\widetilde{Z}$.

Let $\widetilde{\pi} : \mathbb{R} \times \mathbb{P}^1 \longrightarrow \mathbb{R}$ denote the projection. We obtain the set $\widetilde{\pi}(\widetilde{Z}) \subset \mathbb{R}$.

For any $(a, b) \in \mathbb{R}^2$, let $\mathcal{P}_{a,b}(\mathcal{V}) \subset \mathcal{V}$ be the $\mathcal{O}_{\mathbb{P}^1}$-module such that (i) $\mathcal{P}_{a,b}(\mathcal{V})(*\{0, \infty\}) = \mathcal{V}$, (ii) the stalk of $\mathcal{P}_{a,b}(\mathcal{V})$ at 0 is $\mathcal{P}_a(\mathcal{V}_0)$, (iii) the stalk of $\mathcal{P}_{a,b}(\mathcal{V})$ at ∞ is $\mathcal{P}_b(\mathcal{V}_\infty)$.

Let p_i denote the projections of $\mathbb{P}^1_u \times \mathbb{P}^1_{\beta_1}$ onto the i-th components. We obtain the following complex $\mathcal{C}^\bullet_{a,b}$ on $\mathbb{P}^1_u \times \mathbb{P}^1_{\beta_1}$:

$$p_1^*\mathcal{P}_{a,b}\mathcal{V}(*(\mathbb{P}^1 \times \{\infty\})) \xrightarrow{u\mathbb{D}^\lambda_u - \frac{\sqrt{-1}}{2}\beta_1} p_1^*\mathcal{P}_{a,b}\mathcal{V}(*(\mathbb{P}^1 \times \{\infty\}))$$

where the first term sits in the degree 0. We obtain $R^i p_{2*}\mathcal{C}^\bullet_{a,b} = 0$ unless $i = 1$, and $\mathfrak{N}_{a,b}(\mathcal{P}_*\mathcal{V}, \mathbb{D}^\lambda) := R^1 p_{2*}\mathcal{C}^\bullet_{a,b}$ is a locally free $\mathcal{O}_{\mathbb{P}^1}(*\infty)$-module.

Let $t_1 \in \mathbb{R}$. Suppose $t_1 \notin \widetilde{\pi}(\widetilde{Z})$. There exists $\epsilon > 0$ such that $\{t_1 - \epsilon \leq a \leq t_1 + \epsilon\} \cap \widetilde{\pi}(\widetilde{Z}) = \emptyset$. For any $t_1 - \epsilon \leq t_1' \leq t_1 + \epsilon$, we have $\mathcal{P}_{t_1,-t_1}\mathcal{V} = \mathcal{P}_{t_1',-t_1'}\mathcal{V}$, and hence there exists a natural isomorphism

$$\widetilde{\mathfrak{N}}_{t_1,-t_1}(\mathcal{P}_*\mathcal{V},\mathbb{D}^{\lambda}) \simeq \widetilde{\mathfrak{N}}_{t_1',-t_1'}(\mathcal{P}_*\mathcal{V},\mathbb{D}^{\lambda}). \tag{2.50}$$

Suppose $t_1 \in \widetilde{\pi}(\widetilde{Z})$. There exists $\epsilon > 0$ such that $\{t_1-\epsilon \le a \le t_1+\epsilon\}\cap\widetilde{\pi}(\widetilde{Z}) = \{t_1\}$. For any $t_1-\epsilon \le t_1' \le t_1+\epsilon$, we have $\mathcal{P}_{t_1-\epsilon,-t_1-\epsilon}\mathcal{V} \subset \mathcal{P}_{t_1',-t_1'}\mathcal{V}$, and hence there exists the following morphism $\widetilde{\mathfrak{N}}_{t_1-\epsilon,-t_1-\epsilon}(\mathcal{P}_*\mathcal{V},\mathbb{D}^{\lambda}) \longrightarrow \widetilde{\mathfrak{N}}_{t_1',-t_1'}(\mathcal{P}_*\mathcal{V},\mathbb{D}^{\lambda})$.

Lemma 2.10.7 *We set $\widetilde{Z}_{t_1} := \widetilde{\pi}^{-1}(t_1)\cap\widetilde{Z}$, which we naturally regard as a subset of $\mathbb{C}$. Then, the induced morphism*

$$\widetilde{\mathfrak{N}}_{t_1-\epsilon,-t_1-\epsilon}(\mathcal{P}_*\mathcal{V},\mathbb{D}^{\lambda})(*\widetilde{Z}_{t_1}) \longrightarrow \widetilde{\mathfrak{N}}_{t_1',-t_1'}(\mathcal{P}_*\mathcal{V},\mathbb{D}^{\lambda})(*\widetilde{Z}_{t_1}) \tag{2.51}$$

is an isomorphism.

Proof If α is an eigenvalues of $\mathrm{Res}_0(\mathbb{D}^{\lambda})$ on $\mathrm{Gr}^{\mathcal{P}}_{t_1}(\mathcal{V}_0)$, then $\frac{\sqrt{-1}}{2}\alpha$ is contained in $\widetilde{Z}_{t_1}$. If α is an eigenvalues of $\mathrm{Res}_{\infty}(\mathbb{D}^{\lambda})$ on $\mathrm{Gr}^{\mathcal{P}}_{-t_1}(\mathcal{V}_{\infty})$, then $-\frac{\sqrt{-1}}{2}\alpha$ is contained in $\widetilde{Z}_{t_1}$.

Let $\beta_1 \in \mathbb{C}\setminus\widetilde{Z}_{t_1}$. There exists a neighbourhood U of β_1 in $\mathbb{C}\setminus\widetilde{Z}_{t_1}$. By the consideration in the previous paragraph, $\mathrm{Res}_0(\mathbb{D}^{\lambda})-\frac{\sqrt{-1}}{2}\beta_1'$ ($\beta_1' \in U$) are invertible on $\mathrm{Gr}^{\mathcal{P}}_{t_1}(\mathcal{V}_0)$, and $\mathrm{Res}_{\infty}(\mathbb{D}^{\lambda})+\frac{\sqrt{-1}}{2}\beta_1'$ ($\beta_1' \in U$) are invertible on $\mathrm{Gr}^{\mathcal{P}}_{-t_1}(\mathcal{V}_{\infty})$. Hence, we can easily check the claim of the lemma. ■

By Lemma 2.10.7, we obtain the following isomorphism for any $t_1-\epsilon \le t_1' \le t_1+\epsilon$:

$$\widetilde{\mathfrak{N}}_{t_1,-t_1}(\mathcal{P}_*\mathcal{V},\mathbb{D}^{\lambda})_{|\mathbb{P}^1\setminus\widetilde{Z}_{t_1}} \simeq \widetilde{\mathfrak{N}}_{t_1',-t_1'}(\mathcal{P}_*\mathcal{V},\mathbb{D}^{\lambda})_{|\mathbb{P}^1\setminus\widetilde{Z}_{t_1}}. \tag{2.52}$$

By the isomorphisms (2.50) and (2.52), we obtain an $\mathcal{O}_{(\mathbb{R}\times\mathbb{P}^1)\setminus\widetilde{Z}}(*(\mathbb{R}\times\{\infty\}))$-module $\widetilde{\mathfrak{N}}(\mathcal{P}_*\mathcal{V})$ from $\widetilde{\mathfrak{N}}_{t_1,-t_1}(\mathcal{P}_*\mathcal{V})_{|\mathbb{P}^1\setminus\widetilde{Z}_{t_1}}$ ($t_1 \in \mathbb{R}$). The isomorphism f_u on $\mathcal{V}$ induces $\kappa_1^*\widetilde{\mathfrak{N}}(\mathcal{P}_*\mathcal{V},\mathbb{D}^{\lambda}) \simeq \widetilde{\mathfrak{N}}(\mathcal{P}_*\mathcal{V},\mathbb{D}^{\lambda})$. Hence, we obtain an $\mathcal{O}_{\overline{\mathcal{M}}^{\lambda}\setminus Z}(*H^{\lambda}_{\infty})$-module $\mathfrak{N}(\mathcal{P}_*\mathcal{V})$.

Let $(t_1^0,\beta_1^0) \in \widetilde{Z}$. There exist integers $k_1 \ge \ldots \ge k_r$ and a frame $v_1,\ldots,v_r$ of $\widetilde{\mathfrak{N}}(\mathcal{P}_*\mathcal{V},\mathbb{D}^{\lambda})_{(t_1^0-\epsilon,\beta_1)}$ such that the tuple $(\beta_1-\beta_1^0)^{k_i}v_i$ ($i=1,\ldots,r$) is a frame of $\widetilde{\mathfrak{N}}(\mathcal{P}_*\mathcal{V},\mathbb{D}^{\lambda})_{(t_1^0+\epsilon,\beta_1)}$.

We set $\alpha = \frac{\sqrt{-1}}{2}\beta_1^0$. We obtain the filtration W on the spaces $\mathbb{E}_{\alpha}\,\mathrm{Gr}^{\mathcal{P}}_{t_1^0}(\mathcal{V}_0)$ and $\mathbb{E}_{-\alpha}\,\mathrm{Gr}^{\mathcal{P}}_{-t_1^0}(\mathcal{V}_{\infty})$ obtained as the weight filtration of the nilpotent parts of $\mathrm{Res}_0(\mathbb{D}^{\lambda})$ and $\mathrm{Res}_{\infty}(\mathbb{D}^{\lambda})$, respectively. For $k \ge 0$, let $P\,\mathrm{Gr}^W_k\,\mathbb{E}_{\alpha}\,\mathrm{Gr}^{\mathcal{P}}_{t_1^0}(\mathcal{V}_0)$ and $P\,\mathrm{Gr}^W_k\,\mathbb{E}_{-\alpha}\,\mathrm{Gr}^{\mathcal{P}}_{-t_1^0}(\mathcal{V}_{\infty})$ denote the primitive parts.

Proposition 2.10.8 $\dim P\,\mathrm{Gr}^W_k\,\mathbb{E}_\alpha\,\mathrm{Gr}^{\mathcal{P}}_{t_1^0}(\mathcal{V}_0)$ *is equal to the number of* v_i *such that* $k_i = k+1$, *and* $\dim P\,\mathrm{Gr}^W_k\,\mathbb{E}_{-\alpha}\,\mathrm{Gr}^{\mathcal{P}}_{-t_1^0}(\mathcal{V}_\infty)$ *is equal to the number of* v_i *such that* $k_i = -k-1$.

Proof It is enough to consider the case $(t_1^0, \beta_1^0) = (0,0)$. Let $c_1 : \mathcal{P}_{-\epsilon,-\epsilon}\mathcal{V} \longrightarrow \mathcal{P}_{-\epsilon,\epsilon}\mathcal{V}$ and $c_2 : \mathcal{P}_{-\epsilon,-\epsilon}\mathcal{V} \longrightarrow \mathcal{P}_{\epsilon,-\epsilon}\mathcal{V}$ denote the natural inclusions. Let $d_1 : \mathcal{P}_{-\epsilon,\epsilon}\mathcal{V} \longrightarrow \mathcal{P}_{\epsilon,\epsilon}\mathcal{V}$ and $d_2 : \mathcal{P}_{\epsilon,-\epsilon}\mathcal{V} \longrightarrow \mathcal{P}_{\epsilon,\epsilon}\mathcal{V}$ denote the natural inclusions. We obtain the following exact sequence:

$$0 \longrightarrow \mathcal{P}_{-\epsilon,-\epsilon}\mathcal{V} \stackrel{c_1\oplus c_2}{\longrightarrow} \mathcal{P}_{\epsilon,-\epsilon}\mathcal{V} \oplus \mathcal{P}_{-\epsilon,\epsilon}\mathcal{V} \stackrel{d_1-d_2}{\longrightarrow} \mathcal{P}_{\epsilon,\epsilon}\mathcal{V} \longrightarrow 0.$$

It induces the following exact sequence

$$0 \longrightarrow \widetilde{\mathfrak{N}}_{-\epsilon,-\epsilon}(\mathcal{P}_*\mathcal{V}, \mathbb{D}^\lambda) \longrightarrow \widetilde{\mathfrak{N}}_{\epsilon,-\epsilon}(\mathcal{P}_*\mathcal{V}, \mathbb{D}^\lambda) \oplus \widetilde{\mathfrak{N}}_{-\epsilon,\epsilon}(\mathcal{P}_*\mathcal{V}, \mathbb{D}^\lambda) \longrightarrow \widetilde{\mathfrak{N}}_{\epsilon,\epsilon}(\mathcal{P}_*\mathcal{V}, \mathbb{D}^\lambda) \longrightarrow 0. \tag{2.53}$$

On $\mathbb{C}$, we consider the following complex:

$$\mathrm{Gr}^{\mathcal{P}}_0(\mathcal{V}_0) \otimes \mathcal{O}_{\mathbb{C}} \xrightarrow{\mathrm{Res}_0(\mathbb{D}^\lambda) - \frac{\sqrt{-1}}{2}\beta_1} \mathrm{Gr}^{\mathcal{P}}_0(\mathcal{V}_0) \otimes \mathcal{O}_{\mathbb{C}}$$

The quotient is isomorphic to the quotient of

$$\widetilde{\mathfrak{N}}_{-\epsilon,-\epsilon}(\mathcal{P}_*\mathcal{V}, \mathbb{D}^\lambda) \longrightarrow \widetilde{\mathfrak{N}}_{\epsilon,-\epsilon}(\mathcal{P}_*\mathcal{V}, \mathbb{D}^\lambda).$$

We have a similar description of the quotient of

$$\widetilde{\mathfrak{N}}_{-\epsilon,-\epsilon}(\mathcal{P}_*\mathcal{V}, \mathbb{D}^\lambda) \longrightarrow \widetilde{\mathfrak{N}}_{-\epsilon,\epsilon}(\mathcal{P}_*\mathcal{V}, \mathbb{D}^\lambda).$$

Then, we can easily deduce the claim of the proposition from Lemma 2.10.9 below. ■

Lemma 2.10.9 *Let* V *be an* r*-dimensional vector space with a base* $e_1, \ldots, e_r$. *Let* f *be the endomorphism of* V *determined by* $f(e_i) = e_{i+1}$ $(i = 1, \ldots, r-1)$ *and* $f(e_r) = 0$. *Then, the cokernel* $f - z\,\mathrm{id}_V : V \otimes \mathcal{O}_{\mathbb{C}_z} \longrightarrow V \otimes \mathcal{O}_{\mathbb{C}_z}$ *is isomorphic to* $\mathcal{O}_{\mathbb{C}}/z^r\mathcal{O}_{\mathbb{C}}$.

Proof We set $F = f - z\,\mathrm{id}_V$. Let V' denote the vector space $\bigoplus \mathbb{C}_{i=2}^r e_i$, and let $\rho : V \longrightarrow V'$ denote the projection. It is easy to see that $\rho \circ F$ is an epimorphism, and the kernel of $\rho \circ F$ is generated by $\sum_{j=1}^r z^{r-j} e_j$. Then, the claim of the lemma easily follows. ■

Remark 2.10.10 We can consider this kind of transformation for stable good filtered λ-flat bundles of degree 0 which do not necessarily satisfy Condition 2.10.6. It will be explained elsewhere. The transformation naturally appear as the algebraic counterpart of the Nahm transform between periodic monopoles and wild harmonic bundles on $(\mathbb{P}^1, \overline{D})$. Some special cases were studied in [13, 14, 30], and a more general case is studied in [69]. ■

2.11 Filtered Prolongation of Acceptable Bundles

2.11.1 Filtered Bundles on a Neighbourhood of 0 in $\mathbb{C}$

Let Y be a neighbourhood of 0 in $\mathbb{C}$. Let $\mathcal{E}$ be a locally free $\mathcal{O}_Y(*0)$-module of finite rank. A filtered bundle over $\mathcal{E}$ is an increasing sequence of locally free $\mathcal{O}_Y$-submodules $\mathcal{P}_a\mathcal{E} \subset \mathcal{E}$ $(a \in \mathbb{R})$ satisfying the following conditions.

- $\mathcal{P}_a\mathcal{E}$ $(a \in \mathbb{R})$ are lattices of $\mathcal{E}$, i.e., $\mathcal{P}_a\mathcal{E}(*\{0\}) = \mathcal{E}$.
- $\mathcal{P}_{a+n}\mathcal{E} = \mathcal{P}_a\mathcal{E}(n\{0\})$ for any $a \in \mathbb{R}$ and $n \in \mathbb{Z}$.
- For any $a \in \mathbb{R}$, there exists $\epsilon > 0$ such that $\mathcal{P}_{a+\epsilon}\mathcal{E} = \mathcal{P}_a\mathcal{E}$.

In that case, we also say that $\mathcal{P}_*\mathcal{E}$ is a filtered bundle on $(Y, 0)$, for simplicity.

For any $a \in \mathbb{R}$, we set $\mathcal{P}_{<a}\mathcal{E} := \sum_{b<a} \mathcal{P}_b(\mathcal{E})$, and

$$\mathrm{Gr}_a^{\mathcal{P}}(\mathcal{E}) := \mathcal{P}_a(\mathcal{E})/\mathcal{P}_{<a}(\mathcal{E}).$$

We may naturally regard $\mathrm{Gr}_a^{\mathcal{P}}(\mathcal{E})$ as a finite dimensional $\mathbb{C}$-vector space.

A frame $\boldsymbol{v} = (v_1, \ldots, v_{\mathrm{rank}\,\mathcal{E}})$ of $\mathcal{P}_a\mathcal{E}$ is called compatible with the parabolic structure if there exists a decomposition $\boldsymbol{v} = \coprod_{a-1<b\leq a} \boldsymbol{v}_b$ such that the following holds.

- $\boldsymbol{v}_b$ is a tuple of sections of $\mathcal{P}_b(\mathcal{E})$, and $\boldsymbol{v}_b$ induces a base of $\mathrm{Gr}_b^{\mathcal{P}}(\mathcal{E})$.

For any non-zero section s, the number

$$\deg^{\mathcal{P}}(s) := \min\left\{c \in \mathbb{R} \,\middle|\, s \in \mathcal{P}_c\mathcal{E}\right\} \tag{2.54}$$

is called the parabolic degree of s. If $s = 0$, we set $\deg^{\mathcal{P}}(s) = -\infty$.

2.11.1.1 *G*-Equivariance

Let G be a finite group acting on Y preserving 0, which is not necessarily effective. We say that a filtered bundle $\mathcal{P}_*\mathcal{E}$ on $(Y, 0)$ is G-equivariant if $\mathcal{E}$ is G-equivariant, and each $\mathcal{P}_a\mathcal{E}$ is preserved by the G-action.

For a G-equivariant filtered bundle $\mathcal{P}_*\mathcal{E}$ on $(Y, 0)$, each $H^0(Y, \mathcal{P}_a\mathcal{E})$ is naturally a G-representation. The following lemma is obvious because G is finite.

Lemma 2.11.1 *There exist G-invariant subspaces $H_b \subset H^0(Y\, \mathcal{P}_b\mathcal{E})$ $(b \in \mathbb{R})$ such that the natural morphism $H^0(Y, \mathcal{P}_b\mathcal{E}) \longrightarrow \mathrm{Gr}_b^{\mathcal{P}}(\mathcal{E})$ induces an isomorphism $H_b \simeq \mathrm{Gr}_b^{\mathcal{P}}(\mathcal{E})$ of G-representations.* ∎

Corollary 2.11.2 *Let $\mathrm{Gr}_b^{\mathcal{P}}(\mathcal{E}) = \bigoplus_{i=1}^{m(b)} V_{b,i}$ $(a-1 < b \leq a)$ be decompositions of G-representations. Then, there exist a G-invariant neighbourhood Y' of 0 in Y and a G-equivariant decomposition $\mathcal{P}_*\mathcal{E}_{|Y'} = \bigoplus_{a-1<b\leq a} \bigoplus_{i=1}^{m(b)} \mathcal{P}_*\mathcal{E}_{b,i}$ such that the following holds.*

- *We have $\mathrm{Gr}_c^{\mathcal{P}}(\mathcal{E}_{b,i}) = 0$ for $a-1 < c \leq a$ unless $c = b$, and $\mathrm{Gr}_b^{\mathcal{P}}(\mathcal{E}_{b,i}) = V_{b,i}$.*

Proof Let H_b be as in Lemma 2.11.1. We obtain the induced decompositions of G-representations $H_b = \bigoplus_{i=1}^{m(i)} H_{b,i}$. Let $\mathcal{E}'_{b,i}$ denote the $\mathcal{O}_Y(*0)$-submodule of $\mathcal{E}$ generated by $H_{b,i}$. There exists a G-invariant neighbourhood Y' of 0 in Y such that $\mathcal{E}_{|Y'} = \bigoplus \mathcal{E}'_{b,i|Y'}$. Then, the claim of the lemma is easy to check. ∎

Corollary 2.11.3 *If G is cyclic, there exist a G-invariant neighbourhood Y' of 0 in Y and a compatible frame $\boldsymbol{v}$ of $\mathcal{P}_a\mathcal{E}_{|Y'}$ such that $g^*(v_i) = g^{n(i)}v_i$ for some $n(i) \in \mathbb{Z}$.* ∎

The following lemma is obvious.

Lemma 2.11.4 *Suppose that the action of G on Y is trivial. Let $\mathrm{Irr}(G)$ denote the set of the isomorphism classes of the irreducible representations of G. There exists a unique G-equivariant decomposition $\mathcal{P}_*\mathcal{E} = \bigoplus_{\kappa\in\mathrm{Irr}(G)} \mathcal{P}_*\mathcal{E}_\kappa$ such that for any $y \in Y$ and for any $a \in \mathbb{R}$, the G-representation $\mathcal{P}_a(\mathcal{E}_\kappa)_{|y}$ is isomorphic to a direct sum of κ.* ∎

2.11.1.2 Subbundles, Quotient and Splitting

Let $\mathcal{P}_*\mathcal{E}$ be a G-equivariant filtered bundle on $(Y, 0)$. Recall that a G-equivariant locally free $\mathcal{O}_Y(*0)$-submodule $\mathcal{E}' \subset \mathcal{E}$ is called saturated if $\mathcal{E}'' = \mathcal{E}/\mathcal{E}'$ is also locally free. In that case, we define $\mathcal{P}_a(\mathcal{E}') := \mathcal{P}_a(\mathcal{E}) \cap \mathcal{E}'$ for any a, and we obtain the G-equivariant induced filtered bundle over $\mathcal{E}'$. Let $\mathcal{P}_a(\mathcal{E}'')$ be the image of $\mathcal{P}_a(\mathcal{E}) \longrightarrow \mathcal{E}''$, and we obtain the G-equivariant induced filtered bundle over $\mathcal{E}''$.

Lemma 2.11.5 *There exists a G-invariant neighbourhood Y' of 0 in Y and a splitting $\mathcal{E}''_{|Y'} \longrightarrow \mathcal{E}_{|Y'}$ with which we obtain $\mathcal{P}_*(\mathcal{E})_{|Y'} \simeq \mathcal{P}_*(\mathcal{E}')_{|Y'} \oplus \mathcal{P}_*(\mathcal{E}'')_{|Y'}$.*

Proof We may naturally regard $\mathrm{Gr}_b^{\mathcal{P}}(\mathcal{E}')$ as a G-invariant subspace of $\mathrm{Gr}_b^{\mathcal{P}}(\mathcal{E})$. Because G is finite, there exists a G-invariant decomposition $\mathrm{Gr}_b^{\mathcal{P}}(\mathcal{E}) = \mathrm{Gr}_b^{\mathcal{P}}(\mathcal{E}') \oplus V''_b$, and there exists a G-invariant subspace $H''_b \subset H^0(Y, \mathcal{P}_b\mathcal{E})$ such that the map $H^0(Y, \mathcal{P}_b\mathcal{E}) \longrightarrow \mathrm{Gr}_b^{\mathcal{P}}(\mathcal{E})$ induces an isomorphism $H''_b \simeq V''_b$ of G-representations. The composite of the natural morphisms $H^0(Y, \mathcal{P}_b\mathcal{E}) \longrightarrow H^0(Y, \mathcal{P}_b\mathcal{E}'') \longrightarrow \mathrm{Gr}_b^{\mathcal{P}}(\mathcal{E}'')$ induces an isomorphism $H''_b \simeq \mathrm{Gr}_b^{\mathcal{P}}(\mathcal{E}'')$ of G-representations.

Let $\mathcal{E}''_1$ be the $\mathcal{O}_Y(*0)$-submodule of $\mathcal{E}$ generated by $\bigoplus_{a-1<b\leq a} H''_b$. There exists a G-invariant neighbourhood Y' of 0 in Y such that $\mathcal{E}_{|Y'} = \mathcal{E}'_{|Y'} \oplus \mathcal{E}''_{1|Y'}$. Then, it is easy to check the claims of the lemma. ■

2.11.1.3 Basic Functoriality

Let $\mathcal{P}_*\mathcal{E}_i$ $(i = 1, 2)$ be filtered bundles over locally free $\mathcal{O}_Y(*0)$-modules $\mathcal{E}_i$. Let $\mathcal{P}_*\mathcal{E}_1 \oplus \mathcal{P}_*\mathcal{E}_2$ denote the filtered bundle $\mathcal{P}_*(\mathcal{E}_1 \oplus \mathcal{E}_2)$ over $\mathcal{E}_1 \oplus \mathcal{E}_2$ obtained as

$$\mathcal{P}_a(\mathcal{E}_1 \oplus \mathcal{E}_2) = \mathcal{P}_a(\mathcal{E}_1) \oplus \mathcal{P}_a(\mathcal{E}_2).$$

Let $\mathcal{P}_*\mathcal{E}_1 \otimes \mathcal{P}_*\mathcal{E}_2$ denote the filtered bundle $\mathcal{P}_*(\mathcal{E}_1 \otimes \mathcal{E}_2)$ over $\mathcal{E}_1 \otimes \mathcal{E}_2$ defined as

$$\mathcal{P}_a(\mathcal{E}_1 \otimes \mathcal{E}_2) = \sum_{b+c\leq a} \mathcal{P}_b(\mathcal{E}_1) \otimes \mathcal{P}_c(\mathcal{E}_2).$$

Let $\mathcal{H}om(\mathcal{E}_1, \mathcal{E}_2)$ denote the sheaf of $\mathcal{O}_Y(*0)$-morphisms from $\mathcal{E}_1$ to $\mathcal{E}_2$. It is naturally a locally free $\mathcal{O}_Y(*0)$-module. Let $\mathcal{H}om(\mathcal{P}_*\mathcal{E}_1, \mathcal{P}_*\mathcal{E}_2)$ denote the filtered bundle $\mathcal{P}_*\mathcal{H}om(\mathcal{E}_1, \mathcal{E}_2)$ defined as follows for any open subset $U \subset Y$:

$$\begin{aligned}&\mathcal{P}_a\mathcal{H}om(\mathcal{E}_1, \mathcal{E}_2)(U)\\&\quad= \big\{f \in \mathcal{H}om(\mathcal{E}_1, \mathcal{E}_2)(U) \,\big|\, f(\mathcal{P}_b\mathcal{E}_{1|U}) \subset \mathcal{P}_{b+a}(\mathcal{E}_{2|U})\ (\forall b \in \mathbb{R})\big\}. \qquad (2.55)\end{aligned}$$

Remark 2.11.6 If $\mathcal{E}_i$ and $\mathcal{P}_*\mathcal{E}_i$ are G-equivariant, the induced filtered bundles are also G-equivariant by the construction. ■

Note that $\mathcal{O}_Y(*0)$ is equipped with the canonical filtered bundle $\mathcal{P}_*\big(\mathcal{O}_Y(*0)\big)$, i.e., $\mathcal{P}_a(\mathcal{O}_Y(*0)) = \mathcal{O}_Y([a]0)$, where $[a] := \max\{n \in \mathbb{Z} \,|\, n \leq a\}$. For a filtered bundle $\mathcal{P}_*\mathcal{E}$, let $\mathcal{P}_*\mathcal{E}^\vee$ denote the filtered bundle obtained as $\mathcal{H}om(\mathcal{P}_*\mathcal{E}, \mathcal{P}_*\mathcal{O}_Y(*0))$. Note that $\mathcal{P}_*\mathcal{O}_Y(*0)$ is naturally G-equivariant. Hence, if $\mathcal{P}_*\mathcal{E}$ is G-equivariant, then $\mathcal{P}_*\mathcal{E}^\vee$ is also G-equivariant.

2.11.1.4 Pull Back

Let $q \in \mathbb{Z}_{\geq 1}$. Let $\varphi_q : \mathbb{C} \longrightarrow \mathbb{C}$ be the morphism defined by $\varphi_q(\zeta) = \zeta^q$. Let $Y_q := \varphi_q^{-1}(Y)$.

Let $\mathcal{P}_*\mathcal{E}$ be a filtered bundle over a locally free $\mathcal{O}_Y(*0)$-module $\mathcal{E}$ on $(Y, 0)$. We obtain the $\mathcal{O}_{Y_q}(*0)$-module $\varphi_q^*(\mathcal{E})$ and the filtered bundle $\mathcal{P}_*(\varphi_q^*\mathcal{E})$ defined as follows:

$$\mathcal{P}_a(\varphi_q^*\mathcal{E}) = \sum_{\substack{b\in\mathbb{R}, n\in\mathbb{Z}\\ qb+n\leq a}} \zeta^{-n}\varphi_q^*(\mathcal{P}_b\mathcal{E})$$

The filtered bundle $\mathcal{P}_*(\varphi_q^*\mathcal{E})$ is denoted by $\varphi_q^*(\mathcal{P}_*\mathcal{E})$. We can check the following lemma by using Lemma 2.11.5.

Lemma 2.11.7 *Let $\mathcal{E}$ and $\mathcal{P}_*\mathcal{E}$ be as above. Let $\mathcal{E}'$ be a saturated locally free $\mathcal{O}_Y(*0)$-submodule of $\mathcal{E}$. We set $\mathcal{E}'' = \mathcal{E}/\mathcal{E}'$. We have the induced filtered bundles $\mathcal{P}_*\mathcal{E}'$ and $\mathcal{P}_*\mathcal{E}''$ over $\mathcal{E}'$ and $\mathcal{E}''$, respectively. Then, the pull back of the filtered bundle $\varphi_q^*(\mathcal{P}_*\mathcal{E}')$ (resp. $\varphi_q^*(\mathcal{P}_*\mathcal{E}'')$) is equal to the filtered bundle over $\varphi_q^*(\mathcal{E}')$ (resp. $\varphi_q^*(\mathcal{E}'')$) induced by $\varphi_q^*(\mathcal{P}_*\mathcal{E})$ and the inclusion $\varphi_q^*(\mathcal{E}') \subset \varphi_q^*(\mathcal{E})$ (resp. the projection $\varphi_q^*(\mathcal{E}) \longrightarrow \varphi_q^*(\mathcal{E}'')$).* ■

Lemma 2.11.8 *Let $\mathcal{P}_*\mathcal{E}_i$ $(i = 1, 2)$ be filtered bundles on $(Y, 0)$. There exist natural isomorphisms $\varphi_q^*(\mathcal{P}_*\mathcal{E}_1 \oplus \mathcal{P}_*\mathcal{E}_2) \simeq \varphi_q^*(\mathcal{P}_*\mathcal{E}_1) \oplus \varphi_q^*(\mathcal{P}_*\mathcal{E}_2)$, $\varphi_q^*(\mathcal{P}_*\mathcal{E}_1 \otimes \mathcal{P}_*\mathcal{E}_2) \simeq \varphi_q^*(\mathcal{P}_*\mathcal{E}_1) \otimes \varphi_q^*(\mathcal{P}_*\mathcal{E}_2)$, and $\varphi_q^*\mathcal{H}om\big(\mathcal{P}_*\mathcal{E}_1, \mathcal{P}_*\mathcal{E}_2\big) \simeq \mathcal{H}om\big(\varphi_q^*(\mathcal{P}_*\mathcal{E}_1), \varphi_q^*(\mathcal{P}_*\mathcal{E}_2)\big)$.*

Proof By using compatible frames, we can observe that it is enough to check the claims in the case $\operatorname{rank}(\mathcal{E}_i) = 1$. In that case, we can check the claims by direct computations. (For example, see Sect. 2.11.1.7.) ■

2.11.1.5 Push-Forward

Let $\mathcal{P}_*\mathcal{E}_1$ be a filtered bundle over a locally free $\mathcal{O}_{Y_q}(*0)$-module $\mathcal{E}_1$. We obtain the $\mathcal{O}_Y(*0)$-module $\varphi_{q*}(\mathcal{E}_1)$ over which we obtain the following filtered bundle:

$$\mathcal{P}_a(\varphi_{q*}\mathcal{E}_1) = \varphi_{q*}\mathcal{P}_{a/q}(\mathcal{E}_1).$$

The filtered bundle $\mathcal{P}_*(\varphi_{q*}\mathcal{E}_1)$ is denoted by $\varphi_{q*}(\mathcal{P}_*\mathcal{E}_1)$. We can check the following lemma by using Lemma 2.11.5.

Lemma 2.11.9 *Let $\mathcal{E}_1$ be as above. Let $\mathcal{E}_1'$ be a saturated locally free $\mathcal{O}_{Y_q}(*0)$-submodule of $\mathcal{E}_1$. We set $\mathcal{E}_1'' = \mathcal{E}_1/\mathcal{E}_1'$. We obtain the induced filtered bundles $\mathcal{P}_*\mathcal{E}_1'$ and $\mathcal{P}_*\mathcal{E}_1''$ over $\mathcal{E}_1'$ and $\mathcal{E}_1''$, respectively. Then, $\varphi_{q*}(\mathcal{P}_*\mathcal{E}_1')$ (resp. $\varphi_{q*}\mathcal{P}_*\mathcal{E}_1''$) is equal to the filtered bundle over $\varphi_{q*}(\mathcal{E}_1')$ (resp. $\varphi_{q*}(\mathcal{E}_1'')$) induced by $\varphi_{q*}\mathcal{P}_*\mathcal{E}_1$ and the inclusion $\varphi_{q*}\mathcal{E}_1' \subset \varphi_{q*}\mathcal{E}_1$ (resp. the projection $\varphi_{q*}\mathcal{E}_1 \longrightarrow \varphi_{q*}\mathcal{E}_1''$).* ■

Let $\mathcal{P}_*\mathcal{E}$ be a filtered bundle on $(Y, 0)$. Let $\mathcal{P}_*\mathcal{E}_1$ be a filtered bundle on $(Y_q, 0)$. There exist natural isomorphisms $\varphi_{q*}(\mathcal{E}_1 \otimes \varphi_q^*\mathcal{E}) \simeq \varphi_{q*}(\mathcal{E}_1) \otimes \mathcal{E}$ and $\varphi_{q*}\mathcal{H}om\big(\varphi_q^*\mathcal{E}, \mathcal{E}_1\big) \simeq \mathcal{H}om(\mathcal{E}, \varphi_{q*}\mathcal{E}_1)$.

Lemma 2.11.10 *There exist natural isomorphisms*

$$\varphi_{q*}(\mathcal{P}_*\mathcal{E}_1 \otimes \varphi_q^*(\mathcal{P}_*\mathcal{E})) \simeq \varphi_{q*}\mathcal{P}_*\mathcal{E}_1 \otimes \mathcal{P}_*\mathcal{E},$$

$$\varphi_{q*}\mathcal{H}om\big(\varphi_q^*\mathcal{P}_*\mathcal{E}, \mathcal{P}_*\mathcal{E}_1\big) \simeq \mathcal{H}om(\mathcal{P}_*\mathcal{E}, \varphi_{q*}\mathcal{P}_*\mathcal{E}_1).$$

Proof By using compatible frames, we can observe that it is enough to study the case $\operatorname{rank}\mathcal{E} = 1$. Then, we can check the claims by direct computations. ■

2.11.1.6 Descent

We set $\mathrm{Gal}_q := \{\alpha \in \mathbb{C} \,|\, \alpha^q = 1\}$. It acts on Y_q by the multiplication of c to ζ with which we may regard Gal_q as the Galois group of the ramified covering φ_q. Let $\mathcal{P}_*\mathcal{E}_1$ be a Gal_q-equivariant filtered bundle on $(Y_q, 0)$. We have the naturally induced Gal_q-action on $\varphi_{q*}\mathcal{E}_1$. Let $\mathcal{E}_2$ be the Gal_q-invariant part of $\varphi_{q*}\mathcal{E}_1$ which is also a locally free $\mathcal{O}_Y(*0)$-module. Let $\mathcal{P}_a\mathcal{E}_2$ denote the Gal_q-invariant part of $\mathcal{P}_a\varphi_{q*}\mathcal{E}_1$. Thus, we obtain the filtered bundle $\mathcal{P}_*\mathcal{E}_2$, which is called the descent of $\mathcal{P}_*\mathcal{E}_1$.

Lemma 2.11.11 *$\varphi_q^*(\mathcal{P}_*\mathcal{E}_2)$ is naturally isomorphic to $\mathcal{P}_*\mathcal{E}_1$.*

Proof By the adjunction, there exists a natural morphism $\varphi_q^*(\varphi_{q*}\mathcal{E}_1) \longrightarrow \mathcal{E}_1$. By using Corollary 2.11.3, we can check that the induced morphism $\varphi_q^*\mathcal{P}_*\mathcal{E}_2 \longrightarrow \mathcal{P}_*\mathcal{E}_1$ is an isomorphism. ■

We can check the following lemma by using Lemma 2.11.5.

Lemma 2.11.12 *Let $\mathcal{E}_1'$ be a Gal_q-equivariant saturated locally free $\mathcal{O}_Y(*0)$-submodule of $\mathcal{E}_1$. We set $\mathcal{E}_1'' = \mathcal{E}_1/\mathcal{E}_1'$.*

- *We obtain the $\mathcal{O}_Y(*0)$-submodule $\mathcal{E}_2' \subset \mathcal{E}_2$ as the descent of $\mathcal{E}_1'$. Then, the decent $\mathcal{P}_*\mathcal{E}_2'$ of $\mathcal{P}_*\mathcal{E}_1'$ is equal to the filtered bundle over $\mathcal{E}_2'$ induced by $\mathcal{P}_*\mathcal{E}_2$ with the inclusion $\mathcal{E}_2' \subset \mathcal{E}_2$.*
- *We obtain the $\mathcal{O}_Y(*0)$-quotient module $\mathcal{E}_2 \longrightarrow \mathcal{E}_2''$ as the descent of $\mathcal{E}_1''$. Then, the decent $\mathcal{P}_*\mathcal{E}_2''$ of $\mathcal{P}_*\mathcal{E}_1''$ is equal to the filtered bundle over $\mathcal{E}_2''$ induced by $\mathcal{P}_*\mathcal{E}_2$ with the projection $\mathcal{E}_2 \longrightarrow \mathcal{E}_2''$.* ■

We can easily check the following lemma by using a compatible frame.

Lemma 2.11.13 *For a filtered bundle $\mathcal{P}_*\mathcal{E}$ on $(Y, 0)$, $\varphi_q^*(\mathcal{P}_*\mathcal{E})$ is Gal_q-equivariant, and the decent is naturally isomorphic to $\mathcal{P}_*\mathcal{E}$.* ■

2.11.1.7 Some Examples

Let $c \in \mathbb{R}$. Let $\mathcal{O}_Y(*0) \cdot e$ denote the locally free $\mathcal{O}_Y(*0)$-module of rank one with a global frame e. Let $\mathcal{P}_*^{(c)}\big(\mathcal{O}_Y(*0) \cdot e\big)$ denote the filtered bundle over $\mathcal{O}_Y(*0) \cdot e$ defined as $\mathcal{P}_a^{(c)}(\mathcal{O}_Y(*0) \cdot e) = \mathcal{O}_Y([a-c]0) \cdot e$, where $[d] := \max\{n \in \mathbb{Z} \,|\, n \le d\}$ for any $d \in \mathbb{R}$.

Lemma 2.11.14 *For any filtered bundle $\mathcal{P}_*\mathcal{E}$ of rank one, there exist a neighbourhood Y' of 0 in Y and $-1 < c \le 0$ such that $\mathcal{P}_*\mathcal{E}_{|Y'} \simeq \mathcal{P}_*^{(c)}(\mathcal{O}_{Y'}(*0) \cdot e)$.*

Proof There exists a neighbourhood Y' and a frame v of $\mathcal{P}_0\mathcal{E}_{|Y'}$. There exists $-1 < c \le 0$ such that $\deg^{\mathcal{P}}(v) = c$. Then, we obtain $\mathcal{P}_*\mathcal{E}_{|Y'} \simeq \mathcal{P}_*^{(c)}(\mathcal{O}_{Y'}(*0) \cdot e)$. ■

We have the natural isomorphism $\mathcal{O}_Y(*0)\cdot e_1 \otimes \mathcal{O}_Y(*0)\cdot e_2 \simeq \mathcal{O}_Y(*0)\cdot(e_1\otimes e_2)$, under which we have

$$\mathcal{P}_*^{(c_1)}(\mathcal{O}_Y(*0)\cdot e_1)\otimes\mathcal{P}_*^{(c_2)}(\mathcal{O}_Y(*0)\cdot e_2)\simeq\mathcal{P}_*^{(c_1+c_2)}(\mathcal{O}_Y(*0)\cdot(e_1\otimes e_2)).$$

We have the natural isomorphism

$$\mathcal{H}om_{\mathcal{O}_Y(*0)}\big(\mathcal{O}_Y(*0)e_1,\mathcal{O}_Y(*0)e_2\big)\simeq\mathcal{O}_Y(*0)\,e_1^\vee\otimes e_2,$$

where $(e_1^\vee\otimes e_2)$ is determined by $(e_1^\vee\otimes e_2)(e_1)=e_2$. Under the isomorphism, we have

$$\mathcal{H}om\big(\mathcal{P}_*^{(c_1)}(\mathcal{O}_Y(*0)\cdot e_1),\mathcal{P}_*^{(c_2)}(\mathcal{O}_Y(*0)\cdot e_2)\big)\simeq\mathcal{P}_*^{(-c_1+c_2)}(\mathcal{O}_Y(*0)\cdot(e_1^\vee\otimes e_2)).$$

Under the natural isomorphism $\varphi_q^*(\mathcal{O}_Y(*0)\cdot e)\simeq\mathcal{O}_{Y_q}(*0)\cdot\varphi_q^*(e)$, we have

$$\varphi_q^*\big(\mathcal{P}_*^{(c)}(\mathcal{O}_Y(*0)\cdot e)\big)\simeq\mathcal{P}_*^{(qc)}\big(\mathcal{O}_{Y_q}(*0)\cdot\varphi_q^*(e)\big).$$

We have the natural isomorphism

$$\varphi_{q*}(\mathcal{O}_{Y_q}(*0)\cdot e)\simeq\bigoplus_{i=0}^{q-1}\mathcal{O}_Y(*0)\cdot v_i$$

given by $\zeta^i\cdot e\longmapsto v_i$ $(i=0,\ldots,q-1)$. Under the isomorphism, we obtain

$$\varphi_{q*}(\mathcal{P}_*^{(c)}\mathcal{O}_{Y_q}(*0)\cdot e)\simeq\bigoplus_{i=0}^{q-1}\mathcal{P}_*^{(c-i)/q}\big(\mathcal{O}_Y(*0)\cdot v_i\big).$$

Let m be an integer. Let us consider the Gal_q-action on $\mathcal{P}_*(\mathcal{O}_{Y_q}(*0)\,e)$ defined by $\alpha^*(f(\zeta)e)=\alpha^m\cdot f(\alpha\zeta)e$. The naturally induced Gal_q-action on $\varphi_{q*}(\mathcal{O}_{Y_q}(*0)\cdot e)$ is described as $\alpha^*(v_i)=\alpha^{m+i}v_i$. Let i_0 be the unique integer satisfying $0\le i\le q-1$ and $i_0+m\in q\mathbb{Z}$. Then, the descent of $\mathcal{O}_{Y_q}(*0)\cdot e$ is $\mathcal{O}_Y(*0)\cdot v_{i_0}$, and the descent of $\mathcal{P}_*^{(c)}\cdot e$ is $\mathcal{P}^{(c-i_0)/q}\mathcal{O}_Y(*0)\cdot v_{i_0}$.

2.11.2 Acceptable Bundles on a Punctured Disc

Let $(E,\overline{\partial}_E)$ be a holomorphic vector bundle on $Y\setminus\{0\}$ with a Hermitian metric h. For any $a\in\mathbb{R}$ and open subset $U\subset Y$ with $0\in U$, let $\mathcal{P}_a^hE(U)$ denote the space of holomorphic sections s of E on $U\setminus\{0\}$ such that $|s|_h=O\big(|z|^{-a-\epsilon}\big)$ for any $\epsilon>0$, where $|s|_h$ denotes the function on $U\setminus\{0\}$ obtained as the norm of s

with respect to h. For any open subset $U \subset Y$ with $0 \notin U$, let $\mathcal{P}^h_a E(U)$ denote the space of holomorphic sections of s of E on U. Thus, we obtain an $\mathcal{O}_U$-module $\mathcal{P}^h_a E$. Similarly, for any open subset $U \subset Y$ with $0 \in U$, let $\mathcal{P}^h E(U)$ denote the space of holomorphic sections s of E on $U \setminus \{0\}$ such that $|s|_h = O\big(|z|^{-N}\big)$ for some $N > 0$. For any open subset $U \subset Y$ with $0 \notin U$, let $\mathcal{P}^h E(U)$ denote the space of holomorphic sections s of E on U. Thus, we obtain an $\mathcal{O}_U(*0)$-module $\mathcal{P}^h E$.

Suppose that $(E, \overline{\partial}_E, h)$ is acceptable, i.e., the curvature $F(h)$ of the Chern connection satisfies the following estimate with respect to h:

$$F(h) = O\big(|z|^{-2}(-\log|z|)^{-2}\big)\, dz\, d\overline{z}. \tag{2.56}$$

Proposition 2.11.15 ([16, 80]) *Under the assumption of the acceptability, $\mathcal{P}^h_a E$ are locally free $\mathcal{O}_U$-modules, and $\mathcal{P}^h E$ is a locally free $\mathcal{O}_U(*0)$-module.* ■

In this way, we obtain a filtered bundle $\mathcal{P}^h_* E = \big(\mathcal{P}^h_a E \,\big|\, a \in \mathbb{R}\big)$ over $\mathcal{P}^h E$.

Remark 2.11.16 We shall often use the notation $\mathcal{P}_* E$ instead of $\mathcal{P}^h_* E$ to simplify the description if there is no risk of confusion. ■

Suppose that $\boldsymbol{v}$ is a frame of $\mathcal{P}^h_a(E)$ compatible with the parabolic structure. Let h_0 be the Hermitian metric determined as follows:

$$h_0(v_i, v_j) := \begin{cases} |z|^{-2\deg^{\mathcal{P}^h}(v_i)} & (i = j), \\ 0 & (i \neq j). \end{cases}$$

Recall the following lemma.

Lemma 2.11.17 ([64]) *If $(E, \overline{\partial}_E, h)$ is acceptable, there exist $C \geq 1$ and $N > 0$ such that*

$$C^{-1}(-\log|z|)^{-N} h_0 \leq h \leq C(-\log|z|)^{N} h_0.$$

■

2.11.2.1 Basic Functoriality

Let $(E_i, \overline{\partial}_{E_i})$ $(i = 1, 2)$ be holomorphic vector bundle on $Y \setminus \{0\}$ with a Hermitian metric h_i. Suppose that h_i are acceptable. We obtain the filtered bundles $\mathcal{P}^{h_i}_* E_i$. The bundles $E_1 \oplus E_2$, $E_1 \otimes E_2$ and $\mathcal{H}om(E_1, E_2)$ are naturally equipped with the holomorphic structure and the metric $\widetilde{h}$, and they are acceptable. According to [80], the following holds. (It also follows from Lemma 2.11.17.)

Lemma 2.11.18 *There exist natural isomorphisms*

$$\mathcal{P}_*^{\widetilde{h}}(E_1 \oplus E_2) \simeq \mathcal{P}_*^{h_1} E_1 \oplus \mathcal{P}_*^{h_2} E_2,$$

$$\mathcal{P}_*^{\widetilde{h}}(E_1 \otimes E_2) \simeq \mathcal{P}_*^{h_1} E_1 \otimes \mathcal{P}_*^{h_2} E_2,$$

$$\mathcal{P}_*^{\widetilde{h}} \mathcal{H}om(E_1, E_2) \simeq \mathcal{H}om(\mathcal{P}_*^{h_1} E_1, \mathcal{P}_*^{h_2} E_2).$$

■

Let $(E^\vee, \overline{\partial}_{E^\vee})$ denote the dual bundle of $(E, \overline{\partial}_E)$. It is equipped with the induced Hermitian metric $h^\vee$. It is easy to see that $(E^\vee, \overline{\partial}_{E^\vee}, h^\vee)$ is acceptable. As a special case of Lemma 2.11.18, we obtain $\mathcal{P}_*^{h^\vee}(E^\vee) = (\mathcal{P}_*^h E)^\vee$.

2.11.2.2 Pull Back and Descent

Let $(E, \overline{\partial}_E)$ be a holomorphic bundle on $Y \setminus \{0\}$ with a Hermitian metric h such that $(E, \overline{\partial}_E, h)$ is acceptable on $(Y_q, 0)$. We obtain the holomorphic vector bundle $\varphi_q^{-1}(E, \overline{\partial}_E)$ with the induced metric $\varphi_q^{-1}(h)$. Obviously, $\varphi_q^{-1}(E, \overline{\partial}_E, h)$ is acceptable on $(Y_q, 0)$. We can easily check the following by using Lemma 2.11.17:

$$\mathcal{P}_*^{\varphi_q^{-1}(h)}\big(\varphi_q^{-1}(E)\big) \simeq \varphi_q^*(\mathcal{P}_*^h E).$$

Let $(E_1, \overline{\partial}_{E_1})$ be a holomorphic vector bundle on $Y_q \setminus \{0\}$ with a Hermitian metric h_1 such that $(E_1, \overline{\partial}_{E_1}, h_1)$ is acceptable. We naturally obtain a holomorphic vector bundle $\varphi_{q*}(E_1, \overline{\partial}_{E_1})$ with the induced metric h_1. It is easy to see that $\varphi_{q*}(E_1, \overline{\partial}_{E_1}, h_1)$ is acceptable, and

$$\mathcal{P}_*^{\varphi_{q*}h_1}\big(\varphi_q(E_1)\big) \simeq \varphi_{q*}(\mathcal{P}_*^{h_1} E_1).$$

Suppose that $(E_1, \overline{\partial}_{E_1}, h_1)$ is Gal_q-equivariant. We obtain a holomorphic vector bundle $(E_2, \overline{\partial}_{E_2})$ on $Y_q \setminus \{0\}$ with the induced metric h_2 as the descent of $(E_1, \overline{\partial}_{E_1}, h_1)$. It is easy to see that $\mathcal{P}_*^{h_2}(E_2)$ is the descent of $\mathcal{P}_*^{h_1} E_1$.

2.11.3 Global Case

2.11.3.1 Filtered Bundles

Let C be a Riemann surface with a discrete subset $D \subset Y$. Let $\mathcal{E}$ be a locally free $\mathcal{O}_Y(*D)$-module of finite rank. A filtered bundle $\mathcal{P}_*\mathcal{E}$ over $\mathcal{E}$ is a tuple $\big(\mathcal{P}_{\boldsymbol{a}}\mathcal{E} \,\big|\, \boldsymbol{a} \in \mathbb{R}^D\big)$ of locally free $\mathcal{O}_Y$-submodules such that the following holds.

- For $P \in D$, let (C_P, z_P) denote a holomorphic coordinate neighbourhood of P such that $C_P \cap D = \{P\}$ and $z_P(P) = 0$. Then, for each $\boldsymbol{a} = (a(Q) \mid Q \in D) \in \mathbb{R}^D$, $\mathcal{P}_{\boldsymbol{a}}(\mathcal{E})_{|C_P}$ depends only on $a(P)$, denoted by $\mathcal{P}_{a(P)}(\mathcal{E}_{|C_P})$, and $\mathcal{P}_*(\mathcal{E}_{|C_P})$ is a filtered bundle over the $\mathcal{O}_{C_P}(*P)$-module $\mathcal{E}_{|C_P}$.

For any locally free $\mathcal{O}_C(*D)$-submodule $\mathcal{E}' \subset \mathcal{E}$ such that $\mathcal{E}'' = \mathcal{E}/\mathcal{E}'$ is also locally free, we obtain the induced filtered bundles $\mathcal{P}_*(\mathcal{E}')$ and $\mathcal{P}_*(\mathcal{E}'')$ over $\mathcal{E}'$ and $\mathcal{E}''$, respectively, applying the constructions in Sect. 2.11.1.

For filtered bundles $\mathcal{P}_*(\mathcal{E}_i)$ $(i = 1, 2)$ over locally free $\mathcal{O}_Y(*D)$-modules $\mathcal{E}_i$, we obtain the induced filtered bundles $\mathcal{P}_*(\mathcal{E}_1 \oplus \mathcal{E}_2)$ over $\mathcal{E}_1 \oplus \mathcal{E}_2$, $\mathcal{P}_*(\mathcal{E}_1 \otimes \mathcal{E}_2)$ over $\mathcal{E}_1 \otimes \mathcal{E}_2$, and $\mathcal{H}om(\mathcal{P}_*\mathcal{E}_1, \mathcal{P}_*\mathcal{E}_2)$ over $\mathcal{H}om(\mathcal{E}_1, \mathcal{E}_2)$ as in Sect. 2.11.1.3.

If C is a compact Riemann surface, then D is finite, and the degree of a filtered bundle $\mathcal{P}_*\mathcal{E}$ on (C, D) is defined as

$$\deg(\mathcal{P}_*\mathcal{E}) = \deg(\mathcal{P}_{\boldsymbol{a}}\mathcal{E}) - \sum_{P \in D} \sum_{a(P)-1<b\leq a(P)} b \dim \mathrm{Gr}_b^{\mathcal{P}}(\mathcal{E}_{|C_P}). \tag{2.57}$$

The number is independent of the choice of $\boldsymbol{a}$.

2.11.3.2 Acceptable Bundles

Let $(E, \overline{\partial}_E)$ be a holomorphic vector bundle on $C \setminus D$ with a Hermitian metric h. For each $P \in D$, we take a holomorphic coordinate neighbourhood (C_P, z_P) around P such that $C_P \cap D = \{P\}$ and $z_P(P) = 0$. We set $C_P^* = C_P \setminus \{P\}$. By applying the procedure 2.11.2 to the restriction of $(E, \overline{\partial}_E, h)_{|C_P \setminus P}$, we obtain a tuple of sheaves $\mathcal{P}_*^h(E_{|C_P \setminus P}) = \big(\mathcal{P}_a^h(E_{|C_P^*}) \,\big|\, a \in \mathbb{R}\big)$. For $\boldsymbol{a} = (a(P) \mid P \in D) \in \mathbb{R}^D$, we obtain the $\mathcal{O}_C$-module $\mathcal{P}_{\boldsymbol{a}}^h(E)$ from $(E, \overline{\partial}_E)$ and $\mathcal{P}_*^h(E_{|C_P^*})$ $(P \in D)$. Similarly, we obtain the $\mathcal{O}_C(*D)$-module $\mathcal{P}^h(E)$.

We say that $(E, \overline{\partial}_E, h)$ is acceptable if the following holds.

- Let $P \in D$. If (C_P, z_P) is a holomorphic coordinate neighbourhood such that $C_P \cap D = \{P\}$ and $z_P(P) = 0$, then $(E, \overline{\partial}_E, h)_{|C_P \setminus P}$ is acceptable in the sense of Sect. 2.11.2.

If $(E, \overline{\partial}_E, h)$ is acceptable, then $\mathcal{P}^h E$ is a locally free $\mathcal{O}_C(*D)$-module, and $\mathcal{P}_*^h(E)$ is a filtered bundle over $\mathcal{P}^h E$.

Proposition 2.11.19 ([79, 80]) *If C is compact, the following holds.*

$$\deg(\mathcal{P}_*^h E) = \frac{\sqrt{-1}}{2\pi} \int_{C \setminus D} \mathrm{Tr}\, F(h).$$

■

Chapter 3
Formal Difference Modules and Good Parabolic Structure

Abstract We introduce the notion of good parabolic structure of formal difference modules. We also explain the geometrization of formal difference modules, and their relation with formal λ-connections.

3.1 Outline of This Chapter

We recall the classification of formal difference modules [12, 23, 76, 86] in Sect. 3.2. We shall introduce the concept of good filtered bundles in the context of formal difference modules in Sect. 3.3.

We explain a geometrization of difference modules in Sect. 3.4. Namely, we introduce a ringed space $\widehat{H}_{\infty,q}$, and explain an equivalence between formal difference modules and some sheaves on $\widehat{H}_{\infty,q}$. It is the formal version of the equivalence in Sect. 2.10. Then, in Sect. 3.5, we define the notion of good filtered bundles on $\widehat{H}_{\infty,q}$ as the translation of the parabolic structure of difference modules.

In Sect. 3.6, we explain another equivalence between some sheaves on $\widehat{H}_{\infty,q}$ and formal λ-connections, which is the ramified and formal version of the construction in Sect. 2.8.3.4.

3.2 Formal Difference Modules

Let $\mathbb{C}((y^{-1}))$ denote the field of formal Laurent power series of y^{-1}. We fix an algebraic closure of $\mathbb{C}((y^{-1}))$. We also fix a q-th root y_q of y for each $q \in \mathbb{Z}_{\geq 1}$ such that $y_p^{p/q} = y_q$ for any $p \in q\mathbb{Z}_{\geq 1}$.

Let $\varrho \in \mathbb{C}$. Let Φ^* denote the automorphism of the field $\mathbb{C}(y)$ defined by $\Phi^*(f)(y) = f(y+\varrho)$. Note that $\Phi^*(y^{-1}) = (y+\varrho)^{-1} = y^{-1}(1+\varrho y^{-1})^{-1}$. There exists the expansion $(1+\varrho y^{-1})^{-1} = \sum_{j=0}^{\infty}(-\varrho y^{-1})^j$ in $\mathbb{C}((y^{-1}))$. We obtain $\Phi^*\big((y^{-1})^k\big) = (y^{-1})^k\big(\sum_{j=0}^{\infty}(-\varrho y^{-1})^j\big)^k$ in $\mathbb{C}((y^{-1}))$, and $\Phi^*\big((y^{-1})^k\big) - (y^{-1})^k \in (y^{-1})^{k+1}\mathbb{C}[\![y^{-1}]\!]$. Hence, we obtain the induced automorphism of $\mathbb{C}((y^{-1}))$, which

T. Mochizuki, *Periodic Monopoles and Difference Modules*, Lecture Notes in Mathematics 2300, https://doi.org/10.1007/978-3-030-94500-8_3

is also denoted by Φ^*. It induces automorphisms of $\mathbb{C}((y_q^{-1}))$ determined by $\Phi^*(y_q^{-1}) = y_q^{-1}(1+\varrho y^{-1})^{-1/q}$ for any $q \in \mathbb{Z}_{>0}$, where $(1+\varrho y^{-1})^{-1/q}$ denotes the unique power series $1+\sum_{j>0} a_{q,j} y^{-j}$ such that $\left(1+\sum_{j>0} a_{q,j} y^{-j}\right)^q = \sum_{j=0}^{\infty}(-\varrho y)^j$. We regard $y^{-j} = y_q^{-qj}$. In this section, a difference $\mathbb{C}((y_q^{-1}))$-module means a finite dimensional $\mathbb{C}((y_q^{-1}))$-module $\mathcal{N}$ with a $\mathbb{C}$-linear automorphism $\Phi^* : \mathcal{N} \longrightarrow \mathcal{N}$ such that $\Phi^*(fs) = \Phi^*(f) \cdot \Phi^*(s)$. A morphism of $\mathbb{C}((y_q^{-1}))$-difference modules $g : \mathcal{N}_1 \longrightarrow \mathcal{N}_2$ is a $\mathbb{C}((y_q^{-1}))$-linear map such that $\Phi^*(g(v)) = g(\Phi^*(v))$ for any $v \in \mathcal{N}_1$.

3.2.1 Formal Difference Modules of Level ≤1

Let $q \in \mathbb{Z}_{\geq 1}$. Let $\mathcal{N}$ be a difference $\mathbb{C}((y_q^{-1}))$-module. A $\mathbb{C}[\![y_q^{-1}]\!]$-lattice $\mathcal{L}$ of $\mathcal{N}$ is called Φ^*-invariant if $\Phi^*(\mathcal{L}) = \mathcal{L}$. In that case, we obtain the induced automorphism $\Phi^*_{|\infty}$ of $\mathcal{L}_{|\infty} = \mathcal{L}/y_q^{-1}\mathcal{L}$.

By following [12, §1] (see also [12, §6.2]), we say that the level of $\mathcal{N}$ is less than 1 if there exists a Φ^*-invariant lattice $\mathcal{L}$ of $\mathcal{N}$ such that the induced automorphism $\Phi^*_{|\infty}$ of $\mathcal{L}_{|\infty}$ is unipotent.

Proposition 3.2.1 ([12, 23, 76, 86]) *For any difference $\mathbb{C}((y_q^{-1}))$-module $\mathcal{N}$ of level ≤ 1, there exist $p \in q\mathbb{Z}_{>0}$, a finite subset $\mathcal{I}(\mathcal{N}) \subset \left\{\mathfrak{b} \in y_p^{-1}\mathbb{C}[y_p^{-1}] \,\middle|\, \deg_{y_p^{-1}} \mathfrak{b} < p\right\}$ and a decomposition of difference modules*

$$\mathcal{N} \otimes_{\mathbb{C}((y_q^{-1}))} \mathbb{C}((y_p^{-1})) = \bigoplus_{\mathfrak{b} \in \mathcal{I}(\mathcal{N})} \mathcal{N}_{\mathfrak{b}}, \quad (\mathcal{N}_{\mathfrak{b}} \neq 0) \tag{3.1}$$

such that each $\mathcal{N}_{\mathfrak{b}}$ has a Φ^*-invariant lattice $\mathcal{L}_{\mathfrak{b}}$ satisfying $(\Phi^* - (1+\mathfrak{b})\,\mathrm{id}_{\mathcal{N}_{\mathfrak{b}}})\mathcal{L}_{\mathfrak{b}} \subset y_p^{-p}\mathcal{L}_{\mathfrak{b}}$. The set $\mathcal{I}(\mathcal{N})$ and the decomposition (3.1) *are uniquely determined.*

Proof If $\varrho = 0$, the assumption implies that the eigenvalues of the $\mathbb{C}((y_q^{-1}))$-automorphism Φ^* are invertible elements in $\mathbb{C}[[y_p^{-1}]]$ for some $p \in q\mathbb{Z}_{>0}$. Hence, the claim is easily checked. If $\varrho \neq 0$, the claim is known as a part of the classification of difference modules. See [12, Proposition 5]. (It is not difficult to prove it directly by a standard method in the study of formal connections.) ■

By following [12], we say that the level of a difference module $\mathcal{N}$ is 0 if there exists a Φ^*-invariant lattice $\mathcal{L}$ of $\mathcal{N}$ such that $(\Phi^* - \mathrm{id})(\mathcal{L}) = y_q^{-q}\mathcal{L} = y^{-1}\mathcal{L}$. In Proposition 3.2.1, each $\mathcal{N}_{\mathfrak{b}}$ is isomorphic to the tensor product of a difference module of the level 0 and a difference module $\mathbb{C}((y_p^{-1}))\, e$ with the difference operator Φ^* defined as $\Phi^*(e) = (1+\mathfrak{b})e$.

Remark 3.2.2 In Proposition 3.2.1, $\Phi^* - (1+\mathfrak{b})\,\mathrm{id}$ is well defined as a $\mathbb{C}$-linear endomorphism of $\mathcal{N}_{\mathfrak{b}}$. Note that $\big(\Phi^* - (1+\mathfrak{b})\,\mathrm{id}\big)\mathcal{L}_{\mathfrak{b}} \subset y_p^{-p}\mathcal{L}_{\mathfrak{b}}$ holds if and only if there exists a frame $\boldsymbol{v} = (v_1, \dots, v_r)$ of $\mathcal{L}_{\mathfrak{b}}$ such that $\big(\Phi^* - (1+\mathfrak{b})\,\mathrm{id}\big)v_i \in y_p^{-p}\mathcal{L}_{\mathfrak{b}}$. Indeed, the "only if" part of the claim is clear. Suppose that there exists such a frame $\boldsymbol{v}$ of $\mathcal{L}_{\mathfrak{b}}$. For any $f \in \mathbb{C}[\![y_p^{-1}]\!]$, we obtain

$$\big(\Phi^* - (1+\mathfrak{b})\,\mathrm{id}\big)(f v_i) = \Phi^*(f)\big(\Phi^* - (1+\mathfrak{b})\,\mathrm{id}\big)(v_i) + (1+\mathfrak{b})(\Phi^*(f) - f)v_i. \tag{3.2}$$

Because $\Phi^*(f) - f \in y_p^{-p-1}\mathbb{C}[\![y_p^{-1}]\!]$, (3.2) is contained in $y_p^{-p}\mathcal{L}_{\mathfrak{b}}$. ■

The following lemma is standard.

Lemma 3.2.3 *Let $g : \mathcal{N}_1 \longrightarrow \mathcal{N}_2$ be a morphism of $\mathbb{C}(\!(y_q^{-1})\!)$-difference modules of level ≤ 1. Choose $p \in q\mathbb{Z}_{\geq 0}$ such that both $\mathcal{N}_i \otimes \mathbb{C}(\!(y_p^{-1})\!)$ have decompositions of $\mathbb{C}(\!(y_p^{-1})\!)$-difference modules as in (3.1)*

$$\mathcal{N}_i \otimes \mathbb{C}(\!(y_p^{-1})\!) = \bigoplus_{\mathfrak{b} \in y_p^{-1}\mathbb{C}[y_p^{-1}]} \mathcal{N}_{i,\mathfrak{b}}.$$

Then, we have $g\big(\mathcal{N}_{1,\mathfrak{b}}\big) \subset \mathcal{N}_{2,\mathfrak{b}}$, where we set $\mathcal{N}_{i,\mathfrak{b}} = 0$ if $\mathfrak{b} \notin \mathcal{I}(\mathcal{N}_i)$.

Proof We indicate only an outline. We may assume $p = q$. It is enough to prove that $g = 0$ if $\mathcal{N}_1 = \mathcal{N}_{i,\mathfrak{b}_i}$ with $\mathfrak{b}_1 \neq \mathfrak{b}_2$. There exist Φ^*-invariant lattices $\mathcal{L}_i \subset \mathcal{N}_i$ such that $(\Phi^* - (1+\mathfrak{b}_i)\mathcal{N}_i)\mathcal{L}_i \subset y_q^{-q}\mathcal{L}_i$. Let $\boldsymbol{v}_i$ be a frame of $\mathcal{L}_i$. Let A be the matrix representing g with respect to the frames $\boldsymbol{v}_i$, i.e., $g(\boldsymbol{v}_1) = \boldsymbol{v}_2 A$. There exists an expansion $A = \sum_{j \geq N} A_j (y_q^{-1})^j$. Suppose $A \neq 0$. We may assume that $A_N \neq 0$. Let B_i be the matrices determined by $\Phi^*(\boldsymbol{v}_i) = \boldsymbol{v}_i B_i$. We have the expansions $B_i = (1+\mathfrak{b}_i)I_{r(i)} + \sum_{j \geq q} B_{i,j} y_q^{-j}$, where we set $r(i) = \operatorname{rank}\mathcal{N}_i$, and $I_{r(i)}$ denote the $r(i)$-square identity matrices. We also have the expansions $\mathfrak{b}_i = \sum_{j=1}^{q-1} \mathfrak{b}_{i,j} y_q^{-j}$. Let $j(0) := \min\{j \mid \mathfrak{b}_{1,j} - \mathfrak{b}_{2,j}\}$. Because $B_2\Phi^*(A) = AB_1$ and $1 \leq j_0$, we obtain $\mathfrak{b}_{2,j(0)}A_N = \mathfrak{b}_{1,j(0)}A_N$ by taking the coefficients of $y_q^{-N-j_0}$. It implies $A_N = 0$, which contradicts with the assumption. Hence, we obtain $A = 0$. ■

For $p \in q\mathbb{Z}_{>0}$, we set $\mathrm{Gal}_{q,p} := \big\{\gamma \in \mathbb{C} \mid \gamma^{p/q} = 1\big\}$ which acts on $\mathbb{C}(\!(y_p^{-1})\!)$ by $\gamma \bullet f(y_p^{-1}) = f(\gamma^{-1}y_p^{-1})$. It is identified with the Galois group of the extension $\mathbb{C}(\!(y_p^{-1})\!)$ over $\mathbb{C}(\!(y_q^{-1})\!)$. Note that the $\mathrm{Gal}_{q,p}$-action on $\mathbb{C}(\!(y_p^{-1})\!)$ induces a $\mathrm{Gal}_{q,p}$-action on $\mathcal{N} \otimes_{\mathbb{C}(\!(y_q^{-1})\!)} \mathbb{C}(\!(y_p^{-1})\!)$. We obtain the following lemma from the uniqueness of the index set $\mathcal{I}(\mathcal{N})$ and the decomposition (3.1).

Lemma 3.2.4 *The* $\mathrm{Gal}_{q,p}$*-action on* $\mathbb{C}(\!(y_p^{-1})\!)$ *preserves* $\mathcal{I}(\mathcal{N}) \subset \mathbb{C}(\!(y_p^{-1})\!)$. *Moreover, the following holds.*

- *For* $\mathfrak{b} \in \mathcal{I}(\mathcal{N})$, *let* $\mathcal{L}_{\mathfrak{b}}$ *be a lattice of* $\mathcal{N}_{\mathfrak{b}}$ *satisfying* $(\Phi^* - (1+\mathfrak{b})\,\mathrm{id})\mathcal{L}_{\mathfrak{b}} \subset y_p^{-p}\mathcal{L}_{\mathfrak{b}}$. *Then, for any* $\gamma \in \mathrm{Gal}_{q,p}$, $\gamma \bullet \mathcal{L}_{\mathfrak{b}}$ *is a lattice of* $\mathcal{N}_{\gamma\bullet\mathfrak{b}}$ *satisfying* $(\Phi^* - (1+\gamma \bullet \mathfrak{b})\,\mathrm{id})\mathcal{L}_{\gamma\bullet\mathfrak{b}} \subset y_p^{-p}\mathcal{L}_{\gamma\bullet\mathfrak{b}}$. ■

Remark 3.2.5 If $\varrho \neq 0$, for a difference module $\mathcal{N}$ of level 0, there exist frame $\boldsymbol{u} = (u_1, \dots, u_r)$ and a matrix $A \in M_r(\mathbb{C})$ such that $\Phi^*(\boldsymbol{u}) = \boldsymbol{u}(I_r + y_q^{-1}A)$, where $I_r \in M_r(\mathbb{C})$ denotes the identity matrix. (See [12].) ■

Remark 3.2.6 In Sect. 3.6, we shall explain that formal difference modules of level ≤ 1 on $\mathbb{C}(\!(y_q^{-1})\!)$ are equivalent to formal λ-connections on $\mathbb{C}(\!(w_q^{-1})\!)$ whose Poincaré rank is strictly smaller than q. ■

3.2.2 Formal Difference Modules of Pure Slope

Let $\mathcal{N}$ be a difference $\mathbb{C}(\!(y_q^{-1})\!)$-module. For integers m and ℓ, a $\mathbb{C}[\![y_q^{-1}]\!]$-lattice of $\mathcal{N}$ is called $y_q^m(\Phi^*)^\ell$-invariant if $y_q^m(\Phi^*)^\ell(\mathcal{L}) = \mathcal{L}$. Note that the $\mathbb{C}$-linear automorphism of $\mathcal{L}_{|\infty} = \mathcal{L}/y_q^{-1}\mathcal{L}$ is induced by $y_q^m(\Phi^*)^\ell$, which is denoted by $(y_q^m(\Phi^*)^\ell)_{|\infty}$.

For any $\omega \in \mathbb{Q}$, we set $Z(q,\omega) := \{p \in q\mathbb{Z}_{>0} \mid p\omega \in \mathbb{Z}\}$, i.e., $Z(q,\omega) = q\mathbb{Z}_{>0} \cap \omega^{-1}\mathbb{Z}$ if $\omega \neq 0$, and $Z(q,0) = q\mathbb{Z}_{>0}$.

Definition 3.2.7 We say that a difference $\mathbb{C}(\!(y_q^{-1})\!)$-module $\mathcal{N}$ has pure slope $\omega \in \mathbb{Q}$ if there exist $p \in Z(q,\omega)$ and a $y_q^{p\omega}(\Phi^*)^{p/q}$-invariant lattice of $\mathcal{N}$. ■

Lemma 3.2.8

- *Suppose that a difference* $\mathbb{C}(\!(y_q^{-1})\!)$*-module* $\mathcal{N}$ *has a pure slope* ω. *Then, for any* $p \in Z(q,\omega)$, *there exists a* $y_q^{p\omega}(\Phi^*)^{p/q}$*-invariant lattice of* $\mathcal{N}$.
- *Take any* $s \in q\mathbb{Z}_{>0}$. *Then, a difference* $\mathbb{C}(\!(y_q^{-1})\!)$*-module* $\mathcal{N}$ *has a pure slope* ω *if and only if the induced difference* $\mathbb{C}(\!(y_s^{-1})\!)$*-module* $\mathbb{C}(\!(y_s^{-1})\!) \otimes_{\mathbb{C}(\!(y_q^{-1})\!)} \mathcal{N}$ *has a pure slope* ω.
- *Take any* $s \in Z(q,\omega)$. *If a difference* $\mathbb{C}(\!(y_q^{-1})\!)$*-module* $\mathcal{N}$ *has a pure slope* ω, *there exists a* $\mathrm{Gal}_{q,s}$*-invariant lattice* $\mathcal{L}$ *of the induced difference* $\mathbb{C}(\!(y_s^{-1})\!)$*-module* $\mathbb{C}(\!(y_s^{-1})\!) \otimes_{\mathbb{C}(\!(y_q^{-1})\!)} \mathcal{N}$ *such that* $y_s^{s\omega}\Phi^*(\mathcal{L}) = \mathcal{L}$.

Proof Let us study the first claim. By the assumption, there exists $p_1 \in Z(q,\omega)$ such that there exists a $y_q^{p_1\omega}(\Phi^*)^{p_1/q}$-invariant lattice $\mathcal{L}_1$ of $\mathcal{N}$. Let p be any element of $Z(q,\omega)$. If $p \in p_1\mathbb{Z}_{>0}$, it is easy to see that $\mathcal{L}_1$ is $y_q^{p\omega}(\Phi^*)^{p/q}$-invariant. Let us consider the case where there exists $\ell \in \mathbb{Z}_{>0}$ such that $p_1 = p\ell$. We set $\mathcal{L} = \sum_{i=0}^{\ell-1} y_q^{ip\omega} \cdot (\Phi^*)^{ip/q}(\mathcal{L}_1)$. Then, we obtain

$$\begin{aligned} y_q^{p\omega}(\Phi^*)^{p/q}(\mathcal{L}_1) &= \sum_{i=0}^{\ell-1} y_q^{(i+1)p\omega} \cdot (\Phi^*)^{(i+1)p/q}(\mathcal{L}_1) \\ &= \sum_{i=1}^{\ell-1} y_q^{ip\omega}(\Phi^*)^{ip/q}(\mathcal{L}_1) + y_q^{(\ell-1)p\omega} \cdot y_q^{p\omega} y_q^{-p\omega} \mathcal{L}_1 = \mathcal{L}. \end{aligned} \tag{3.3}$$

The first claim in the general case immediately follows.

Let us study the second claim. The "only if" part of the claim is obvious. Let us study the "if" part of the claim. It is enough to study the case $s \in Z(q, \omega)$. Suppose that $\mathbb{C}((y_s^{-1})) \otimes_{\mathbb{C}((y_q^{-1}))} \mathcal{N}$ has a pure slope ω. Let $\mathcal{L}$ be a $y_s^{s\omega}\Phi^*$-invariant lattice of $\mathbb{C}((y_s^{-1})) \otimes_{\mathbb{C}((y_q^{-1}))} \mathcal{N}$. It is also a $(y_s^{s\omega})^{s/q}(\Phi^*)^{s/q}$-invariant lattice of $\mathbb{C}((y_s^{-1})) \otimes_{\mathbb{C}((y_q^{-1}))} \mathcal{N}$. Note that $(y_s^{s\omega})^{s/q} = y_q^{s\omega}$. We set $\mathcal{L}_1 := \mathcal{L} \cap \mathcal{N}$ for the natural inclusion $\mathcal{N} \longrightarrow \mathbb{C}((y_s^{-1})) \otimes_{\mathbb{C}((y_q^{-1}))} \mathcal{N}$. Because both $\mathcal{L}$ and $\mathcal{N}$ are preserved by the action of $y_q^{s\omega}(\Phi^*)^{s/q}$ on $\mathbb{C}((y_s^{-1})) \otimes_{\mathbb{C}((y_q^{-1}))} \mathcal{N}$, we obtain that $\mathcal{L}_1$ is $y_q^{s\omega}(\Phi^*)^{s/q}$-invariant. Hence, $\mathcal{N}$ has pure slope ω.

As for the third claim, we have already proved that there exists a lattice $\mathcal{L}_1$ of $\mathbb{C}((y_s^{-1})) \otimes_{\mathbb{C}((y_q^{-1}))} \mathcal{N}$, such that $y_s^{s\omega}\Phi^*(\mathcal{L}_1) = \mathcal{L}_1$. Then, $\mathcal{L} = \sum_{\gamma \in \mathrm{Gal}_{q,s}} \gamma \bullet \mathcal{L}_1$ has the desired property. ■

Example 3.2.9 A difference $\mathbb{C}((y_q^{-1}))$-module of level ≤ 1 has pure slope 0. ■

Example 3.2.10 For any $\ell \in \mathbb{Z}$ and $\alpha \in \mathbb{C}^*$, let $\boldsymbol{L}_q(\ell, \alpha)$ denote the difference $\mathbb{C}((y_q^{-1}))$-module $\boldsymbol{L}_q(\ell, \alpha) = \mathbb{C}((y_q^{-1}))\, e$ with the difference operator Φ^* defined as

$$\Phi^*(e) = y_q^{-\ell} \alpha \cdot e.$$

Then, $\boldsymbol{L}_q(\ell, \alpha)$ has pure slope ℓ/q. ■

Lemma 3.2.11 *Let $\mathcal{N}_i$ $(i = 1, 2)$ be difference $\mathbb{C}((y_q^{-1}))$-modules with pure slopes ω_i. Any morphism of difference $\mathbb{C}((y_q^{-1}))$-modules $\mathcal{N}_1 \longrightarrow \mathcal{N}_2$ is 0 unless the following condition is satisfied.*

- $\omega_1 = \omega_2$.
- *For any $p \in Z(q, \omega)$, and for any $y_q^{p\omega_1}(\Phi^*)^{p/q}$-invariant lattices $\mathcal{L}_i$ of $\mathcal{N}_i$, the induced automorphisms on $\mathcal{L}_{i|\infty}$ have a common eigenvalue.*

Proof Let $f : \mathcal{N}_1 \longrightarrow \mathcal{N}_2$ be a morphism of difference $\mathbb{C}((y_q^{-1}))$-modules. Take any $p \in Z(q, \omega_1) \cap Z(q, \omega_2)$. There exist $y_q^{p\omega_i}(\Phi^*)^{p/q}$-invariant lattices $\mathcal{L}_i$ of $\mathcal{N}_i$. Let $\boldsymbol{v}_i$ be frames of $\mathcal{L}_i$. Let $r_i := \operatorname{rank} \mathcal{N}_i$. We obtain the matrices $A_i \in M_{r_i}(\mathbb{C}[\![y_q^{-1}]\!])$ such $y_q^{p\omega}(\Phi^*)^{p/q}\boldsymbol{v}_i = \boldsymbol{v}_i A_i$. For the expansion $A_i = \sum_{j=0}^{\infty} A_{i,j} y_q^{-j}$, $A_{i,0}$ are invertible.

Let B be the matrix determined by $f(\boldsymbol{v}_1) = \boldsymbol{v}_2 \cdot B$. If $B \neq 0$, there exists the expansion $B = \sum_{j \le j_0} B_j y_q^j$ such that $B_{j_0} \neq 0$. Because f is compatible with the difference operators, we obtain

$$B(y_q^{-1}) \cdot A_1 \cdot y_q^{-p\omega_1} = A_2 \cdot B\big((1 + (p/q)\varrho y_q^{-q})^{-1/q}\big) \cdot y_q^{-p\omega_2}. \tag{3.4}$$

By comparing the expansions of the both sides of (3.4), we obtain $B_{j_0} = 0$ unless $\omega_1 = \omega_2$ holds and $A_{i,0}$ have a common eigenvalue. Then, the claim of the lemma follows. ■

We can classify difference modules with pure slope by the following proposition and Lemma 3.2.8.

Proposition 3.2.12 ([12, 23, 76, 86]) *Let $\mathcal{N}$ be a difference $\mathbb{C}(\!(y_q^{-1})\!)$-module with pure slope ω. Suppose that $q\omega \in \mathbb{Z}$. Then, there exists a unique decomposition of difference $\mathbb{C}(\!(y_q^{-1})\!)$-modules*

$$\mathcal{N} = \bigoplus_{\alpha \in \mathbb{C}^*} \mathcal{N}_\alpha \tag{3.5}$$

such that each $\mathcal{N}_\alpha$ is isomorphic to the tensor product of $\mathbf{L}_q(q\omega, \alpha)$ and a difference $\mathbb{C}(\!(y_q^{-1})\!)$-module of level ≤ 1.

Proof Let $\mathcal{L}$ be a $y_q^{q\omega}\Phi^*$-invariant lattice of $\mathcal{N}$. We obtain the generalized eigen decomposition

$$\mathcal{L}_{|\infty} = \bigoplus_{\alpha \in \mathbb{C}^*} \mathbb{E}_\alpha(\mathcal{L}_{|\infty}) \tag{3.6}$$

of $(y_q^{q\omega}\Phi^*)_{|\infty}$, i.e., the restriction of $\alpha^{-1}(y_q^{q\omega}\Phi^*)_{|\infty}$ to $\mathbb{E}_\alpha(\mathcal{L}_{|\infty})$ are unipotent. It is a standard fact that there exists a decomposition $\mathcal{L} = \bigoplus \mathcal{L}_\alpha$ of $\mathbb{C}[\![y_q^{-1}]\!]$-lattice such that (i) $y_q^{q\omega}\Phi^*(\mathcal{L}_\alpha) = \mathcal{L}_\alpha$, (ii) $\mathcal{L}_{\alpha|\infty} = \mathbb{E}_\alpha(\mathcal{L}_{|\infty})$. (See [12, §3], for example.) Then, the induced decomposition $\mathcal{N} = \bigoplus \mathbb{C}(\!(y_q^{-1})\!) \otimes_{\mathbb{C}[\![y_q^{-1}]\!]} \mathcal{L}_\alpha$ has the desired property. The uniqueness follows from Lemma 3.2.11. ■

Let $\mathcal{N}_i$ $(i = 1, 2)$ be difference $\mathbb{C}(\!(y_q^{-1})\!)$-modules with pure slope ω. Suppose $q\omega \in \mathbb{Z}$. Each $\mathcal{N}_i$ has a decomposition $\mathcal{N}_i = \bigoplus_{\alpha \in \mathbb{C}^*} \mathcal{N}_{i,\alpha}$ as in (3.5). We obtain the following lemma from Lemma 3.2.11.

Lemma 3.2.13 *Any morphism $F : \mathcal{N}_1 \longrightarrow \mathcal{N}_2$ preserves the decompositions $\mathcal{N}_i = \bigoplus_{\alpha \in \mathbb{C}^*} \mathcal{N}_{i,\alpha}$.* ■

Let $\mathcal{N}$ be a difference $\mathbb{C}(\!(y_q^{-1})\!)$-difference of pure slope $\omega \in \mathbb{Q}$. Take $p \in q\mathbb{Z}_{>0}$ such that $p\omega \in \mathbb{Z}$. Then, there exists a decomposition

$$\mathbb{C}(\!(y_p^{-1})\!) \otimes_{\mathbb{C}(\!(y_q^{-1})\!)} \mathcal{N} = \bigoplus \mathcal{N}_\alpha. \tag{3.7}$$

The following lemma is easy to see.

Lemma 3.2.14 *For any $\gamma \in \mathrm{Gal}_{q,p}$, we obtain $\gamma \bullet \mathcal{N}_\alpha = \mathcal{N}_{\alpha\gamma^{-p\omega}}$. As a result, the following holds.*

- *Let F be the automorphism of $\mathbb{C}((y_p^{-1}))\otimes_{\mathbb{C}((y_q^{-1}))}\mathcal{N}$ obtained by the multiplication of $\alpha y_p^{-p\omega}$ on $\mathcal{N}_\alpha$ in (3.7). Then, F is equivariant with respect to the $\mathrm{Gal}_{q,p}$-action* ■

3.2.3 Slope Decomposition of Formal Difference Modules

We can understand the structure of general formal difference modules by the slope decomposition in the following theorem.

Theorem 3.2.15 (see [12, 23, 76, 86]) *For any difference $\mathbb{C}((y_q^{-1}))$-module $\mathcal{N}$, there exists a unique decomposition of difference $\mathbb{C}((y_q^{-1}))$-modules*

$$\mathcal{N} = \bigoplus_{\omega \in \mathbb{Q}} \mathcal{S}_\omega(\mathcal{N})$$

such that each $\mathcal{S}_\omega(\mathcal{N})$ has pure slope ω. It is called the slope decomposition. ■

Lemma 3.2.11 implies the following lemma.

Lemma 3.2.16 *Any morphism of difference $\mathbb{C}((y_q^{-1}))$-modules $F : \mathcal{N}_1 \longrightarrow \mathcal{N}_2$ preserves the slope decompositions $\mathcal{N}_i = \bigoplus \mathcal{S}_\omega(\mathcal{N}_i)$.* ■

Corollary 3.2.17 *Let $\mathcal{N}$ be any difference $\mathbb{C}((y_q^{-1}))$-module. For any difference $\mathbb{C}((y_q^{-1}))$-submodule $\mathcal{N}_1 \subset \mathcal{N}$, we have $\mathcal{S}_\omega(\mathcal{N}_1) \subset \mathcal{S}_\omega(\mathcal{N})$. For any difference $\mathbb{C}((y_q^{-1}))$-quotient module $\mathcal{N}_2$ of $\mathcal{N}$, $\mathcal{S}_\omega(\mathcal{N}_2)$ is a quotient of $\mathcal{S}_\omega(\mathcal{N})$* ■

3.3 Good Filtered Bundles of Formal Difference Modules

3.3.1 Filtered Bundles over $\mathbb{C}((y_q^{-1}))$-Modules

Let $\mathcal{E}$ be any $\mathbb{C}((y_q^{-1}))$-module of finite rank. A filtered bundle over $\mathcal{E}$ is defined to be an increasing sequence $\mathcal{P}_*\mathcal{E} = \big(\mathcal{P}_a\mathcal{E} \,\big|\, a \in \mathbb{R}\big)$ of $\mathbb{C}[\![y_q^{-1}]\!]$-lattices of $\mathcal{E}$ such that (i) for any $a \in \mathbb{R}$, there exists $\epsilon > 0$ such that $\mathcal{P}_a\mathcal{E} = \mathcal{P}_{a+\epsilon}\mathcal{E}$, (ii) $\mathcal{P}_{a+n}\mathcal{E} = y_q^n\mathcal{P}_a\mathcal{E}$ for any $a \in \mathbb{R}$ and $n \in \mathbb{Z}$. This is the formal version of the notion of filtered bundle in Sect. 2.11.1. We set

$$\mathrm{Gr}_a^{\mathcal{P}}(\mathcal{E}) := \mathcal{P}_a(\mathcal{E})\big/\mathcal{P}_{<a}(\mathcal{E}), \tag{3.8}$$

where $\mathcal{P}_{<a}(\mathcal{E}) = \sum_{b<a} \mathcal{P}_b(\mathcal{E})$. A frame $\boldsymbol{v}$ of $\mathcal{P}_a\mathcal{E}$ is called compatible if there exists a decomposition $\boldsymbol{v} = \coprod_{a-1<b\leq a} \boldsymbol{v}_b$ such that $\boldsymbol{v}_b$ is a tuple of elements of $\mathcal{P}_b\mathcal{E}$ and induces a base of $\mathrm{Gr}^{\mathcal{P}}_b(\mathcal{E})$. The parabolic degrees $\deg^{\mathcal{P}}(s)$ for $s \in \mathcal{E}$ are also defined as in (2.54).

3.3.1.1 G-Equivariance

Let G be a finite group acting on the field $\mathbb{C}(\!(y_q^{-1})\!)$ such that $g^*(y_q^{-1}) \in y_q^{-1}\mathbb{C}[\![y_q^{-1}]\!]$ for any $g \in G$. A $\mathbb{C}(\!(y_q^{-1})\!)$-module $\mathcal{E}$ is called G-equivariant if $\mathcal{E}$ is a G-representation over $\mathbb{C}$ such that $g^*(fv) = g^*(f) \cdot g^*(v)$ for any $g \in G$. We say that a filtered bundle $\mathcal{P}_*\mathcal{E}$ over $\mathcal{E}$ is G-equivariant if each $\mathcal{P}_a\mathcal{E}$ $(a \in \mathbb{R})$ is preserved by the G-action. Because G is finite, the following lemma is obvious.

Lemma 3.3.1 *If $\mathcal{E}$ and $\mathcal{P}_*\mathcal{E}$ are G-equivariant, then there exist G-invariant subspaces $H_b \subset \mathcal{P}_b\mathcal{E}$ $(b \in \mathbb{R})$ such that the natural morphisms $\mathcal{P}_b\mathcal{E} \longrightarrow \mathrm{Gr}^{\mathcal{P}}_b(\mathcal{E})$ induce isomorphisms of the G-representations $H_b \simeq \mathrm{Gr}^{\mathcal{P}}_b(\mathcal{E})$.* ■

As in the case of Corollary 2.11.2, we obtain the following.

Corollary 3.3.2 *Let $\mathrm{Gr}^{\mathcal{P}}_b(\mathcal{E}) = \bigoplus_{i=1}^{m(b)} V_{b,i}$ $(a-1 < b \leq a)$ be decompositions of G-representations. Then, there exists a G-equivariant decomposition $\mathcal{P}_*\mathcal{E} = \bigoplus_{a-1<b\leq a} \bigoplus_{i=1}^{m(b)} \mathcal{P}_*\mathcal{E}_{b,i}$ such that the following holds.*

- *We have $\mathrm{Gr}^{\mathcal{P}}_c(\mathcal{E}_{b,i}) = 0$ for $a-1 < c \leq a$ unless $c = b$, and $\mathrm{Gr}^{\mathcal{P}}_b(\mathcal{E}_{b,i}) = V_{b,i}$.* ■

Corollary 3.3.3 *If G is cyclic, there exists a compatible frame $\boldsymbol{v}$ of $\mathcal{P}_a\mathcal{E}$ such that $g^*(v_i) = g^{n(i)}v_i$ for some $n(i) \in \mathbb{Z}$.* ■

3.3.1.2 Submodules, Quotient Modules and Splittings

Let $\mathcal{E}$ be a G-equivariant $\mathbb{C}(\!(y_q^{-1})\!)$-module equipped with a G-equivariant filtered bundle $\mathcal{P}_*\mathcal{E}$. For any G-equivariant $\mathbb{C}(\!(y_q^{-1})\!)$-submodule $\mathcal{E}' \subset \mathcal{E}$, we set $\mathcal{P}_a(\mathcal{E}') = \mathcal{P}_a(\mathcal{E}) \cap \mathcal{E}'$ $(a \in \mathbb{R})$ and we obtain the G-equivariant induced filtered bundle over $\mathcal{E}'$. Put $\mathcal{E}'' = \mathcal{E}/\mathcal{E}'$. Let $\mathcal{P}_a(\mathcal{E}'')$ $(a \in \mathbb{R})$ denote the image of $\mathcal{P}_a(\mathcal{E}) \longrightarrow \mathcal{E}''$, and we obtain the G-equivariant induced filtered bundle over $\mathcal{E}''$. The following lemma is similar to Lemma 2.11.5.

Lemma 3.3.4 *There exists a G-equivariant splitting $\mathcal{E}'' \longrightarrow \mathcal{E}$ which induces $\mathcal{P}_*\mathcal{E} \simeq \mathcal{P}_*\mathcal{E}' \oplus \mathcal{P}_*\mathcal{E}''$.* ■

3.3.1.3 Basic Functoriality

Let $\mathcal{E}_i$ $(i = 1, 2)$ be $\mathbb{C}(\!(y_q^{-1})\!)$-modules of finite rank. Let $\mathcal{P}_*\mathcal{E}_i$ be filtered bundles over $\mathcal{E}_i$. By setting $\mathcal{P}_a(\mathcal{E}_1 \oplus \mathcal{E}_2) = \mathcal{P}_a\mathcal{E}_1 \oplus \mathcal{P}_a\mathcal{E}_2$, we obtain the filtered bundle

$\mathcal{P}_*(\mathcal{E}_1 \oplus \mathcal{E}_2)$ over $\mathcal{E}_1 \oplus \mathcal{E}_2$, which is denoted by $\mathcal{P}_*(\mathcal{E}_1) \oplus \mathcal{P}_*(\mathcal{E}_2)$. By setting $\mathcal{P}_a(\mathcal{E}_1 \otimes \mathcal{E}_2) = \sum_{b+c \leq a} \mathcal{P}_b(\mathcal{E}_1) \otimes \mathcal{P}_c(\mathcal{E}_2)$, we obtain the filtered bundle $\mathcal{P}_*(\mathcal{E}_1 \otimes \mathcal{E}_2)$ over $\mathcal{E}_1 \otimes \mathcal{E}_2$, which is denoted by $\mathcal{P}_*(\mathcal{E}_1) \otimes \mathcal{P}_*(\mathcal{E}_2)$. By setting

$$\mathcal{P}_a \operatorname{Hom}(\mathcal{E}_1, \mathcal{E}_2) = \left\{ f \in \operatorname{Hom}(\mathcal{E}_1, \mathcal{E}_2) \,\middle|\, f(\mathcal{P}_b \mathcal{E}_1) \subset \mathcal{P}_{b+a} \mathcal{E}_2 \ (\forall b \in \mathbb{R}) \right\},$$

we obtain the filtered bundle $\mathcal{P}_* \operatorname{Hom}(\mathcal{E}_1, \mathcal{E}_2)$ over $\operatorname{Hom}(\mathcal{E}_1, \mathcal{E}_2)$, which is denoted by $\operatorname{Hom}(\mathcal{P}_*\mathcal{E}_1, \mathcal{P}_*\mathcal{E}_2)$.

Note that $\mathbb{C}(\!(y_q^{-1})\!)$ is equipped with the canonical filtered bundle $\mathcal{P}_*\mathbb{C}(\!(y_q^{-1})\!)$ defined by $\mathcal{P}_a(\mathbb{C}(\!(y_q^{-1})\!)) = y_q^{[a]}\mathbb{C}[[y_q^{-1}]]$, where $[a] := \max\{n \in \mathbb{Z} \,|\, n \leq a\}$. For a filtered bundle $\mathcal{P}_*\mathcal{E}$, the dual $(\mathcal{P}_*\mathcal{E})^\vee$ is defined as $\operatorname{Hom}(\mathcal{P}_*\mathcal{E}, \mathcal{P}_*(\mathbb{C}(\!(y_q^{-1})\!)))$.

Remark 3.3.5 If $\mathcal{P}_*\mathcal{E}_i$ are G-equivariant, the induced filtered bundles are also G-equivariant. Because $\mathcal{P}_*\mathbb{C}(\!(y_q^{-1})\!)$ is naturally G-equivariant, $\mathcal{P}_*\mathcal{E}^\vee$ is G-equivariant if $\mathcal{P}_*\mathcal{E}$ is G-equivariant. ■

3.3.1.4 Pull Back

Let $p \in q\mathbb{Z}_{>0}$. For any $\mathbb{C}(\!(y_q^{-1})\!)$-module $\mathcal{E}$ of finite rank, we set $\varphi_{q,p}^*(\mathcal{E}) := \mathbb{C}(\!(y_p^{-1})\!) \otimes_{\mathbb{C}(\!(y_q^{-1})\!)} \mathcal{E}$. (See also Sect. 3.4.2 below for this notation.) Let $\mathcal{P}_*\mathcal{E}$ be a filtered bundle over a $\mathbb{C}(\!(y_q^{-1})\!)$-module $\mathcal{E}$ of finite rank. For any $a \in \mathbb{R}$, we define

$$\mathcal{P}_a(\varphi_{q,p}^*\mathcal{E}) = \sum_{n+b(p/q) \leq a} y_p^n \mathcal{P}_b(\mathcal{E}) \otimes_{\mathbb{C}[\![y_q^{-1}]\!]} \mathbb{C}[\![y_p^{-1}]\!].$$

Thus, we obtain a filtered bundle $\mathcal{P}_*\varphi_{q,p}^*\mathcal{E}$ over $\varphi_{q,p}^*\mathcal{E}$, which we denote by $\varphi_{q,p}^*(\mathcal{P}_*\mathcal{E})$, and called the pull back of $\mathcal{P}_*\mathcal{E}$. The filtered bundle $\mathcal{P}_*\mathcal{E}$ is naturally $\mathrm{Gal}_{q,p}$-equivariant. (See Sect. 3.2.1 for $\mathrm{Gal}_{q,p}$.) By the induced $\mathrm{Gal}_{q,p}$-action on $\mathrm{Gr}_a^{\mathcal{P}}(\mathcal{E})$, we obtain the canonical decomposition

$$\mathrm{Gr}_a^{\mathcal{P}}(\varphi_{q,p}^*\mathcal{E}) = \bigoplus_{m=0}^{(p/q)-1} \mathcal{G}_m \,\mathrm{Gr}_a^{\mathcal{P}}(\varphi_{q,p}^*\mathcal{E}), \tag{3.9}$$

where $\alpha \in \mathrm{Gal}_{q,p}$ acts on $\mathcal{G}_m \,\mathrm{Gr}_a^{\mathcal{P}}(\mathcal{E})$ by the multiplication of α^m. There exists the natural isomorphism

$$\mathcal{G}_0 \,\mathrm{Gr}_a^{\mathcal{P}}(\varphi_{q,p}^*\mathcal{E}) \simeq \mathrm{Gr}_{aq/p}^{\mathcal{P}}(\mathcal{E}). \tag{3.10}$$

We can check the following lemma by using Lemma 3.3.4.

Lemma 3.3.6 *Let $\mathcal{E}$ and $\mathcal{P}_*\mathcal{E}$ be as above. Let $\mathcal{E}' \subset \mathcal{E}$ be a $\mathbb{C}(\!(y_q^{-1})\!)$-submodule. We set $\mathcal{E}'' = \mathcal{E}/\mathcal{E}''$. We have the induced filtered bundles $\mathcal{P}_*\mathcal{E}'$ and $\mathcal{P}_*\mathcal{E}''$ over $\mathcal{E}'$ and*

$\mathcal{E}''$, respectively. Then, $\varphi^*_{q,p}(\mathcal{P}_*\mathcal{E}')$ (resp. $\varphi^*_{q,p}(\mathcal{P}_*\mathcal{E}'')$) is equal to the filtered bundle over $\varphi^*_{q,p}\mathcal{E}'$ (resp. $\varphi^*_{q,p}\mathcal{E}''$) induced by $\varphi^*_{q,p}(\mathcal{P}_*\mathcal{E})$ and the inclusion $\varphi^*_{q,p}(\mathcal{E}') \subset \varphi^*_{q,p}(\mathcal{E})$ (resp. the projection $\varphi^*_{q,p}(\mathcal{E}) \longrightarrow \varphi^*_{q,p}(\mathcal{E}'')$). ■

We can check the following lemma by using compatible frames as in the case of Lemma 2.11.8.

Lemma 3.3.7 *Let $\mathcal{E}_i$ be $\mathbb{C}((y_q^{-1}))$-modules of finite rank equipped with a filtered bundle $\mathcal{P}_*\mathcal{E}_i$. Then, there exist natural isomorphisms $\varphi^*_{q,p}(\mathcal{P}_*\mathcal{E}_1 \oplus \mathcal{P}_*\mathcal{E}_2) \simeq \varphi^*_{q,p}(\mathcal{P}_*\mathcal{E}_1) \oplus \varphi^*_{q,p}(\mathcal{P}_*\mathcal{E}_2)$, $\varphi^*_{q,p}(\mathcal{P}_*\mathcal{E}_1 \otimes \mathcal{P}_*\mathcal{E}_2) \simeq \varphi^*_{q,p}(\mathcal{P}_*\mathcal{E}_1) \otimes \varphi^*_{q,p}(\mathcal{P}_*\mathcal{E}_2)$, and $\varphi^*_{q,p}\operatorname{Hom}\big(\mathcal{P}_*\mathcal{E}_1, \mathcal{P}_*\mathcal{E}_2\big) \simeq \operatorname{Hom}\big(\varphi^*_{q,p}(\mathcal{P}_*\mathcal{E}_1), \varphi^*_{q,p}(\mathcal{P}_*\mathcal{E}_2)\big)$.* ■

3.3.1.5 Push-Forward

Let $p \in q\mathbb{Z}_{\geq 1}$. Any $\mathbb{C}((y_p^{-1}))$-module $\mathcal{E}_1$ of finite rank is naturally a $\mathbb{C}((y_q^{-1}))$-module of finite rank, which is denoted by $\varphi_{q,p*}(\mathcal{E}_1)$. Let $\mathcal{P}_*\mathcal{E}_1$ be a filtered bundle over a $\mathbb{C}((y_p^{-1}))$-module $\mathcal{E}_1$ of finite rank. For any $a \in \mathbb{R}$, we obtain the following $\mathbb{C}[\![y_q^{-1}]\!]$-submodule

$$\mathcal{P}_a\big(\varphi_{q,p*}(\mathcal{E}_1)\big) = \mathcal{P}_{aq/p}\mathcal{E}_1 \subset \varphi_{q,p*}(\mathcal{E}_1).$$

Thus, we obtain a filtered bundle $\mathcal{P}_*(\varphi_{q,p*}\mathcal{E}_1)$ over $\varphi_{q,p*}\mathcal{E}_1$, which is denoted by $\varphi_{q,p*}(\mathcal{P}_*\mathcal{E}_1)$, and called the push-forward of $\mathcal{P}_*\mathcal{E}_1$.

Lemma 3.3.8 *Let $\mathcal{E}_1$ and $\mathcal{P}_*\mathcal{E}_1$ be as above. Let $\mathcal{E}'_1 \subset \mathcal{E}_1$ be a $\mathbb{C}((y_q^{-1}))$-submodule. We set $\mathcal{E}''_1 := \mathcal{E}_1/\mathcal{E}'_1$. They are equipped with the induced filtered bundles $\mathcal{P}_*\mathcal{E}'_1$ and $\mathcal{P}_*\mathcal{E}''_1$, respectively. Then, $\varphi_{q,p*}(\mathcal{P}_*\mathcal{E}'_1)$ (resp. $\varphi_{q,p*}(\mathcal{P}_*\mathcal{E}''_1)$) is equal to the filtered bundle induced by $\varphi_{q,p*}(\mathcal{P}_*\mathcal{E}_1)$ and the inclusion $\varphi_{q,p*}(\mathcal{E}'_1) \subset \varphi_{q,p*}(\mathcal{E}_1)$ (resp. the projection $\varphi_{q,p*}(\mathcal{E}_1) \longrightarrow \varphi_{q,p*}(\mathcal{E}''_1)$).* ■

Let $\mathcal{E}$ be a $\mathbb{C}((y_q^{-1}))$-module of finite rank equipped with a filtered bundle $\mathcal{P}_*\mathcal{E}$. Let $\mathcal{E}_1$ and $\mathcal{P}_*\mathcal{E}_1$ be as above. We obtain the following lemma by using a compatible frame of $\mathcal{P}_*\mathcal{E}$.

Lemma 3.3.9 *The natural isomorphism $\varphi_{q,p*}(\mathcal{E}_1 \otimes \varphi^*_{q,p}\mathcal{E}) \simeq \varphi_{q,p*}(\mathcal{E}_1) \otimes \mathcal{E}$ induces*

$$\varphi_{q,p*}(\mathcal{P}_*\mathcal{E}_1 \otimes \varphi^*_{q,p}\mathcal{P}_*\mathcal{E}) \simeq \varphi_{q,p*}(\mathcal{P}_*\mathcal{E}_1) \otimes \mathcal{P}_*\mathcal{E}.$$

The natural isomorphism $\varphi_{q,p*}(\operatorname{Hom}(\varphi^*_{q,p}\mathcal{E}, \mathcal{E}_1)) \simeq \operatorname{Hom}(\mathcal{E}, \varphi_{q,p*}\mathcal{E}_1)$ induces

$$\varphi_{q,p*}(\operatorname{Hom}(\varphi^*_{q,p}\mathcal{P}_*\mathcal{E}, \mathcal{P}_*\mathcal{E}_1)) \simeq \operatorname{Hom}(\mathcal{P}_*\mathcal{E}, \varphi_{q,p*}\mathcal{P}_*\mathcal{E}_1).$$

■

3.3.1.6 Descent

Let $\mathcal{E}_1$ be a $\mathbb{C}((y_p^{-1}))$-module of finite rank equipped with a filtered bundle $\mathcal{P}_*\mathcal{E}_1$. Suppose that $\mathcal{E}_1$ and $\mathcal{P}_*\mathcal{E}_1$ are $\mathrm{Gal}_{q,p}$-equivariant. We obtain the $\mathbb{C}((y_q^{-1}))$-module of finite rank $\mathcal{E}_2$ as the $\mathrm{Gal}_{q,p}$-invariant part of $\varphi_{q,p*}(\mathcal{E}_1)$, equipped with the induced filtered bundle $\mathcal{P}_*\mathcal{E}_2$. It is called the descent of $\mathcal{P}_*\mathcal{E}_1$. The following lemma is similar to Lemma 2.11.11.

Lemma 3.3.10 *$\varphi_{q,p}^*\mathcal{P}_*\mathcal{E}_2$ is naturally isomorphic to $\mathcal{P}_*\mathcal{E}_1$.* ■

The following lemma is similar to Lemma 2.11.12.

Lemma 3.3.11 *Let $\mathcal{E}_1'$ be a $\mathrm{Gal}_{q,p}$-equivariant $\mathbb{C}((y_p^{-1}))$-submodule of $\mathcal{E}_1$. We set $\mathcal{E}_1'' := \mathcal{E}_1/\mathcal{E}_1'$, which is naturally $\mathrm{Gal}_{q,p}$-equivariant. We obtain the $\mathbb{C}((y_q^{-1}))$-submodule $\mathcal{E}_2' \subset \mathcal{E}_2$ as the descent of $\mathcal{E}_1'$. We also obtain the $\mathbb{C}((y_q^{-1}))$-quotient module $\mathcal{E}_2 \longrightarrow \mathcal{E}_2''$ as the descent of $\mathcal{E}_1''$. Then, the decent $\mathcal{P}_*\mathcal{E}_2'$ of $\mathcal{P}_*\mathcal{E}_1'$ is equal to the filtered bundle over $\mathcal{E}_2'$ induced by $\mathcal{P}_*\mathcal{E}_2$ with the inclusion $\mathcal{E}_2' \subset \mathcal{E}_2$. The descent of $\mathcal{P}_*\mathcal{E}_2''$ of $\mathcal{P}_*\mathcal{E}_1''$ is equal to the filtered bundle over $\mathcal{E}_2''$ induced by $\mathcal{P}_*\mathcal{E}_2$ with the projection $\mathcal{E}_2 \longrightarrow \mathcal{E}_2''$.* ■

The following lemma is similar to Lemma 2.11.13, and easy to check.

Lemma 3.3.12 *For a filtered bundle $\mathcal{P}_*\mathcal{E}$ over a $\mathbb{C}((y_q^{-1}))$-module $\mathcal{E}$, the pull back $\varphi_{q,p}^*(\mathcal{P}_*\mathcal{E})$ is naturally equipped with the $\mathrm{Gal}_{q,p}$-action, and the descent of $\varphi_{q,p}^*(\mathcal{P}_*\mathcal{E})$ is naturally isomorphic to $\mathcal{P}_*\mathcal{E}$.* ■

3.3.2 Good Filtered Bundles over Formal Difference Modules

Let $(\mathcal{N}, \Phi^*)$ be a difference $\mathbb{C}((y_q^{-1}))$-module. We set $S(q) := \big\{ \sum_{j=1}^{q-1} \mathfrak{b}_j y_q^{-j} \,\big|\, \mathfrak{b}_j \in \mathbb{C} \big\}$.

Definition 3.3.13 We say that $(\mathcal{N}, \Phi^*)$ is unramified if there exists a decomposition of difference $\mathbb{C}((y_q^{-1}))$-modules

$$\mathcal{N} = \bigoplus_{\ell \in \mathbb{Z}} \bigoplus_{\alpha \in \mathbb{C}^*} \bigoplus_{\mathfrak{b} \in S(q)} \mathcal{N}_{\ell,\alpha,\mathfrak{b}} \tag{3.11}$$

such that each $\mathcal{N}_{\ell,\alpha,\mathfrak{b}}$ has a lattice $\mathcal{L}_{\ell,\alpha,\mathfrak{b}}$ satisfying

$$(y_q^{\ell}\alpha^{-1}\Phi - (1+\mathfrak{b})\,\mathrm{id})\mathcal{L}_{\ell,\alpha,\mathfrak{b}} \subset y_q^{-q}\mathcal{L}_{\ell,\alpha,\mathfrak{b}}.$$

Such a lattice $\bigoplus \mathcal{L}_{\ell,\alpha,\mathfrak{b}}$ is called an unramifiedly good lattice of $(\mathcal{N}, \Phi^*)$. ■

Definition 3.3.14 If $(\mathcal{N}, \Phi^*)$ is unramified, a filtered bundle $\mathcal{P}_*\mathcal{N}$ over $\mathcal{N}$ is called unramifiedly good for $(\mathcal{N}, \Phi^*)$ if each $\mathcal{P}_a\mathcal{N}$ is unramifiedly good lattices of $(\mathcal{N}, \Phi^*)$. Such an object $(\mathcal{P}_*\mathcal{N}, \Phi^*)$ is called an unramifiedly good filtered difference $\mathbb{C}(\!(y_q^{-1})\!)$-module. ■

By Propositions 3.2.1, 3.2.12, and Theorem 3.2.15, the following holds.

Proposition 3.3.15 *For any $\mathbb{C}(\!(y_q^{-1})\!)$-difference module $(\mathcal{N}, \Phi^*)$, there exist $p \in q\mathbb{Z}_{>0}$ such that $\varphi_{q,p}^*(\mathcal{N}, \Phi)$ is unramified.* ■

Definition 3.3.16 A filtered bundle $\mathcal{P}_*\mathcal{N}$ over $\mathcal{N}$ is called good with respect to Φ^* if there exists $p \in q\mathbb{Z}_{>0}$ such that $\varphi_{q,p}^*(\mathcal{P}_*\mathcal{N})$ is unramifiedly good for $(\mathcal{N}, \Phi^*)$. Such $(\mathcal{P}_*\mathcal{N}, \Phi^*)$ is called a good filtered difference $\mathbb{C}(\!(y_q^{-1})\!)$-module. ■

The following lemma is obvious.

Lemma 3.3.17 *Let $(\mathcal{P}_*\mathcal{N}, \Phi^*)$ be a good filtered difference module. Then, the filtered bundle $\mathcal{P}_*\mathcal{N}$ is compatible with the slope decomposition in Theorem 3.2.15, i.e., $\mathcal{P}_*\mathcal{N} = \bigoplus_{\omega\in\mathbb{Q}} \mathcal{P}_*(\mathcal{S}_\omega\mathcal{N})$. Moreover, for any $p \in Z(q,\omega)$, each $\mathcal{P}_a\mathcal{S}_\omega\mathcal{N}$ is a $y_q^{p\omega}(\Phi^*)^{p/q}$-invariant lattice.* ■

Lemma 3.3.18 *Let $(\mathcal{P}_*\mathcal{N}, \Phi^*)$ be a good filtered difference $\mathbb{C}(\!(y_q^{-1})\!)$-module. If $(\mathcal{N}, \Phi^*)$ is unramified, then $(\mathcal{P}_*\mathcal{N}, \Phi^*)$ is an unramifiedly good filtered difference $\mathbb{C}(\!(y_q^{-1})\!)$-module.*

Proof Because $(\mathcal{N}, \Phi^*)$ is assumed to be unramified, there exists a decomposition (3.11) of $\mathcal{N}$ as in Definition 3.3.13. There exists $p \in q\mathbb{Z}_{>0}$ such that $\varphi_{q,p}^*(\mathcal{P}_*\mathcal{N}, \Phi^*)$ is unramifiedly good. There exists a decomposition

$$\varphi_{q,p}^*(\mathcal{N}) = \bigoplus_{\ell\in\mathbb{Z}} \bigoplus_{\alpha\in\mathbb{C}^*} \bigoplus_{\mathfrak{b}\in S(p)} (\varphi_{q,p}^*(\mathcal{N}))_{\ell,\alpha,\mathfrak{b}} \tag{3.12}$$

as in Definition 3.3.13. It is easy to observe that $\varphi_{q,p}^*\mathcal{N}_{\ell,\alpha,\mathfrak{b}} = \mathcal{N}_{\ell(p/q),\alpha,\mathfrak{b}}$. The decomposition (3.12) is preserved by the natural action of $\mathrm{Gal}_{q,p}$. Because the filtered bundle $\varphi^*(\mathcal{P}_*\mathcal{N})$ is compatible with the decomposition (3.12), we obtain that the filtered bundle $\mathcal{P}_*\mathcal{N}$ is compatible with the decomposition (3.11), i.e., $\mathcal{P}_*\mathcal{N} = \bigoplus \mathcal{P}_*\mathcal{N}_{\ell,\alpha,\mathfrak{b}}$. Because

$$y_p^p\big(y_p^{p\ell/q}\alpha^{-1}\Phi^* - (1+\mathfrak{b})\,\mathrm{id}\big)\mathcal{P}_b(\varphi_{q,p}^*\mathcal{N})_{p\ell/q,\alpha,\mathfrak{b}} \subset \mathcal{P}_b(\varphi_{q,p}^*\mathcal{N})_{p\ell/q,\alpha,\mathfrak{b}},$$

for any $b \in \mathbb{R}$, we easily obtain $\big(y_q^\ell\alpha^{-1}\Phi^* - (1+\mathfrak{b})\,\mathrm{id}\big)\mathcal{P}_b\mathcal{N}_{\ell,\alpha,\mathfrak{b}} \subset y_q^{-q}\mathcal{P}_b\mathcal{N}_{\ell,\alpha,\mathfrak{b}}$. Hence, $\mathcal{P}_*\mathcal{N}$ is good with respect to Φ^*. ■

Lemma 3.3.19 *Let $(\mathcal{P}_*\mathcal{N}, \Phi^*)$ be a good filtered difference $\mathbb{C}(\!(y_q^{-1})\!)$-module. Let $\mathcal{N}' \subset \mathcal{N}$ be a difference $\mathbb{C}(\!(y_q^{-1})\!)$-submodule. We set $\mathcal{N}'' = \mathcal{N}/\mathcal{N}'$. They are equipped with the induced filtered bundle $\mathcal{P}_*\mathcal{N}'$ and $\mathcal{P}_*\mathcal{N}''$, respectively. Then, $(\mathcal{P}_*\mathcal{N}', \Phi^*)$ and $(\mathcal{P}_*\mathcal{N}'', \Phi^*)$ are good filtered difference $\mathbb{C}(\!(y_q^{-1})\!)$-modules.*

Proof We explain the proof for $\mathcal{N}'$. The other case can be argued similarly. By Lemma 3.3.6 and Lemma 3.3.11, it is enough to study the case where both $(\mathcal{N}, \Phi^*)$ and $(\mathcal{N}', \Phi^*)$ are unramified, i.e., there exist decompositions

$$\mathcal{N} = \bigoplus_{\ell\in\mathbb{Z}} \bigoplus_{\alpha\in\mathbb{C}^*} \bigoplus_{\mathfrak{b}\in S(q)} \mathcal{N}_{\ell,\alpha,\mathfrak{b}}, \tag{3.13}$$

$$\mathcal{N}' = \bigoplus_{\ell\in\mathbb{Z}} \bigoplus_{\alpha\in\mathbb{C}^*} \bigoplus_{\mathfrak{b}\in S(q)} \mathcal{N}'_{\ell,\alpha,\mathfrak{b}} \tag{3.14}$$

as in (3.11). By Lemma 3.2.3 and Lemma 3.2.11, the decompositions (3.13) and (3.14) are compatible with the inclusion $\mathcal{N}' \subset \mathcal{N}$. Hence, the induced filtration $\mathcal{P}_*\mathcal{N}'$ is compatible with the decomposition (3.14). It is easy to check that each $\mathcal{P}_a(\mathcal{N}')$ is an unramifiedly good lattice. ■

Lemma 3.3.20 *If* $(\mathcal{P}_*\mathcal{N}_i, \Phi^*)$ $(i = 1, 2)$ *be good filtered difference* $\mathbb{C}((y_q^{-1}))$*-modules, the induced objects* $(\mathcal{P}_*(\mathcal{N}_1) \oplus \mathcal{P}_*(\mathcal{N}_2), \Phi^*)$, $(\mathcal{P}_*(\mathcal{N}_1) \otimes \mathcal{P}_*(\mathcal{N}_2), \Phi^*)$ *and* $(\mathrm{Hom}(\mathcal{P}_*\mathcal{N}_1, \mathcal{P}_*\mathcal{N}_2), \Phi^*)$ *are also good filtered difference* $\mathbb{C}((y_q^{-1}))$*-modules.*

Proof By Lemma 3.3.7 and Lemma 3.3.12, it is enough to consider the case where $(\mathcal{P}_*\mathcal{N}_i, \Phi^*)$ are unramified. Then, it is easy to check. ■

3.3.3 The Induced Endomorphisms on the Graded Pieces

Let $(\mathcal{P}_*\mathcal{N}, \Phi^*)$ be a good filtered difference $\mathbb{C}((y_q^{-1}))$-module. If $\mathcal{N}$ is unramified, there exists the decomposition (3.11) as in Definition 3.3.13. By the condition, $y_q^q(y_q^\ell\alpha^{-1}\Phi^* - (1 + \mathfrak{b})\,\mathrm{id})$ induces the $\mathbb{C}$-linear endomorphism $\mathfrak{F}_{a,\ell,\alpha,\mathfrak{b}}$ of $\mathrm{Gr}^{\mathcal{P}}_a(\mathcal{N}_{\ell,\alpha,\mathfrak{b}})$. We obtain the endomorphism $\bigoplus_{\ell,\alpha,\mathfrak{b}} \mathfrak{F}_{a,\ell,\alpha,\mathfrak{b}}$ on $\mathrm{Gr}^{\mathcal{P}}_a(\mathcal{N})$ which we denote by $\mathrm{Res}(\Phi^*)$. We obtain the monodromy weight filtration W on $\mathrm{Gr}^{\mathcal{P}}_a(\mathcal{N})$ with respect to the nilpotent part of $\mathrm{Res}(\Phi^*)$. It is compatible with the decomposition $\mathrm{Gr}^{\mathcal{P}}_a(\mathcal{N}) = \bigoplus_{\ell,\alpha,\mathfrak{b}} \mathrm{Gr}^{\mathcal{P}}_a(\mathcal{N}_{\ell,\alpha,\mathfrak{b}})$.

In general, there exists $p \in q\mathbb{Z}_{>0}$ such that $\varphi^*_{q,p}(\mathcal{N}, \Phi^*)$ is unramified. We obtain the endomorphism $\mathrm{Res}(\Phi^*)$ on $\mathrm{Gr}^{\mathcal{P}}_a(\varphi^*_{q,p}\mathcal{N})$.

Lemma 3.3.21 $\mathrm{Res}(\Phi^*)$ *is equivariant with respect to the natural* $\mathrm{Gal}_{q,p}$*-action on* $\mathrm{Gr}^{\mathcal{P}}_a(\varphi^*_{q,p}\mathcal{N})$.

Proof It follows from Lemma 3.2.4 and Lemma 3.2.14. ■

By Lemma 3.3.21, $\mathrm{Res}(\Phi^*)$ is compatible with the canonical decomposition (3.9). By using the identification $\mathrm{Gr}^{\mathcal{P}}_a(\mathcal{N}) \simeq \mathcal{G}_0\,\mathrm{Gr}^{\mathcal{P}}_{pa/q}(\varphi^*_{q,p}\mathcal{N})$, we obtain an endomorphism on $\mathrm{Gr}^{\mathcal{P}}_a(\mathcal{N})$, denoted by $\mathrm{Res}(\Phi^*)$. We also obtain the monodromy weight filtration W on $\mathrm{Gr}^{\mathcal{P}}_a(\mathcal{N})$ with respect to the nilpotent part of $\mathrm{Res}(\Phi^*)$.

3.4 Geometrization of Formal Difference Modules

3.4.1 *Ringed Spaces*

A ringed space is defined to be a topological space X equipped with a sheaf of algebras A_X on X, called the structure sheaf. In this monograph, the structure sheaf of any ringed space is assumed to be commutative. A morphism of ringed spaces $F : (X, A_X) \longrightarrow (Y, A_Y)$ is a continuous map $F : X \longrightarrow Y$ with a morphism of sheaves of algebras $F^{-1}(A_Y) \longrightarrow A_X$. For an A_Y-module $\mathcal{F}$, we obtain an A_X-module $F^*(\mathcal{F}) := A_X \otimes_{F^{-1}A_Y} F^{-1}(\mathcal{F})$. If X is a subspace of Y and if $F^{-1}(A_Y) = A_X$, $F^*(\mathcal{F})$ is denoted as $\mathcal{F}_{|X}$. Even if $F^{-1}(A_Y) \neq A_X$, we shall often denote $F^*(\mathcal{F})$ as $\mathcal{F}_{|X}$ to simplify the description if there is no risk of confusion.

For an A_X-module $\mathcal{F}'$, we obtain an A_Y-module $F_*(\mathcal{F}')$ as the standard push-forward.

3.4.2 *Some Formal Spaces*

For any $q \in \mathbb{Z}_{\geq 1}$, let $\widehat{\infty}_{y,q}$ denote the ringed space which consists of the point $\infty_{y,q}$ with the ring $\mathcal{O}_{\widehat{\infty}_{y,q}} := \mathbb{C}[\![y_q^{-1}]\!]$. In other words, $\widehat{\infty}_{y,q}$ is the completion of the projective line $\mathbb{P}^1_{y_q}$ along $y_q = \infty$. We also set $\mathcal{O}_{\widehat{\infty}_{y,q}}(*\infty_{y,q}) := \mathbb{C}(\!(y_q^{-1})\!)$. If $p \in q\mathbb{Z}_{\geq 1}$, we obtain the naturally defined morphism of the ringed spaces $\varphi_{q,p} : \widehat{\infty}_{y,p} \longrightarrow \widehat{\infty}_{y,q}$ induced by the inclusions $\varphi_{q,p}^* : \mathbb{C}[\![y_q^{-1}]\!] \longrightarrow \mathbb{C}[\![y_p^{-1}]\!]$. There also exists the natural ring homomorphism $\varphi_{q,p}^* : \mathcal{O}_{\widehat{\infty}_{y,q}}(*\infty) \longrightarrow \mathcal{O}_{\widehat{\infty}_{y,p}}(*\infty)$. The space $\widehat{\infty}_{y,1}$ is also denoted by $\widehat{\infty}_y$.

We set $H^{\mathrm{cov}}_{\infty,q} := \mathbb{R}_t \times \{\infty_{y,q}\}$. Let $\mathcal{O}_{\widehat{H}^{\mathrm{cov}}_{\infty,q}}$ (resp. $\mathcal{O}_{\widehat{H}^{\mathrm{cov}}_{\infty,q}}(*H^{\mathrm{cov}}_{\infty,q})$) denote the sheaf of locally constant functions on $H^{\mathrm{cov}}_{\infty,q}$ to $\mathbb{C}[\![y_q^{-1}]\!]$ (resp. $\mathbb{C}(\!(y_q^{-1})\!)$). Let $\widehat{H}^{\mathrm{cov}}_{\infty,q}$ denote the ringed space which consists of the topological space $H^{\mathrm{cov}}_{\infty,q}$ with the sheaf of algebras $\mathcal{O}_{\widehat{H}^{\mathrm{cov}}_{\infty,q}}$.

Let $T \in \mathbb{R}_{>0}$. Let κ denote the natural $\mathbb{Z}$-action on $H^{\mathrm{cov}}_{\infty,q}$ induced by $\kappa_n(t) = t+nT$ $(n \in \mathbb{Z})$. We obtain the isomorphism $\kappa_n^*\mathcal{O}_{\widehat{H}^{\mathrm{cov}}_{\infty,q}} \simeq \mathcal{O}_{\widehat{H}^{\mathrm{cov}}_{\infty,q}}$ induced by $\kappa_n^*(y) = y+2\sqrt{-1}nT\lambda$. In this sense, the ringed space $\widehat{H}^{\mathrm{cov}}_{\infty,q}$ is equipped with a $\mathbb{Z}$-action κ. The sheaf $\mathcal{O}_{\widehat{H}^{\mathrm{cov}}_{\infty,q}}(*H^{\mathrm{cov}}_{\infty,q})$ is also naturally $\mathbb{Z}$-equivariant.

Let $H_{\infty,q}$ denote the quotient space of $H^{\mathrm{cov}}_{\infty,q}$ by the action κ. We obtain the sheaves $\mathcal{O}_{\widehat{H}_{\infty,q}}$ and $\mathcal{O}_{\widehat{H}_{\infty,q}}(*H_{\infty,q})$ as the descents of $\mathcal{O}_{\widehat{H}^{\mathrm{cov}}_{\infty,q}}$ and $\mathcal{O}_{\widehat{H}^{\mathrm{cov}}_{\infty,q}}(*H^{\mathrm{cov}}_{\infty,q})$, respectively. Let $\widehat{H}_{\infty,q}$ denote the ringed space obtained as the topological space $H_{\infty,q}$ with the sheaf of algebras $\mathcal{O}_{\widehat{H}_{\infty,q}}$.

The projections $\widehat{H}^{\mathrm{cov}}_{\infty,q} \longrightarrow \widehat{H}_{\infty,q}$ and $H^{\mathrm{cov}}_{\infty,q} \longrightarrow H_{\infty,q}$ are denoted by ϖ_q. We also denote $\widehat{H}^{\mathrm{cov}}_{\infty,q}$, $\widehat{H}_{\infty,q}$, etc., by $\widehat{H}^{\mathrm{cov}}_{\infty,q,T}$, $\widehat{H}_{\infty,q,T}$, etc., respectively, when we emphasize the dependence on T.

Remark 3.4.1 The space $\widehat{H}_\infty$ and $\widehat{H}_\infty^{\mathrm{cov}}$ are the formal completion of the spaces $\overline{\mathcal{M}}^\lambda$ and $\overline{M}^\lambda$ along H_∞^λ and $H_\infty^{\lambda\,\mathrm{cov}}$. (See Sect. 2.7 for the notation.) ■

The ringed space $\mathbb{R}$ with the sheaf of $\mathbb{C}$-valued locally constant functions is also denoted by $\mathbb{R}$. We set $S_T^1 := \mathbb{R}/T\mathbb{Z}$. The projection $\widehat{H}_{\infty,q}^{\mathrm{cov}} \longrightarrow \mathbb{R}$ induces a morphism $\pi_q : \widehat{H}_{\infty,q} \longrightarrow S_T^1$. It also induces an isomorphism $H_{\infty,q} \simeq S_T^1$. For $t \in S_T^1$, let $\widehat{H}_{\infty,q}\langle t\rangle$ denote the formal space obtained as the point set $\{t\} \subset H_{\infty,q} \simeq S_T^1$ with the ring $\mathcal{O}_{\widehat{H}_{\infty,q},t}$. For any $\mathcal{O}_{\widehat{H}_{\infty,q}}(*H_{\infty,q})$-module $\mathcal{E}$, let $\mathcal{E}_{|\widehat{H}_{\infty,q}\langle t\rangle}$ denote the pull back of $\mathcal{E}$ by the natural morphism $\widehat{H}_{\infty,q}\langle t\rangle \longrightarrow \widehat{H}_{\infty,q}$. Note that if we choose $\widetilde{t} \in \mathbb{R}$ which is mapped to $t \in S_T^1$, there exists the induced isomorphism $\{\widetilde{t}\} \times \widehat{\infty}_{y,q} \simeq \widehat{H}_{\infty,q}\langle t\rangle$. Hence, we may regard $\mathcal{E}_{|\widehat{H}_{\infty,q}\langle t\rangle}$ as $\mathbb{C}(\!(y_q^{-1})\!)$-module in a way depending on the lift $\widetilde{t}$.

Let $\mathcal{E}^{\mathrm{cov}}$ be a locally free $\mathcal{O}_{\widehat{H}_{\infty,q}^{\mathrm{cov}}}(*H_{\infty,q}^{\mathrm{cov}})$-module of finite rank. For any $t \in \mathbb{R}$, let $\mathcal{E}^{\mathrm{cov}}_{|\{t\}\times\widehat{\infty}_{y,q}}$ denote the pull back of $\mathcal{E}^{\mathrm{cov}}$ by $\{t\} \times \widehat{\infty}_{y,q} \longrightarrow \widehat{H}_{\infty,q}^{\mathrm{cov}}$. For any $t_1, t_2 \in \mathbb{R}$, there exists the natural isomorphism $\mathcal{E}^{\mathrm{cov}}_{|\{t_1\}\times\widehat{\infty}_{y,q}} \longrightarrow \mathcal{E}^{\mathrm{cov}}_{|\{t_2\}\times\widehat{\infty}_{y,q}}$.

3.4.3 Difference Modules and $\mathcal{O}_{\widehat{H}_{\infty,q}}(*H_{\infty,q})$-Modules

Let $\mathcal{E}$ be a locally free $\mathcal{O}_{\widehat{H}_{\infty,q}}(*H_{\infty,q})$-module of finite rank. We obtain $\mathcal{E}^{\mathrm{cov}} := \varpi_q^{-1}\mathcal{E}$ on $\widehat{H}_{\infty,q}^{\mathrm{cov}}$. There exists the isomorphism

$$\Pi_{T,0} : \mathcal{E}^{\mathrm{cov}}_{|\{0\}\times\widehat{\infty}_{y,q}} \longrightarrow \mathcal{E}^{\mathrm{cov}}_{|\{T\}\times\widehat{\infty}_{y,q}}$$

induced by the parallel transport. There also exists the natural identification

$$\mathcal{E}^{\mathrm{cov}}_{|\{T\}\times\widehat{\infty}_{y,q}} = \mathcal{E}_{|\widehat{H}_{\infty,q}\langle 0\rangle} = \mathcal{E}^{\mathrm{cov}}_{|\{0\}\times\widehat{\infty}_{y,q}}.$$

We define the difference operator $\Phi^* : \mathcal{E}_{|\widehat{H}_{\infty,q}\langle 0\rangle} \longrightarrow \mathcal{E}_{|\widehat{H}_{\infty,q}\langle 0\rangle}$ as the composite of the following morphisms:

$$\mathcal{E}_{|\widehat{H}_{\infty,q}\langle 0\rangle} = \mathcal{E}^{\mathrm{cov}}_{|\{0\}\times\widehat{\infty}_{y,q}} \xrightarrow{\Pi_{T,0}} \mathcal{E}^{\mathrm{cov}}_{|\{T\}\times\widehat{\infty}_{y,q}} = \mathcal{E}_{|\widehat{H}_{\infty,q}\langle 0\rangle}.$$

It is also equivalent to the following constructions. Let $H^0\big(H_{\infty,q}^{\mathrm{cov}}, \mathcal{E}^{\mathrm{cov}}\big)$ denote the space of the sections of $\mathcal{E}^{\mathrm{cov}}$ on $H_{\infty,q}^{\mathrm{cov}}$. We define Φ^* on $H^0\big(H_{\infty,q}^{\mathrm{cov}}, \mathcal{E}^{\mathrm{cov}}\big)$ as the composite of the following maps:

$$H^0\big(H_{\infty,q}^{\mathrm{cov}}, \mathcal{E}^{\mathrm{cov}}\big) \xrightarrow{\kappa_1^*} H^0\big(H_{\infty,q}^{\mathrm{cov}}, \kappa_1^*\mathcal{E}^{\mathrm{cov}}\big) = H^0\big(H_{\infty,q}^{\mathrm{cov}}, \mathcal{E}^{\mathrm{cov}}\big).$$

Here, we obtain the second equality by the identifications $\kappa_1^*\mathcal{E}^{\mathrm{cov}} = (\varpi_q \circ \kappa_1)^*\mathcal{E} = \varpi_q^*\mathcal{E} = \mathcal{E}^{\mathrm{cov}}$. Under the natural isomorphism $H^0\big(H^{\mathrm{cov}}_{\infty,q}, \mathcal{E}^{\mathrm{cov}}\big) \simeq \mathcal{E}^{\mathrm{cov}}_{|\{0\}\times\widehat{\infty}_{y,q}} = \mathcal{E}_{|\widehat{H}_{\infty,q}\langle 0\rangle}$, the two constructions are the same. Note that the constructions are compatible with direct sums, tensor products and inner homomorphisms.

Proposition 3.4.2 *The above construction induces an equivalence of the categories of locally free $\mathcal{O}_{\widehat{H}_{\infty,q}}(*H_{\infty,q})$-modules of finite rank and $(2\sqrt{-1}\lambda T)$-difference $\mathbb{C}(\!(y_q^{-1})\!)$-modules.*

Proof Let $(\mathcal{N}, \Phi^*)$ be a $(2\sqrt{-1}\lambda T)$-difference $\mathbb{C}(\!(y_q^{-1})\!)$-module. Let $p_2 : \widehat{H}^{\mathrm{cov}}_{\infty,q} \to \widehat{\infty}_{y,q}$ denote the projection. We obtain the $\mathcal{O}_{\widehat{H}^{\mathrm{cov}}_{\infty,q}}(*H^{\mathrm{cov}}_{\infty,q})$-module $p_2^{-1}\mathcal{N}$ which is naturally $\mathbb{Z}$-equivariant by the action $\kappa_n^*(p_2^{-1}(s)) = p_2^{-1}\big((\Phi^*)^n(s)\big)$. We obtain an $\mathcal{O}_{\widehat{H}_{\infty,q}}(*H_{\infty,q})$-module as the descent. This is a converse construction of the previous one. ■

The properties for difference $\mathbb{C}(\!(y_q^{-1})\!)$-modules are translated to the properties of $\mathcal{O}_{\widehat{H}_{\infty,q}}(*H_{\infty,q})$-modules.

Definition 3.4.3 Let $\mathcal{E}$ be a locally free $\mathcal{O}_{\widehat{H}_{\infty,q}}(*H_{\infty,q})$-module.

- The level of $\mathcal{E}$ is 0 (resp. less than 1) if the level of the difference $\mathbb{C}(\!(y_q^{-1})\!)$-module $\mathcal{E}_{|\widehat{H}_{\infty,q}\langle 0\rangle}$ is 0 (resp. less than 1). (See Sect. 3.2.1.)
- $\mathcal{E}$ has pure slope $\omega \in \mathbb{Q}$ if the difference $\mathbb{C}(\!(y_q^{-1})\!)$-module $\mathcal{E}_{|\widehat{H}_{\infty,q}\langle 0\rangle}$ has pure slope ω. (See Sect. 3.2.2)
- $\mathcal{E}$ is unramified if the difference $\mathbb{C}(\!(y_q^{-1})\!)$-module $\mathcal{E}_{|\widehat{H}_{\infty,q}\langle 0\rangle}$ is unramified. (See Definition 3.3.13.) ■

Any $\mathcal{O}_{\widehat{H}_{\infty,q}}(*H_{\infty,q})$-module $\mathcal{E}$ of finite rank has the unique slope decomposition

$$\mathcal{E} = \bigoplus_{\omega\in\mathbb{Q}} \mathcal{S}_\omega\mathcal{E}, \tag{3.15}$$

where each $\mathcal{S}_\omega\mathcal{E}$ has pure slope ω.

3.4.4 Lattices and the Induced Local Systems

Let $\mathcal{E}$ be a locally free $\mathcal{O}_{\widehat{H}_{\infty,q}}(*H_{\infty,q})$-module of finite rank. A lattice of $\mathcal{E}$ means a locally free $\mathcal{O}_{\widehat{H}_{\infty,q}}$-submodule $\mathcal{E}_0 \subset \mathcal{E}$ such that $\mathcal{O}_{\widehat{H}_{\infty,q}}(*H_{\infty,q})\otimes\mathcal{E}_0 = \mathcal{E}$. Similarly, for a locally free $\mathcal{O}_{\widehat{H}^{\mathrm{cov}}_{\infty,q}}(*H^{\mathrm{cov}}_{\infty,q})$-module $\mathcal{E}^{\mathrm{cov}}$ of finite rank, a lattice of $\mathcal{E}^{\mathrm{cov}}$ means a locally free $\mathcal{O}_{\widehat{H}^{\mathrm{cov}}_{\infty,q}}$-submodule $\mathcal{E}^{\mathrm{cov}}_0 \subset \mathcal{E}^{\mathrm{cov}}$ such that $\mathcal{O}_{\widehat{H}^{\mathrm{cov}}_{\infty,q}}(*H^{\mathrm{cov}}_{\infty,q})\otimes\mathcal{E}^{\mathrm{cov}}_0 = \mathcal{E}^{\mathrm{cov}}$.

Let $\mathcal{E}$ be a locally free $\mathcal{O}_{\widehat{H}_{\infty,q}}(*H_{\infty,q})$-module of finite rank. We set $\mathcal{E}^{\mathrm{cov}} := \varpi_q^{-1}(\mathcal{E})$. Any lattice $\mathcal{E}_0$ of $\mathcal{E}$ induces a lattice $\mathcal{E}^{\mathrm{cov}}_0 = \varpi_q^{-1}(\mathcal{E}_0)$ of $\mathcal{E}^{\mathrm{cov}}$. For any

$k \in \mathbb{Z}_{>0}$, we obtain a $\mathbb{Z}$-equivariant local system $\mathcal{E}_0^{\mathrm{cov}}/y_q^{-k}\mathcal{E}_0^{\mathrm{cov}}$ on $H_{\infty,q}^{\mathrm{cov}}$. It induces a local system $\mathrm{Loc}_k(\mathcal{E}_0)$ on $H_{\infty,q}$. The following lemma is obvious.

Lemma 3.4.4 *Let $\mathcal{E}$ be an $\mathcal{O}_{\widehat{H}_{\infty,q}}(*H_{\infty,q})$-module.*

- *There exists a lattice $\mathcal{E}_0$ of $\mathcal{E}$ if and only if $\mathcal{E}$ has pure slope 0.*
- *The level of $\mathcal{E}$ is less than 1 if and only if there exists a lattice $\mathcal{E}_0$ of $\mathcal{E}$ such that the monodromy of $\mathrm{Loc}_1(\mathcal{E}_0)$ is unipotent.*
- *The level of $\mathcal{E}$ is 0 if and only if there exists a lattice $\mathcal{E}_0$ of $\mathcal{E}$ such that the monodromy of $\mathrm{Loc}_q(\mathcal{E}_0)$ is the identity.* ■

3.5 Filtered Bundles in the Formal Case

3.5.1 Pull Back and Descent of $\mathcal{O}_{\widehat{H}_{\infty,p}}(*H_{\infty,p})$-Modules

For any $p \in q\mathbb{Z}_{\geq 1}$, there exists the naturally induced morphism $\mathcal{R}_{q,p} : \widehat{H}_{\infty,p} \longrightarrow \widehat{H}_{\infty,q}$, whose underlying map $\mathcal{R}_{q,p} : H_{\infty,p} \longrightarrow H_{\infty,q}$ is the identity, and the morphism of sheaves $\mathcal{R}_{q,p}^*\mathcal{O}_{\widehat{H}_{\infty,q}} \longrightarrow \mathcal{O}_{\widehat{H}_{\infty,p}}$ is induced by the extension $\mathbb{C}[\![y_q^{-1}]\!] \longrightarrow \mathbb{C}[\![y_p^{-1}]\!]$. The $\mathrm{Gal}_{q,p}$-action on $\mathbb{C}[\![y_p^{-1}]\!]$ induces a $\mathrm{Gal}_{q,p}$-action on $\widehat{H}_{\infty,p}$, which we may regard as the Galois groups action for the ramified covering $\widehat{H}_{\infty,p} \longrightarrow \widehat{H}_{\infty,q}$.

For any $\mathcal{O}_{\widehat{H}_{\infty,q}}$-module $\mathcal{E}$, we obtain the $\mathcal{R}_{q,p}^*\mathcal{E}$ as the pull back as in Sect. 3.4.1. If $\mathcal{E}$ is an $\mathcal{O}_{\widehat{H}_{\infty,q}}(*H_{\infty,q})$-module, then $\mathcal{R}_{q,p}^*\mathcal{E}$ is naturally an $\mathcal{O}_{\widehat{H}_{\infty,p}}(*H_{\infty,p})$-module.

Let $\mathcal{E}_1$ be an $\mathcal{O}_{\widehat{H}_{\infty,p}}$-module. We obtain the $\mathcal{O}_{\widehat{H}_{\infty,q}}$-module $\mathcal{R}_{q,p*}\mathcal{E}_1$ obtained as the push-forward. If $\mathcal{E}_1$ is an $\mathcal{O}_{\widehat{H}_{\infty,p}}(*H_{\infty,p})$-module, then $\mathcal{R}_{q,p*}\mathcal{E}_1$ is also an $\mathcal{O}_{\widehat{H}_{\infty,q}}(*H_{\infty,q})$-module. If $\mathcal{E}_1$ is a $\mathrm{Gal}_{q,p}$-equivariant $\mathcal{O}_{\widehat{H}_{\infty,p}}$-module, $\mathcal{R}_{q,p*}\mathcal{E}_1$ is equipped with the induced $\mathrm{Gal}_{q,p}$-action. The invariant part of $\mathcal{R}_{q,p*}\mathcal{E}_1$ is called the descent of $\mathcal{E}_1$ with respect to the $\mathrm{Gal}_{q,p}$-action.

3.5.2 Filtered Bundles

We consider the family version of filtered bundles in Sect. 3.3.1 in a straightforward way.

Definition 3.5.1 A filtered bundle over a locally free $\mathcal{O}_{\widehat{H}_{\infty,q}}(*H_{\infty,q})$-module $\mathcal{E}$ of finite rank is defined to be a family of filtered bundles $\mathcal{P}_*(\mathcal{E}_{|\widehat{H}_{\infty,q}\langle t\rangle})$ $(t \in S_T^1)$ over $\mathcal{E}_{|\widehat{H}_{\infty,q}\langle t\rangle}$. The family is also denoted as $\mathcal{P}_*\mathcal{E}$, for simplicity. We also say that $\mathcal{P}_*\mathcal{E}$ is a filtered bundle on $(\widehat{H}_{\infty,q}, H_{\infty,q})$. ■

Remark 3.5.2 In Definition 3.5.1, we do not impose any condition between the filtrations $\mathcal{P}_*(\mathcal{E}_{|\widehat{H}_{\infty,q}\langle t_1\rangle})$ and $\mathcal{P}_*(\mathcal{E}_{|\widehat{H}_{\infty,q}\langle t_2\rangle})$ under the natural isomorphisms $\mathcal{E}_{|\widehat{H}_{\infty,q}\langle t_1\rangle} \simeq \mathcal{E}_{|\widehat{H}_{\infty,q}\langle t_2\rangle}$. We shall explain what condition should be imposed in Sects. 3.5.3–3.5.5, and we shall eventually introduce the goodness condition in Definition 3.5.15. ■

3.5.2.1 Subbundles and Quotient Bundles

Let $\mathcal{P}_*\mathcal{E}$ be a filtered bundle on $(\widehat{H}_{\infty,q}, H_{\infty,q})$. For any locally free $\mathcal{O}_{\widehat{H}_{\infty,q}}(*H_{\infty,q})$-submodule $\mathcal{E}' \subset \mathcal{E}$, we obtain the filtered bundle $\mathcal{P}_*\mathcal{E}'$ over $\mathcal{E}'$ by setting $\mathcal{P}_a(\mathcal{E}'_{|\widehat{H}_{\infty,q}\langle t\rangle}) := \mathcal{P}_a(\mathcal{E}_{|\widehat{H}_{\infty,q}\langle t\rangle}) \cap \mathcal{E}'_{|\widehat{H}_{\infty,q}\langle t\rangle}$. We also obtain the filtered bundle over $\mathcal{E}'' = \mathcal{E}/\mathcal{E}'$ by setting $\mathcal{P}_a(\mathcal{E}''_{|\widehat{H}_{\infty,q}\langle t\rangle})$ as the image of $\mathcal{P}_a(\mathcal{E}_{|\widehat{H}_{\infty,q}\langle t\rangle}) \longrightarrow \mathcal{E}''_{|\widehat{H}_{\infty,q}\langle t\rangle}$.

3.5.2.2 Basic Functoriality

Let $\mathcal{E}_i$ $(i = 1, 2)$ be locally free $\mathcal{O}_{\widehat{H}_{\infty,q}}(*H_{\infty,q})$-modules. Let

$$\mathcal{P}_*(\mathcal{E}_i) = \big(\mathcal{P}_*(\mathcal{E}_{i|\widehat{H}_{\infty,q}\langle t\rangle}) \,\big|\, t \in S^1_T\big)$$

be filtered bundles over $\mathcal{E}_i$. The induced filtered bundles $\Big(\mathcal{P}_*(\mathcal{E}_{1|\widehat{H}_{\infty,q}\langle t\rangle}) \oplus \mathcal{P}_*(\mathcal{E}_{2|\widehat{H}_{\infty,q}\langle t\rangle})\Big| t \in S^1_T\Big)$ over $\mathcal{E}_1 \oplus \mathcal{E}_2$ is denoted by $\mathcal{P}_*(\mathcal{E}_1 \oplus \mathcal{E}_2)$. The filtered bundle $\Big(\mathcal{P}_*(\mathcal{E}_{2|\widehat{H}_{\infty,q}\langle t\rangle}) \otimes \mathcal{P}_*(\mathcal{E}_{2|\widehat{H}_{\infty,q}\langle t\rangle})\Big| t \in S^1_T\Big)$ over $\mathcal{E}_1 \otimes \mathcal{E}_2$ is denoted by $\mathcal{P}_*(\mathcal{E}_1 \otimes \mathcal{E}_2)$. Let $\mathcal{H}om(\mathcal{E}_1, \mathcal{E}_2)$ denote the sheaf of $\mathcal{O}_{\widehat{H}_{\infty,q}}(*H_{\infty,q})$-homomorphisms from $\mathcal{E}_1$ to $\mathcal{E}_2$. Similarly, we obtain a naturally defined filtered bundle over $\mathcal{H}om(\mathcal{E}_1, \mathcal{E}_2)$, which is denoted by $\mathcal{H}om(\mathcal{P}_*\mathcal{E}_1, \mathcal{P}_*\mathcal{E}_2)$. Let $\mathcal{P}_*\big(\mathcal{O}_{\widehat{H}_{\infty,q}}(*H_{\infty,q})\big)$ denote the filtered bundle over $\mathcal{O}_{\widehat{H}_{\infty,q}}(*H_{\infty,q})$ defined by $\mathcal{P}_a\big(\mathcal{O}_{\widehat{H}_{\infty,q}}(*H_{\infty,q})\big) = y_q^{[a]}\mathcal{O}_{\widehat{H}_{\infty,q}}$. We define $\mathcal{P}_*\mathcal{E}^\vee = \mathcal{H}om\big(\mathcal{P}_*\mathcal{E}, \mathcal{P}_*(\mathcal{O}_{\widehat{H}_{\infty,q}}(*H_{\infty,q}))\big)$.

3.5.2.3 Pull Back

Let $\mathcal{P}_*\mathcal{E}$ be a filtered bundle over a locally free $\mathcal{O}_{\widehat{H}_{\infty,q}}(*H_{\infty,q})$-module $\mathcal{E}$. Because $\mathcal{R}^*_{q,p}(\mathcal{E})_{|\widehat{H}_{\infty,p}\langle t\rangle}$ $(t \in S^1_T)$ are the pull back of $\mathcal{E}_{|\widehat{H}_{\infty,q}\langle t\rangle}$ by $\varphi_{q,p} : \widehat{\infty}_{y,p} \longrightarrow \widehat{\infty}_{y,q}$, we obtain the filtered bundles $\mathcal{P}_*\big(\mathcal{R}^*_{q,p}(\mathcal{E})_{|\widehat{H}_{\infty,p}\langle t\rangle}\big) = \varphi^*_{q,p}\big(\mathcal{P}_*(\mathcal{E}_{|\widehat{H}_{\infty,q}\langle t\rangle})\big)$ as in Sect. 3.3.1. The tuple $\big(\mathcal{P}_*(\mathcal{R}^*_{q,p}(\mathcal{E})_{|\widehat{H}_{\infty,p}\langle t\rangle}) \,\big|\, t \in S^1_T\big)$ is denoted by $\mathcal{R}^*_{q,p}(\mathcal{P}_*\mathcal{E})$, and called the pull back of $\mathcal{P}_*\mathcal{E}$. We obtain the following lemma from Lemma 3.3.6.

Lemma 3.5.3 *Let $\mathcal{E}$ and $\mathcal{P}_*\mathcal{E}$ be as above. Let $\mathcal{E}'$ be a free $\mathcal{O}_{\widehat{H}_{\infty,q}}(*H_{\infty,q})$-submodule of $\mathcal{E}$. We put $\mathcal{E}'' = \mathcal{E}/\mathcal{E}'$. They are equipped with the induced filtered bundle $\mathcal{P}_*\mathcal{E}'$ and $\mathcal{P}_*\mathcal{E}''$, respectively. Then, $\mathcal{R}^*_{q,p}(\mathcal{P}_*\mathcal{E}')$ (resp. $\mathcal{R}^*_{q,p}(\mathcal{P}_*\mathcal{E}'')$) is equal to the filtered bundle over $\mathcal{R}^*_{q,p}\mathcal{E}'$ (resp. $\mathcal{R}^*_{q,p}\mathcal{E}'$) induced by $\mathcal{R}^*_{q,p}(\mathcal{P}_*\mathcal{E})$ and the inclusion $\mathcal{R}^*_{q,p}\mathcal{E}' \subset \mathcal{R}^*_{q,p}\mathcal{E}$ (the projection $\mathcal{R}^*_{q,p}\mathcal{E} \longrightarrow \mathcal{R}^*_{q,p}\mathcal{E}''$).* ■

We obtain the following lemma from Lemma 3.3.7.

Lemma 3.5.4 *Let $\mathcal{P}_*\mathcal{E}_i$ be filtered bundles over $(\widehat{H}_{\infty,q}, H_{\infty,q})$. Then, there exist natural isomorphisms $\mathcal{R}^*_{q,p}(\mathcal{P}_*\mathcal{E}_1 \oplus \mathcal{P}_*\mathcal{E}_2) \simeq \mathcal{R}^*_{q,p}(\mathcal{P}_*\mathcal{E}_1) \oplus \mathcal{R}^*_{q,p}(\mathcal{P}_*\mathcal{E}_2)$, $\mathcal{R}^*_{q,p}(\mathcal{P}_*\mathcal{E}_1 \otimes \mathcal{P}_*\mathcal{E}_2) \simeq \mathcal{R}^*_{q,p}(\mathcal{P}_*\mathcal{E}_1) \otimes \mathcal{R}^*_{q,p}(\mathcal{P}_*\mathcal{E}_2)$, and $\mathcal{R}^*_{q,p}\mathcal{H}om(\mathcal{P}_*\mathcal{E}_1, \mathcal{P}_*\mathcal{E}_2) \simeq \mathcal{H}om\big(\mathcal{R}^*_{q,p}(\mathcal{P}_*\mathcal{E}_1), \mathcal{R}^*_{q,p}(\mathcal{P}_*\mathcal{E}_2)\big)$.* ■

3.5.2.4 Push-Forward

Let $\mathcal{P}_*\mathcal{E}_1$ be a filtered bundle over a locally free $\mathcal{O}_{\widehat{H}_{\infty,p}}(*H_{\infty,p})$-module $\mathcal{E}_1$. Because $\mathcal{R}_{q,p*}(\mathcal{E}_1)_{|\widehat{H}_{\infty,q}\langle t\rangle} = \varphi_{q,p*}(\mathcal{E}_{1|\widehat{H}_{\infty,p}\langle t\rangle})$, we obtain the induced filtered bundle $\mathcal{P}_*\big(\mathcal{R}_{q,p*}(\mathcal{E}_1)_{|\widehat{H}_{\infty,q}\langle t\rangle}\big) = \varphi_{q,p*}\big(\mathcal{P}_*(\mathcal{E}_{1|\widehat{H}_{\infty,p}\langle t\rangle})\big)$. The tuple

$$\big(\mathcal{P}_*(\mathcal{R}_{q,p*}(\mathcal{E}_1)_{|\widehat{H}_{\infty,q}\langle t\rangle}) \,\big|\, t \in S^1_T\big)$$

is denoted by $\mathcal{R}_{q,p*}(\mathcal{P}_*\mathcal{E}_1)$, and called the push-forward of $\mathcal{P}_*\mathcal{E}_1$. We obtain the following lemma from Lemma 3.3.8.

Lemma 3.5.5 *Let $\mathcal{E}_1$ and $\mathcal{P}_*\mathcal{E}_1$ be as above. Let $\mathcal{E}'_1 \subset \mathcal{E}_1$ be a locally free $\mathcal{O}_{\widehat{H}_{\infty,p}}(*H_{\infty,p})$-submodule. We set $\mathcal{E}''_1 = \mathcal{E}_1/\mathcal{E}''_1$. They are equipped with the induced filtered bundles $\mathcal{P}_*(\mathcal{E}'_1)$ and $\mathcal{P}_*(\mathcal{E}''_1)$, respectively. Then, $\mathcal{R}_{q,p*}(\mathcal{P}_*\mathcal{E}'_1)$ (resp. $\mathcal{R}_{q,p*}(\mathcal{P}_*\mathcal{E}''_1)$) is equal to the filtered bundle induced by $\mathcal{R}_{q,p*}(\mathcal{P}_*\mathcal{E}_1)$ and the inclusion $\mathcal{R}_{q,p*}(\mathcal{E}'_1) \subset \mathcal{R}_{q,p*}(\mathcal{E}_1)$ (the projection $\mathcal{R}_{q,p*}(\mathcal{E}_1) \longrightarrow \mathcal{R}_{q,p*}(\mathcal{E}''_1)$).* ■

Let $\mathcal{P}_*\mathcal{E}$ be a filtered bundle on $(\widehat{H}_{\infty,q}, H_{\infty,q})$. Let $\mathcal{P}_*\mathcal{E}_1$ be a filtered bundle on $(\widehat{H}_{\infty,p}, H_{\infty,p})$. We obtain the following lemma from Lemma 3.3.9.

Lemma 3.5.6 *The natural isomorphism $\mathcal{R}_{q,p*}(\mathcal{E}_1 \otimes \mathcal{R}^*_{q,p}\mathcal{E}) \simeq \mathcal{R}_{q,p*}(\mathcal{E}_1) \otimes \mathcal{E}$ induces*

$$\mathcal{R}_{q,p*}(\mathcal{P}_*\mathcal{E}_1 \otimes \mathcal{R}^*_{q,p}\mathcal{P}_*\mathcal{E}) \simeq \mathcal{R}_{q,p*}(\mathcal{P}_*\mathcal{E}_1) \otimes \mathcal{P}_*\mathcal{E}.$$

The natural isomorphism $\mathcal{R}_{q,p*}(\mathcal{H}om(\mathcal{R}^*_{q,p}\mathcal{E}, \mathcal{E}_1)) \simeq \mathcal{H}om(\mathcal{E}, \mathcal{R}_{q,p*}\mathcal{E}_1)$ induces

$$\mathcal{R}_{q,p*}(\mathcal{H}om(\mathcal{R}^*_{q,p}\mathcal{P}_*\mathcal{E}, \mathcal{P}_*\mathcal{E}_1)) \simeq \mathcal{H}om(\mathcal{P}_*\mathcal{E}, \mathcal{R}_{q,p*}\mathcal{P}_*\mathcal{E}_1).$$

■

3.5.2.5 Descent

If $\mathcal{P}_*\mathcal{E}_1$ is $\mathrm{Gal}_{q,p}$-equivariant, then we obtain the induced filtered bundle $\mathcal{P}_*\mathcal{E}_2$ over the descent $\mathcal{E}_2$ of $\mathcal{E}_1$ by taking the $\mathrm{Gal}_{q,p}$-invariant part of $\mathcal{P}_*\big(\mathcal{R}_{q,p*}(\mathcal{E}_1)_{|\widehat{H}_{q,\infty}\langle t\rangle}\big)$ $(t \in S^1_T)$ as in Sect. 3.3.1. We obtain the following lemma from Lemma 3.3.10.

Lemma 3.5.7 *$\mathcal{R}^*_{q,p}(\mathcal{P}_*\mathcal{E}_2)$ is naturally isomorphic to $\mathcal{P}_*\mathcal{E}_1$.* ■

We obtain the following lemma from Lemma 3.3.11.

Lemma 3.5.8 *Let $\mathcal{E}'_1$ be a $\mathrm{Gal}_{q,p}$-equivariant locally free $\mathcal{O}_{\widehat{H}_{\infty,p}}(*H_{\infty,p})$-submodule of $\mathcal{E}_1$. We set $\mathcal{E}''_1 := \mathcal{E}_1/\mathcal{E}'_1$. We obtain the locally free $\mathcal{O}_{\widehat{H}_{\infty,q}}(*H_{\infty,q})$-submodule $\mathcal{E}'_2 \subset \mathcal{E}_2$ as the descent of $\mathcal{E}'_1$. We also obtain the locally free $\mathcal{O}_{\widehat{H}_{\infty,q}}(*H_{\infty,q})$-quotient module $\mathcal{E}_2 \longrightarrow \mathcal{E}''_2$ as the descent of $\mathcal{E}''_1$. Then, the decent $\mathcal{P}_*\mathcal{E}'_2$ of $\mathcal{P}_*\mathcal{E}'_1$ is equal to the filtered bundle over $\mathcal{E}'_2$ induced by $\mathcal{P}_*\mathcal{E}_2$ with the inclusion $\mathcal{E}'_2 \subset \mathcal{E}_2$. Similarly, the decent $\mathcal{P}_*\mathcal{E}''_2$ of $\mathcal{P}_*\mathcal{E}''_1$ is equal to the filtered bundle over $\mathcal{E}''_2$ induced by $\mathcal{P}_*\mathcal{E}_2$ with the projection $\mathcal{E}_2 \longrightarrow \mathcal{E}''_2$.* ■

We obtain the following lemma from Lemma 3.3.12.

Lemma 3.5.9 *For a filtered bundle $\mathcal{P}_*\mathcal{E}$ on $(\widehat{H}_{\infty,q}, H_{\infty,q})$, $\mathcal{R}^*_{q,p}(\mathcal{P}_*\mathcal{E})$ is naturally $\mathrm{Gal}_{q,p}$-equivariant, and $\mathcal{P}_*\mathcal{E}$ is isomorphic to the descent of $\mathcal{R}^*_{q,p}\mathcal{P}_*\mathcal{E}$.* ■

3.5.3 Basic Filtered Objects with Pure Slope

Let $q \in \mathbb{Z}_{>0}$, $\ell \in \mathbb{Z}$ and $\alpha \in \mathbb{C}^*$. We consider the $\mathbb{Z}$-action on $\widehat{\mathbb{L}}^{\lambda\,\mathrm{cov}}_q(\ell,\alpha) := \mathcal{O}_{\widehat{H}^{\mathrm{cov}}_{\infty,q}}(*H^{\mathrm{cov}}_{\infty,q})\, e_{q,\ell,\alpha}$ given by

$$\kappa_1^*(e_{q,\ell,\alpha})$$
$$= \alpha\,(1+|\lambda|^2)^{\ell/q}\, y_q^{-\ell}(1+2\sqrt{-1}\lambda T y^{-1})^{-\ell/q} \exp\Big(\frac{\ell}{q}G(2\sqrt{-1}T\lambda y^{-1})\Big)\, e_{q,\ell,\alpha}. \tag{3.16}$$

Here, $G(x) = 1 - x^{-1}\log(1+x)$. We obtain the induced $\mathcal{O}_{\widehat{H}_{\infty,q}}(*H_{\infty,q})$-module, which is denoted by $\widehat{\mathbb{L}}^{\lambda}_q(\ell,\alpha)$. It has pure slope ℓ/q.

Let $a \in \mathbb{R}$. Let $e^t_{q,\ell,\alpha}$ denote the restriction of $e_{q,\ell,\alpha}$ to $\{t\} \times \widehat{\infty}_{y,q}$. We define the filtrations

$$\mathcal{P}^{(a)}_*\Big(\widehat{\mathbb{L}}^{\lambda\,\mathrm{cov}}_q(\ell,\alpha)_{|\{t\}\times\widehat{\infty}_{y,q}}\Big)$$

by setting

$$\deg^{\mathcal{P}}(e^t_{q,\ell,\alpha}) = a - \frac{\ell t}{T}.$$

Because the filtrations are preserved by the $\mathbb{Z}$-action, we obtain a tuple of the filtrations $\mathcal{P}_*^{(a)}\widehat{\mathbb{L}}_q^{\lambda}(\ell,\alpha) = \big(\mathcal{P}_*^{(a)}(\widehat{\mathbb{L}}_q^{\lambda}(\ell,\alpha)_{|\widehat{H}_{\infty,q}\langle t\rangle}) \,\big|\, t \in S_T^1\big)$.

Remark 3.5.10 The filtered bundles in this section naturally appear in the context of periodic monopoles. See Sect. 5.1. ■

3.5.4 Good Filtered Bundles over $\mathcal{O}_{\widehat{H}_{\infty,q}}(*H_{\infty,q})$-Modules with Level ≤1

Let $\mathcal{E}$ be a locally free $\mathcal{O}_{\widehat{H}_{\infty,q}}(*H_{\infty,q})$-module of finite rank with level ≤ 1. (See Definition 3.4.3.) Note that any Φ^*-invariant lattice $\mathcal{L}$ of the difference $\mathbb{C}((y_q^{-1}))$-module $\mathcal{N} := \mathcal{E}_{|\widehat{H}_{\infty,q}\langle 0\rangle}$ induces a lattice $\Upsilon(\mathcal{L})$ of $\mathcal{E}$. In particular, it induces lattices $\Upsilon(\mathcal{L})_{|\widehat{H}_{\infty,q}\langle t\rangle}$ of $\mathcal{E}_{|\widehat{H}_{\infty,q}\langle t\rangle}$ $(t \in S_T^1)$.

Definition 3.5.11 Suppose that the level of $\mathcal{E}$ is less than 1. A filtered bundle $\mathcal{P}_*(\mathcal{E}_{|\widehat{H}_{\infty,q}\langle t\rangle} \mid t \in S_T^1)$ is called good if there exists a filtered bundle $\mathcal{P}_*(\mathcal{N})$ over $\mathcal{N}$ which is good with respect to Φ^* such that $\mathcal{P}_*(\mathcal{E}_{|\widehat{H}_{\infty,q}\langle t\rangle}) = \Upsilon(\mathcal{P}_*\mathcal{N})_{|\widehat{H}_{\infty,q}\langle t\rangle}$. ■

Lemma 3.5.12 *Let $\mathcal{P}_*\mathcal{E}$ be a good filtered bundle over $\mathcal{E}$. Let $\mathcal{E}'$ be a locally free $\mathcal{O}_{\widehat{H}_{\infty,q}}(*H_{\infty,q})$-submodule of $\mathcal{E}$. We set $\mathcal{E}'' = \mathcal{E}/\mathcal{E}'$. They are equipped with the induced filtered bundles $\mathcal{P}_*\mathcal{E}'$ and $\mathcal{P}_*\mathcal{E}''$, respectively. Then, $\mathcal{P}_*\mathcal{E}'$ and $\mathcal{P}_*\mathcal{E}''$ are also good in the sense of Definition* 3.5.11.

Proof We explain the proof for $\mathcal{E}'$. The other case can be argued similarly. Note that the level of $\mathcal{E}'$ is also less than 1, which follows from Lemma 3.2.13 and Lemma 3.2.16. We set $\mathcal{N}' = \mathcal{E}'_{|\widehat{H}_{\infty,q}\langle 0\rangle}$, which is a $\mathbb{C}((y_q^{-1}))$-difference submodule of $\mathcal{N}$, which is equipped with the induced filtration $\mathcal{P}_*\mathcal{N}'$. By Lemma 3.3.19, $\mathcal{P}_*\mathcal{N}'$ is good. It is easy to see that $\mathcal{P}_*(\mathcal{E}'_{|\widehat{H}_{\infty,q}\langle t\rangle}) = \Upsilon(\mathcal{P}_*\mathcal{N})_{|\widehat{H}_{\infty,q}\langle t\rangle}$. Hence, $\mathcal{P}_*\mathcal{E}'$ is good. ■

If moreover $\mathcal{E}$ is unramified, there exists a decomposition

$$\mathcal{E} = \bigoplus_{\mathfrak{b}\in S(q)} \mathcal{E}_{\mathfrak{b}} \tag{3.17}$$

such that for a lattice $\mathcal{E}_{\mathfrak{b},0}$ of $\mathcal{E}_{\mathfrak{b}}$, the monodromy of $\mathrm{Loc}_q(\mathcal{E}_{\mathfrak{b},0})$ is equal to the multiplication of $1+\mathfrak{b}$. The following lemma is easy to see.

Lemma 3.5.13 *Suppose that the level of $\mathcal{E}$ is less than 1 and that $\mathcal{E}$ is unramified. Then, a filtered bundle $\mathcal{P}_*(\mathcal{E}) = \big(\mathcal{P}_*(\mathcal{E}_{|\widehat{H}_{\infty,q}\langle t\rangle}) \,\big|\, t \in S_T^1\big)$ is good if and only if the following condition is satisfied.*

- *For each $a \in \mathbb{R}$, there exists a lattice $\mathcal{P}_a(\mathcal{E})$ of $\mathcal{E}$ such that $\mathcal{P}_a(\mathcal{E}_{|\widehat{H}_{\infty,q}\langle t\rangle}) = \mathcal{P}_a(\mathcal{E})_{|\widehat{H}_{\infty,q}\langle t\rangle}$.*

- *The lattice* $\mathcal{P}_a(\mathcal{E})$ *is compatible with the decomposition (3.17), i.e., by setting* $\mathcal{P}_a(\mathcal{E}_{\mathfrak{b}}) = \mathcal{P}_a\mathcal{E} \cap \mathcal{E}_{\mathfrak{b}}$*, we obtain* $\mathcal{P}_a(\mathcal{E}) = \bigoplus \mathcal{P}_a\mathcal{E}_{\mathfrak{b}}$.
- *The monodromy of* $\mathrm{Loc}_q(\mathcal{P}_a\mathcal{E}_{\mathfrak{b}})$ *are equal to the multiplication of* $1 + \mathfrak{b}$. ■

3.5.5 Good Filtered Bundles over $\mathcal{O}_{\widehat{H}_{\infty,q}}(*H_{\infty,q})$-Modules

Let $\mathcal{E}$ be a locally free $\mathcal{O}_{\widehat{H}_{\infty,q}}(*H_{\infty,q})$-module of finite rank.

Definition 3.5.14 We say that $\mathcal{E}$ is unramified modulo level ≤ 1 if there exists a decomposition

$$\mathcal{E} = \bigoplus_{\ell \in \mathbb{Z}} \bigoplus_{\alpha \in \mathbb{C}^*} \mathcal{E}_{\ell,\alpha} \tag{3.18}$$

such that each $\mathcal{E}_{\ell,\alpha}$ is the tensor product of $\widehat{\mathbb{L}}_q^{\lambda}(\ell, \alpha)$ and a locally free $\mathcal{O}_{\widehat{H}_{\infty,q}}(*H_{\infty,q})$-module of level ≤ 1. ■

Note that the decomposition (3.18) is uniquely determined.

Definition 3.5.15 If $\mathcal{E}$ is unramified modulo level ≤ 1, a filtered bundle $\mathcal{P}_*\mathcal{E}$ is called good if the following conditions are satisfied.

- $\mathcal{P}_*\mathcal{E}$ is compatible with the decomposition (3.18), i.e., $\mathcal{P}_*(\mathcal{E}) = \bigoplus_{\ell,\alpha} \mathcal{P}_*(\mathcal{E}_{\ell,\alpha})$.
- There exists a good filtered bundle $\mathcal{P}_*\big(\widehat{\mathbb{L}}_q^{\lambda}(\ell, \alpha)^{-1} \otimes \mathcal{E}_{\ell,\alpha}\big)$ over $\widehat{\mathbb{L}}_q^{\lambda}(\ell, \alpha)^{-1} \otimes \mathcal{E}_{\ell,\alpha}$ (Definition 3.5.11) such that

$$\mathcal{P}_*(\mathcal{E}_{\ell,\alpha|\widehat{H}_{\infty,q}\langle t\rangle}) = \mathcal{P}_*^{(0)}\big(\widehat{\mathbb{L}}_q^{\lambda}(\ell, \alpha)_{|\widehat{H}_{\infty,q}\langle t\rangle}\big) \otimes \mathcal{P}_*\big(\widehat{\mathbb{L}}_q^{\lambda}(\ell, \alpha)^{-1} \otimes \mathcal{E}_{\ell,\alpha}\big)_{|\widehat{H}_{\infty,q}\langle t\rangle}.$$

■

Recall that for any locally free $\mathcal{O}_{\widehat{H}_{\infty,q}}(*H_{\infty,q})$-module $\mathcal{E}$ of finite rank there exists $p \in q\mathbb{Z}_{>0}$ such that $\mathcal{R}_{q,p}^*\mathcal{E}$ is unramified modulo level ≤ 1. (See Proposition 3.2.12 and Theorem 3.2.15.)

Definition 3.5.16 For a locally free $\mathcal{O}_{\widehat{H}_{\infty,q}}(*H_{\infty,q})$-module $\mathcal{E}$ of finite rank, a filtered bundle $\mathcal{P}_*\mathcal{E}$ over $\mathcal{E}$ is good if there exist $p \in q\mathbb{Z}_{>0}$ such that (i) $\mathcal{R}_{q,p}^*(\mathcal{E})$ is unramified modulo level ≤ 1, (ii) $\mathcal{R}_{q,p}^*(\mathcal{P}_*\mathcal{E})$ is a good filtered bundle over $\mathcal{R}_{q,p}^*(\mathcal{E})$ (Definition 3.5.15). ■

3.5.5.1 An Equivalence

For a good filtered bundle $\mathcal{P}_*\mathcal{E}$ over a locally free $\mathcal{O}_{\widehat{H}_{\infty,q}}(*H_{\infty,q})$-module $\mathcal{E}$, by taking the restriction to $\widehat{H}_{\infty,q}\langle 0\rangle$, we obtain a filtered bundle $\mathcal{P}_*(\mathcal{E}_{|\widehat{H}_{\infty,q}\langle 0\rangle})$

over $\mathcal{E}_{|\widehat{H}_{\infty,q}\langle 0\rangle}$. By the construction, it is good for the difference $\mathbb{C}(\!(y_q^{-1})\!)$-module $\mathcal{E}_{|\widehat{H}_{\infty,q}\langle 0\rangle}$. The following proposition is obvious by the definitions.

Proposition 3.5.17 *By the restriction from $\mathcal{P}_*\mathcal{E}$ to $\mathcal{P}_*(\mathcal{E}_{|\widehat{H}_{\infty,q}\langle 0\rangle})$, we obtain an equivalence between good filtered bundles on locally free $\mathcal{O}_{\widehat{H}_{\infty,q}}(*H_{\infty,q})$-modules and good filtered difference $\mathbb{C}(\!(y_q^{-1})\!)$-modules.* ■

3.5.5.2 Some Properties

The following lemma is obvious.

Lemma 3.5.18 *Let $\mathcal{P}_*\mathcal{E}$ be a good filtered bundle on $(\widehat{H}_{\infty,q}, H_{\infty,q})$. Then, it is compatible with the slope decomposition* (3.15), *i.e., $\mathcal{P}_*\mathcal{E} = \bigoplus \mathcal{P}_*\mathcal{S}_\omega\mathcal{E}$.* ■

Lemma 3.5.19 *Let $\mathcal{P}_*\mathcal{E}$ be a good filtered bundle on $(\widehat{H}_{\infty,q}, H_{\infty,q})$. Let $\mathcal{E}'$ be a locally free $\mathcal{O}_{\widehat{H}_{\infty,q}}(*H_{\infty,q})$-submodule of $\mathcal{E}$. We set $\mathcal{E}'' = \mathcal{E}/\mathcal{E}'$. Then, the induced filtered bundle $\mathcal{P}_*\mathcal{E}'$ (resp. $\mathcal{P}_*\mathcal{E}''$) over $\mathcal{E}'$ (resp. $\mathcal{E}''$) is good.*

Proof By Lemma 3.5.3 and Lemma 3.5.9, it is enough to study the case where $\mathcal{E}$ and $\mathcal{E}'$ are unramified modulo ≤ 1. Because $\mathcal{P}_*\mathcal{E}$ is compatible with the decomposition (3.18), it is enough to study the case where the level of $\mathcal{E}$ is less than 1, which we have already studied in Lemma 3.5.12. ■

Lemma 3.5.20 *Let $\mathcal{P}_*\mathcal{E}_i$ $(i = 1, 2)$ be good filtered bundles over locally free $\mathcal{O}_{\widehat{H}_{\infty,q}}(*H_{\infty,q})$-modules $\mathcal{E}_i$. Then, the induced objects $\mathcal{P}_*(\mathcal{E}_1) \oplus \mathcal{P}_*(\mathcal{E}_2)$, $\mathcal{P}_*(\mathcal{E}_1) \otimes \mathcal{P}_*(\mathcal{E}_2)$, and $\mathcal{H}om(\mathcal{P}_*\mathcal{E}_1, \mathcal{P}_*\mathcal{E}_2)$ are also good filtered bundles. In particular, for a good filtered bundle $\mathcal{P}_*\mathcal{E}$, its dual $(\mathcal{P}_*\mathcal{E})^\vee$ is also a good filtered bundle.*

Proof By Lemmas 3.5.4 and 3.5.9, it is enough to consider the case where $\mathcal{E}_i$ are unramified modulo level ≤ 1. Because the good filtered bundles $\mathcal{P}_*\mathcal{E}_i$ are compatible with the decompositions as in (3.18), we have only to consider the case where the levels of $\mathcal{E}_i$ are less than 1. Then, it follows from the claims in the case of $\mathbb{C}(\!(y_q^{-1})\!)$-difference modules (Lemma 3.3.20). ■

3.5.6 Global Lattices on the Covering Space

Let $\mathcal{P}_*\mathcal{E}$ be a good filtered bundle over a locally free $\mathcal{O}_{\widehat{H}_{\infty,q}}(*H_{\infty,q})$-module $\mathcal{E}$. We set $\mathcal{E}^{\mathrm{cov}} := \varpi_q^{-1}(\mathcal{E})$. We obtain the induced filtrations $\mathcal{P}_*\big(\mathcal{E}^{\mathrm{cov}}_{|\{t\}\times\widehat{\infty}_{y,q}}\big)$ $(t \in H^{\mathrm{cov}}_{\infty,q})$. The slope decomposition $\mathcal{E} = \bigoplus \mathcal{S}_\omega(\mathcal{E})$ induces a decomposition $\mathcal{E}^{\mathrm{cov}} = \bigoplus \mathcal{S}_\omega(\mathcal{E})^{\mathrm{cov}}$. For each $t \in H^{\mathrm{cov}}_{\infty,q}$, we obtain the decomposition $\mathcal{P}_*(\mathcal{E}^{\mathrm{cov}}_{|\{t\}\times\widehat{\infty}_{y,q}}) = \bigoplus \mathcal{P}_*(\mathcal{S}_\omega(\mathcal{E})^{\mathrm{cov}}_{|\{t\}\times\widehat{\infty}_{y,q}})$.

Lemma 3.5.21 *For any* $t_1, t_2 \in H^{\mathrm{cov}}_{\infty,q}$*, the isomorphism* $\mathcal{S}_\omega(\mathcal{E})^{\mathrm{cov}}_{|\{t_1\}\times\widehat{\infty}_{y,q}} \simeq \mathcal{S}_\omega(\mathcal{E})^{\mathrm{cov}}_{|\{t_2\}\times\widehat{\infty}_{y,q}}$ *induces the following isomorphisms for any* $a \in \mathbb{R}$*:*

$$\mathcal{P}_{a-t_1 q\omega/T}\mathcal{S}_\omega(\mathcal{E})^{\mathrm{cov}}_{|\{t_1\}\times\widehat{\infty}_{y,q}} \simeq \mathcal{P}_{a-t_2 q\omega/T}\mathcal{S}_\omega(\mathcal{E})^{\mathrm{cov}}_{|\{t_2\}\times\widehat{\infty}_{y,q}}. \tag{3.19}$$

Proof If $\mathcal{E}$ is unramified, the claim is clear by the definition of good filtrations. In general, there exists $p \in q\mathbb{Z}_{>0}$ such that $\mathcal{R}^*_{q,p}(\mathcal{E})$ is unramified. We obtain the isomorphism

$$\mathcal{P}_{a-t_1 p\omega/T}\mathcal{S}_\omega(\mathcal{R}^*_{q,p}\mathcal{E})^{\mathrm{cov}}_{|\{t_1\}\times\widehat{\infty}_{y,q}} \simeq \mathcal{P}_{a-t_2 p\omega/T}\mathcal{S}_\omega(\mathcal{R}^*_{q,p}\mathcal{E})^{\mathrm{cov}}_{|\{t_2\}\times\widehat{\infty}_{y,q}}.$$

It is compatible with the natural $\mathrm{Gal}_{q,p}$-action. By using the natural isomorphisms (3.10), we obtain (3.19). ∎

By Lemma 3.5.21, for each $a \in \mathbb{R}$, there uniquely exists a locally free $\mathcal{O}_{\widehat{H}^{\mathrm{cov}}_{\infty,q}}$-submodule $\mathfrak{P}_a(\mathcal{E}^{\mathrm{cov}})$ of $\mathcal{E}^{\mathrm{cov}}$ satisfying

$$\mathfrak{P}_a(\mathcal{E}^{\mathrm{cov}})_{|\{t\}\times\widehat{\infty}_{y,q}} := \bigoplus_{\omega\in\mathbb{Q}} \mathcal{P}_{a-tq\omega/T}(\mathcal{S}_\omega\mathcal{E}^{\mathrm{cov}}_{|\{t\}\times\widehat{\infty}_{y,q}}).$$

Thus, we obtain the global filtration $\mathfrak{P}_*$ of $\mathcal{E}^{\mathrm{cov}}$. The following lemma is obvious by the construction.

Lemma 3.5.22 *Let* $(\mathcal{P}_*\mathcal{N}, \Phi^*)$ *be a good filtered difference* $\mathbb{C}(\!(y_q^{-1})\!)$*-module obtained as the restriction of* $\mathcal{P}_*\mathcal{E}$ *to* $\widehat{H}_{\infty,q}\langle 0\rangle$*. Let* $p_2 : \widehat{H}^{\mathrm{cov}}_{\infty,q} \longrightarrow \widehat{\infty}_{y,q}$ *denote the projection. Then, under the natural isomorphism* $\mathcal{E}^{\mathrm{cov}} \simeq p_2^{-1}(\mathcal{N})$*, we obtain* $\mathfrak{P}_*(\mathcal{E}^{\mathrm{cov}}) = p_2^{-1}(\mathcal{P}_*\mathcal{N})$*.* ∎

We set $\mathrm{Gr}^{\mathfrak{P}}_a(\mathcal{E}^{\mathrm{cov}}) := \mathfrak{P}_a(\mathcal{E}^{\mathrm{cov}})/\mathfrak{P}_{<a}(\mathcal{E}^{\mathrm{cov}})$. We obtain the decomposition $\mathrm{Gr}^{\mathfrak{P}}_a(\mathcal{E}^{\mathrm{cov}}) = \bigoplus \mathrm{Gr}^{\mathfrak{P}}_a(\mathcal{S}_\omega\mathcal{E}^{\mathrm{cov}})$ induced by the slope decomposition. By the construction, there exist the natural isomorphisms:

$$\mathrm{Gr}^{\mathfrak{P}}_a(\mathcal{S}_\omega\mathcal{E}^{\mathrm{cov}})_{\{t\}\times\widehat{\infty}_{y,q}} \simeq \mathrm{Gr}^{\mathcal{P}}_{a-tq\omega/T}(\mathcal{S}_\omega\mathcal{E}^{\mathrm{cov}}_{|\{t\}\times\widehat{\infty}_{y,q}}).$$

By Lemma 3.5.22 and the construction in Sect. 3.3.3, there exist the endomorphism $\mathfrak{F}_{\mathcal{E}}$ of $\mathrm{Gr}^{\mathfrak{P}}_a(\mathcal{E}^{\mathrm{cov}})$ induced by $\mathrm{Res}(\Phi^*)$ for the corresponding good filtered difference $\mathbb{C}(\!(y_q^{-1})\!)$-module. We also obtain the monodromy weight filtration W of the nilpotent part of $\mathfrak{F}_{\mathcal{E}}$ on each $\mathrm{Gr}^{\mathfrak{P}}_a(\mathcal{E}^{\mathrm{cov}})$. They are compatible with the slope decomposition.

3.5.7 Local Lattices

Let $\mathcal{P}_*\mathcal{E} = \big(\mathcal{P}_*\mathcal{E}_{|\widehat{H}_{\infty,q}\langle t\rangle} \,\big|\, t \in S^1_T\big)$ denote a good filtered bundle over a locally free $\mathcal{O}_{\widehat{H}_{\infty,q}}(*H_{\infty,q})$-module $\mathcal{E}$. For any $t \in S^1_T$, we set

$$\mathcal{P}ar(\mathcal{P}_*\mathcal{E}, t) := \big\{b \in \mathbb{R} \,\big|\, \mathrm{Gr}^{\mathcal{P}}_b(\mathcal{E}_{|\widehat{H}_{\infty,q}\langle t\rangle}) \neq 0\big\}.$$

We clearly have $\mathcal{P}ar(\mathcal{P}_*\mathcal{E}, t) = \bigcup_{\omega\in\mathbb{Q}} \mathcal{P}ar\big(\mathcal{P}_*(\mathcal{S}_\omega\mathcal{E}), t\big)$. The following lemma is clear by the construction.

Lemma 3.5.23 *For $t_0, t_1 \in S^1_T$, there exists the following relation:*

$$\mathcal{P}ar(\mathcal{P}_*\mathcal{S}_\omega\mathcal{E}, t_1) = \big\{b - q\omega T^{-1}(\widetilde{t}_1 - \widetilde{t}_0) \,\big|\, b \in \mathcal{P}ar(\mathcal{P}_*\mathcal{S}_\omega\mathcal{E}, t_0)\big\}.$$

Here, $\widetilde{t}_i \in \mathbb{R}$ are any lifts of $t_i \in S^1_T = \mathbb{R}/T\mathbb{Z}$. ■

Take $t_0 \in S^1_T \simeq H_{\infty,q}$. For $0 < \epsilon < T/10$, we set

$$H_{\infty,q}\langle t_0, \epsilon\rangle :=]t_0 - \epsilon, t_0 + \epsilon[\subset H_{\infty,q},$$

and let $\widehat{H}_{\infty,q}\langle t_0, \epsilon\rangle$ denote the ringed space obtained as $H_{\infty,q}\langle t_0, \epsilon\rangle$ with

$$\mathcal{O}_{\widehat{H}_{\infty,q}\langle t_0,\epsilon\rangle} := \mathcal{O}_{\widehat{H}_{\infty,q}|\widehat{H}_{\infty,q}\langle t_0,\epsilon\rangle}.$$

We set $\mathcal{O}_{\widehat{H}_{\infty,q}\langle t_0,\epsilon\rangle}(*H_{\infty,q}\langle t_0, \epsilon\rangle) = \mathcal{O}_{\widehat{H}_{\infty,q}}(*H_{\infty,q})_{|\widehat{H}_{\infty,q}\langle t_0,\epsilon\rangle}$. The following lemma is clear by the condition of good filtered bundles.

Lemma 3.5.24 *For each $a \in \mathbb{R}$, there uniquely exists a locally free $\mathcal{O}_{\widehat{H}_{\infty,q}\langle t_0,\epsilon\rangle}$-submodule $\boldsymbol{P}^{(t_0)}_a(\mathcal{E}_{|\widehat{H}_{\infty,q}\langle t_0,\epsilon\rangle}) \subset \mathcal{E}_{|\widehat{H}_{\infty,q}\langle t_0,\epsilon\rangle}$ satisfying the following condition for any $t \in H_{\infty,q}\langle t_0, \epsilon\rangle$:*

$$\boldsymbol{P}^{(t_0)}_a(\mathcal{E}_{|\widehat{H}_{\infty,q}\langle t_0,\epsilon\rangle})_{|\widehat{H}_{\infty,q}\langle t\rangle} = \bigoplus_{\omega} \mathcal{P}_{a-q\omega(\widetilde{t}-\widetilde{t}_0)/T}(\mathcal{S}_\omega\mathcal{E}_{|\widehat{H}_{\infty,q}\langle t\rangle}).$$

Here, $\widetilde{t}_0 \in \mathbb{R}$ denotes any lift of t_0, and $\widetilde{t}$ denotes the lift of t such that $|\widetilde{t} - \widetilde{t}_0| < \epsilon$. It satisfies the following.

- $\boldsymbol{P}^{(t_0)}_a(\mathcal{E}_{|\widehat{H}_{\infty,q}\langle t_0,\epsilon\rangle}) = \bigoplus_\omega \boldsymbol{P}^{(t_0)}_a(\mathcal{S}_\omega\mathcal{E}_{|\widehat{H}_{\infty,q}\langle t_0,\epsilon\rangle})$.
- $\boldsymbol{P}^{(t_0)}_a(\mathcal{E}_{|\widehat{H}_{\infty,q}\langle t_0,\epsilon\rangle}) \otimes_{\mathcal{O}_{\widehat{H}_{\infty,q}\langle t_0,\epsilon\rangle}} \mathcal{O}_{\widehat{H}_{\infty,q}\langle t_0,\epsilon\rangle}(*H_{\infty,q}\langle t_0, \epsilon\rangle) = \mathcal{E}_{|\widehat{H}_{\infty,q}\langle t_0,\epsilon\rangle}$. ■

We set $\boldsymbol{P}^{(t_0)}_{<a}(\mathcal{E}_{|\widehat{H}_{\infty,q}\langle t_0,\epsilon\rangle}) = \sum_{b<a} \boldsymbol{P}^{(t_0)}_b(\mathcal{E}_{|\widehat{H}_{\infty,q}\langle t_0,\epsilon\rangle})$. We obtain the following locally constant sheaves on $H_{\infty,q}\langle t_0, \epsilon\rangle$:

$$\boldsymbol{G}^{(t_0)}_a(\mathcal{E}_{|\widehat{H}_{\infty,q}\langle t_0,\epsilon\rangle}) := \boldsymbol{P}^{(t_0)}_a(\mathcal{E}_{|\widehat{H}_{\infty,q}\langle t_0,\epsilon\rangle}) \Big/ \boldsymbol{P}^{(t_0)}_{<a}(\mathcal{E}_{|\widehat{H}_{\infty,q}\langle t_0,\epsilon\rangle}).$$

There exists the decomposition:

$$G_*^{(t_0)}(\mathcal{E}_{|\widehat{H}_{\infty,q}\langle t_0,\epsilon\rangle}) = \bigoplus_{\omega\in\mathbb{Q}} G_*^{(t_0)}(\mathcal{S}_\omega\mathcal{E}_{|\widehat{H}_{\infty,q}\langle t_0,\epsilon\rangle}).$$

The following lemma is clear by the construction.

Lemma 3.5.25 *Let $\widetilde{t}_0 \in \mathbb{R}$ which is a lift of t_0. Let $\widehat{H}^{\mathrm{cov}}_{\infty,q}\langle\widetilde{t}_0,\epsilon\rangle :=]\widetilde{t}_0-\epsilon, \widetilde{t}_0+\epsilon[\times\widehat{\infty}_{y,q}$. For $\omega\in\mathbb{Q}$, there exists a natural isomorphism*

$$G_b^{(t_0)}(\mathcal{S}_\omega\mathcal{E}_{|\widehat{H}_{\infty,q}\langle t_0,\epsilon\rangle}) \simeq \mathrm{Gr}^{\mathfrak{P}}_{b+\omega q\widetilde{t}_0/T}(\mathcal{S}_\omega\mathcal{E}^{\mathrm{cov}})_{|\widehat{H}^{\mathrm{cov}}_{\infty,q}\langle\widetilde{t}_0,\epsilon\rangle}.$$

As a result, there exist the nilpotent endomorphism $\mathfrak{N}$ and the monodromy weight filtration W on $G_b^{(t_0)}(\mathcal{E}_{|\widehat{H}_{\infty,q}\langle t_0,\epsilon\rangle})$. ■

We set $\nu(\mathcal{E}) := \max\{|\omega_1|+|\omega_2| \mid \mathcal{S}_{\omega_i}(\mathcal{E})\neq 0\}$. We assume that ϵ satisfies

$$10q\nu(\mathcal{E})\epsilon/T < \min\{|a-b| \mid a,b\in \mathcal{P}ar(\mathcal{P}_*\mathcal{E},t_0),\ a\neq b\}. \tag{3.20}$$

Let v be a non-zero section of $\boldsymbol{P}_a^{(t_0)}(\mathcal{E}_{|\widehat{H}_{\infty,q}\langle t_0,\epsilon\rangle})$. According to the decomposition

$$\boldsymbol{P}_a^{(t_0)}(\mathcal{E}_{|\widehat{H}_{\infty,q}\langle t_0,\epsilon\rangle}) = \bigoplus \boldsymbol{P}_a^{(t_0)}(\mathcal{S}_\omega\mathcal{E}_{|\widehat{H}_{\infty,q}\langle t_0,\epsilon\rangle}),$$

we obtain the decomposition $v=\sum v_\omega$. For any v_ω and for any $t\in H_{\infty,q}\langle t_0,\epsilon\rangle$, we set

$$\deg^{\mathcal{P}}(v_{\omega|\widehat{H}_{\infty,q}\langle t\rangle}) := \begin{cases} \inf\{b\in\mathbb{R} \mid v_\omega\in\mathcal{P}_b(\mathcal{E}_{|\widehat{H}_{\infty,q}\langle t\rangle})\} & (v_\omega\neq 0)\\ -\infty & (v_\omega=0).\end{cases}$$

Lemma 3.5.26 *Suppose that there exists ω_0 such that*

$$\deg^{\mathcal{P}}(v_{\omega|\widehat{H}_{\infty,q}\langle t_0\rangle}) < \deg^{\mathcal{P}}(v_{\omega_0|\widehat{H}_{\infty,q}\langle t_0\rangle})$$

for any $\omega\neq\omega_0$. Then, we obtain

$$\deg^{\mathcal{P}}(v_{\omega|\widehat{H}_{\infty,q}\langle t\rangle}) < \deg^{\mathcal{P}}(v_{\omega_0|\widehat{H}_{\infty,q}\langle t\rangle})$$

for $t\in H_{\infty,q}\langle t_0,\epsilon\rangle$ and any $\omega\neq\omega_0$.

Proof It follows from Lemma 3.5.23 and (3.20). ■

3.5.8 *Complement for Good Filtered Bundles with Level ≤1*

Let $\mathcal{E}$ be a locally free $\mathcal{O}_{\widehat{H}_{\infty,q}}(*H_{\infty,q})$-module of level ≤ 1. Let $\mathcal{P}_*\mathcal{E} = \big(\mathcal{P}_*(\mathcal{E}_{|\widehat{H}_{\infty,q}\langle t\rangle}) \,\big|\, t \in S^1_T\big)$ be a good filtered bundle over $\mathcal{E}$. In this case, for each $a \in \mathbb{R}$, there exist $\mathcal{O}_{\widehat{H}_{\infty,q}}$-lattices $\mathcal{P}_a\mathcal{E} \subset \mathcal{E}$ such that $\mathcal{P}_a(\mathcal{E})_{|\widehat{H}_{\infty,q}\langle t\rangle} = \mathcal{P}_a\big(\mathcal{E}_{|\widehat{H}_{\infty,q}\langle t\rangle}\big)$. The other filtrations are also simply described in terms of $\mathcal{P}_a\mathcal{E}$. Indeed, we obtain $\mathfrak{P}_a(\mathcal{E}^{\mathrm{cov}}) = \varpi_q^{-1}(\mathcal{P}_a\mathcal{E})$. For any $t_0 \in S^1_T$, we also obtain $\boldsymbol{P}^{(t_0)}_a(\mathcal{E}_{|\widehat{H}_{\infty,q}\langle t_0,\epsilon\rangle}) = \mathcal{P}_a(\mathcal{E})_{|\widehat{H}_{\infty,q}\langle t_0,\epsilon\rangle}$.

3.6 Formal Difference Modules of Level ≤1 and Formal λ-Connections

Let $q \in \mathbb{Z}_{\geq 1}$. Let us explain that $(2\sqrt{-1}\lambda T)$-difference $\mathbb{C}(\!(y_q^{-1})\!)$-modules of level ≤ 1 are equivalent to formal λ-connections whose Poincaré rank are strictly smaller than q.

3.6.1 *Formal λ-Connections*

We introduce a formal variable x, and we fix a q-th root x_q of x. Let $\mathcal{V}$ be a $\mathbb{C}(\!(x_q^{-1})\!)$-module with a λ-connection ∇^λ. A $\mathbb{C}[\![x_q^{-1}]\!]$-lattice $\mathcal{L}$ of $\mathcal{V}$ is called ∇^λ_x-invariant if $\nabla^\lambda_x\mathcal{L} \subset \mathcal{L}$ is satisfied, where $\nabla^\lambda_x = q^{-1}x^{-1}x_q\nabla^\lambda_{x_q} = q^{-1}x_q^{-q+1}\nabla^\lambda_{x_q}$. For any ∇^λ_x-invariant lattice $\mathcal{L}$, we obtain the endomorphism F_{∇^λ} of $\mathcal{L}_{|\infty_{x,q}} = \mathcal{L}/x_q^{-1}\mathcal{L}$ as follows; for any $v \in \mathcal{L}_{|\infty_{x,q}}$, we take $\widetilde{v} \in \mathcal{L}$ which induces v, then we put $F_{\nabla^\lambda}(v) = \nabla^\lambda(\widetilde{v})_{|\infty_{x,q}}$.

Recall that there exists $p \in q\mathbb{Z}_{>0}$ and a decomposition

$$\varphi^*_{q,p}(\mathcal{V}, \nabla^\lambda) = \bigoplus_{\mathfrak{a}\in x_p\mathbb{C}[x_p]} (\mathcal{V}_{\mathfrak{a}}, \nabla^\lambda_{\mathfrak{a}}) \tag{3.21}$$

such that there exist $\mathbb{C}[\![x_q^{-1}]\!]$-lattices $\mathcal{L}_{\mathfrak{a}}$ of $\mathcal{V}_{\mathfrak{a}}$ such that $\nabla^\lambda_{\mathfrak{a}} - d\mathfrak{a}\,\mathrm{id}$ are logarithmic with respect to $\mathcal{L}_{\mathfrak{a}}$. (See Sect. 3.3.1 or Sect. 3.4.2 for $\varphi^*_{q,p}$.) The set $\mathcal{I}(\mathcal{V}, \nabla^\lambda) = \{\mathfrak{a} \mid \mathcal{V}_{\mathfrak{a}} \neq 0\}$ is uniquely determined. If $\mathcal{I}(\mathcal{V}, \nabla^\lambda)$ is contained in $x_q\mathbb{C}[x_q]$, then $\mathcal{I}(\mathcal{V}, \nabla^\lambda)$ is called unramified. In that case, $(\mathcal{V}, \nabla^\lambda)$ is called unramified. The Poincaré rank of $(\mathcal{V}, \nabla^\lambda)$ is defined to be

$$\max\Big\{\frac{q}{p}\deg_{x_p}\mathfrak{a}(x_p) \,\Big|\, \mathfrak{a} \in \mathcal{I}(\mathcal{V}, \nabla^\lambda)\Big\}.$$

We say that the $(\mathcal{V}, \nabla^{\lambda})$ is regular if the Poincaré rank of $(\mathcal{V}, \nabla^{\lambda})$ is 0. It is equivalent to the existence of a $\mathbb{C}[\![x_q^{-1}]\!]$-lattice $\mathcal{L}$ of $\mathcal{V}$ for which ∇^{λ} is logarithmic, i.e., $x\nabla_x^{\lambda}\mathcal{L} \subset \mathcal{L}$.

The following lemma is standard.

Lemma 3.6.1 *Poincaré rank of $(\mathcal{V}, \nabla^{\lambda})$ is less than q if and only if there exists a ∇_x^{λ}-invariant lattice of $(\mathcal{V}, \nabla^{\lambda})$. If Poincaré rank of $(\mathcal{V}, \nabla^{\lambda})$ is strictly less than q, for any ∇_x^{λ}-invariant lattice $\mathcal{L}$, the induced endomorphism $F_{\nabla^{\lambda}}$ of $\mathcal{L}_{|\infty_{x,q}}$ is nilpotent.* ∎

3.6.2 Some Sheaves of Algebras on $\widehat{H}_{\infty,q}$

For any open subset $U \subset H^{\mathrm{cov}}_{\infty,q}$, let $\mathcal{K}_{\widehat{H}^{\mathrm{cov}}_{\infty,q}}(U)$ denote the space of formal power series with the variable y_q^{-1} over $C^{\infty}(U)$, i.e.,

$$\mathcal{K}_{\widehat{H}^{\mathrm{cov}}_{\infty,q}}(U) := \Big\{ \sum_{j=0}^{\infty} a_j(t) y_q^{-j} \,\Big|\, a_j(t) \in C^{\infty}(U) \Big\}.$$

Similarly, for any open subset $U \subset H^{\mathrm{cov}}_{\infty,q}$, let $\mathcal{K}_{\widehat{H}^{\mathrm{cov}}_{\infty,q}}(*H^{\mathrm{cov}}_{\infty,q})(U)$ denote the space of formal Laurent power series with the variable y_q^{-1} over $C^{\infty}(U)$. Thus, we obtain sheaves of algebras $\mathcal{K}_{\widehat{H}^{\mathrm{cov}}_{\infty,q}}$ and $\mathcal{K}_{\widehat{H}^{\mathrm{cov}}_{\infty,q}}(*H^{\mathrm{cov}}_{\infty,q})$ on $H^{\mathrm{cov}}_{\infty,q}$. Note that $\mathcal{K}_{\widehat{H}^{\mathrm{cov}}_{\infty,q}}(*H^{\mathrm{cov}}_{\infty,q})$ is naturally isomorphic to $\mathcal{O}_{\widehat{H}^{\mathrm{cov}}_{\infty,q}}(*H_{\infty,q}) \otimes_{\mathcal{O}_{\widehat{H}^{\mathrm{cov}}_{\infty}}} \mathcal{K}_{\widehat{H}^{\mathrm{cov}}_{\infty,q}}$. We obtain the operators $\partial_t : \mathcal{K}_{\widehat{H}^{\mathrm{cov}}_{\infty,q}} \longrightarrow \mathcal{K}_{\widehat{H}^{\mathrm{cov}}_{\infty,q}}$ and $\partial_t : \mathcal{K}_{\widehat{H}^{\mathrm{cov}}_{\infty,q}}(*H^{\mathrm{cov}}_{\infty,q}) \longrightarrow \mathcal{K}_{\widehat{H}^{\mathrm{cov}}_{\infty,q}}(*H^{\mathrm{cov}}_{\infty,q})$ defined by $\partial_t \sum a_j(t) y_q^{-j} = \sum \partial_t a_j(t) y_q^{-j}$. The kernels are $\mathcal{O}_{\widehat{H}^{\mathrm{cov}}_{\infty,q}}$ and $\mathcal{O}_{\widehat{H}^{\mathrm{cov}}_{\infty,q}}(*H^{\mathrm{cov}}_{\infty,q})$, respectively.

The sheaves $\mathcal{K}_{\widehat{H}^{\mathrm{cov}}_{\infty,q}}$ and $\mathcal{K}_{\widehat{H}^{\mathrm{cov}}_{\infty,q}}(*H^{\mathrm{cov}}_{\infty,q})$ are equivariant with respect to the $\mathbb{Z}$-action κ. The operators ∂_t on $\mathcal{K}_{\widehat{H}^{\mathrm{cov}}_{\infty,q}}$ and $\mathcal{K}_{\widehat{H}^{\mathrm{cov}}_{\infty,q}}(*H^{\mathrm{cov}}_{\infty,q})$ are also equivariant.

We obtain the sheaves $\mathcal{K}_{\widehat{H}_{\infty,q}}$ and $\mathcal{K}_{\widehat{H}_{\infty,q}}(*H_{\infty,q})$ on $H_{\infty,q}$ as the descents of $\mathcal{K}_{\widehat{H}^{\mathrm{cov}}_{\infty,q}}$ and $\mathcal{K}_{\widehat{H}^{\mathrm{cov}}_{\infty,q}}(*H^{\mathrm{cov}}_{\infty,q})$, respectively. We obtain the induced operators ∂_t on $\mathcal{K}_{\widehat{H}_{\infty,q}}$ and $\mathcal{K}_{\widehat{H}_{\infty,q}}(*H_{\infty,q})$. The kernels are $\mathcal{O}_{\widehat{H}_{\infty,q}}$ and $\mathcal{O}_{\widehat{H}_{\infty,q}}(*H_{\infty,q})$, respectively.

3.6.3 From Formal λ-Connections to Formal Difference Modules

Let Ψ denote the map $H_{\infty} \longrightarrow \{\infty\}$. There exists the morphism of sheaves $\Psi^* : \Psi^{-1}\mathcal{O}_{\widehat{\infty}_x} \longrightarrow \mathcal{K}_{\widehat{H}_{\infty}}$ induced by $\Psi^*(x^{-1}) = y^{-1}(1 - 2\sqrt{-1}\lambda t/y)^{-1}$. We regard (Ψ, Ψ^*) as a "C^{∞}"-map $\Psi : \widehat{H}_{\infty} \longrightarrow \widehat{\infty}_x$. Similarly, we obtain the map Ψ_q :

$H_{\infty,q} \longrightarrow \{\infty_{x,q}\}$ and the morphism of sheaves $\Psi_q^* : \Psi_q^{-1}\mathcal{O}_{\widehat{\infty}_{x,q}} \longrightarrow \mathcal{K}_{\widehat{H}_{\infty,q}}$ induced by $\Psi_q^*(x_q^{-1}) = y_q^{-1}(1 - 2\sqrt{-1}\lambda t/y_q^q)^{-1/q}$. We regard the pair (Ψ_q, Ψ_q^*) as a C^∞-map $\Psi_q : \widehat{H}_{\infty,q} \longrightarrow \widehat{\infty}_{x,q}$. Let $\widehat{\infty}_{x,q} \longrightarrow \widehat{\infty}_x$ be the ramified covering given by $x_q \longmapsto x = x_q^q$. Then, the following diagram is commutative:

$$\begin{array}{ccc} \widehat{H}_{\infty,q} & \longrightarrow & \widehat{H}_\infty \\ \Psi_q \downarrow & & \Psi \downarrow \\ \widehat{\infty}_{x,q} & \longrightarrow & \widehat{\infty}_x . \end{array}$$

Let $(\mathcal{V}, \nabla^\lambda)$ be a $\mathbb{C}(\!(x_q^{-1})\!)$-module with a λ-connection. We obtain a locally free $\mathcal{K}_{\widehat{H}_{\infty,q}}(*H_{\infty,q})$-module

$$\widetilde{\mathcal{V}}^\infty := \mathcal{K}_{\widehat{H}_{\infty,q}}(*H_{\infty,q}) \otimes_{\Psi_q^{-1}\mathcal{O}_{\widehat{\infty}_{x,q}}(*\infty_{x,q})} \Psi_q^{-1}(\mathcal{V}) \simeq \mathcal{K}_{\widehat{H}_{\infty,q}} \otimes_{\Psi_q^{-1}\mathcal{O}_{\widehat{\infty}_{x,q}}} \Psi_q^{-1}(\mathcal{V}). \tag{3.22}$$

There uniquely exists an operator $\partial_{\widetilde{\mathcal{V}}^\infty,t} : \widetilde{\mathcal{V}}^\infty \longrightarrow \widetilde{\mathcal{V}}^\infty$ determined by the following condition for any local sections f and s of $\mathcal{K}_{\widehat{H}_{\infty,q}}$ and $\mathcal{V}$:

$$\partial_{\widetilde{\mathcal{V}}^\infty,t}\big(f \cdot \Psi_q^{-1}(s)\big) = \partial_t(f)\Psi_q^{-1}(s) + f\Psi_q^{-1}(-2\sqrt{-1}\nabla_x^\lambda s). \tag{3.23}$$

Let $\Psi_q^*(\mathcal{V}, \nabla^\lambda)$ denote the $\mathcal{O}_{\widehat{H}_{\infty,q}}(*H_{\infty,q})$-module obtained as the kernel of $\partial_{\widetilde{\mathcal{V}}^\infty,t}$: $\widetilde{\mathcal{V}}^\infty \longrightarrow \widetilde{\mathcal{V}}^\infty$.

If ∇_x^λ-invariant lattice $\mathcal{L}$ of $\mathcal{V}$ exists, we obtain a $\mathcal{K}_{\widehat{H}_{\infty,q}}$-submodule

$$\widetilde{\mathcal{L}}^\infty := \mathcal{K}_{\widehat{H}_{\infty,q}} \otimes_{\Psi_q^{-1}\mathcal{O}_{\widehat{\infty}_{x,q}}} \Psi_q^{-1}(\mathcal{L}) \subset \widetilde{\mathcal{V}}^\infty \tag{3.24}$$

such that $\partial_{\widetilde{\mathcal{V}}^\infty,t}\widetilde{\mathcal{L}}^\infty \subset \widetilde{\mathcal{L}}^\infty$. We set $\Psi_q^*(\mathcal{L}, \nabla^\lambda) := \widetilde{\mathcal{L}}^\infty \cap \Psi_q^*(\mathcal{V}, \nabla^\lambda)$ which is an $\mathcal{O}_{\widehat{H}_{\infty,q}}$-module.

Proposition 3.6.2 *Suppose that the Poincaré rank of $(\mathcal{V}, \nabla^\lambda)$ is less than q.*

- *$\Psi_q^*(\mathcal{V}, \nabla^\lambda)$ is a locally free $\mathcal{O}_{\widehat{H}_{\infty,q}}(*H_{\infty,q})$-module of rank $\mathcal{V}$ such that $\Psi_q^*(\mathcal{V}, \nabla^\lambda)$ has pure slope 0. If the Poincaré rank of $(\mathcal{V}, \nabla^\lambda)$ is strictly less than q, then the level of $\Psi_q^*(\mathcal{V}, \nabla^\lambda)$ is less than 1. If $(\mathcal{V}, \nabla^\lambda)$ is regular, the level of $\Psi_q^*(\mathcal{V}, \nabla^\lambda)$ is 0.*
- *For any ∇_x^λ-invariant lattice $\mathcal{L}$ of $\mathcal{V}$, $\Psi_q^*(\mathcal{L}, \nabla^\lambda)$ is an $\mathcal{O}_{\widehat{H}_{\infty,q}}$-lattice of $\Psi_q^*(\mathcal{V}, \nabla^\lambda)$. If the induced endomorphism F_{∇^λ} of $\mathcal{L}_{|\infty_{x,q}}$ is nilpotent, then the monodromy of* $\mathrm{Loc}_1\big(\Psi_q^*(\mathcal{L}, \nabla^\lambda)\big)$ *is unipotent. If ∇^λ is logarithmic with respect to $\mathcal{L}$, then the monodromy of* $\mathrm{Loc}_q\big(\Psi_q^*(\mathcal{L}, \nabla^\lambda)\big)$ *is the identity.*

Proof Let $\mathcal{L}$ be a ∇_x^λ-invariant lattice of $(\mathcal{V}, \nabla^\lambda)$. Set $r := \operatorname{rank}(\mathcal{V})$. Let $\boldsymbol{v} = (v_1, \dots, v_r)$ be a frame of $\mathcal{L}$. We obtain $A \in M_r(\mathbb{C}[\![x_q^{-1}]\!])$ determined by $\nabla_x^\lambda \boldsymbol{v} = \boldsymbol{v}A$. We obtain the induced frame $\widetilde{\boldsymbol{v}} = (\widetilde{v}_1, \dots, \widetilde{v}_r) = \Psi_q^{-1}(\boldsymbol{v})$ of $\widetilde{\mathcal{L}}^\infty$. The action of $\partial_{\widetilde{\mathcal{V}}^\infty,t}$ is expressed as follows with respect to $\widetilde{\boldsymbol{v}}$:

$$\partial_{\widetilde{\mathcal{V}}^\infty,t}\widetilde{\boldsymbol{v}} = \widetilde{\boldsymbol{v}} \cdot \Big(-2\sqrt{-1}\Psi_q^* A \Big).$$

Note that for the expansion $A = \sum_{j=0}^\infty A_j x_q^{-j}$, where $A_j \in M_r(\mathbb{C})$, we obtain

$$\Psi_q^*(A)(t, y_q) = \sum_{j=0}^\infty A_j y_q^{-j}(1 - 2\sqrt{-1}\lambda t / y_q^q)^{-j/q}.$$

Let $P \in H_{\infty,q}$ be any point, and let U be a simply connected neighbourhood of P in $H_{\infty,q}$. We put $\widetilde{\boldsymbol{v}}_P^{(1)} := \boldsymbol{v}_{|U} \exp(2\sqrt{-1}A_0 t)$. Let $\widetilde{A}^{(1)} \in M_r(\mathcal{K}_{\widehat{H}_{\infty,q}}(U))$ be determined by $\partial_{\widetilde{\mathcal{V}}^\infty,t}\widetilde{\boldsymbol{v}}_P^{(1)} = \widetilde{\boldsymbol{v}}_P^{(1)}\widetilde{A}^{(1)}$. It is easy to check that $\widetilde{A}^{(1)} \in y_q^{-1}M_r(\mathcal{K}_{\widehat{H}_{\infty,q}}(U))$.

Lemma 3.6.3 *There exist matrices* $G^{(i)} \in M_r(C^\infty(U))$ $(i = 2, 3, \dots)$ *such that the following condition is satisfied.*

- *Let* $I_r \in M_r(\mathbb{C})$ *denote the identity matrix. We set*

$$\widetilde{\boldsymbol{v}}_P^{(i)} = \widetilde{\boldsymbol{v}}_P^{(1)}(I_r - y_q^{-1}G^{(2)}) \cdots (I_r - y_q^{-i+1}G^{(i)}).$$

  Let $\widetilde{A}^{(i)} \in M_r(\mathcal{K}_{\widehat{H}_{\infty,q}}(U))$ *be determined by* $\partial_{\widetilde{\mathcal{V}}^\infty,t}\widetilde{\boldsymbol{v}}_P^{(i)} = \widetilde{\boldsymbol{v}}_P^{(i)}\widetilde{A}^{(i)}$. *Then,* $\widetilde{A}^{(i)}$ *is contained in* $y_q^{-i}M_r(\mathcal{K}_{\widehat{H}_{\infty,q}}(U))$.

Proof We construct $G^{(i)}$ inductively. Suppose that $G^{(i)}$ $(i \le \ell)$ are constructed. For the expansion $A^{(\ell)} = \sum_{j\ge\ell} A_j^{(\ell)} y_q^{-j}$, there exists $G^{(\ell+1)} \in M_r(C^\infty(U))$ such that $\partial_t G^{(\ell+1)} = A_\ell^{(\ell)}$. We can check that the condition is satisfied for $i \le \ell + 1$. ■

By Lemma 3.6.3, there exists a frame $\widetilde{\boldsymbol{v}}_P^{(\infty)}$ of $\widetilde{\mathcal{L}}_{|U}^\infty$ such that $\partial_{\widetilde{\mathcal{V}}^\infty,t}\widetilde{\boldsymbol{v}}_P^{(\infty)} = 0$. Hence, $\Psi^*(\mathcal{L}, \nabla^\lambda)$ is a locally free $\mathcal{O}_{\widehat{H}_{\infty,q}}$-module whose rank is equal to $\operatorname{rank}\mathcal{V}$. Note that the conjugacy class of the monodromy of the local system $\operatorname{Loc}_1(\Psi^*(\mathcal{L}, \nabla^\lambda))$ is equal to the conjugacy class of $\exp(-2\sqrt{-1}\lambda A_0 T)$. Hence, if A_0 is nilpotent, the monodromy of $\operatorname{Loc}_1(\Psi^*(\mathcal{L}, \nabla^\lambda))$ is unipotent. If ∇^λ is logarithmic with respect to $\mathcal{L}$, we obtain $A_j = 0$ for $j = 0, \dots, q-1$, and hence the monodromy of $\operatorname{Loc}_q(\Psi^*(\mathcal{L}, \nabla^\lambda))$ is the identity. Thus, we obtain the claims for $\Psi^*(\mathcal{L}, \nabla^\lambda)$. The claims for $\Psi^*(\mathcal{V}, \nabla^\lambda)$ immediately follow. ■

Remark 3.6.4 Suppose that the Poincaré rank of $(\mathcal{V}, \nabla^\lambda)$ is less than q. Let $\mathcal{L}$ be a lattice such that $\nabla_x^\lambda \mathcal{L} \subset \mathcal{L}$. If any eigenvalues α of F_∇ on $\mathcal{L}_{|\infty_{x,q}}$ satisfy $\exp(2\sqrt{-1}T\alpha) = 1$, then the level of $\Psi^*(\mathcal{V}, \nabla^\lambda)$ is strictly less than 1. ■

Let $P \in H_{\infty,q}$. By evaluating C^{∞}-functions at t, we obtain the morphisms of algebras $\mathcal{K}_{\widehat{H}_{\infty,q},P} \longrightarrow \mathcal{O}_{\widehat{H}_{\infty,q},P}$ and $\mathcal{K}_{\widehat{H}_{\infty,q}}(*H_{\infty,q})_P \longrightarrow \mathcal{O}_{\widehat{H}_{\infty,q}}(*H_{\infty,q})_P$. The natural morphism $\Psi_q^*(\mathcal{V}, \nabla^{\lambda}) \longrightarrow \widetilde{\mathcal{V}}^{\infty}$ induces

$$\Psi_q^*(\mathcal{V}, \nabla^{\lambda})_P \xrightarrow{c_{\mathcal{V},1}} \widetilde{\mathcal{V}}_P^{\infty} \xrightarrow{c_{\mathcal{V},2}} \widetilde{\mathcal{V}}_P^{\infty} \otimes_{\mathcal{K}_{\widehat{H}_{\infty,q}}(*H_{\infty,q})_P} \mathcal{O}_{\widehat{H}_{\infty,q}}(*H_{\infty,q})_P. \tag{3.25}$$

Similarly, for a ∇_x^{λ}-invariant lattice $\mathcal{L}$ of $(\mathcal{V}, \nabla^{\lambda})$, we obtain

$$\Psi_q^*(\mathcal{L}, \nabla^{\lambda})_P \xrightarrow{c_{\mathcal{L},1}} \widetilde{\mathcal{L}}_P^{\infty} \xrightarrow{c_{\mathcal{L},2}} \widetilde{\mathcal{L}}_P^{\infty} \otimes_{\mathcal{K}_{\widehat{H}_{\infty,q},P}} \mathcal{O}_{\widehat{H}_{\infty,q},P}. \tag{3.26}$$

Lemma 3.6.5 *The morphisms $c_{2,\mathcal{V}} \circ c_{1,\mathcal{V}}$ and $c_{2,\mathcal{L}} \circ c_{1,\mathcal{L}}$ are isomorphisms.*

Proof Because $\widetilde{\mathcal{V}}^{\infty} = \Psi_q^*(\mathcal{V}, \nabla^{\lambda}) \otimes_{\mathcal{O}_{\widehat{H}_{\infty,q}}} \mathcal{K}_{\widehat{H}_{\infty,q}}$ and $\widetilde{\mathcal{L}}^{\infty} = \Psi_q^*(\mathcal{L}, \nabla^{\lambda}) \otimes_{\mathcal{O}_{\widehat{H}_{\infty,q}}} \mathcal{K}_{\widehat{H}_{\infty,q}}$, the claim is obvious. ■

Lemma 3.6.6 *Let $(\mathcal{V}_i, \nabla^{\lambda})$ $(i = 1, 2)$ be finite dimensional $\mathbb{C}((x_q^{-1}))$-vector spaces with λ-connection. Then, there exist natural isomorphisms*

$$\Psi_q^*\big((\mathcal{V}_1, \nabla^{\lambda}) \oplus (\mathcal{V}_2, \nabla^{\lambda})\big) \simeq \Psi_q^*\big((\mathcal{V}_1, \nabla^{\lambda})\big) \oplus \Psi_q^*\big((\mathcal{V}_2, \nabla^{\lambda})\big) \tag{3.27}$$

$$\Psi_q^*\big((\mathcal{V}_1, \nabla^{\lambda}) \otimes (\mathcal{V}_2, \nabla^{\lambda})\big) \simeq \Psi_q^*\big((\mathcal{V}_1, \nabla^{\lambda})\big) \otimes \Psi_q^*\big((\mathcal{V}_2, \nabla^{\lambda})\big) \tag{3.28}$$

$$\Psi_q^*\Big(\mathrm{Hom}\,\big((\mathcal{V}_1, \nabla^{\lambda}), (\mathcal{V}_2, \nabla^{\lambda})\big)\Big) \simeq \mathcal{H}om\Big(\Psi_q^*\big((\mathcal{V}_1, \nabla^{\lambda})\big), \Psi_q^*\big((\mathcal{V}_2, \nabla^{\lambda})\big)\Big). \tag{3.29}$$

In particular, for a finite dimensional $\mathbb{C}((x_q^{-1}))$-vector space $\mathcal{V}$ with a λ-connection ∇^{λ}, we obtain

$$\Psi_q^*\big((\mathcal{V}, \nabla^{\lambda})^{\vee}\big) \simeq \Psi_q^*\big((\mathcal{V}, \nabla^{\lambda})\big)^{\vee}.$$

Proof We explain only the claim for the tensor product. The others are similar. We set $\mathcal{V}_3 := \mathcal{V}_1 \otimes \mathcal{V}_2$. By the construction, there exists a natural isomorphism

$$\widetilde{\mathcal{V}}_1^{\infty} \otimes_{\mathcal{K}_{\widehat{H}_{\infty,q}}(*H_{\infty,q})} \widetilde{\mathcal{V}}_2^{\infty} \simeq \widetilde{\mathcal{V}}_3^{\infty}.$$

The natural morphisms $\Psi_q^*(\mathcal{V}_i, \nabla^{\lambda}) \longrightarrow \widetilde{\mathcal{V}}_i^{\infty}$ $(i = 1, 2)$ induce the following:

$$\Psi_q^*(\mathcal{V}_1, \nabla^{\lambda}) \otimes \Psi_q^*(\mathcal{V}_2, \nabla^{\lambda}) \longrightarrow \widetilde{\mathcal{V}}_3^{\infty}.$$

It factors through $\Psi_q^*(\mathcal{V}_3, \nabla^{\lambda})$, and it induces the desired isomorphism (3.28). ■

Remark 3.6.7 We can compare the constructions in this section and Sect. 2.8.3.4 by replacing (t, y, x) with $(t_1, \beta_1, (1 + |\lambda|^2)w)$. See Remark 2.8.9. ■

3.6.4 Equivalence

The following proposition is fundamental in our study.

Proposition 3.6.8 *The functor Ψ^* induces an equivalence of the categories of the following objects.*

- *finite dimensional $\mathbb{C}((x_q^{-1}))$-modules with λ-connection of Poincaré rank $< q$.*
- *locally free $\mathcal{O}_{\widehat{H}_{\infty,q}}(*H_{\infty,q})$-modules of finite rank with level ≤ 1.*

It also induces an equivalence of the categories of the following objects.

- *finite dimensional $\mathbb{C}((x_q^{-1}))$-modules with regular λ-connection.*
- *locally free $\mathcal{O}_{\widehat{H}_{\infty,q}}(*H_{\infty,q})$-modules of finite rank whose level is 0.*

Proof Let us prove that the functor is essentially surjective. (See also Sect. 3.6.4.1 for a simplified proof in easy cases.) Let $\mathcal{E}$ be a locally free $\mathcal{O}_{\widehat{H}_{\infty,q}}(*H_{\infty,q})$-module of rank r whose level is less than 1. Let $\mathcal{E}_0$ be an $\mathcal{O}_{\widehat{H}_{\infty,q}}$-lattice of $\mathcal{E}$ such that the monodromy of $\mathrm{Loc}_1(\mathcal{E}_0)$ is unipotent. We set

$$\mathcal{E}^{\infty} := \mathcal{K}_{\widehat{H}_{\infty,q}} \otimes_{\mathcal{O}_{\widehat{H}_{\infty,q}}} \mathcal{E}, \qquad \mathcal{E}_0^{\infty} := \mathcal{K}_{\widehat{H}_{\infty,q}} \otimes_{\mathcal{O}_{\widehat{H}_{\infty,q}}} \mathcal{E}_0.$$

We obtain the naturally defined operators $\partial_{\mathcal{E}^\infty,t} : \mathcal{E}^\infty \longrightarrow \mathcal{E}^\infty$ and $\partial_{\mathcal{E}^\infty,t} : \mathcal{E}_0^\infty \longrightarrow \mathcal{E}_0^\infty$.

Note that $\Psi_q^*(x_q^{-1}) = y_q^{-1}(1-2\sqrt{-1}\lambda t/y_q^q)^{-1/q}$ is a global section of $\mathcal{K}_{\widehat{H}_{\infty,q}}$ on $H_{\infty,q}$, which is denoted by x_q^{-1} to simplify the description. Any section f of $\mathcal{K}_{\widehat{H}_{\infty,q}}$ on $H_{\infty,q}$ is expanded as $f = \sum_{j=0}^{\infty} f_j(t)x_q^{-j}$ for $f_j(t) \in C^\infty(H_{\infty,q})$.

There exists a global frame $\boldsymbol{u}$ of $\mathcal{E}_0^\infty$. We obtain $A \in M_r(\mathcal{K}_{\widehat{H}_{\infty,q}}(H_{\infty,q}))$ determined by $\partial_{\mathcal{E}^\infty,t}\boldsymbol{u} = \boldsymbol{u}A$. We have the expansion $A = \sum_{j=0}^{\infty} A_j x_q^{-j}$.

Lemma 3.6.9 *We may assume that A_0 is constant and nilpotent.*

Proof Let $C^\infty_{H_{\infty,q}}$ denote the sheaf of C^∞-functions on $H_{\infty,q}$. We obtain the locally free $C^\infty_{H_{\infty,q}}$-module $\mathrm{Loc}_1(\mathcal{E}_0) \otimes C^\infty_{H_{\infty,q}}$. It is equipped with the frame $\boldsymbol{u}_0$ induced by $\boldsymbol{u}$, and the connection ∂_t induced by $\partial_{\mathcal{E}^\infty,t}$. We obtain $\partial_t\boldsymbol{u}_0 = \boldsymbol{u}_0A_0$. Because the monodromy of $\mathrm{Loc}_1(\mathcal{E}_0)$ is unipotent, there exists a frame $\boldsymbol{u}_0'$ of $\mathrm{Loc}_1(\mathcal{E}_0) \otimes C^\infty_{H_{\infty,q}}$ such that $\partial_t\boldsymbol{u}_0' = \boldsymbol{u}_0' \cdot N$ for a constant nilpotent matrix N. There exists $G \in \mathrm{GL}_r(C^\infty(H_{\infty,q}))$ such that $\boldsymbol{u}_0' = \boldsymbol{u}_0 \cdot G$. By considering $\boldsymbol{u} \cdot G$ instead of $\boldsymbol{u}$, we may assume that A_0 is a constant nilpotent matrix. ■

Lemma 3.6.10 *There exist matrices $G^{(i)} \in M_r(C^\infty(H_{\infty,q}))$ $(i = 1, 2, \dots)$ such that the following condition is satisfied.*

- *We set $\boldsymbol{u}^{(i)} = \boldsymbol{u}(I_r + x_q^{-1}G^{(1)})(I_r + x_q^{-2}G^{(2)}) \cdots (I_r + x_q^{-i}G^{(i)})$. Let $A^{(i)}$ be the matrix determined by $\partial_{\mathcal{E}^\infty,t}\boldsymbol{u}^{(i)} = \boldsymbol{u}^{(i)}A^{(i)}$. Then, for the expansion $A^{(i)} = \sum_{j=0}^{\infty} A_j^{(i)} x_q^{-j}$, $A_j^{(i)}$ $(j \leq i)$ are constant.*

Proof We shall construct such $G^{(i)}$ inductively. Suppose that we have already constructed $G^{(i)}$ $(i \leq \ell - 1)$. By the assumption, $A_j^{(\ell-1)}$ $(j < \ell)$ are constant. Moreover, by the construction, $A_0^{(\ell-1)} = A_0$ is a constant nilpotent matrix. There exists $G^{(\ell)}(t) \in M_r(C^\infty(H_{\infty,q}))$ such that

$$\partial_t G^{(\ell)}(t) + \big(-G^{(\ell)}(t)A_0 + A_0 G^{(\ell)}(t)\big) + A_\ell^{(\ell-1)}(t)$$

is constant. We set $\boldsymbol{u}^{(\ell)} = \boldsymbol{u}^{(\ell-1)}(I_r + x_q^{-\ell}G^{(\ell)})$. We obtain

$$A^{(\ell)} = (I_r + x_q^{-\ell}G^{(\ell)})^{-1}\Big(\partial_t G^{(\ell)} x_q^{-\ell} + G^{(\ell)}\partial_t x_q^{-\ell} + A^{(\ell-1)}(I_r + x_q^{-\ell}G^{(\ell)})\Big).$$

Note that $\partial_t x_q^{-j_0} = 2\sqrt{-1}\lambda j_0 q^{-1} x_q^{-j_0-q} = -2\sqrt{-1}\lambda\partial_x(x_q^{-j_0})$. Because

$$A_\ell^{(\ell)} = \partial_t G^{(\ell)}(t) + \big(-G^{(\ell)}(t)A_0 + A_0 G^{(\ell)}(t)\big) + A_\ell^{(\ell-1)}(t),$$

$A_\ell^{(\ell)}$ is constant. Hence, the induction can proceed. ■

By Lemma 3.6.10, there exists a global frame $\boldsymbol{u}^{(\infty)}$ of $\mathcal{E}_0^\infty$ with the following property.

- Let $A^{(\infty)}$ be the matrix determined by $\partial_{\mathcal{E}^\infty,t}\boldsymbol{u}^{(\infty)} = \boldsymbol{u}^{(\infty)}A^{(\infty)}$. Then, for the expansion $A^{(\infty)} = \sum_{j=0}^{\infty} A_j^{(\infty)} x_q^{-j}$, $A_j^{(\infty)}$ are constant. Moreover, $A_0^{(\infty)}$ is nilpotent.

We set $\mathcal{V} = \bigoplus_{i=1}^r \mathbb{C}((x_q^{-1}))e_i$. We define the λ-connection by

$$\nabla_x^\lambda \boldsymbol{e} = \boldsymbol{e}(-2\sqrt{-1})^{-1}A^{(\infty)}.$$

Then, Poincaré rank of $(\mathcal{V}, \nabla^\lambda)$ is strictly smaller than q, and there exists an isomorphism $\Psi_q^*(\mathcal{V}, \nabla^\lambda) \simeq \mathcal{E}$. If the monodromy of $\mathrm{Loc}_q(\mathcal{E}_0)$ is the identity, we can easily observe that $\widetilde{A}_j^{(\infty)} = 0$ $(0 \leq j < q)$, i.e., ∇_x^λ is regular. Hence, Ψ_q^* is essentially surjective.

The functor Ψ_q^* is clearly faithful. Let us prove that Ψ_q^* is full. Let $(\mathcal{V}_i, \nabla^\lambda)$ $(i = 1, 2)$ be $\mathbb{C}((x_q^{-1}))$-modules with λ-connection whose Poincaré rank are strictly smaller than q. Let $F : \Psi_q^*(\mathcal{V}_1, \nabla^\lambda) \longrightarrow \Psi_q^*(\mathcal{V}_2, \nabla^\lambda)$ be a morphism of $\mathcal{O}_{\widehat{H}_{\infty,q}}(*H_{\infty,q})$-modules. Let us prove that F is induced by a morphism $(\mathcal{V}_1, \nabla^\lambda) \longrightarrow (\mathcal{V}_2, \nabla^\lambda)$.

There exist ∇_x^λ-invariant lattices $\mathcal{L}_i \subset \mathcal{V}_i$ such that the induced endomorphisms F_{∇^λ} of $\mathcal{L}_{i|\infty_{x,q}}$ are nilpotent. Let $\boldsymbol{v}_i$ be a frame of $\mathcal{L}_i$. Let $A_i(x_q^{-1})$ be determined by $\nabla_x^\lambda \boldsymbol{v}_i = \boldsymbol{v}_i A_i(x_q^{-1})$. Note that $A_i(0)$ are nilpotent.

We obtain frames $\widetilde{\boldsymbol{v}}_i := \Psi_q^{-1}(\boldsymbol{v}_i)$ of $\widetilde{\mathcal{V}}_i^{\infty} := \Psi_q^{-1}(\mathcal{V}_i) \otimes_{\Psi_q^{-1}(\mathcal{O}_{\widehat{\infty}_{x,q}}(*\infty_{x,q}))} \mathcal{K}_{\widehat{H}_{\infty,q}}(*H_{\infty,q})$. The sheaves are equipped with the operators $\partial_{\widetilde{\mathcal{V}}_i^{\infty},t}$. We set $\widetilde{A}_i := -2\sqrt{-1}\Psi_q^* A_i$, and then $\partial_{\widetilde{\mathcal{V}}_i^{\infty},t}\widetilde{\boldsymbol{v}}_i = \widetilde{\boldsymbol{v}}_i \widetilde{A}_i(x_q^{-1})$ holds.

Let B be the matrix determined by $F(\boldsymbol{v}_1) = \boldsymbol{v}_2 B$ whose entries are sections of $\mathcal{K}_{\widehat{H}_{\infty,q}}(*H_{\infty,q})$ on $H_{\infty,q}$. Because $F \circ \partial_{\widetilde{\mathcal{V}}_1^{\infty},t} = \partial_{\widetilde{\mathcal{V}}_2^{\infty},t} \circ F$, we obtain

$$\partial_t B - B\widetilde{A}_1 + \widetilde{A}_2 B = 0. \tag{3.30}$$

There exists the expansion $B = \sum_{j \geq -N}^{\infty} B_j(t) x_q^{-j}$, where $B_j(t)$ are matrices whose entries are C^∞-functions on $H_{\infty,q}$.

Lemma 3.6.11 *$B_j(t)$ are constant.*

Proof Suppose that there exists j such that $B_j(t)$ are non-constant. Let j_0 be the minimum of such j. We have the expansions $\widetilde{A}_i = \sum_{j=0}^{\infty} \widetilde{A}_{i,j} x_q^{-j}$. We obtain that $\partial_t B_{j_0}(t) + \widetilde{A}_{2,0} B_{j_0} - B_{j,0}\widetilde{A}_{1,0}$ is constant. Because $\widetilde{A}_{i,0}$ are nilpotent, we obtain that B_{j_0} is constant, which contradicts our choice of j_0. Hence, we obtain that B_j are constant for any j. ■

By Lemma 3.6.11, F is induced by the morphism $f : \mathcal{V}_1 \longrightarrow \mathcal{V}_2$ defined as $f(\boldsymbol{v}_1) = \boldsymbol{v}_2 B$. The relation (3.30) implies that f is compatible with the λ-connections. Hence, Ψ_q^* is full, and Proposition 3.6.8 is proved. ■

Remark 3.6.12 Proposition 3.6.8 allows us to understand the formal structure of $\mathcal{O}_{\widehat{H}_{\infty,q}}(*H_{\infty,q})$-modules with level ≤ 1 in terms of λ-flat bundles. It is useful for our understanding of the asymptotic behaviour of monopoles. See Sect. 2.9.5.2. ■

Remark 3.6.13 Proposition 3.4.2 and Proposition 3.6.8 provides us with an equivalence between some classes of λ-connections and $2\sqrt{-1}\lambda T$-difference modules. See Sects. 3.6.5 and 3.6.6 for some explicit examples. ■

We also obtained the following from the above proof.

Corollary 3.6.14 *Let $(\mathcal{V}, \nabla^\lambda)$ be a $\mathbb{C}((x_q^{-1}))$-module with a λ-connection whose Poincaré rank is strictly less than q. Then, the constructions in* Sects. 3.4.3 *and* 3.6.3 *induce natural bijective correspondences between the following objects.*

- *∇_x^λ-invariant lattices $\mathcal{L}$ of $(\mathcal{V}, \nabla^\lambda)$ such that the induced endomorphism of $\mathcal{L}_{|\infty}$ is nilpotent.*
- *$\mathcal{O}_{\widehat{H}_{\infty,q}}$-lattices $\mathcal{E}_0$ of $\Psi_q^*(\mathcal{V}, \nabla^\lambda)$ such that the monodromy of* $\mathrm{Loc}_1(\mathcal{E}_0)$ *is unipotent.*
- *$\mathbb{C}[\![y_q^{-1}]\!]$-lattices $\mathcal{N}_0$ of the $(2\sqrt{-1}\lambda T)$-difference module $\Psi_q^*(\mathcal{V}, \nabla^\lambda)_{|\widehat{H}_{\infty,q}\langle 0\rangle}$ such that $\Phi^*(\mathcal{N}_0) = \mathcal{N}_0$ and that the induced automorphism of $\mathcal{N}_{0|\infty}$ is unipotent.*

■

3.6.4.1 Simpler Cases of Proposition 3.6.8

Let us explain a simplified proof of the essential surjectivity in Proposition 3.6.8 in some special cases, as an explanation of basic ideas of the proof. Let $\mathcal{E}$ and $\mathcal{E}_0$ be as in the proof of Proposition 3.6.8.

Let us consider the case $\lambda = 0$. We obtain the $\mathbb{C}[\![y_q^{-1}]\!]$-module $\mathcal{L} = \mathcal{E}_{0|\widehat{H}_{\infty,q}\langle 0\rangle}$. If $\lambda = 0$, we obtain the $\mathbb{C}[\![y_q^{-1}]\!]$-automorphism M of $\mathcal{L}$. By the assumption, the induced automorphism $M_{|y_q^{-1}=0}$ of $\mathcal{L}/y_q^{-1}\mathcal{L}$ is unipotent. Hence, there exists an $\mathbb{C}[\![y_q^{-1}]\!]$-endomorphism N of $\mathcal{L}$ such that (i) $\exp(2\sqrt{-1}TN) = M$, (ii) the induced endomorphism $N_{|y_q^{-1}=0}$ of $\mathcal{L}/y_q^{-1}\mathcal{L}$ is nilpotent. We obtain the meromorphic Higgs field ∇^0 of $\mathcal{L}$ by $\nabla^0_x = N$. Then, it is easy to see that $\Psi_q^*(\mathcal{L},\nabla^0) \simeq \mathcal{E}_0$, and we obtain the essential surjectivity in this case. We also remark that we can obtain a similar equivalence even in the analytic case, not only in the formal case, if $\lambda = 0$.

Let us consider the case where $\lambda \neq 0$ but $\operatorname{rank}\mathcal{E} = 1$. Let v be a frame of $\mathcal{E}_0$. There exists a global section $a = \sum_{j=1}^{\infty} a_j(t)x_q^{-j}$ of $\mathcal{K}_{\widehat{H}_{\infty,q}}$ such that $\partial_t v = va$. We would like to find a section $b = \sum_{j=1}^{\infty} b_j(t)x_q^{-j}$ of $\mathcal{K}_{\widehat{H}_{\infty,q}}$ such that $\partial_t(b) + a \in x_q^{-1}\mathbb{C}[\![x_q^{-1}]\!]$. Indeed, if there exists such b, we put $v' = v\exp(b)$ for which $\partial_t v' = v'(\partial_t b + a)$ holds, and hence we obtain that $(\mathcal{E}_0^{\infty}, \partial_t)$ comes from a λ-connection of rank one. To find b such that $\partial_t b + a \in x_q^{-1}\mathbb{C}[\![x_q^{-1}]\!]$, we note the following.

- For a C^{∞}-function f on $H_{\infty,q} \simeq S^1_T$, there exists a function g such that $f - \partial_t g$ is constant.
- We have $\partial_t x_q^{-j} = 2\sqrt{-1}\lambda j q^{-1} x_q^{-j-q}$.

Then, by using a standard inductive argument, we can easily construct b such that $\partial_t b + a \in x_q^{-1}\mathbb{C}[\![x_q^{-1}]\!]$.

3.6.5 *Example 1*

Let $\mathfrak{a} \in x_q\mathbb{C}[x_q]$ such that $\deg_{x_q}\mathfrak{a} < q$. Let us consider the $\mathbb{C}(\!(x_q^{-1})\!)$-module $\mathcal{V} = \mathbb{C}(\!(x_q^{-1})\!)\cdot v$ with the λ-connection $\nabla^{\lambda}v = v\cdot d\mathfrak{a}$. We obtain an $\mathcal{K}_{\widehat{H}_{\infty,q}}(*H_{\infty,q})$-module $\widetilde{\mathcal{V}}^{\infty}$ as in (3.22), which is equipped with the induced frame $\widetilde{v} = \Psi_q^{-1}(v)$ and the operator $\partial_{\widetilde{\mathcal{V}}^{\infty},t}$ defined as in (3.23). We have

$$\partial_{\widetilde{\mathcal{V}}^{\infty},t}\widetilde{v} = \widetilde{v}\Big(-2\sqrt{-1}\big(\partial_x\mathfrak{a}(x_q)\big)_{|x_q=y_q(1-2\sqrt{-1}\lambda t y^{-1})^{1/q}}\Big).$$

Let us compute the corresponding difference module explicitly.

We set $\widetilde{\mathcal{V}}^{\infty,\mathrm{cov}} := \varpi_q^{-1}\widetilde{\mathcal{V}}^{\infty}$ which is an $\mathcal{K}_{\widehat{H}^{\mathrm{cov}}_{\infty,q}}(*H^{\mathrm{cov}}_{\infty,q})$-module. It is equipped with the induced frame $\widetilde{v}^{\mathrm{cov}} = \varpi_q^{-1}(\widetilde{v})$. It is also equipped with the operator induced by $\partial_{\widetilde{\mathcal{V}}^{\infty\,\mathrm{cov}},t}$, which we denote by ∂_t to simplify the description.

3.6.5.1

Let us study the case $\lambda \neq 0$. Note that

$$\partial_t \widetilde{v}^{\mathrm{cov}} = \widetilde{v}^{\mathrm{cov}} \partial_t \Big(\lambda^{-1} \mathfrak{a}\big(y_q (1 - 2\sqrt{-1}\lambda t y^{-1})^{1/q} \big) \Big).$$

Because $\deg_{x_q} \mathfrak{a} < q$,

$$-\lambda^{-1} \mathfrak{a}\big(y_q (1 - 2\sqrt{-1}\lambda t y_q^{-q})^{1/q} \big) + \lambda^{-1} \mathfrak{a}(y_q)$$

is a section of $y_q^{-1} \mathcal{K}_{\widehat{H}^{\mathrm{cov}}_{\infty,q}} (*H^{\mathrm{cov}}_{\infty,q})$ on $H^{\mathrm{cov}}_{\infty,q}$. Hence,

$$\widetilde{u} = \widetilde{v}^{\mathrm{cov}} \cdot \exp\Big(- \lambda^{-1} \mathfrak{a}(y_q (1 - 2\sqrt{-1}\lambda t y^{-1})^{1/q}) + \lambda^{-1} \mathfrak{a}(y_q) \Big)$$

is a section of $\widetilde{\mathcal{V}}^{\infty\ \mathrm{cov}}$ such that $\partial_t \widetilde{u} = 0$. Note that $\kappa_1^*(\widetilde{v}^{\mathrm{cov}}) = \widetilde{v}^{\mathrm{cov}}$. We also note that $\kappa_1^*(x_q) = x_q$ which implies that $\kappa_1^*(y_q(1 - 2\lambda\sqrt{-1} t y^{-1})^{1/q}) = y_q(1 - 2\lambda\sqrt{-1} t y^{-1})^{1/q}$. Hence, we obtain

$$\kappa_1^*(\widetilde{u}) = \widetilde{u} \cdot \exp\Big(\lambda^{-1} \mathfrak{a}\big(y_q (1 + 2\sqrt{-1}\lambda T y^{-1})^{1/q} \big) - \lambda^{-1} \mathfrak{a}(y_q) \Big).$$

Therefore, the corresponding difference module is the $\mathbb{C}(\!(y_q^{-1})\!)$-module $\mathbb{C}(\!(y_q^{-1})\!) \cdot u$ with the difference operator defined as

$$\Phi^* u = u \cdot \exp\Big(\lambda^{-1} \mathfrak{a}\big(y_q (1 + 2\sqrt{-1}\lambda T y^{-1})^{1/q} \big) - \lambda^{-1} \mathfrak{a}(y_q) \Big).$$

3.6.5.2

Let us study the case $\lambda = 0$. Note that

$$\partial_t \widetilde{v}^{\mathrm{cov}} = \widetilde{v}^{\mathrm{cov}} \big(- 2\sqrt{-1} (\partial_x \mathfrak{a})(y_q) \big).$$

We set $\widetilde{u} = \widetilde{v}^{\mathrm{cov}} \exp\big(2t\sqrt{-1} (\partial_x \mathfrak{a})(y_q) \big)$, which is a frame of $\widetilde{\mathcal{V}}^{\infty\ \mathrm{cov}}$ such that $\partial_t \widetilde{u} = 0$. Because $\kappa_1^* \widetilde{v} = \widetilde{v}$, we obtain

$$\kappa_1^*(\widetilde{u}) = \widetilde{u} \cdot \exp\big(2T\sqrt{-1} (\partial_x \mathfrak{a})(y_q) \big).$$

Therefore, the corresponding difference module is the $\mathbb{C}(\!(y_q^{-1})\!)$-module $\mathbb{C}(\!(y_q^{-1})\!) \cdot u$ with the difference operator Φ^* defined as

$$\Phi^* u = u \cdot \exp\big(2T\sqrt{-1} (\partial_x \mathfrak{a})(y_q) \big).$$

3.6.6 Example 2

We set $\mathcal{V} = \bigoplus_{i=1}^{r} \mathbb{C}((x_q^{-1}))v_i$. Let $A \in M_r(\mathbb{C})$. We consider the λ-connection ∇^λ on $\mathcal{V}$ defined by $\nabla_x^\lambda \boldsymbol{v} = \boldsymbol{v} \cdot \big(x^{-1}A\big)$. We obtain the $\mathcal{K}_{\widehat{H}_{\infty,q}}(*H_{\infty,q})$-module $\widetilde{\mathcal{V}}^\infty$ as in (3.22) with the operator $\partial_{\widetilde{\mathcal{V}}^\infty,t}$ and the frame $\widetilde{\boldsymbol{v}} = \Psi_q^{-1}(\boldsymbol{v})$. We obtain

$$\partial_{\widetilde{\mathcal{V}}^\infty,t}\widetilde{\boldsymbol{v}} = \widetilde{\boldsymbol{v}}\Big(-2\sqrt{-1}(y-2\sqrt{-1}\lambda t)^{-1}A\Big).$$

Let us compute the corresponding difference module explicitly. As in Sect. 3.6.5, we set $\widetilde{\mathcal{V}}^{\infty,\mathrm{cov}} = \varpi_q^{-1}(\widetilde{\mathcal{V}}^\infty)$, which is equipped with the induced frame $\widetilde{\boldsymbol{v}}^{\mathrm{cov}} = \varpi_q^{-1}(\widetilde{\boldsymbol{v}})$ and the induced differential operator ∂_t.

3.6.6.1

Let us study the case $\lambda \neq 0$. We obtain a frame $\widetilde{\boldsymbol{u}}$ of $\widetilde{\mathcal{V}}^{\infty\ \mathrm{cov}}$ satisfying $\partial_t \widetilde{\boldsymbol{u}} = 0$ as follows:

$$\widetilde{\boldsymbol{u}} := \widetilde{\boldsymbol{v}} \cdot \exp\Big(-\lambda^{-1}A\log\big((1-2\sqrt{-1}\lambda t y^{-1})\big)\Big).$$

The following holds:

$$\kappa_1^*(\widetilde{\boldsymbol{u}}) = \widetilde{\boldsymbol{u}} \cdot \exp\Big(\lambda^{-1}A\log\big(y^{-1}(y+2\sqrt{-1}\lambda T)\big)\Big).$$

Hence, the corresponding difference module is $\bigoplus_{i=1}^{r}\mathbb{C}((y_q^{-1}))u_i$ with the difference operator Φ^* defined as

$$\Phi^*\boldsymbol{u} = \boldsymbol{u}\exp\Big(\lambda^{-1}A\log\big(y^{-1}(y+2\sqrt{-1}\lambda T)\big)\Big).$$

3.6.6.2

Let us study the case $\lambda = 0$. We obtain a frame $\widetilde{\boldsymbol{u}}$ of $\widetilde{\mathcal{V}}^{\infty\,\mathrm{cov}}$ such that $\partial_t \widetilde{\boldsymbol{u}} = 0$ as follows:

$$\widetilde{\boldsymbol{u}} = \widetilde{\boldsymbol{v}}\exp\big(2\sqrt{-1}ty^{-1}A\big).$$

Note that $\kappa_1^*(\widetilde{\boldsymbol{u}}) = \widetilde{\boldsymbol{u}}\exp\Big(2\sqrt{-1}Ty^{-1}A\Big)$. Hence, the corresponding difference module is $\bigoplus_{i=1}^{r}\mathbb{C}((y_q^{-1}))u_i$ with the difference operator Φ^* defined as $\Phi^*\boldsymbol{u} = \boldsymbol{u}\exp\Big(2\sqrt{-1}Ty^{-1}A\Big)$.

3.6.7 Comparison of Good Filtered Bundles

Let $(\mathcal{V}, \nabla^{\lambda})$ be a $\mathbb{C}((x_q^{-1}))$-module of finite rank with a λ-connection. Let $\mathcal{P}_*\mathcal{V}$ be a filtered bundle over $\mathcal{V}$. Recall that $(\mathcal{P}_*\mathcal{V}, \nabla^{\lambda})$ is called good if there exist $p \in q\mathbb{Z}_{>0}$ and a decomposition

$$\varphi_{q,p}^*(\mathcal{P}_*\mathcal{V}) = \bigoplus_{\mathfrak{a}\in x_q\mathbb{C}[x_q]} \mathcal{P}_*\mathcal{V}_{\mathfrak{a}}$$

such that $(\nabla^{\lambda} - d\mathfrak{a}\,\mathrm{id})\mathcal{P}_a\mathcal{V}_{\mathfrak{a}} \subset \mathcal{P}_a\mathcal{V}_{\mathfrak{a}} \cdot dx_q/x_q$ for any $\mathfrak{a}$ and a.

Proposition 3.6.15 *Suppose that Poincaré rank of $(\mathcal{V}, \nabla^{\lambda})$ is strictly less than q, and suppose that each $\mathcal{P}_a\mathcal{V}$ is ∇_x^{λ}-invariant. Then, the induced filtered bundle $\Psi_q^*(\mathcal{P}_*\mathcal{V}, \nabla^{\lambda})$ is good if and only if $(\mathcal{P}_*\mathcal{V}, \nabla^{\lambda})$ is good.*

Proof Let us consider the case where $(\mathcal{V}, \nabla^{\lambda})$ is unramified, i.e., there exists

$$(\mathcal{V}, \nabla^{\lambda}) = \bigoplus_{\mathfrak{a}\in\mathcal{I}(\mathcal{V},\nabla^{\lambda})} (\mathcal{V}_{\mathfrak{a}}, \nabla_{\mathfrak{a}}^{\lambda}) \tag{3.31}$$

such that $(\mathcal{V}, \nabla^{\lambda} - d\mathfrak{a}\,\mathrm{id})$ are regular. We obtain the induced decomposition

$$\Psi_q^*(\mathcal{V}, \nabla^{\lambda}) = \bigoplus_{\mathfrak{a}\in\mathcal{I}(\mathcal{V},\nabla^{\lambda})} \Psi_q^*(\mathcal{V}_{\mathfrak{a}}, \nabla_{\mathfrak{a}}^{\lambda}). \tag{3.32}$$

By the equivalence in Proposition 3.6.8, the filtered bundle $\mathcal{P}_*\mathcal{V}$ is compatible with the decomposition (3.31) if and only if the induced filtered bundle $\Psi_q^*(\mathcal{P}_*\mathcal{V}, \nabla^{\lambda})$ is compatible with the decomposition (3.32), according to Corollary 3.6.14. By the computation in Sect. 3.6.5, it is enough to consider the case where $\mathcal{I}(\mathcal{V}, \nabla^{\lambda}) = \{0\}$, which is reduced to the following lemma.

Lemma 3.6.16 *Let $\mathcal{E}_0$ be a locally free $\mathcal{O}_{\widehat{H}_{\infty,q}}$-module of rank r. We set $\mathcal{E}_0^{\infty} := \mathcal{K}_{\widehat{H}_{\infty,q}} \otimes_{\mathcal{O}_{\widehat{H}_{\infty,q}}} \mathcal{E}_0$. Then, the monodromy of $\mathrm{Loc}_q(\mathcal{E}_0)$ is the identity if and only if there exists a frame $\boldsymbol{v}$ of $\mathcal{E}_0^{\infty}$ such that $\partial_t \boldsymbol{v} = \boldsymbol{v} \cdot x_q^{-q} A$ for $A \in M_r(\mathbb{C}[\![x_q^{-1}]\!])$.*

Proof Let $C_{H_{\infty,q}}^{\infty}$ denote the sheaf of C^{∞}-functions on $H_{\infty,q}$. Note that

$$\mathrm{Loc}_q(\mathcal{E}_0)^{\infty} := \mathrm{Loc}_q(\mathcal{E}_0) \otimes C_{H_{\infty,q}}^{\infty}$$

is naturally isomorphic to $\mathcal{E}_0^{\infty}/x_q^{-q}\mathcal{E}_0^{\infty}$. We also remark that the tuple y_q^{-j} $(j = 0, \dots, q-1)$ is a frame of the local system $\mathrm{Loc}_q(\mathcal{O}_{\widehat{H}_{\infty,q}})$, and equal to the tuple induced by x_q^{-j} $(j = 0, \dots, q-1)$.

Let $\boldsymbol{v} = (v_1, \ldots, v_r)$ be a frame of $\mathcal{E}_0^{\infty}$. Then, the tuple $\boldsymbol{v}_0 = \big(y_q^{-j} v_i \,\big|\, j = 0, \ldots, q-1,\, i = 1, \ldots, r\big)$ is a frame of $\mathcal{E}_0^{\infty}$ over $\mathcal{C}^{\infty}_{H_{\infty,q}}$. It is called the induced frame.

Assume that there exists a frame $\boldsymbol{v}$ of $\mathcal{E}_0^{\infty}$ satisfying the condition in the lemma. It induces a frame $\boldsymbol{v}_0$ of $\mathrm{Loc}_q(\mathcal{E}_0)^{\infty}$ as above. We obtain $\partial_t \boldsymbol{v}_0 = 0$. Hence, the monodromy of $\mathrm{Loc}_q(\mathcal{E}_0)$ is the identity.

Conversely, suppose that the monodromy of $\mathrm{Loc}_q(\mathcal{E}_0)$ is the identity. It implies that $\mathrm{Loc}_q(\mathcal{E}_0)$ is a free module over the sheaf of algebras $\mathrm{Loc}_q(\mathcal{O}_{\widehat{H}_{\infty,q}})$. There exists a frame $\boldsymbol{u} = (u_1, \ldots, u_r)$ of $\mathrm{Loc}_q(\mathcal{E}_0)$ over $\mathrm{Loc}_q(\mathcal{O}_{\widehat{H}_{\infty,q}})$.

Let $\boldsymbol{v}$ be any frame of $\mathcal{E}_0^{\infty}$ which induces $\boldsymbol{u}$ as a tuple of sections of $\mathrm{Loc}_q(\mathcal{E}_0)^{\infty}$. Then, we obtain $\partial_t \boldsymbol{u} = \boldsymbol{u} \cdot x_q^{-q} A$ for a global section A of $M_r(\mathcal{K}_{\widehat{H}_{\infty,q}})$ on $H_{\infty,q}$. As in the proof of Proposition 3.6.8, after an appropriate gauge transform of the frame, we obtain $A \in M_r(\mathbb{C}[\![x_q^{-1}]\!])$. Thus, we obtain Lemma 3.6.16 and the claim of Proposition 3.6.15 in the unramified case. ■

For $p \in q\mathbb{Z}_{>0}$, there exists the following natural commutative diagram:

$$\begin{array}{ccc} \widehat{H}_{p,\infty} & \xrightarrow{\mathcal{R}_{q,p}} & \widehat{H}_{\infty,q} \\ {\scriptstyle\Psi_p}\big\downarrow & & {\scriptstyle\Psi_q}\big\downarrow \\ \widehat{\infty}_{x,p} & \xrightarrow{\varphi_{q,p}} & \widehat{\infty}_{x,q}. \end{array}$$

Hence, there exists the following natural isomorphism:

$$\Psi_p^* \varphi_{q,p}^* (\mathcal{P}_* \mathcal{V}, \nabla^{\lambda}) \simeq \mathcal{R}_{q,p}^* \Psi_q^* (\mathcal{P}_* \mathcal{V}, \nabla^{\lambda}).$$

By the consideration in the unramified case, $\varphi_{q,p}^*(\mathcal{P}_*\mathcal{V}, \nabla^{\lambda})$ is good if and only if $\Psi_p^* \varphi_{q,p}^*(\mathcal{P}_*\mathcal{V}, \nabla^{\lambda})$ is good. Thus, we obtain the claim in the general case. ■

3.6.8 Comparison of the Associated Graded Pieces

Let $(\mathcal{P}_*\mathcal{V}, \nabla^{\lambda})$ be a good filtered λ-flat bundle whose Poincaré rank is strictly less than 1. We obtain the good filtered bundle $\Psi_q^*(\mathcal{P}_*\mathcal{V}, \nabla^{\lambda})$ on $(\widehat{H}_{\infty,q}, H_{\infty,q})$.

Take any $P \in H_{\infty,q}$. For any $v \in \mathcal{V}$, we obtain an element $\Psi_q^{-1}(v)_P$ of $\widetilde{\mathcal{V}}_P^{\infty}$. By using Lemma 3.6.5, we obtain

$$(c_{2,\mathcal{V}} \circ c_{1,\mathcal{V}})^{-1}\Big(c_{2,\mathcal{V}}\big(\Psi_q^{-1}(v)_P\big)\Big) \in \Psi_q^*(\mathcal{V}, \nabla^{\lambda})_P.$$

Thus, we obtain a $\mathbb{C}$-linear map $\mathcal{V} \longrightarrow \Psi_q^*(\mathcal{V}, \nabla^{\lambda})_P$.

Proposition 3.6.17 *The induced morphism* $\mathrm{Gr}_a^{\mathcal{P}}(\mathcal{V}) \longrightarrow \mathrm{Gr}_a^{\mathcal{P}}(\Psi_q^*(\mathcal{V}, \nabla^\lambda))_P$ *is a* $\mathbb{C}$*-linear isomorphism. Moreover, under the isomorphism,* $q^{-1}(2\sqrt{-1})T\,\mathrm{Res}(\nabla^\lambda)$ *is equal to* $\mathfrak{F}_{\Psi_q^*(\mathcal{V},\nabla^\lambda)}$. *(See* Sect. 3.5.6 *for the endomorphism* $\mathfrak{F}_{\Psi_q^*(\mathcal{V},\nabla^\lambda)}$.*) In particular, the weight filtrations of the nilpotent parts of* $\mathrm{Res}(\nabla^\lambda)$ *and* $\mathfrak{F}_{\Psi_q^*(\mathcal{V},\nabla^\lambda)}$ *are equal.*

Proof The first claim is clear. Let us study the case where $(\mathcal{V}, \nabla^\lambda)$ is regular. Let $\boldsymbol{v}$ be a frame of $\mathcal{P}_a\mathcal{V}$ compatible with the parabolic structure as in Sect. 2.11.1, i.e., there is a decomposition $\boldsymbol{v} = \coprod_{a-1<b\leq a} \boldsymbol{v}_b$ such that each $\boldsymbol{v}_b$ is a tuple of sections of $\mathcal{P}_b(\mathcal{V})$ and induces a base of $\mathrm{Gr}_b^{\mathcal{P}}(\mathcal{V})$. For $j \in \mathbb{Z}_{\geq 0}$ and $a-1 < b \leq a$, we obtain tuples of sections $\Psi_q^{-1}(x_q^{-j}\boldsymbol{v}_b)$ of $\widetilde{\mathcal{P}_a\mathcal{V}}^\infty$. Let $[\Psi_q^{-1}(x_q^{-j}\boldsymbol{v}_b)]$ denote the induced sections of a locally free $\mathcal{C}^\infty_{H_{\infty,q}}$-module

$$\widetilde{\mathcal{P}_a\mathcal{V}}^\infty / \widetilde{\mathcal{P}_{<a-q}\mathcal{V}}^\infty. \tag{3.33}$$

Then, the union of $[\Psi_q^{-1}(x_q^{-j}\boldsymbol{v}_b)]$ $(a-1 < b < a,\ 0 \leq j \leq q-1)$ and $[\Psi_q^{-1}(x_q^{-j}\boldsymbol{v}_a)]$ $(0 \leq j \leq q)$ is a frame of (3.33) over $\mathcal{C}^\infty_{H_{\infty,q}}$.

We set $r(a) := \dim_{\mathbb{C}} \mathrm{Gr}_a^{\mathcal{P}}(\mathcal{V})$. Let $[\boldsymbol{v}_a]$ denote the base of $\mathrm{Gr}_a^{\mathcal{P}}(\mathcal{V})$ induced by $\boldsymbol{v}_a$. We obtain the matrix $A \in M_{r(a)}(\mathbb{C})$ by $\mathrm{Res}\,\nabla^\lambda[\boldsymbol{v}_a] = [\boldsymbol{v}_a]$. Namely, for $\boldsymbol{v}_a = (v_{a,1}, \dots, v_{a,r(a)})$, $\mathrm{Res}(\nabla)[v_{a,j}] = \sum A_{i,j}[v_{a,i}]$.

For any $a-1 < b < a$, and for any $j = 0, \dots, q-1$, we obtain $\partial_t[\Psi_q^{-1}(x_q^{-j}\boldsymbol{v}_b)] = 0$. For $j = 1, \dots, q$, we obtain $\partial_t[\Psi_q^{-1}(x_q^{-j}\boldsymbol{v}_a)] = 0$. Because $\partial_t[\Psi_q^{-1}(\boldsymbol{v}_a)] = [\Psi_q^{-1}(x_q^{-q}\boldsymbol{v}_a)]q^{-1}(-2\sqrt{-1}A)$,

$$\varpi_q^{-1}\Psi_q^{-1}[v_{a,j}] + q^{-1}2\sqrt{-1}t\sum A_{i,j}\Psi_q^{-1}[x_q^{-q}v_{a,i}] \quad (i = 1, \dots, r(a))$$

are flat sections of the pull back of (3.33) by ϖ_q. Thus, we obtain the claim of the proposition in the case where $(\mathcal{V}, \nabla^\lambda)$ is regular.

As in the proof of Proposition 3.6.15, by using the computation in Sect. 3.6.5 and the pull back by $\varphi_{q,p}$ such that $\varphi_{q,p}^*(\mathcal{V}, \nabla^\lambda)$ is unramified, we can deduce the claim in the general case. ■

3.6.9 Some Functoriality

Let $(\mathcal{V}_i, \nabla^\lambda)$ $(i = 1, 2)$ be finite dimensional $\mathbb{C}((x_q^{-1}))$-vector spaces with λ-connection. Let $(\mathcal{P}_*\mathcal{V}_i, \nabla^\lambda)$ be good filtered λ-flat bundles over $(\mathcal{V}_i, \nabla^\lambda)$.

Proposition 3.6.18 *There exist the following natural isomorphisms:*

$$\Psi_q^*(\mathcal{P}_*\mathcal{V}_1, \nabla^\lambda) \oplus \Psi_q^*(\mathcal{P}_*\mathcal{V}_2, \nabla^\lambda) \simeq \Psi_q^*\big((\mathcal{P}_*\mathcal{V}_1, \nabla^\lambda) \oplus (\mathcal{P}_*\mathcal{V}_2, \nabla^\lambda)\big) \qquad (3.34)$$

$$\Psi_q^*(\mathcal{P}_*\mathcal{V}_1, \nabla^\lambda) \otimes \Psi_q^*(\mathcal{P}_*\mathcal{V}_2, \nabla^\lambda) \simeq \Psi_q^*\big((\mathcal{P}_*\mathcal{V}_1, \nabla^\lambda) \otimes (\mathcal{P}_*\mathcal{V}_2, \nabla^\lambda)\big) \qquad (3.35)$$

$$\mathcal{H}om\Big(\Psi_q^*(\mathcal{P}_*\mathcal{V}_1, \nabla^\lambda), \Psi_q^*(\mathcal{P}_*\mathcal{V}_2, \nabla^\lambda)\Big) \simeq \Psi_q^*\Big(\operatorname{Hom}\big((\mathcal{P}_*\mathcal{V}_1, \nabla^\lambda), (\mathcal{P}_*\mathcal{V}_2, \nabla^\lambda)\big)\Big). \qquad (3.36)$$

In particular, for a good filtered λ-flat bundle $(\mathcal{P}_*\mathcal{V}, \nabla^\lambda)$ over a finite dimensional $\mathbb{C}(\!(x_q^{-1})\!)$-vector space $\mathcal{V}$ with a λ-connection ∇^λ, we obtain $\Psi_q^*(\mathcal{V}, \nabla^\lambda)^\vee \simeq \Psi_q^*\big((\mathcal{V}, \nabla^\lambda)^\vee\big)$. ■

3.7 Appendix: Pull Back and Descent in the ℝ-Direction

To emphasize the dependence of $\widehat{H}_{\infty,q}$ on T, we denote it by $\widehat{H}_{\infty,q,T}$. We denote $\pi_q : \widehat{H}_{\infty,q,T} \longrightarrow S_T^1$ by $\pi_{q,T}$. As explained, we use the covering maps $\mathcal{R}_{q,p} : \widehat{H}_{\infty,p,T} \longrightarrow \widehat{H}_{\infty,q,T}$ for $p \in q\mathbb{Z}_{>0}$ to define the notion of good filtered bundles. There also exists the naturally defined covering map $\mathcal{T}_{T,mT} : \widehat{H}_{\infty,q,mT} \longrightarrow \widehat{H}_{\infty,q,T}$, which we may use to define the notion of good filtered bundles in another way. Although we will not use it in this monograph, we explain it because it is also a natural way.

For any $m \in \mathbb{Z}_{\geq 1}$, there exists the following natural commutative diagram:

$$\begin{array}{ccc} \widehat{H}_{\infty,q,mT} & \xrightarrow{\mathcal{T}_{T,mT}} & \widehat{H}_{\infty,q,T} \\ {\scriptstyle \pi_{q,mT}}\downarrow & & \downarrow{\scriptstyle \pi_{q,T}} \\ S_{mT}^1 & \xrightarrow{\varphi} & S_T^1 \end{array}$$

Note that $\mathcal{T}_{T,mT}^{-1}\mathcal{O}_{\widehat{H}_{\infty,q,T}} = \mathcal{O}_{\widehat{H}_{\infty,q,mT}}$. There exits the pull back $\mathcal{T}_{T,mT}^* = \mathcal{T}_{T,mT}^{-1}$ of $\mathcal{O}_{\widehat{H}_{\infty,q,T}}$-modules and the push-forward $\mathcal{T}_{T,mT*}$ of $\mathcal{O}_{\widehat{H}_{\infty,q,mT}}$-modules.

We obtain the action of $\mathbb{Z}/m\mathbb{Z}$ on $\widehat{H}_{\infty,q,mT}$ induced by $n(t, y_q) = (t + nT, y_q(1+2\sqrt{-1}\lambda Tny^{-1})^{1/q})$. Let $\mathcal{E}_1$ be an $\mathcal{O}_{\widehat{H}_{\infty,q,mT}}$-module which is equivariant with respect to the $\mathbb{Z}/m\mathbb{Z}$-action. Then, the push-forward $\mathcal{T}_{T,mT*}\mathcal{E}_1$ is equipped with the naturally induced $\mathbb{Z}/m\mathbb{Z}$-action. The invariant part is called the descent of $\mathcal{E}_1$ with respect to the $\mathbb{Z}/m\mathbb{Z}$-action.

Let us consider a filtered bundle $\big(\mathcal{P}_*\mathcal{E}_{|\widehat{H}_{\infty,q,T}\langle t\rangle} \,\big|\, t \in S_T^1\big)$ over a locally free $\mathcal{O}_{\widehat{H}_{\infty,q,T}}(H_{\infty,q,T})$-module $\mathcal{E}$. Because $\mathcal{T}_{T,mT}^*(\mathcal{E})_{|\widehat{H}_{\infty,q,mT}\langle t_1\rangle} = \mathcal{E}_{|\widehat{H}_{\infty,q,T}\langle\varphi(t_1)\rangle}$, we obtain a filtered bundle over $\mathcal{T}_{T,m_T}^*(\mathcal{E})$ as follows:

$$\mathcal{P}_*\big(\mathcal{T}_{T,mT}^*(\mathcal{E})_{|\widehat{H}_{\infty,q,mT}\langle t_1\rangle}\big) := \mathcal{P}_*\mathcal{E}_{|\widehat{H}_{\infty,q,T}\langle\varphi(t_1)\rangle} \quad (t_1 \in S_{mT}^1).$$

Let $\left(\mathcal{P}_*\mathcal{E}_{1|\widehat{H}_{\infty,q,mT}\langle t_1\rangle} \,\middle|\, t_1 \in S^1_{mT}\right)$ be a filtered bundle over a locally free $\mathcal{O}_{\widehat{H}_{\infty,q,mT}}(*H_{\infty,q,mT})$-module $\mathcal{E}_1$. We obtain the filtered bundle over $\mathcal{T}_{T,mT*}\mathcal{E}_1$ given as follows:

$$\mathcal{P}_*\mathcal{T}_{T,mT*}(\mathcal{E}_1)_{|\widehat{H}_{\infty,q,T}\langle t\rangle} = \bigoplus_{t_1\in\varphi^{-1}(t)} \mathcal{P}_*\mathcal{E}_{1|\widehat{H}_{\infty,q,mT}\langle t_1\rangle}. \tag{3.37}$$

If $\mathcal{E}_1$ and $\mathcal{P}_*\mathcal{E}_1$ are $\mathbb{Z}/m\mathbb{Z}$-equivariant, then we obtain the filtered bundle over the descent of $\mathcal{E}_1$ by taking the invariant part of (3.37). The following lemma is clear by the construction.

Lemma 3.7.1 *Let $\mathcal{E}$ be a locally free $\mathcal{O}_{\widehat{H}_{\infty,q}}(*H_{\infty,q})$-module. Let $\mathcal{P}_*\mathcal{E}$ be a filtered bundle over $\mathcal{E}$. Let $\mathcal{P}_*\mathcal{T}^*_{T,mT}\mathcal{E}$ be the induced filtered bundle over $\mathcal{T}^*_{T,mT}\mathcal{E}$. Then, $\mathcal{P}_*\mathcal{E}$ is the descent of $\mathcal{P}_*\mathcal{T}^*_{T,mT}\mathcal{E}$.* ■

Proposition 3.7.2 *Let $\mathcal{E}$ be a locally free $\mathcal{O}_{\widehat{H}_{\infty,q,T}}(*H_{\infty,q,T})$-module, and let $\left(\mathcal{P}_*\mathcal{E}_{|\widehat{H}_{\infty,q,T}\langle t\rangle} \,\middle|\, t \in S^1_T\right)$ be a filtered bundle over $\mathcal{E}$. Then, it is good if and only if the induced filtered bundle over $\mathcal{T}^*_{T,mT}\mathcal{E}$ is good.*

Proof We set $\mathcal{E}' = \mathcal{T}^*_{T,mT}\mathcal{E}$. Let us consider the case where the level of $\mathcal{E}$ is less than 1. We use the notation $\Psi_{q,T}$ to emphasize the dependence on T. There exists a $\mathbb{C}((x_q^{-1}))$-module with λ-connection $(\mathcal{V}, \nabla^\lambda)$ such that $\mathcal{E} = \Psi^*_{q,T}(\mathcal{V}, \nabla)$. By construction, we obtain the natural isomorphism $\mathcal{E}' = \mathcal{T}^*_{T,mT} \circ \Psi^*_{q,T}(\mathcal{V}, \nabla_\lambda) \simeq \Psi^*_{q,mT}(\mathcal{V}, \nabla^\lambda)$. If $\mathcal{P}_*\mathcal{E}$ is good, there exists a good filtered λ-flat bundle $(\mathcal{P}_*\mathcal{V}, \nabla^\lambda)$ over $(\mathcal{V}, \nabla^\lambda)$ such that $\mathcal{P}_*\mathcal{E} = \Psi^*_{q,T}(\mathcal{P}_*\mathcal{V})$. Because $\mathcal{P}_*\mathcal{E}' = \Psi^*_{q,mT}(\mathcal{P}_*\mathcal{V}, \nabla^\lambda)$, it is also good. The converse can be proved similarly.

We use the notation $\widehat{\mathbb{L}}^\lambda_{q,T}(\ell, \alpha)$ to emphasize the dependence on T. According to the computation in Sect. 3.7.1, there exists the natural isomorphism $\mathcal{T}^*_{T,mT}\widehat{\mathbb{L}}^\lambda_{q,T}(\ell, \alpha) \simeq \widehat{\mathbb{L}}^\lambda_{q,mT}(m\ell, \alpha^m)$. Under the isomorphism, we obtain

$$\mathcal{P}_*\mathcal{T}^*_{T,mT}\widehat{\mathbb{L}}^\lambda_{q,T}(\ell, \alpha) = \mathcal{P}_*\widehat{\mathbb{L}}^\lambda_{q,mT}(m\ell, \alpha^m)$$

by the construction of the filtrations.

Let $\mathcal{E}$ be a locally free $\mathcal{O}_{\widehat{H}_{\infty,q,T}}(*H_{\infty,q,T})$-module. We have $p \in q\mathbb{Z}_{\geq 1}$ such that $\mathcal{E}_1 := \mathcal{R}^*_{p,q}\mathcal{E}$ is unramified. Let $\mathcal{E}' := \mathcal{T}^*_{T,mT}\mathcal{E}$ and $\mathcal{E}'_1 := \mathcal{T}^*_{T,mT}\mathcal{E}_1$. It is easy to observe that $\mathcal{E}'$ is the descent of $\mathcal{E}'_1$ with respect to the ramified covering $\mathcal{R}_{p,q,mT}$: $\widehat{H}_{\infty,p,mT} \longrightarrow \widehat{H}_{\infty,q,mT}$.

Let $\mathcal{P}_*\mathcal{E}$ be a good filtered bundle over $\mathcal{E}$. The induced filtered bundle $\mathcal{P}_*\mathcal{E}_1$ over $\mathcal{E}_1$ is good. Let $\mathcal{P}_*\mathcal{E}'_1$ be the filtered bundle obtained as the pull back of $\mathcal{P}_*\mathcal{E}_1$ by $\mathcal{T}_{T,mT}$. Then, by the previous consideration, $\mathcal{P}_*\mathcal{E}'_1$ is good. We can easily observe that $\mathcal{P}_*\mathcal{E}'_1$ is $\mathrm{Gal}_{q,p}$-equivariant, and $\mathcal{P}_*\mathcal{E}'$ is the descent of $\mathcal{P}_*\mathcal{E}'_1$. Hence, $\mathcal{P}_*\mathcal{E}'$ is good.

Conversely, suppose that $\mathcal{P}_*\mathcal{E}'$ is good. The pull back $\mathcal{P}_*\mathcal{E}'_1$ over $\mathcal{E}'_1$ is good. We can easily observe that $\mathcal{P}_*\mathcal{E}'_1$ is equivariant with respect to $\mathbb{Z}/m\mathbb{Z}$-action and

$\mathrm{Gal}_{q,p}$-action. Let $\mathcal{P}_*\mathcal{E}_1$ be the descent of $\mathcal{P}_*\mathcal{E}_1'$. It is good, $\mathrm{Gal}_{q,p}$-equivariant, and the descent is $\mathcal{P}_*\mathcal{E}$. Hence, we obtain the claim of the proposition. ■

Corollary 3.7.3 *Let $\mathcal{E}'$ be a locally free $\mathcal{O}_{\widehat{H}_{\infty,q,mT}}(*H_{\infty,q,mT})$-module. Let $\mathcal{P}_*\mathcal{E}'$ be a $\mathbb{Z}/m\mathbb{Z}$-equivariant good filtered bundle over $\mathcal{E}'$. Then, the descent of $\mathcal{P}_*\mathcal{E}'$ is also good.* ■

3.7.1 Examples

Let $q \in \mathbb{Z}_{\geq 1}$ and $\ell \in \mathbb{Z}$. We consider the automorphism Φ^* of $\mathbb{C}((y_q^{-1}))$ induced by $\Phi^*(y) = y + \varrho$. We consider the difference $\mathbb{C}((y_q^{-1}))$-module $\widehat{\mathbb{L}}_q^{\lambda}(\ell)$ given by $\widehat{\mathbb{L}}_q^{\lambda}(\ell) = \mathbb{C}((y_q^{-1}))\, e$ with

$$\Phi^*(e) = \Big(1 + \frac{|\varrho|^2}{4}\Big)^{\ell/q} y_q^{-\ell}(1 + \varrho y^{-1})^{-\ell/q} \exp\Big(\frac{\ell}{q} G(\varrho y^{-1})\Big) \cdot e.$$

Here, $G(x) = 1 - x^{-1}\log(1+x)$. Because this example has the following property, the corresponding $\mathcal{O}_{\widehat{H}_{\infty,q,T}}(*H_{\infty,q,T})$-modules are convenient with respect to the pull back and push-forward with respect to $\mathcal{T}_{T,mT}$.

Lemma 3.7.4 *For any $m \in \mathbb{Z}_{\geq 1}$, we have*

$$(\Phi^m)^*(e) = \Big(1 + \frac{|\varrho|^2}{4}\Big)^{m\ell/q} y_q^{-m\ell}(1 + \varrho m y^{-1})^{-m\ell/q} \exp\Big(\frac{m\ell}{q} G(\varrho m y^{-1})\Big) \cdot e.$$

Proof We give only a formal argument. to simplify the description. We have

$$(\Phi^m)^* e = \Big(1 + \frac{|\varrho|^2}{4}\Big)^{m\ell/q} \prod_{j=1}^{m} (y + j\varrho)^{-\ell/q} \cdot \exp\Big(\frac{\ell}{q} \sum_{j=0}^{m-1} G\big(\varrho(y + j\varrho)^{-1}\big)\Big) \cdot e.$$

We have

$$\begin{aligned}
&\frac{\ell}{q}\sum_{j=0}^{m-1}\Big(1 - \frac{y+\varrho}{\varrho}\log\Big(\frac{y+(j+1)\varrho}{y+j\varrho}\Big)\Big) \\
&= \frac{\ell m}{q} - \frac{\ell}{q}\sum_{j=0}^{m-1}\frac{y}{\varrho}\log\Big(\frac{y+(j+1)\varrho}{y+j\varrho}\Big) - \frac{\ell}{q}\sum_{j=0}^{m-1} j \log\Big(\frac{y+(j+1)\varrho}{y+j\varrho}\Big) \\
&= \frac{\ell m}{q} - \frac{\ell m}{q}\frac{y}{\varrho m}\log\Big(\frac{y+m\varrho}{y}\Big)
\end{aligned}$$

$$-\frac{\ell}{q}\Big(\sum_{j=0}^{m-1}(j+1)\log(y+(j+1)\varrho)-\sum_{j=0}^{m-1}\log(y+(j+1)\varrho)$$
$$-\sum_{j=0}^{m-1} j\log(y+j\varrho)\Big)=\frac{\ell m}{q}G(\varrho m y^{-1})-\frac{\ell}{q}\Big(m\log(y+m\varrho)-\sum_{j=1}^{m}\log(y+j\varrho)\Big). \tag{3.38}$$

Then, we obtain the claim of the lemma. ∎

Chapter 4
Filtered Bundles

Abstract There are two purposes in this chapter. One is to formulate the notion of stable good filtered bundle with Dirac type singularity on $(\overline{\mathcal{M}}^{\lambda}; Z, H^{\lambda}_{\infty})$. They are equivalent to stable parabolic difference modules, and we shall later establish that they are the algebraic counterpart of monopoles. The other is to introduce some basic notions related to good filtered bundles on a ramified covering, as a preparation for the study of the asymptotic behaviour of monopoles.

4.1 Outline of This Chapter

In Sect. 4.2, we introduce the notion of filtered bundles over locally free $\mathcal{O}_{\overline{\mathcal{M}}^{\lambda}\setminus Z}(*H^{\lambda}_{\infty})$-modules of Dirac type. We also define the stability condition for such filtered bundles. Moreover, we explain the comparison with parabolic difference modules. In Sect. 4.3, we introduce the notion of filtered bundles on a ramified covering space $\mathcal{B}^{\lambda}_{q}$ of a neighbourhood of H^{λ}_{∞}. In Sect. 4.4, we explain an extension of a mini-holomorphic bundle $\mathcal{B}^{\lambda}_{q} \setminus H^{\lambda}_{\infty,q}$ with a Hermitian metric to a filtered object on $(\mathcal{B}^{\lambda}_{q}, H^{\lambda}_{\infty,q})$. We introduce the conditions called norm estimates and strong adaptedness in this context. In Sect. 4.5, we study the analytification of the construction in Sect. 3.6, which is also a ramified version of the construction in Sect. 2.8.3.4. It is our purpose to compare the norm estimate for λ-connections and the norm estimate for $\mathcal{O}_{\mathcal{B}^{\lambda}_{q}}(*H^{\lambda}_{\infty,q})$-modules (Proposition 4.5.7).

4.2 Filtered Bundles in the Global Case

We use the notation in Sect. 2.7. Let Z be a finite subset of $\mathcal{M}^{\lambda}$. Let $\pi^{\lambda} : \overline{\mathcal{M}}^{\lambda} \longrightarrow S^1_T$ denote the projection induced by $(t_1, \beta_1) \longmapsto t_1$. We set $\overline{\mathcal{M}}^{\lambda}\langle t_1\rangle := (\pi^{\lambda})^{-1}(t_1)$ and $H^{\lambda}_{\infty}\langle t_1\rangle := (\pi^{\lambda})^{-1}(t_1) \cap H^{\lambda}_{\infty}$ for any $t_1 \in S^1_T$. If there is no risk of confusion, the point $H^{\lambda}_{\infty}\langle t_1\rangle \in \overline{\mathcal{M}}^{\lambda}\langle t_1\rangle$ is also denoted by ∞.

T. Mochizuki, *Periodic Monopoles and Difference Modules*, Lecture Notes in Mathematics 2300, https://doi.org/10.1007/978-3-030-94500-8_4

Let $\mathcal{E}$ be a locally free $\mathcal{O}_{\overline{\mathcal{M}}^{\lambda}\setminus Z}(*H^{\lambda}_{\infty})$-module of Dirac type. Let $\mathcal{E}_{|\overline{\mathcal{M}}^{\lambda}\langle t_1\rangle\setminus Z}$ denote the pull back of $\mathcal{E}$ by $\overline{\mathcal{M}}^{\lambda}\langle t_1\rangle \setminus Z \longrightarrow \overline{\mathcal{M}}^{\lambda} \setminus Z$, which is naturally a locally free $\mathcal{O}_{\overline{\mathcal{M}}^{\lambda}\langle t_1\rangle\setminus Z}(*\infty)$-module.

Definition 4.2.1 A filtered bundle over $\mathcal{E}$ is defined to be a family of filtered bundles $\mathcal{P}_*(\mathcal{E}_{|\overline{\mathcal{M}}^{\lambda}\langle t_1\rangle\setminus Z})$ $(t_1 \in S^1_T)$ over $\mathcal{E}_{|\overline{\mathcal{M}}^{\lambda}\langle t_1\rangle\setminus Z}$. (See Sect. 2.11.3.1 for the notion of filtered bundles on punctured Riemann surfaces.) Such a family is also denoted by $\mathcal{P}_*\mathcal{E}$. ■

For any $P \in H^{\lambda}_{\infty}$, let $\mathcal{O}_{\widehat{H}^{\lambda}_{\infty},P}$ denote the completion of the local ring $\mathcal{O}_{\overline{\mathcal{M}}^{\lambda}_{\infty},P}$ with respect to the maximal ideal. Thus, we obtain the sheaf of algebras $\mathcal{O}_{\widehat{H}^{\lambda}_{\infty}}$ on H^{λ}_{∞}. Let $\widehat{H}^{\lambda}_{\infty}$ denote the ringed space obtained as the topological space H^{λ}_{∞} with the sheaf of algebras $\mathcal{O}_{\widehat{H}^{\lambda}_{\infty}}$. We also set $\mathcal{O}_{\widehat{H}^{\lambda}_{\infty}}(*H^{\lambda}_{\infty})_P := \mathcal{O}_{\overline{\mathcal{M}}^{\lambda}}(*H^{\lambda}_{\infty})_P \otimes_{\mathcal{O}_{\overline{\mathcal{M}}^{\lambda},P}} \mathcal{O}_{\widehat{H}^{\lambda}_{\infty},P}$, and we obtain the sheaf of algebras $\mathcal{O}_{\widehat{H}^{\lambda}_{\infty}}(*H^{\lambda}_{\infty})$.

Remark 4.2.2 The ringed space H^{λ}_{∞} is naturally isomorphic to the formal space $\widehat{H}_{\infty,1}$ in Sect. 3.4.2 by the change of the variables from (t_1, β_1) to (t, y). The sheaf of algebras $\mathcal{O}_{\widehat{H}^{\lambda}_{\infty}}(*H^{\lambda}_{\infty})$ is naturally isomorphic to $\mathcal{O}_{\widehat{H}_{\infty,1}}(*H_{\infty,1})$. ■

Let $k^{\lambda} : \widehat{H}^{\lambda}_{\infty} \longrightarrow \overline{\mathcal{M}}^{\lambda} \setminus Z$ denote the morphism of ringed spaces obtained as the inclusion $H^{\lambda}_{\infty} \longrightarrow \overline{\mathcal{M}}^{\lambda} \setminus Z$ with the natural morphism of sheaves of algebras $\mathcal{O}_{\overline{\mathcal{M}}^{\lambda}\setminus Z|H^{\lambda}_{\infty}} \longrightarrow \mathcal{O}_{\widehat{H}^{\lambda}_{\infty}}$. We set $\mathcal{E}_{|\widehat{H}^{\lambda}_{\infty}} := (k^{\lambda})^*(\mathcal{E})$, which is naturally a locally free $\mathcal{O}_{\widehat{H}^{\lambda}_{\infty}}(*H^{\lambda}_{\infty})$-module. A filtered bundle over $\mathcal{E}$ naturally induces a filtered bundle over $\mathcal{E}_{|\widehat{H}^{\lambda}_{\infty}}$ in the sense of Definition 3.5.1.

Definition 4.2.3 $\mathcal{P}_*\mathcal{E}$ is called good if the induced filtered bundle over $\mathcal{E}_{|\widehat{H}^{\lambda}_{\infty}}$ is good in the sense of Definition 3.5.16. ■

Definition 4.2.4 A (good) filtered bundle of Dirac type on $(\overline{\mathcal{M}}^{\lambda}; Z, H^{\lambda}_{\infty})$ is a locally free $\mathcal{O}_{\overline{\mathcal{M}}^{\lambda}\setminus Z}(*H^{\lambda}_{\infty})$-module of Dirac type $\mathcal{E}$ equipped with a (good) filtered bundle $\mathcal{P}_*\mathcal{E}$. ■

Remark 4.2.5 For filtered bundles $\mathcal{P}_*(\mathcal{E}_i)$ $(i = 1, 2)$ over locally free $\mathcal{O}_{\overline{\mathcal{M}}^{\lambda}\setminus Z}(*H^{\lambda}_{\infty})$-modules $\mathcal{E}_i$, we naturally define the filtered bundles $\mathcal{P}_*(\mathcal{E}_1)\oplus\mathcal{P}_*(\mathcal{E}_2)$ over $\mathcal{E}_1\oplus\mathcal{E}_2$, $\mathcal{P}_*(\mathcal{E}_1)\otimes\mathcal{P}_*(\mathcal{E}_2)$ over $\mathcal{E}_1\otimes\mathcal{E}_2$ and $\mathcal{H}om(\mathcal{P}_*\mathcal{E}_1, \mathcal{P}_*\mathcal{E}_2)$ over $\mathcal{H}om(\mathcal{E}_1, \mathcal{E}_2)$ by applying the constructions in Sect. 2.11.1 to $\mathcal{P}_*(\mathcal{E}_{i|\overline{\mathcal{M}}^{\lambda}\langle t_1\rangle\setminus Z})$. If $\mathcal{P}_*\mathcal{E}_i$ $(i = 1, 2)$ are good, then $\mathcal{P}_*(\mathcal{E}_1) \oplus \mathcal{P}_*(\mathcal{E}_2)$, $\mathcal{P}_*(\mathcal{E}_1) \otimes \mathcal{P}_*(\mathcal{E}_2)$ and $\mathcal{H}om(\mathcal{P}_*\mathcal{E}_1, \mathcal{P}_*\mathcal{E}_2)$ are also good, which follows from Lemma 3.5.20. Similarly, for a filtered bundle $\mathcal{P}_*\mathcal{E}$ over $\mathcal{E}$, we naturally obtain the dual $(\mathcal{P}_*\mathcal{E})^{\vee}$ over $\mathcal{E}^{\vee}$. If $\mathcal{P}_*\mathcal{E}$ is good, then $(\mathcal{P}_*\mathcal{E})^{\vee}$ is also good. ■

For any $t_1 \in \mathbb{R}$, we set $\overline{M}^{\lambda}\langle t_1\rangle := \{t_1\} \times \mathbb{P}^1_{\beta_1} \subset \overline{M}^{\lambda}$. The projection $\varpi^{\lambda} : \overline{M}^{\lambda} \longrightarrow \overline{\mathcal{M}}^{\lambda}$ induces an isomorphism $\overline{M}^{\lambda}\langle\widetilde{t}_1\rangle \simeq \overline{\mathcal{M}}^{\lambda}\langle t_1\rangle$ for any lift $\widetilde{t}_1 \in \mathbb{R}$ of $t_1 \in S^1_T$. We set $Z^{\mathrm{cov}} := (\varpi^{\lambda})^{-1}(Z)$. We obtain the locally free

$\mathcal{O}_{\overline{M}^{\lambda}\setminus Z^{\mathrm{cov}}}(*H_{\infty}^{\lambda\,\mathrm{cov}})$-module $\mathcal{E}^{\mathrm{cov}} := (\varpi^{\lambda})^{-1}(\mathcal{E})$. As the restriction, we obtain the locally free $\mathcal{O}_{\overline{M}^{\lambda}\langle t_1\rangle\setminus Z^{\mathrm{cov}}}(*H_{\infty}^{\lambda\,\mathrm{cov}})$-modules $\mathcal{E}^{\mathrm{cov}}_{|\overline{M}^{\lambda}\langle t_1\rangle\setminus Z^{\mathrm{cov}}}$ $(t_1 \in \mathbb{R})$. There exists the induced filtered bundles $\mathcal{P}_*(\mathcal{E}^{\mathrm{cov}}_{|\overline{M}^{\lambda}\langle t_1\rangle\setminus Z^{\mathrm{cov}}})$ $(t_1 \in \mathbb{R})$. By choosing $t_1^0, t_1^1 \in \mathbb{R}$, we obtain the isomorphism

$$\mathcal{E}^{\mathrm{cov}}_{|\{t_1^0\}\times\widehat{\infty}} \simeq \mathcal{E}^{\mathrm{cov}}_{|\{t_1^1\}\times\widehat{\infty}}. \tag{4.1}$$

In Definition 4.2.1, we do not impose any relation between the filtered bundles $\mathcal{P}_*(\mathcal{E}^{\mathrm{cov}}_{|\{t_1^0\}\times\widehat{\infty}})$ and $\mathcal{P}_*(\mathcal{E}^{\mathrm{cov}}_{|\{t_1^1\}\times\widehat{\infty}})$ under the isomorphism (4.1). Recall that there exists the slope decomposition $\mathcal{E}_{|\widehat{H}_{\infty}^{\lambda}} = \bigoplus_{\omega\in\mathbb{Q}} \mathcal{S}_{\omega}(\mathcal{E}_{|\widehat{H}_{\infty}^{\lambda}})$, which induces the decomposition $\mathcal{E}^{\mathrm{cov}}_{|\widehat{H}_{\infty}^{\lambda\,\mathrm{cov}}} = \bigoplus_{\omega\in\mathbb{Q}} \mathcal{S}_{\omega}(\mathcal{E}^{\mathrm{cov}}_{|\widehat{H}_{\infty}^{\lambda}})$. If the goodness condition in Definition 4.2.3 is satisfied, there exist the decompositions

$$\mathcal{P}_*(\mathcal{E}^{\mathrm{cov}}_{|\{t_1\}\times\widehat{\infty}}) = \bigoplus_{\omega\in\mathbb{Q}} \mathcal{P}_*\mathcal{S}_{\omega}(\mathcal{E}^{\mathrm{cov}}_{|\{t_1\}\times\widehat{\infty}}) \quad (t_1 \in \mathbb{R}),$$

and there exist the following isomorphisms as in Lemma 3.5.21:

$$\mathcal{P}_{a-t_1^0\omega/T}\mathcal{S}_{\omega}(\mathcal{E}^{\mathrm{cov}}_{|\{t_1^0\}\times\widehat{\infty}}) \simeq \mathcal{P}_{a-t_1^1\omega/T}\mathcal{S}_{\omega}(\mathcal{E}^{\mathrm{cov}}_{|\{t_1^1\}\times\widehat{\infty}}). \tag{4.2}$$

We shall later discuss more detailed coherence property in the ramified situation (See Sect. 4.3.4.)

4.2.1 Subbundles and Quotient Bundles

Let $\mathcal{E}_1 \subset \mathcal{E}$ be a locally free $\mathcal{O}_{\overline{\mathcal{M}}^{\lambda}\setminus Z}(*H_{\infty}^{\lambda})$-submodule of $\mathcal{E}$. We say that $\mathcal{E}_1$ is saturated if $\mathcal{E}_2 = \mathcal{E}/\mathcal{E}_1$ is torsion-free. In that case, we obtain filtered bundles $\mathcal{P}_*\mathcal{E}_i$ over $\mathcal{E}_i$ by applying the constructions in Sect. 2.11.1.2. Namely, by setting

$$\mathcal{P}_b(\mathcal{E}_{1|\overline{\mathcal{M}}^{\lambda}\langle t_1\rangle\setminus Z}) := \mathcal{P}_b(\mathcal{E}_{|\overline{\mathcal{M}}^{\lambda}\langle t_1\rangle\setminus Z}) \cap \mathcal{E}_{1|\overline{\mathcal{M}}^{\lambda}\langle t_1\rangle\setminus Z},$$

we obtain the induced filtered bundle over $\mathcal{E}_1$, which we denote by $\mathcal{P}_*\mathcal{E}_1$. By setting

$$\mathcal{P}_b(\mathcal{E}_{2|\overline{\mathcal{M}}^{\lambda}\langle t_1\rangle\setminus Z}) := \mathrm{Im}\Big(\mathcal{P}_b(\mathcal{E}_{|\overline{\mathcal{M}}^{\lambda}\langle t_1\rangle\setminus Z}) \longrightarrow \mathcal{E}_{2|\overline{\mathcal{M}}^{\lambda}\langle t_1\rangle\setminus Z}\Big),$$

we obtain a filtered bundle $\mathcal{P}_*\mathcal{E}_2$ over $\mathcal{E}_2$. We obtain the following lemma from Lemma 3.5.19.

Lemma 4.2.6 *If $\mathcal{P}_*\mathcal{E}$ is good, $\mathcal{P}_*\mathcal{E}_i$ are also good.* ■

4.2.2 Degree and Slope

Let $\mathcal{P}_*\mathcal{E}$ be a good filtered bundle of Dirac type on $(\overline{\mathcal{M}}^{\lambda}; Z, H^{\lambda}_{\infty})$. Note that the numbers $\deg(\mathcal{P}_*\mathcal{E}_{|\overline{\mathcal{M}}^{\lambda}\langle t_1\rangle})$ are well defined for any $t_1 \in S^1_T \setminus \pi^{\lambda}(Z)$, and they induce a function on $S^1_T \setminus \pi^{\lambda}(Z)$. We have the slope decomposition $\mathcal{E}_{|\widehat{H}^{\lambda}_{\infty}} = \bigoplus \mathcal{S}_{\omega}(\mathcal{E}_{|\widehat{H}^{\lambda}_{\infty}})$. We set $r(\omega) := \operatorname{rank} \mathcal{S}_{\omega}(\mathcal{E}_{|\widehat{H}^{\lambda}_{\infty}})$.

Lemma 4.2.7 *Let $0 \leq \widetilde{t}^0_1 < \widetilde{t}^1_1 < T$ $(i = 0, 1)$. Suppose that the induced points $t^i_1 \in S^1_T$ are contained in a connected component of $S^1_T \setminus \pi^{\lambda}(Z)$. Then, the following holds.*

$$\deg(\mathcal{P}_*\mathcal{E}_{|\overline{\mathcal{M}}^{\lambda}\langle t^1_1\rangle}) - \deg(\mathcal{P}_*\mathcal{E}_{|\overline{\mathcal{M}}^{\lambda}\langle t^0_1\rangle}) = \sum_{\omega\in\mathbb{Q}} \frac{r(\omega)\omega}{T}(\widetilde{t}^1_1 - \widetilde{t}^0_1).$$

Proof It follows from (4.2) and the definition $\deg(\mathcal{P}_*\mathcal{E}_{|\overline{\mathcal{M}}^{\lambda}\langle t_1\rangle})$ (see the formula (2.57)). ∎

By Lemma 4.2.7, on each connected component of $S^1_T \setminus \pi^{\lambda}(Z)$, the function is affine with respect to the local coordinate t_1, and hence it is integrable over S^1_T. The degree of $\mathcal{P}_*\mathcal{E}$ is defined as

$$\deg(\mathcal{P}_*\mathcal{E}) := \frac{1}{T}\int_0^T \deg\left(\mathcal{P}_*\mathcal{E}_{|\overline{\mathcal{M}}^{\lambda}\langle t_1\rangle\setminus Z}\right) dt_1.$$

We define $\mu(\mathcal{P}_*\mathcal{E}) := \deg(\mathcal{P}_*\mathcal{E})/\operatorname{rank}\mathcal{E}$.

Let us rewrite $\deg(\mathcal{P}_*\mathcal{E})$. For each $P \in Z$, we obtain the tuple of integers $(\ell_1(P), \dots, \ell_{\operatorname{rank}\mathcal{E}}(P))$ as in Lemma 2.3.5.

Lemma 4.2.8 *The following holds.*

$$\begin{aligned}&\deg(\mathcal{P}_*\mathcal{E})\\ &= \lim_{\substack{\epsilon>0\\ \epsilon\to 0}} \deg\left(\mathcal{P}_*\mathcal{E}_{|\overline{\mathcal{M}}^{\lambda}\langle -\epsilon\rangle}\right) + \sum_{\omega\in\mathbb{Q}} \frac{r(\omega)\cdot\omega}{2} + \sum_{P\in Z}\left(1 - \frac{\pi^{\lambda}(P)}{T}\right)\sum_{k=1}^{\operatorname{rank}(\mathcal{E})} \ell_k(P).\end{aligned} \tag{4.3}$$

Here, we naturally regard $0 \leq \pi^{\lambda}(P) < T$.

Proof Let $0 \leq \tilde{t}_1^0 < T$ such that the induced point $t_1^0 \in S_T^1$ is contained in $\pi^\lambda(Z)$. Then, the following holds:

$$\lim_{\substack{\epsilon>0 \\ \epsilon\to 0}} \Big(\deg(\mathcal{P}_*\mathcal{E}_{|\overline{\mathcal{M}}^\lambda\langle t_1^0+\epsilon\rangle}) - \deg(\mathcal{P}_*\mathcal{E}_{|\overline{\mathcal{M}}^\lambda\langle t_1^0-\epsilon\rangle}) \Big) = \sum_{P\in\overline{\mathcal{M}}^\lambda\langle t_1\rangle} \sum_{k=1}^{\operatorname{rank}\mathcal{E}} \ell_k(P).$$

Then, the claim follows from Lemma 4.2.7. ■

Lemma 4.2.9 *Let $\mathcal{P}_*\mathcal{E}$ be a filtered bundle over $\mathcal{E}$. Let $0 \longrightarrow \mathcal{E}_1 \longrightarrow \mathcal{E} \longrightarrow \mathcal{E}_2 \longrightarrow 0$ be an exact sequence of locally free $\mathcal{O}_{\overline{\mathcal{M}}^\lambda\setminus Z}(*H_\infty^\lambda)$-modules of Dirac type. Let $\mathcal{P}_*\mathcal{E}_i$ be the induced filtered bundles over $\mathcal{E}_i$. Then, we obtain* $\deg(\mathcal{P}_*\mathcal{E}_1) + \deg(\mathcal{P}_*\mathcal{E}_2) = \deg(\mathcal{P}_*\mathcal{E})$. *As a result, we obtain .*

$$\mu(\mathcal{P}_*\mathcal{E}) - \mu(\mathcal{P}_*\mathcal{E}_1) = \frac{r_2}{r_1}\big(\mu(\mathcal{P}_*\mathcal{E}_2) - \mu(\mathcal{P}_*\mathcal{E})\big).$$

Proof For each $t_1 \in S_T^1 \setminus \pi^\lambda(Z)$, we obtain

$$\deg(\mathcal{P}_*\mathcal{E}_{1|\overline{\mathcal{M}}^\lambda\langle t_1\rangle}) + \deg(\mathcal{P}_*\mathcal{E}_{2|\overline{\mathcal{M}}^\lambda\langle t_1\rangle}) = \deg(\mathcal{P}_*\mathcal{E}_{|\overline{\mathcal{M}}^\lambda\langle t_1\rangle}).$$

Then, the claim of the lemma follows. ■

Corollary 4.2.10 *Let $\mathcal{P}_*\mathcal{E}$ and $\mathcal{P}_*\mathcal{E}_i$ $(i = 1, 2)$ be as in Lemma* 4.2.9. *Then, one of the following holds: (i) $\mu(\mathcal{P}_*\mathcal{E}_1) < \mu(\mathcal{P}_*\mathcal{E}) < \mu(\mathcal{P}_*\mathcal{E}_2)$, (ii) $\mu(\mathcal{P}_*\mathcal{E}_1) > \mu(\mathcal{P}_*\mathcal{E}) > \mu(\mathcal{P}_*\mathcal{E}_2)$, (iii) $\mu(\mathcal{P}_*\mathcal{E}_1) = \mu(\mathcal{P}_*\mathcal{E}) = \mu(\mathcal{P}_*\mathcal{E}_2)$.* ■

4.2.3 Stability Condition

Definition 4.2.11 Let $\mathcal{P}_*\mathcal{E}$ be a good filtered bundle of Dirac type over $(\overline{\mathcal{M}}^\lambda; Z, H_\infty^\lambda)$. It is called stable (semistable) if the following holds for any saturated $\mathcal{O}_{\overline{\mathcal{M}}^\lambda\setminus Z}(*H_\infty^\lambda)$-submodule $\mathcal{E}_1$ of Dirac type with $0 < \operatorname{rank}\mathcal{E}_1 < \operatorname{rank}\mathcal{E}$:

$$\mu(\mathcal{P}_*\mathcal{E}_1) < (\leq)\mu(\mathcal{P}_*\mathcal{E})$$

It is called polystable if it is the direct sum of stable filtered bundles $\mathcal{P}_*(\mathcal{E}) = \bigoplus \mathcal{P}_*\mathcal{E}_i$ with $\mu(\mathcal{P}_*\mathcal{E}_i) = \mu(\mathcal{P}_*\mathcal{E})$. ■

We have some standard lemmas.

Lemma 4.2.12 *Let $f : \mathcal{P}_*\mathcal{E}_1 \longrightarrow \mathcal{P}_*\mathcal{E}_2$ be a morphism of stable good filtered bundles of Dirac type on $(\overline{\mathcal{M}}^\lambda; Z, H_\infty^\lambda)$ with $\mu(\mathcal{P}_*\mathcal{E}_1) = \mu(\mathcal{P}_*\mathcal{E}_2)$. Then, f is an isomorphism or* 0.

Proof Let $\mathcal{E}_3$ be the image of f. Suppose that $\mathcal{E}_3 \neq 0$. We obtain filtered bundles $\mathcal{P}_*^{(i)}\mathcal{E}_3$ $(i = 1, 2)$ over $\mathcal{E}_3$ induced by $\mathcal{P}_*\mathcal{E}_i$. By Corollary 4.2.10, we obtain

$$\mu(\mathcal{P}_*\mathcal{E}_1) \leq \mu(\mathcal{P}_*^{(1)}\mathcal{E}_3) \leq \mu(\mathcal{P}_*^{(2)}\mathcal{E}_3) \leq \mu(\mathcal{P}_*\mathcal{E}_2).$$

Because $\mu(\mathcal{P}_*\mathcal{E}_1) = \mu(\mathcal{P}_*\mathcal{E}_2)$, the equalities hold. Hence, we obtain that $\mathcal{P}_*\mathcal{E}_1 \simeq \mathcal{P}_*^{(1)}\mathcal{E}_3 \simeq \mathcal{P}_*^{(2)}\mathcal{E}_3 \simeq \mathcal{P}_*\mathcal{E}_2$. ■

Corollary 4.2.13 *Let $\mathcal{P}_*\mathcal{E}$ be a stable good filtered bundle of Dirac type on $(\overline{\mathcal{M}}^{\lambda}; Z, H_{\lambda}^{\infty})$. Any endomorphism f of $\mathcal{P}_*\mathcal{E}$ is the multiplication of a complex number.* ■

As a result, we obtain the following proposition.

Proposition 4.2.14 *Any polystable good filtered bundle of Dirac type $\mathcal{P}_*\mathcal{E}$ on $(\overline{\mathcal{M}}^{\lambda}; Z, H_{\lambda}^{\infty})$ has a unique decomposition*

$$\mathcal{P}_*\mathcal{E} = \bigoplus_{i \in \Lambda} \mathcal{P}_*\mathcal{E}_i \otimes_{\mathbb{C}} U_i, \tag{4.4}$$

where $\mathcal{P}_*\mathcal{E}_i$ are stable good filtered bundles of Dirac type on $(\overline{\mathcal{M}}^{\lambda}; Z, H_{\infty}^{\lambda})$ such that $\mathcal{P}_*\mathcal{E}_i \not\simeq \mathcal{P}_*\mathcal{E}_j$ $(i \neq j)$, and U_i denotes $\mathbb{C}$-vector spaces. ■

4.2.4 Good Filtered Bundles of Dirac Type and Parabolic Difference Modules

4.2.4.1 Polystable Parabolic Difference Modules

Let Φ^* be an automorphism of the field $\mathbb{C}(\beta_1)$ defined by $\Phi^*(f)(\beta_1) = f(\beta_1 + 2\sqrt{-1}\lambda T)$. Let $\boldsymbol{V}$ be a $2\sqrt{-1}\lambda T$-difference module as in Definition 1.6.1. We set $\boldsymbol{V}_{|\widehat{\infty}} := \mathbb{C}(\!(y^{-1})\!) \otimes_{\mathbb{C}(y)} \boldsymbol{V}$. We recall that a filtered bundle $\mathcal{P}_*(\boldsymbol{V}_{|\widehat{\infty}})$ over $\boldsymbol{V}_{|\widehat{\infty}}$ is called a good parabolic structure of the difference module $\boldsymbol{V}$ at ∞ (Definition 1.6.6) if $(\mathcal{P}_*\boldsymbol{V}_{|\widehat{\infty}}, \Phi^*)$ is a good filtered difference module in the sense of Definition 3.3.16. We also recall that a parabolic structure of the difference module $\boldsymbol{V}$ consists of a parabolic structure at finite place $(V, m, (\boldsymbol{\tau}_x, \boldsymbol{L}_x)_{x \in \mathbb{C}})$ (see Definition 2.10.1) and a good parabolic structure at infinity $\mathcal{P}_*(\boldsymbol{V}_{|\widehat{\infty}})$, and it is called a parabolic difference module (Definition 1.6.7). Such a tuple $(\boldsymbol{V}, V, m, (\boldsymbol{\tau}_x, \boldsymbol{L}_x)_{x \in \mathbb{C}}, \mathcal{P}_*(\boldsymbol{V}_{|\widehat{\infty}}))$ is denoted by $\boldsymbol{V}_*$.

Let $\boldsymbol{V}' \subset \boldsymbol{V}$ be a difference submodule. By setting $V' = V \cap \boldsymbol{V}'$, $L'_{x,i} = L_{x,i} \cap (\boldsymbol{V}' \otimes_{\mathbb{C}(\beta_1)} \mathbb{C}(\!(\beta_1 - x)\!))$, and $\mathcal{P}_a(\boldsymbol{V}'_{|\widehat{\infty}}) = \mathcal{P}_a(\boldsymbol{V}_{|\widehat{\infty}}) \cap \boldsymbol{V}'_{|\widehat{\infty}}$ $(a \in \mathbb{R})$, we obtain a parabolic structure of $\boldsymbol{V}'$. (See Lemma 3.3.19.) The induced parabolic difference module is denoted by $\boldsymbol{V}'_*$.

Let $V'' = V/V'$ be the quotient difference module. Let V'' denote the image of $V \longrightarrow V''$, which is isomorphic to V/V'. Let $L''_{x,i}$ denote the image of $L_{x,i} \longrightarrow V'' \otimes_{\mathbb{C}(\beta_1)} \mathbb{C}((\beta_1 - x))$, which is isomorphic to $L_{x,i}/L'_{x,i}$. Let $\mathcal{P}_a(V''_{|\widehat{\infty}})$ ($a \in \mathbb{R}$) denote the image of $\mathcal{P}_a(V_{|\widehat{\infty}}) \longrightarrow V''_{|\widehat{\infty}}$, which is isomorphic to $\mathcal{P}_a(V_{|\widehat{\infty}})/\mathcal{P}_a(V'_{|\widehat{\infty}})$. Note that it is a good parabolic structure of V'' at infinity (see Lemma 3.3.19). Thus, we obtain a parabolic structure of the difference module V''. The induced parabolic difference module is denoted by V''_*.

For a parabolic difference module V_*, we define the degree $\deg(V_*)$ and the slope $\mu(V_*)$ by the formulas (1.9) and (1.10), respectively.

Lemma 4.2.15 *Let $V' \subset V$ be a difference submodule, and we set $V'' = V/V'$. Then, we have* $\deg(V_*) = \deg(V'_*) + \deg(V''_*)$.

Proof It is enough to check that

$$\deg(L_{i,x}, L_{i-1,x}) = \deg(L'_{i,x}, L'_{i-1,x}) + \deg(L''_{i,x}, L''_{i-1,x})$$

for any $x \in \mathbb{C}$ and $i = 1, \ldots, m(x)$. Note that for any $N \subset L_{i,x} \cap L_{i-1,x}$ such that $(L_{i,x} \cap L_{i-1,x})/N$ is torsion, we have

$$\deg(L_{i,x}, L_{i-1,x}) = \dim_{\mathbb{C}}(L_{i,x}/N) - \dim_{\mathbb{C}}(L_{i-1,x}/N).$$

We set $N' := N \cap (V' \otimes \mathbb{C}((\beta_1 - x)))$ and $N'' = N/N'$. Then, $N' \subset L'_{i,x} \cap L'_{i-1,x}$ and $N'' \subset L''_{i,x} \cap L''_{i-1,x}$, and the quotients are torsion. Hence, we obtain

$$\begin{aligned} &\deg(L'_{i,x}, L'_{i-1,x}) + \deg(L''_{i,x}, L''_{i-1,x}) \\ &= \dim(L'_{i,x}/N') - \dim(L'_{i-1,x}/N') + \dim(L''_{i,x}/N'') - \dim(L''_{i-1,x}/N'') \\ &= \dim(L_{i,x}/N) - \dim(L_{i-1,x}/N) = \deg(L_{i,x}, L_{i-1,x}). \end{aligned} \quad (4.5)$$

Thus, we are done. ■

As a consequence of Lemma 4.2.15, a similar statement to Corollary 4.2.10 holds. The stability, semistability and polystability conditions are defined in the standard way as in Definition 1.6.8. We obtain standard statements similar to Lemma 4.2.12, Corollary 4.2.13 and Proposition 4.2.14.

4.2.4.2 Equivalence

Let $\mathcal{P}_*\mathcal{E}$ be a good filtered bundle of Dirac type on $(\overline{\mathcal{M}}^\lambda; Z, H^\infty_\lambda)$. We use the notation in Sect. 2.10.3. From a locally free $\mathcal{O}_{\overline{\mathcal{M}}^\lambda \setminus Z}(*H^\lambda_\infty)$-module $\mathcal{E}$, we obtained a difference module $V(\mathcal{E})$ equipped with a parabolic structure at finite place

$$(V(\mathcal{E}), m_Z, (\boldsymbol{\tau}_{Z,x}, \boldsymbol{L}_{Z,x}(\mathcal{E}))_{x \in \mathbb{C}}). \quad (4.6)$$

There exist natural isomorphisms

$$V(\mathcal{E})_{|\widehat{\infty}} \simeq \mathcal{E}^{\rm cov}_{|\{0\}\times\widehat{\infty}} \simeq \mathcal{E}_{|\widehat{H}^{\lambda}_{\infty}\langle 0\rangle}. \tag{4.7}$$

Hence, we obtain the filtered bundle $\mathcal{P}_*(V(\mathcal{E})_{|\widehat{\infty}}) := \mathcal{P}_*(\mathcal{E}_{|\widehat{H}^{\lambda}_{\infty}\langle 0\rangle})$ under the isomorphism (4.7). By the definition of good filtered bundle over $\mathcal{E}$, $\mathcal{P}_*(V(\mathcal{E})_{|\widehat{\infty}})$ is a good parabolic structure of $V(\mathcal{E})$ at infinity. Let $V_*(\mathcal{P}_*\mathcal{E})$ denote the difference module $V(\mathcal{E})$ with the parabolic structure at finite place (4.6) and the good parabolic structure at infinity $\mathcal{P}_*(V(\mathcal{E})_{|\widehat{\infty}})$. By Lemma 4.2.8, we obtain

$$\deg(\mathcal{P}_*\mathcal{E}) = \deg(V_*(\mathcal{P}_*\mathcal{E})). \tag{4.8}$$

The following proposition is obvious by Lemma 2.10.3 and Proposition 3.5.17.

Proposition 4.2.16 *The above construction induces an equivalence between good filtered bundles $\mathcal{E}$ of Dirac type on $(\overline{\mathcal{M}}^{\lambda}; Z, H^{\lambda}_{\infty})$ and parabolic difference modules $V_* = (V, V, m_Z, (\boldsymbol{\tau}_{Z,x}, \boldsymbol{L}_{Z,x})_{x\in\mathbb{C}}, \mathcal{P}_*(V_{|\widehat{\infty}}))$, where m_Z (resp. $\boldsymbol{\tau}_{Z,x}$ $(x \in \mathbb{C})$) are determined by Z as in (2.47) (resp. (2.48) and (2.49)). The construction is functorial with respect to direct sums, tensor products and inner homomorphisms.* ■

For any locally free $\mathcal{O}_{\overline{\mathcal{M}}^{\lambda}\setminus Z}(*H^{\lambda}_{\infty})$-module $\mathcal{E}' \subset \mathcal{E}$ such that $\mathcal{E}'' = \mathcal{E}/\mathcal{E}'$ is also locally free, we obtain the induced difference submodule $V(\mathcal{E}') \subset V(\mathcal{E})$ for which $V(\mathcal{E})/V(\mathcal{E}') \simeq V(\mathcal{E}'')$. By the constructions, the parabolic difference module $V(\mathcal{P}_*\mathcal{E}')$ is equal to the difference module $V(\mathcal{E}')$ equipped with the parabolic structure induced by $V_*(\mathcal{P}_*\mathcal{E})$ and $V(\mathcal{E}') \subset V(\mathcal{E})$. The parabolic difference module $V_*(\mathcal{P}_*\mathcal{E}'')$ is equal to the difference module $V(\mathcal{E}'')$ equipped with the parabolic structure induced by $V_*(\mathcal{P}_*\mathcal{E})$ and $V(\mathcal{E}) \longrightarrow V(\mathcal{E}'')$.

Conversely, let $V' \subset V(\mathcal{E})$ be a difference submodule. We set $V'' = V(\mathcal{E})/V'$. We obtain the induced parabolic difference modules V'_* and V''_*. We have good filtered bundle $\mathcal{P}_*\mathcal{E}'$ (resp. $\mathcal{P}_*\mathcal{E}''$) on $(\overline{\mathcal{M}}^{\lambda}; Z, H^{\lambda}_{\infty})$ such that $V_*(\mathcal{P}_*\mathcal{E}') \simeq V'_*$ (resp. $V_*(\mathcal{P}_*\mathcal{E}'') \simeq V''_*$). By the construction, it is easy to see that there exists an exact sequence $0 \longrightarrow \mathcal{E}' \longrightarrow \mathcal{E} \longrightarrow \mathcal{E}'' \longrightarrow 0$ which induces the exact sequences

$$0 \longrightarrow \mathcal{P}_*(\mathcal{E}'_{|\overline{\mathcal{M}}^{\lambda}\langle t_1\rangle\setminus Z}) \longrightarrow \mathcal{P}_*(\mathcal{E}_{|\overline{\mathcal{M}}^{\lambda}\langle t_1\rangle\setminus Z}) \longrightarrow \mathcal{P}_*(\mathcal{E}''_{|\overline{\mathcal{M}}^{\lambda}\langle t_1\rangle\setminus Z}) \longrightarrow 0.$$

As a result, we also obtain the following proposition.

Proposition 4.2.17 *The above construction preserves the stability, semistability and polystability conditions.* ■

4.3 Filtered Bundles on Ramified Coverings

We generalize the notion of filtered bundles to the ramified situation.

4.3.1 The Case $\lambda = 0$

We use the notation in Sect. 2.7. We set $Y^{0,\mathrm{cov}} := \overline{M}^0 \setminus (\mathbb{R}\times\{0\}) = \mathbb{R}_t \times (\mathbb{P}^1_w \setminus \{0\})$. We also set $Y^{0,\mathrm{cov},*} := Y^{0,\mathrm{cov}} \setminus H^0_\infty = \mathbb{R}_t \times \mathbb{C}^*_w$. They are preserved by the $\mathbb{Z}$-action κ. The quotient spaces are denoted by Y^0 and Y^{0*}. They are naturally open subsets of $\overline{\mathcal{M}}^0$.

For any positive integer q, we fix a q-th root w_q of w such that $w_{qm}^m = w_q$ for any $m \in \mathbb{Z}_{\geq 1}$. There exists the ramified covering $\mathbb{R}_t \times \mathbb{P}^1_{w_q} \longrightarrow \overline{M}^0 = \mathbb{R}_t \times \mathbb{P}^1_w$. Let $Y^{0\,\mathrm{cov}}_q$, $Y^{0\,\mathrm{cov}\,*}_q$ and $H^{0\,\mathrm{cov}}_{\infty,q}$ denote the inverse images of $Y^{0\,\mathrm{cov}}$, $Y^{0\,\mathrm{cov}\,*}$ and $H^{0\,\mathrm{cov}}_\infty$, respectively. Let κ_q be the $\mathbb{Z}$-action on $\mathbb{R}_t \times \mathbb{P}^1_{w_q}$ defined by $\kappa_{q,n}(t, w_q) = (t + nT, w_q)$, which preserves the subsets Y^0_q, Y^{0*}_q and $H^0_{\infty,q}$. The quotient spaces are denoted by Y^0_q, Y^{0*}_q and $H^0_{\infty,q}$, respectively. The projection $Y^{0\,\mathrm{cov}}_q \longrightarrow Y^0_q$ is denoted by ϖ^0_q. Note that $Y^{0\,\mathrm{cov}}_q$ and Y^0_q are naturally equipped with mini-complex structures induced by the mini-complex structure of $\mathbb{R}_t \times \mathbb{P}^1_{w_q}$. Clearly, $Y^{0\,\mathrm{cov}}_1$ and Y^0_1 are equal to $Y^{0\,\mathrm{cov}}$ and Y^0, respectively.

Let $\Psi^0_q : Y^0_q \longrightarrow \mathbb{P}^1_{w_q} \setminus \{0\}$ denote the projection. The restriction $Y^{0*}_q \longrightarrow \mathbb{C}_{w_q} \setminus \{0\}$ is denoted by Ψ^0_q or simply Ψ_q.

4.3.2 Ramified Coverings for General λ

Let $Y^{\lambda\ \mathrm{cov}}$ denote the open subset of $\overline{M}^\lambda$ defined as follows:

$$Y^{\lambda,\mathrm{cov}} := \overline{M}^\lambda \setminus \left\{(t_1, \beta_1) \,\middle|\, \beta_1 - 2\sqrt{-1}\lambda t_1 = 0\right\}.$$

We also put $Y^{\lambda,\mathrm{cov}\,*} := Y^{\lambda,\mathrm{cov}} \setminus H^{\lambda,\mathrm{cov}}_\infty$. The $\mathbb{Z}$-action κ on $\overline{M}^\lambda$ preserves $Y^{\lambda\ \mathrm{cov}}$ and $Y^{\lambda,\mathrm{cov}\,*}$. The quotient spaces are denoted by Y^λ and $Y^{\lambda *}$, respectively, which are naturally open subsets of $\overline{\mathcal{M}}^\lambda$. Clearly, $Y^{\lambda *} = Y^\lambda \setminus H^\lambda_\infty$. Note that $Y^{\lambda,\mathrm{cov},*} = Y^{0,\mathrm{cov},*}$ and $Y^{\lambda,*} = Y^{0,*}$ under the C^∞-identifications $M^\lambda = M^0$ and $\mathcal{M}^\lambda = \mathcal{M}^0$. Recall the relation $w = (1 + |\lambda|^2)^{-1}(\beta_1 - 2\sqrt{-1}\lambda t_1)$. By using the morphism $\Psi^\lambda : \overline{M}^\lambda \longrightarrow \mathbb{P}^1_w$ in Sect. 2.8.3 and the projection $\varpi^\lambda : \overline{M}^\lambda \longrightarrow \overline{\mathcal{M}}^\lambda$, we have $Y^\lambda = (\Psi^\lambda)^{-1}(\mathbb{P}^1_w \setminus \{0\})$ and $Y^{\lambda\ \mathrm{cov}} = (\varpi^\lambda)^{-1}(Y^\lambda)$.

Let us construct ramified coverings of $Y^{\lambda\ \mathrm{cov}}$. Let $\iota_\lambda : \mathbb{R}_u \times \mathbb{P}^1_x \simeq \mathbb{R}_{t_1} \times \mathbb{P}^1_{\beta_1}$ denote the diffeomorphism defined by $\iota_\lambda(u, x) := (u, x + 2\sqrt{-1}\lambda u)$. We have $\iota_\lambda^{-1}(t_1, \beta_1) = (t_1, \beta_1 - 2\sqrt{-1}\lambda t_1)$. Let x_q be a q-th root of x, and let $\mathbb{R}_u \times \mathbb{P}^1_{x_q} \longrightarrow \mathbb{R}_{t_1} \times \mathbb{P}^1_{\beta_1}$ be the C^∞-map obtained as the composite of the ramified covering $(u, x_q) \longmapsto (u, x_q^q)$ and the diffeomorphism ι_λ. Let $Y^{\lambda,\mathrm{cov}}_q$, $Y^{\lambda,\mathrm{cov}\,*}_q$ and $H^{\lambda,\mathrm{cov}}_{\infty,q}$ be the inverse image of $Y^{\lambda,\mathrm{cov}}$, $Y^{\lambda,\mathrm{cov}\,*}$, and $H^{\lambda,\mathrm{cov}}_\infty$, respectively. Let κ_q denote the $\mathbb{Z}$-action on $\mathbb{R}_u \times \mathbb{P}^1_{x_q}$ defined by $\kappa_{q,n}(u, x_q) = (u + nT, x_q)$. It induces the

$\mathbb{Z}$-action on $Y_q^{\lambda,\mathrm{cov}}$, $Y_q^{\lambda,\mathrm{cov}*}$, and $H_{\infty,q}^{\lambda,\mathrm{cov}}$. The quotient spaces are denoted by Y_q^{λ}, $Y_q^{\lambda*}$, and $H_{\infty,q}^{\lambda}$. The projection $Y_q^{\lambda\,\mathrm{cov}} \longrightarrow Y_q^{\lambda}$ is denoted by ϖ_q^{λ}. The function x_q is well defined on $Y_q^{\lambda*}$, which is equal to $(1+|\lambda|^2)^{1/q}w_q$.

Because $Y_q^{\lambda\,\mathrm{cov}*} \longrightarrow Y^{\lambda\,\mathrm{cov}*}$ and $Y_q^{\lambda*} \longrightarrow Y^{\lambda*}$ are covering spaces, we obtain the induced mini-complex structures on $Y_q^{\lambda\,\mathrm{cov}*}$ and $Y_q^{\lambda*}$.

Lemma 4.3.1 *The mini-complex structures on $Y_q^{\lambda\,\mathrm{cov}*}$ and $Y_q^{\lambda*}$ uniquely extend to mini-complex structures on $Y_q^{\lambda\,\mathrm{cov}}$ and Y_q^{λ}, respectively.*

Proof The uniqueness is clear. Let P be any point of $H_{\infty,q}^{\lambda\,\mathrm{cov}}$. There exists a neighbourhood U_P of P in $Y_q^{\lambda\,\mathrm{cov}*}$ such that

$$\beta_{1,q}^{-1} := x_q^{-1}(1+2\sqrt{-1}\lambda u x^{-1})^{-1/q}$$

is a well defined function on U_P. It is a q-th root of β_1^{-1}. We obtain the mini-complex coordinate system $(t_1, \beta_{1,q}^{-1})$ on U_P, which is compatible with the mini-complex structure of $U_P \cap Y_q^{\lambda\,\mathrm{cov},*}$. By varying $P \in H_{\infty,q}^{\lambda\,\mathrm{cov}}$, we obtain a mini-complex structure of $Y_q^{\lambda,\mathrm{cov}}$. It induces a mini-complex structure on Y_q^{λ} which is the extension of the mini-complex structure on $Y_q^{\lambda*}$. ■

We emphasize $Y_q^{\lambda\,\mathrm{cov}*} = Y_q^{0\,\mathrm{cov}*}$ and $Y_q^{\lambda*} = Y_q^{0*}$ as Riemannian manifolds. We also note that Y_q^{λ} (resp. $Y_q^{\lambda\,\mathrm{cov}}$) is the fiber product of Y^{λ} (resp. $Y^{\lambda\,\mathrm{cov}}$) and $\mathbb{P}^1_{w_q} \setminus \{0\}$ over $\mathbb{P}^1_w \setminus \{0\}$.

Let $\Psi_q^{\lambda} : Y_q^{\lambda} \longrightarrow \mathbb{P}^1_{w_q} \setminus \{0\}$ denote the projection induced by $(t_1, \beta_{1,q}) \longmapsto (1+|\lambda|^2)^{-1/q}(\beta_{1,q}^q - 2\sqrt{-1}\lambda t_1)^{1/q}$. It is proper. The restriction $Y_q^{\lambda*} \longrightarrow \mathbb{C}_{w_q} \setminus \{0\}$ is also denoted by Ψ_q because it is identified with the projection $\Psi_q : Y_q^{0*} \longrightarrow \mathbb{C}_{w_q} \setminus \{0\}$.

For any $P \in H_{\infty,q}^{\lambda\,\mathrm{cov}}$, let $\mathcal{O}_{\widehat{H}_{\infty,q}^{\lambda\,\mathrm{cov}},P}$ denote the completion of the local ring $\mathcal{O}_{Y_q^{\lambda\,\mathrm{cov}},P}$ with respect to the maximal ideal. Thus, we obtain the sheaf of algebras $\mathcal{O}_{\widehat{H}_{\infty,q}^{\lambda\,\mathrm{cov}}}$ on $H_{\infty,q}^{\lambda\,\mathrm{cov}}$. The ringed space $H_{\infty,q}^{\lambda\,\mathrm{cov}}$ with $\mathcal{O}_{\widehat{H}_{\infty,q}^{\lambda\,\mathrm{cov}}}$ is denoted by $\widehat{H}_{\infty,q}^{\lambda\,\mathrm{cov}}$. Similarly, for any $P \in H_{\infty,q}^{\lambda}$, let $\mathcal{O}_{\widehat{H}_{\infty,q}^{\lambda},P}$ denote the completion of the local ring $\mathcal{O}_{Y_q^{\lambda},P}$ with respect to the maximal ideal. We obtain the sheaf of algebras $\mathcal{O}_{\widehat{H}_{\infty,q}^{\lambda}}$ on $H_{\infty,q}^{\lambda}$, and the ringed space $\widehat{H}_{\infty,q}^{\lambda}$. The ringed space $\widehat{H}_{\infty,q}^{\lambda\,\mathrm{cov}}$ is equipped with the induced $\mathbb{Z}$-action κ, and $\widehat{H}_{\infty,q}^{\lambda}$ is the quotient space.

For any $P \in H_{\infty,q}^{\lambda}$, we set $\mathcal{O}_{\widehat{H}_{\infty,q}^{\lambda}}(*H_{\infty,q}^{\lambda})_P := \mathcal{O}_{Y_q^{\lambda}}(*H_{\infty,q}^{\lambda})_P \otimes_{\mathcal{O}_{Y_q^{\lambda},P}} \mathcal{O}_{\widehat{H}_{\infty,q}^{\lambda},P}$. Thus, we obtain the sheaf of algebras $\mathcal{O}_{\widehat{H}_{\infty,q}^{\lambda}}(*H_{\infty,q}^{\lambda})$ on $H_{\infty,q}^{\lambda}$. Similarly, we obtain the sheaf of algebras $\mathcal{O}_{\widehat{H}_{\infty,q}^{\lambda\,\mathrm{cov}}}(*H_{\infty,q}^{\lambda\,\mathrm{cov}})$ as the formal completion of $\mathcal{O}_{Y_q^{\lambda\,\mathrm{cov}}}(*H_{\infty,q}^{\lambda\,\mathrm{cov}})$ along $H_{\infty,q}^{\lambda\,\mathrm{cov}}$.

Remark 4.3.2 The ringed space $\widehat{H}_{\infty,q}^{\lambda\,\mathrm{cov}}$ with the $\mathbb{Z}$-action κ is isomorphic to $\widehat{H}_{\infty,q}^{\mathrm{cov}}$ with the $\mathbb{Z}$-action in Sect. 3.4.2 by the change of the variables $(t_1, \beta_{1,q}^{-1}, x_q^{-1}) =$

(t, y_q^{-1}, x_q^{-1}). The ringed space $\widehat{H}^{\lambda}_{\infty,q}$ is isomorphic to $\widehat{H}_{\infty,q}$ with the $\mathbb{Z}$-action in Sect. 3.4.2. ■

Let $k_q^{\lambda} : \widehat{H}^{\lambda}_{\infty,q} \longrightarrow Y_q^{\lambda}$ denote the morphism of ringed spaces obtained as the inclusion $H^{\lambda}_{\infty,q} \longrightarrow Y_q^{\lambda}$ and the natural morphism of sheaves of algebras $\mathcal{O}_{Y_q^{\lambda}|H^{\lambda}_{\infty,q}} \longrightarrow \mathcal{O}_{\widehat{H}^{\lambda}_{\infty,q}}$. For any $\mathcal{O}_{Y_q^{\lambda}}$-module $\mathcal{F}$, we set $\mathcal{F}_{|\widehat{H}^{\lambda}_{\infty,q}} := (k_q^{\lambda})^*(\mathcal{F})$. Similarly, we obtain the morphism of ringed spaces $\widehat{H}^{\lambda\,\mathrm{cov}}_{\infty,q} \longrightarrow Y_q^{\lambda\,\mathrm{cov}}$. The pull back of $\mathcal{O}_{Y_q^{\lambda\,\mathrm{cov}}}$-module $\mathcal{F}$ denoted by $\mathcal{F}_{|\widehat{H}^{\lambda\,\mathrm{cov}}_{\infty,q}}$.

For any $p \in q\mathbb{Z}_{\geq 1}$, let $\mathcal{R}_{q,p} : Y_p^{\lambda,\mathrm{cov}} \longrightarrow Y_q^{\lambda,\mathrm{cov}}$ denote the $\mathbb{Z}$-equivariant map induced by $x_p \longmapsto x_p^{p/q}$. Note that the pull back of functions induces $\mathcal{R}_{q,p}^{-1}\mathcal{O}_{Y_q^{\lambda\,\mathrm{cov}}} \longrightarrow \mathcal{O}_{Y_p^{\lambda\,\mathrm{cov}}}$. We define the pull back of $\mathcal{O}_{Y_q^{\lambda,\mathrm{cov}}}(*H^{\lambda,\mathrm{cov}}_{\infty,q})$-modules and the push-forward of $\mathcal{O}_{Y_p^{\lambda,\mathrm{cov}}}(*H^{\lambda,\mathrm{cov}}_{\infty,p})$-modules in natural ways. Let $\mathrm{Gal}_{q,p}$ denote the Galois group of the ramified covering $\mathcal{R}_{q,p}$. We also define the descent of $\mathrm{Gal}_{q,p}$-equivariant $\mathcal{O}_{Y_p^{\lambda,\mathrm{cov}}}(*H^{\lambda,\mathrm{cov}}_{\infty,p})$-modules in a natural way.

We obtain the induced ramified covering $\mathcal{R}_{q,p} : Y_p^{\lambda} \longrightarrow Y_q^{\lambda}$. We define the pull back of $\mathcal{O}_{Y_q^{\lambda}}(*H^{\lambda}_{\infty,q})$-modules and the push-forward of $\mathcal{O}_{Y_p^{\lambda}}(*H^{\lambda}_{\infty,p})$-modules. We also define the descent of $\mathrm{Gal}_{q,p}$-equivariant $\mathcal{O}_{Y_p^{\lambda}}(*H^{\lambda}_{\infty,p})$-modules.

4.3.3 Filtered Bundles

Let $\mathcal{B}_q^{\lambda}$ be a neighbourhood of $H^{\lambda}_{\infty,q}$ in Y_q^{λ}. Let $\pi_q^{\lambda} : \mathcal{B}_q^{\lambda} \longrightarrow S_T^1$ be the map induced by $\pi_q^{\lambda}(u, x_q) = u$. In terms of the local coordinate systems $(t_1, \beta_{1,q}^{-1})$, it is described as $\pi_q^{\lambda}(t_1, \beta_{1,q}^{-1}) = t_1$. We set $\mathcal{B}_q^{\lambda}\langle t_1\rangle := (\pi_q^{\lambda})^{-1}(t_1)$. We also set $H^{\lambda}_{\infty,q}\langle t_1\rangle := (\pi_q^{\lambda})^{-1}(t_1) \cap H^{\lambda}_{\infty,q}$. The point $H^{\lambda}_{\infty,q}\langle t_1\rangle \in \mathcal{B}_q^{\lambda}\langle t_1\rangle$ is also denoted by ∞ if there is no risk of confusion. For $t_1 \in S_T^1 \simeq H^{\lambda}_{\infty,q}$, let $\widehat{H}^{\lambda}_{\infty,q}\langle t_1\rangle$ denote the ringed space $\left(H^{\lambda}_{\infty,q}, \mathcal{O}_{\widehat{H}^{\lambda}_{\infty,q},t_1}\right)$. It is the formal completion of $\mathcal{B}_q^{\lambda}\langle t_1\rangle$ along $H^{\lambda}_{\infty,q}\langle t_1\rangle$.

Let $\mathcal{E}$ be a locally free $\mathcal{O}_{\mathcal{B}_q^{\lambda}}(*H^{\lambda}_{\infty,q})$-module of finite rank. We obtain the locally free $\mathcal{O}_{\mathcal{B}_q^{\lambda}\langle t_1\rangle}(*\infty)$-module $\mathcal{E}_{|\mathcal{B}_q^{\lambda}\langle t_1\rangle}$ as the pull back of $\mathcal{E}$ by the inclusion $\mathcal{B}_q^{\lambda}\langle t_1\rangle \longrightarrow \mathcal{B}_q^{\lambda}$.

Definition 4.3.3 A family of filtered bundles $\mathcal{P}_*\mathcal{E} = \left(\mathcal{P}_*(\mathcal{E}_{|\mathcal{B}_q^{\lambda}\langle t_1\rangle}) \,\middle|\, t_1 \in S_T^1\right)$ is called a filtered bundle over $\mathcal{E}$. ■

Note that a filtered bundle $\mathcal{P}_*\mathcal{E} = \left(\mathcal{P}_*(\mathcal{E}_{|\mathcal{B}_1^{\lambda}\langle t_1\rangle}) \,\middle|\, t_1 \in S_T^1\right)$ over $\mathcal{E}$ induces a filtered bundle $\mathcal{P}_*\left(\mathcal{E}_{|\widehat{H}^{\lambda}_{\infty,q}}\right) = \left(\mathcal{P}_*(\mathcal{E}_{|\widehat{H}^{\lambda}_{\infty,q}\langle t_1\rangle}) \,\middle|\, t_1 \in S_T^1\right)$ over $\mathcal{E}_{|\widehat{H}^{\lambda}_{\infty,q}}$.

Definition 4.3.4 A filtered bundle $\mathcal{P}_*\mathcal{E}$ over $\mathcal{E}$ is called good if the induced filtered bundle over $\mathcal{E}_{|\widehat{H}^{\lambda}_{\infty,q}}$ is good in the sense of Definition 3.5.16. ■

Definition 4.3.5 A (good) filtered bundle on $(\mathcal{B}_q^\lambda, H_{\infty,q}^\lambda)$ is defined to be a locally free $\mathcal{O}_{\mathcal{B}_q^\lambda}(*H_{\infty,q}^\lambda)$-module $\mathcal{E}$ of finite rank equipped with a (good) filtered bundle $\mathcal{P}_*\mathcal{E}$. ■

Let $p \in q\mathbb{Z}_{>0}$. We set $\mathcal{B}_p^\lambda := \mathcal{R}_{q,p}^{-1}(\mathcal{B}_q^\lambda) \subset Y_p^\lambda$, which is a neighbourhood of $H_{\infty,p}^\lambda$. Let $\mathcal{P}_*\mathcal{E}$ be a filtered bundle on $(\mathcal{B}_q^\lambda, H_{\infty,q}^\lambda)$. We obtain a locally free $\mathcal{O}_{\mathcal{B}_p^\lambda}(*H_{\infty,p}^\lambda)$-module $\mathcal{R}_{q,p}^*(\mathcal{E})$. We also obtain a family of filtered bundles $\mathcal{R}_{q,p}^*\big(\mathcal{P}_*(\mathcal{E}_{|\mathcal{B}_q^\lambda\langle t_1\rangle})\big)$ $(t_1 \in S_T^1)$ over $\mathcal{R}_{q,p}^*(\mathcal{E})_{|\mathcal{B}_q^\lambda\langle t_1\rangle}$. Thus, we obtain a filtered bundle $\mathcal{R}_{q,p}^*(\mathcal{P}_*\mathcal{E})$ on $(\mathcal{B}_p^\lambda, H_{\infty,p}^\lambda)$.

Let $\mathcal{P}_*\mathcal{E}'$ be a filtered bundle on $(\mathcal{B}_p^\lambda, H_{\infty,p}^\lambda)$. By the push-forward, we obtain a locally free $\mathcal{O}_{\mathcal{B}_q^\lambda}(*H_{\infty,q}^\lambda)$-module $\mathcal{R}_{q,p*}(\mathcal{E}')$. Each $\mathcal{R}_{q,p*}(\mathcal{E}')_{|\mathcal{B}_q^\lambda\langle t_1\rangle}$ is equipped with a filtered bundle $\mathcal{R}_{q,p*}\big(\mathcal{P}_*(\mathcal{E}'_{|\mathcal{B}_p^\lambda\langle t_1\rangle})\big)$. Thus, we obtain a filtered bundle $\mathcal{R}_{q,p*}(\mathcal{P}_*\mathcal{E}')$. If $\mathcal{P}_*\mathcal{E}'$ is $\mathrm{Gal}_{q,p}$-equivariant, then the descent of $\mathcal{P}_*\mathcal{E}'$ is defined to be the $\mathrm{Gal}_{q,p}$-invariant part of $\mathcal{R}_{q,p*}(\mathcal{P}_*\mathcal{E}')$. The following lemma is obvious by definition.

Lemma 4.3.6 *Let $\mathcal{P}_*\mathcal{E}$ be a filtered bundle on $(\mathcal{B}_q^\lambda, H_{\infty,q}^\lambda)$.*

- *$\mathcal{R}_{q,p}^*(\mathcal{P}_*\mathcal{E})$ is naturally $\mathrm{Gal}_{q,p}$-equivariant, and $\mathcal{P}_*\mathcal{E}$ is the descent of $\mathcal{R}_{q,p}^*(\mathcal{P}_*\mathcal{E})$.*
- *$\mathcal{P}_*\mathcal{E}$ is good if and only if $\mathcal{R}_{q,p}^*(\mathcal{P}_*\mathcal{E})$ is good.* ■

Remark 4.3.7 For filtered bundles $\mathcal{P}_*(\mathcal{E}_i)$ $(i = 1, 2)$ over locally free $\mathcal{O}_{\mathcal{B}_q^\lambda}(*H_\infty^\lambda)$-modules $\mathcal{E}_i$, we naturally define the filtered bundles $\mathcal{P}_*(\mathcal{E}_1) \oplus \mathcal{P}_*(\mathcal{E}_2)$ over $\mathcal{E}_1 \oplus \mathcal{E}_2$, $\mathcal{P}_*(\mathcal{E}_1) \otimes \mathcal{P}_*(\mathcal{E}_2)$ over $\mathcal{E}_1 \otimes \mathcal{E}_2$ and $\mathcal{H}om(\mathcal{P}_*\mathcal{E}_1, \mathcal{P}_*\mathcal{E}_2)$ over $\mathcal{H}om(\mathcal{E}_1, \mathcal{E}_2)$ by applying the constructions in Sect. 2.11.1 to $\mathcal{P}_*(\mathcal{E}_{i|\mathcal{B}_q^\lambda\langle t_1\rangle})$. If $\mathcal{P}_*\mathcal{E}_i$ $(i = 1, 2)$ are good, then $\mathcal{P}_*(\mathcal{E}_1) \oplus \mathcal{P}_*(\mathcal{E}_2)$, $\mathcal{P}_*(\mathcal{E}_1) \otimes \mathcal{P}_*(\mathcal{E}_2)$ and $\mathcal{H}om(\mathcal{P}_*\mathcal{E}_1, \mathcal{P}_*\mathcal{E}_2)$ are also good. Similarly, for a filtered bundle $\mathcal{P}_*\mathcal{E}$ over $\mathcal{E}$, we naturally obtain the dual $(\mathcal{P}_*\mathcal{E})^\vee$ over $\mathcal{E}^\vee$. If $\mathcal{P}_*\mathcal{E}$ is good, then $(\mathcal{P}_*\mathcal{E})^\vee$ is also good. ■

4.3.4 Local Lattices and the Weight Filtration on the Graded Pieces

Take $t_1^0 \in S_T^1$ and $0 < \epsilon < T/10$. Let $\mathcal{B}_q^\lambda\langle t_1^0, \epsilon\rangle$ denote the inverse image of $]t_1^0 - \epsilon, t_1^0 + \epsilon[$ by $\pi_q^\lambda : \mathcal{B}_q^\lambda \longrightarrow S_T^1$. We set

$$H_{\infty,q}^\lambda\langle t_1^0, \epsilon\rangle := (\pi_q^\lambda)^{-1}\big(]t_1^0 - \epsilon, t_1^0 + \epsilon[\big) \cap H_{\infty,q}^\lambda.$$

As in Sect. 3.5.7, let $\widehat{H}_{\infty,q}^\lambda\langle t_1^0, \epsilon\rangle$ denote the ringed space $H_{\infty,q}^\lambda\langle t_1^0, \epsilon\rangle$ with $\mathcal{O}_{\widehat{H}_{\infty,q}^\lambda\langle t_1^0,\epsilon\rangle} := \mathcal{O}_{\widehat{H}_{\infty,q}^\lambda|H_{\infty,q}^\lambda\langle t_1^0,\epsilon\rangle}$. For any $\mathcal{O}_{\mathcal{B}_q^\lambda}$-module $\mathcal{F}$, let $\mathcal{F}_{|\widehat{H}_{\infty,q}^\lambda\langle t_1^0,\epsilon\rangle}$ denote the pull back of $\mathcal{F}$ by the morphism of the ringed spaces $\widehat{H}_{\infty,q}^\lambda\langle t_1^0, \epsilon\rangle \longrightarrow \mathcal{B}_q^\lambda$.

Similarly, for any $\mathcal{O}_{\widehat{H}^{\lambda}_{\infty,q}}$-module $\mathcal{F}$, let $\mathcal{F}_{|\widehat{H}^{\lambda}_{\infty,q}\langle t_1^0,\epsilon\rangle}$ denote the pull back of $\mathcal{F}$ by the morphism of the ringed spaces $\widehat{H}^{\lambda}_{\infty,q}\langle t_1^0,\epsilon\rangle \longrightarrow \widehat{H}^{\lambda}_{\infty,q}$.

Let $\mathcal{P}_*\mathcal{E}$ be a good filtered bundle on $(\mathcal{B}^{\lambda}_q, H^{\lambda}_{\infty,q})$. For each $a \in \mathbb{R}$, there exists the lattice $\boldsymbol{P}^{(t_1^0)}_a(\mathcal{E}_{|\widehat{H}^{\lambda}_{\infty,q}\langle t_1^0,\epsilon\rangle})$ of $\mathcal{E}_{|\widehat{H}^{\lambda}_{\infty,q}\langle t_1^0,\epsilon\rangle}$ as in Sect. 3.5.7. Let $k^{\lambda}_q : \widehat{H}^{\lambda}_{\infty,q}\langle t_1^0,\epsilon\rangle \longrightarrow \mathcal{B}^{\lambda}_q\langle t_1^0,\epsilon\rangle$ denote the naturally induced morphism. We obtain the following morphisms of sheaves:

$$\mathcal{E}_{|\mathcal{B}^{\lambda}_q\langle t_1^0,\epsilon\rangle} \xrightarrow{b_1} k^{\lambda}_{q*}\Big(\mathcal{E}_{|\widehat{H}^{\lambda}_{\infty,q}\langle t_1^0,\epsilon\rangle}\Big) \xrightarrow{b_2} k^{\lambda}_{q*}\Big(\mathcal{E}_{|\widehat{H}^{\lambda}_{\infty,q}\langle t_1^0,\epsilon\rangle}\Big/\boldsymbol{P}^{(t_1^0)}_a\big(\mathcal{E}_{|\widehat{H}^{\lambda}_{\infty,q}\langle t_1^0,\epsilon\rangle}\big)\Big).$$

We define $\boldsymbol{P}^{(t_1^0)}_a(\mathcal{E}_{|\mathcal{B}^{\lambda}_q\langle t_1^0,\epsilon\rangle})$ as the kernel of the epimorphism $b_2 \circ b_1$. Thus, we obtain the locally free $\mathcal{O}_{\mathcal{B}^{\lambda}_q\langle t_1^0,\epsilon\rangle}$-lattice $\boldsymbol{P}^{(t_1^0)}_a(\mathcal{E}_{|\mathcal{B}^{\lambda}_q\langle t_1^0,\epsilon\rangle})$ of $\mathcal{E}_{|\mathcal{B}^{\lambda}_q\langle t_1^0,\epsilon\rangle}$ satisfying the condition $\boldsymbol{P}^{(t_1^0)}_a(\mathcal{E}_{|\mathcal{B}^{\lambda}_q\langle t_1^0,\epsilon\rangle})_{|\widehat{H}^{\lambda}_{\infty,q}\langle t_1^0,\epsilon\rangle} = \boldsymbol{P}^{(t_1^0)}_a(\mathcal{E}_{|\widehat{H}^{\lambda}_{\infty,q}\langle t_1^0,\epsilon\rangle})$. We obtain the following local system on $H^{\lambda}_{\infty,q}\langle t_1^0,\epsilon\rangle$:

$$\boldsymbol{G}^{(t_1^0)}_a(\mathcal{E}_{|\mathcal{B}^{\lambda}_q\langle t_1^0,\epsilon\rangle}) := \boldsymbol{P}^{(t_1^0)}_a(\mathcal{E}_{|\mathcal{B}^{\lambda}_q\langle t_1^0,\epsilon\rangle})\Big/\boldsymbol{P}^{(t_1^0)}_{<a}(\mathcal{E}_{|\mathcal{B}^{\lambda}_q\langle t_1^0,\epsilon\rangle}).$$

There exists the natural isomorphism.

$$\boldsymbol{G}^{(t_1^0)}_a(\mathcal{E}_{|\mathcal{B}^{\lambda}_q\langle t_1^0,\epsilon\rangle})_{|H^{\lambda}_{\infty,q}\langle t_1^0,\epsilon\rangle} \simeq \boldsymbol{G}^{(t_1^0)}_a(\mathcal{E}_{|\widehat{H}^{\lambda}_{\infty,q}\langle t_1^0,\epsilon\rangle}) = \bigoplus_{\omega} \boldsymbol{G}^{(t_1^0)}_a\Big(\mathcal{S}_{\omega}(\mathcal{E}_{|\widehat{H}^{\lambda}_{\infty,q}})_{|\widehat{H}^{\lambda}_{\infty,q}\langle t_1^0,\epsilon\rangle}\Big). \tag{4.9}$$

There exists the filtration W on $\boldsymbol{G}^{(t_1^0)}_a(\mathcal{E}_{|\widehat{H}^{\lambda}_{\infty,q}\langle t_1^0,\epsilon\rangle})$ as in Sect. 3.5.7. It induces the filtration W on $\boldsymbol{G}^{(t_1^0)}_a(\mathcal{E}_{|\mathcal{B}^{\lambda}_q\langle t_1^0,\epsilon\rangle})$.

4.3.5 Convenient Frame

Definition 4.3.8 A frame $\boldsymbol{v}$ of $\boldsymbol{P}^{(t_1^0)}_a(\mathcal{E}_{|\mathcal{B}^{\lambda}_q\langle t_1^0,\epsilon\rangle})$ is called a convenient frame around P if the following conditions are satisfied.

- There exists a decomposition $\boldsymbol{v} = \coprod_{a-1<b\leq a} \boldsymbol{v}_b$ such that $\boldsymbol{v}_b$ is a tuple of sections of $\boldsymbol{P}^{(t_1^0)}_b(\mathcal{E}_{|\mathcal{B}^{\lambda}_q\langle t_1^0,\epsilon\rangle})$, and that $\boldsymbol{v}_b$ induces a frame of $\boldsymbol{G}^{(t_1^0)}_b(\mathcal{E}_{|\mathcal{B}^{\lambda}_q\langle t_1^0,\epsilon\rangle})$.

- There exists a decomposition $\boldsymbol{v}_b = \coprod_{\omega \in \mathbb{Q}} \boldsymbol{v}_{b,\omega}$ such that $\boldsymbol{v}_{b,\omega}$ induces a base of
$$\boldsymbol{G}_b^{(t_1^0)}\Big(\mathcal{S}_\omega\big(\mathcal{E}_{|\widehat{H}^\lambda_{\infty,q}}\big)_{|\widehat{H}^\lambda_q\langle t_1^0,\epsilon\rangle}\Big)$$
in the decomposition (4.9).
- There exists a decomposition $\boldsymbol{v}_{b,\omega} = \coprod_{k\in\mathbb{Z}} \boldsymbol{v}_{b,\omega,k}$ such that $\coprod_{k\le\ell} \boldsymbol{v}_{b,\omega,k}$ induces a base of
$$W_\ell \boldsymbol{G}_b^{(t_1^0)}\Big(\mathcal{S}_\omega\big(\mathcal{E}_{|\widehat{H}^\lambda_{\infty,q}}\big)_{|\widehat{H}^\lambda_{\infty,q}\langle t_1^0,\epsilon\rangle}\Big).$$

It is easy to see the existence of a convenient frame. ■

We set $\nu(\mathcal{E}) := \max\big\{|\omega_1| + |\omega_2| \,\big|\, \mathcal{S}_{\omega_i}\mathcal{E}_{|\widehat{H}^\lambda_{\infty,q}} \neq 0\big\}$. Suppose that $\epsilon > 0$ satisfies
$$10q\nu(\mathcal{E})\epsilon/T < \min\big\{|a-b| \,\big|\, a,b \in \mathcal{P}ar\big(\mathcal{P}_*\mathcal{E}_{|\mathcal{B}^\lambda_q\langle t_1^0\rangle}\big),\ a \neq b\big\}. \tag{4.10}$$

Lemma 4.3.9 *Let* $\boldsymbol{v} = \coprod_{b,\omega,k} \boldsymbol{v}_{b,\omega,k}$ *be a convenient frame of* $\boldsymbol{P}_a^{(t_1^0)}\big(\mathcal{E}_{|\mathcal{B}^\lambda\langle t_1^0,\epsilon\rangle}\big)$. *Then, the following holds.*

- *For each* $t_1 \in H^\lambda_{\infty,q}\langle t_1^0,\epsilon\rangle$, $\boldsymbol{v}_{b,\omega,k}$ *induces a tuple of sections of*
$$\mathcal{P}_{b-(t_1-t_1^0)q\omega/T}\big(\mathcal{S}_\omega(\mathcal{E}_{|\widehat{H}^\lambda_{\infty,q}})_{|\widehat{H}^\lambda_{\infty,q}\langle t_1\rangle}\big).$$
- *The induced tuple* $[\boldsymbol{v}_{b,\omega,k}]$ *of elements of* $\mathrm{Gr}^{\mathcal{P}}_{b-(t_1-t_1^0)q\omega/T}\big(\mathcal{S}_\omega(\mathcal{E}_{|\widehat{H}^\lambda_{\infty,q}})_{|\widehat{H}^\lambda_{\infty,q}\langle t_1\rangle}\big)$ *are contained in*
$$W_k\,\mathrm{Gr}^{\mathcal{P}}_{b-(t_1-t_1^0)q\omega/T}\big(\mathcal{S}_\omega(\mathcal{E}_{|\widehat{H}^\lambda_{\infty,q}})_{|\widehat{H}^\lambda_{\infty,q}\langle t_1\rangle}\big). \tag{4.11}$$
- $\coprod_{k\le\ell}[\boldsymbol{v}_{b,\omega,k}]$ *is a base of* (4.11).

Proof It follows from Lemma 3.5.26. ■

4.4 Hermitian Metrics and Filtrations

4.4.1 Prolongation by Growth Conditions

Let $\mathcal{B}^\lambda_q$ denote an open neighbourhood of $H^\lambda_{\infty,q}$ in Y^λ_q as in Sect. 4.3.3. We set $\mathcal{B}^{\lambda*}_q := \mathcal{B}^\lambda_q \setminus H^\lambda_{\infty,q}$. Let $(E,\overline{\partial}_E)$ be a mini-holomorphic bundle on $\mathcal{B}^{\lambda*}_q$. Let h be a Hermitian metric of E. We explain a general procedure to construct an $\mathcal{O}_{\mathcal{B}^\lambda_q}(*H^\lambda_{\infty,q})$-module $\mathcal{P}^h E$, although it is not necessarily locally free.

Let U be any open subset in $\mathcal{B}_q^{\lambda}$. Let $\mathcal{P}^h E(U)$ denote the space of mini-holomorphic sections s of E on $U \setminus H_{\infty,q}^{\lambda}$ such that the following holds.

- For any point P of $U \cap H_{\infty,q}^{\lambda}$, we take a relatively compact neighbourhood U_P of P in $\mathcal{B}_q^{\lambda}$. Then, there exists $N(P) > 0$ such that

$$\left|s_{|U_P \setminus H_{\infty,q}^{\lambda}}\right|_h = O\big(|x_q|^{N(P)}\big). \tag{4.12}$$

See Sect. 4.3.2 for the function x_q. We may replace it with w_q or $\beta_{1,q}$ in (4.12). Thus, we obtain an $\mathcal{O}_{\mathcal{B}_q^{\lambda}}(*H_{\infty,q}^{\lambda})$-module $\mathcal{P}^h E$.

For each $t_1 \in S_T^1$, we set $\mathcal{B}_q^{\lambda *}\langle t_1\rangle := \mathcal{B}_q^{\lambda}\langle t_1\rangle \setminus H_{\infty,q}^{\lambda}$. We obtain the holomorphic vector bundle $(E, \overline{\partial}_E)_{|\mathcal{B}_1^{\lambda *}\langle t_1\rangle}$ with the Hermitian metric $h_{|\mathcal{B}_q^{\lambda *}\langle t_1\rangle}$. By applying the construction in Sect. 2.11.2 to $(E, \overline{\partial}_E, h)_{|\mathcal{B}_1^{\lambda *}\langle t_1\rangle}$ we obtain the families of sheaves $\mathcal{P}_*^h(E_{|\mathcal{B}_q^{\lambda *}\langle t_1\rangle}) = \big(\mathcal{P}_b^h(E_{|\mathcal{B}_q^{\lambda *}\langle t_1\rangle}) \,\big|\, b \in \mathbb{R}\big)$. The tuple $\big(\mathcal{P}_*^h(E_{|\mathcal{B}_q^{\lambda *}\langle t_1\rangle}) \,\big|\, t_1 \in S_T^1\big)$ is denoted by $\mathcal{P}_*^h(E)$.

Remark 4.4.1 $\mathcal{P}_*^h(E_{|\mathcal{B}_q^{\lambda *}\langle t_1\rangle})$ are not necessarily filtered bundles. ∎

Remark 4.4.2 Set $\mathcal{P}^h\big(E_{|\mathcal{B}_q^{\lambda *}(t_1)}\big) := \bigcup_{a \in \mathbb{R}} \mathcal{P}_a^h(E_{|\mathcal{B}_q^{\lambda *}(t_1)})$. We have the naturally defined morphism $(\mathcal{P}^h E)_{|\mathcal{B}_q^{\lambda}(t_1)} \longrightarrow \mathcal{P}^h(E_{|\mathcal{B}_q^{\lambda *}(t_1)})$, but it is not necessarily surjective. ∎

Remark 4.4.3 We shall often use the notation $\mathcal{P}E$ and $\mathcal{P}_*E$ instead of $\mathcal{P}^h E$ and $\mathcal{P}_*^h E$, respectively, if there is no risk of confusion. ∎

4.4.2 Norm Estimate for Good Filtered Bundles

Let $\mathcal{P}_*\mathcal{E}$ be a good filtered bundle over a locally free $\mathcal{O}_{\mathcal{B}_q^{\lambda}}(*H_{\infty,q}^{\lambda})$-module $\mathcal{E}$. Let $(E, \overline{\partial}_E)$ be the mini-holomorphic bundle obtained as the restriction of $\mathcal{E}$ to $\mathcal{B}_q^{\lambda *}$.

Let $P \in H_{\infty,q}^{\lambda}$. We set $t_1^0 = \pi_q^{\lambda}(P) \in S_T^1 \simeq H_{\infty,q}^{\lambda}$. Take ϵ satisfying (4.10). Let $\boldsymbol{v} = \coprod \boldsymbol{v}_{b,\omega,k}$ be a convenient frame on $\mathcal{B}_q^{\lambda}\langle t_1^0, \epsilon\rangle$ as in Sect. 4.3.5. We set $\mathcal{B}_q^{\lambda *}\langle t_1^0, \epsilon\rangle := \mathcal{B}_q^{\lambda}\langle t_1^0, \epsilon\rangle \setminus H_{\infty,q}^{\lambda}$. For $v_i \in \boldsymbol{v}_{b,\omega,k}$, we set $\omega(v_i) := \omega$, $b(v_i) = b$ and $k(v_i) = k$. Let h_P be a Hermitian metric of $E_{|\mathcal{B}_q^{\lambda *}\langle t_1^0, \epsilon\rangle}$ determined by the following condition:

$$h_P(v_i, v_j) := \begin{cases} |x_q|^{2b(v_i) - 2\omega(v_i) q (t_1 - t_1^0)/T} (\log |x_q|)^{k(v_i)} & (i = j) \\ 0 & (i \neq j). \end{cases}$$

The following lemma is clear by the construction.

Lemma 4.4.4 *Let h'_P be the Hermitian metric of $E_{|\mathcal{B}_q^{\lambda*}\langle t_1^0,\epsilon\rangle}$ constructed in the same way from another convenient frame. Then, for any relatively compact neighbourhood U_P of P in $\mathcal{B}_q^{\lambda}\langle t_1^0,\epsilon\rangle$, h_P and h'_P are mutually bounded on $U_P \setminus H^{\lambda}_{\infty,q}$.* ■

Definition 4.4.5 Let h be a Hermitian metric of E. We say that $(E, \overline{\partial}_E, h)$ satisfies the norm estimate with respect to $\mathcal{P}_*\mathcal{E}$ if the following holds.

- For any $P \in H^{\lambda}_{\infty,q}$, let $t_1^0 = \pi_q^{\lambda}(P)$, and we take $\epsilon > 0$ satisfying (4.10). We construct a Hermitian metric h_P of $E_{|\mathcal{B}_q^{\lambda*}\langle t_1^0,\epsilon\rangle}$ from a convenient frame as above. Then, for any relatively compact neighbourhood U_P of P in $\mathcal{B}_q^{\lambda}\langle t_1^0,\epsilon\rangle$, h and h_P are mutually bounded on $U_P \setminus H^{\lambda}_{\infty,q}$. ■

If $(E, \overline{\partial}_E, h)$ satisfies the norm estimate with respect to $\mathcal{P}_*\mathcal{E}$, then we obtain $\mathcal{P}^h E = \mathcal{E}$ and $\mathcal{P}^h(E_{|\mathcal{B}_q^{\lambda*}\langle t_1\rangle}) = \mathcal{P}_*(\mathcal{E}_{|\mathcal{B}_q^{\lambda}\langle t_1\rangle})$ $(t_1 \in S^1_T)$.

Take $p \in q\mathbb{Z}_{\geq 1}$. We set $\mathcal{B}_p^{\lambda} := \mathcal{R}_{q,p}^{-1}(\mathcal{B}_q^{\lambda})$. We obtain $\mathcal{R}_{q,p}^{-1}(E, \overline{\partial}_E, h)$ on $\mathcal{B}_p^{\lambda*} = \mathcal{B}_p^{\lambda} \setminus H^{\lambda}_{\infty,p}$ and a good filtered bundle $\mathcal{R}^*_{q,p}(\mathcal{P}_*\mathcal{E})$ on $(\mathcal{B}_p^{\lambda}, H^{\lambda}_{\infty,p})$.

Lemma 4.4.6 *$(E, \overline{\partial}_E, h)$ satisfies the norm estimate with respect to $\mathcal{P}_*\mathcal{E}$ if and only if $\mathcal{R}_{q,p}^{-1}(E, \overline{\partial}_E, h)$ satisfies the norm estimate with respect to $\mathcal{R}^*_{q,p}(\mathcal{P}_*\mathcal{E})$.*

Proof Set $m := p/q$. Let P be any point of $H^{\lambda}_{\infty,q}$, and set $t_1^0 := \pi_q^{\lambda}(P)$. Take $\epsilon > 0$ satisfying (4.10) for $\mathcal{P}_*(\mathcal{E}_{|\mathcal{B}_q^{\lambda*}\langle t_1^0\rangle})$ and $\mathcal{R}^*_{q,p}\mathcal{P}_*(\mathcal{E}_{|\mathcal{B}_q^{\lambda*}\langle t_1^0\rangle})$.

Let $\boldsymbol{v} = \bigcup \boldsymbol{v}_{b,\omega,k}$ be a convenient frame of $\boldsymbol{P}_a^{(t_1^0)}\mathcal{E}$ on $\mathcal{B}_q^{\lambda}\langle t_1^0,\epsilon\rangle$. We construct a Hermitian metric h_P of $E_{|\mathcal{B}_q^{\lambda*}\langle t_1^0,\epsilon\rangle}$ from $\boldsymbol{v}_P$ as above.

We set $P' = \mathcal{R}_{q,p}^{-1}(P) \in H^{\lambda}_{\infty,p}$. For each $a-1 < b \leq a$, let $n(b) \in \mathbb{Z}$ be determined by $ma - 1 < mb + n(b) \leq ma$. We set

$$\boldsymbol{v}'_{b',\omega,k} := \coprod_{mb+n(b)=b'} \beta_{1,p}^{n(b)} \cdot \mathcal{R}^*_{q,p}(\boldsymbol{v}_{b,\omega,k}).$$

Then, $\boldsymbol{v}' = \bigcup_{b',\omega,k} \boldsymbol{v}'_{b',\omega,k}$ is a convenient frame of $\boldsymbol{P}_{ma}^{(t_1^0)}(\mathcal{R}^*_{q,p}\mathcal{E})$ on $\mathcal{B}_p^{\lambda}\langle t_1^0,\epsilon\rangle$. We construct a Hermitian metric $h_{P'}$ of $\mathcal{R}_{q,p}^{-1}(E)_{|\mathcal{B}_p^{\lambda*}\langle t_1^0,\epsilon\rangle}$ from $\boldsymbol{v}'$ as above. We can easily observe that $h_{P'}$ and $\mathcal{R}_{q,p}^{-1}h_P$ are mutually bounded. Then, the claim of the lemma is clear. ■

4.4.3 Strong Adaptedness

We introduce a condition called strong adaptedness, which is weaker than the norm estimate. Let $\mathcal{P}_*\mathcal{E}$ and $(E, \overline{\partial}_E)$ be as in Sect. 4.4.2.

Definition 4.4.7 Let h be a Hermitian metric of E. We say that h is strongly adapted to $\mathcal{P}_*\mathcal{E}$ if the following condition is satisfied.

- For any $P \in H^{\lambda}_{\infty,q}$, we take a neighbourhood U_P and a Hermitian metric h_P of $E_{|U_P \setminus H^{\lambda}_{\infty,q}}$ as in Sect. 4.4.2. Then, for any $\delta > 0$, there exists a constant $C_\delta \geq 1$ such that $C_\delta^{-1}|w_q|^{-\delta}h_P \leq h \leq C_\delta |w_q|^{\delta} h_P$. ■

We obtain the following as in Lemma 4.4.6.

Lemma 4.4.8 *Take $p \in q\mathbb{Z}_{\geq 1}$. Then, h is strongly adapted to $\mathcal{P}_*\mathcal{E}$ if and only if $\mathcal{R}^{-1}_{q,p}(h)$ is strongly adapted to $\mathcal{R}^*_{q,p}(\mathcal{P}_*\mathcal{E})$.* ■

Remark 4.4.9 The strong adaptedness in Definition 4.4.7 is stronger than the adaptedness in the context of harmonic bundles, i.e., it is a locally uniform estimate in the S^1_T-direction. ■

4.5 Comparison with λ-Connections

4.5.1 Some Sheaves on $Y^{\lambda\,\mathrm{cov}}_q$ and Y^{λ}_q

There exists the complex vector field $\partial_{\overline{\beta}_1}$ on Y^{λ}_1. By taking the lift with respect to the covering map $\mathcal{R}_{1,q} : Y^{\lambda\,\mathrm{cov}*}_q \longrightarrow Y^{\lambda\,\mathrm{cov}*}_1$, we obtain the complex vector field $\partial_{\overline{\beta}_1}$ on $Y^{\lambda\,\mathrm{cov}*}_q$ such that $\mathcal{R}_{1,q*}\partial_{\overline{\beta}_1} = \partial_{\overline{\beta}_1}$. Locally around any point of $H^{\lambda\,\mathrm{cov}}_{\infty,q}$, $\mathcal{R}_{1,q}$ is described as $(t_1, \beta^{-1}_{1,q}) \longmapsto (t_1, \beta^{-q}_{1,q})$, and we have $\partial_{\overline{\beta}_1} = \overline{\beta}^{-q}_{1,q}(q^{-1}\overline{\beta}_{1,q}\partial_{\overline{\beta}_{1,q}})$. Hence $\partial_{\overline{\beta}_1}$ extends to a complex vector field on $Y^{\lambda\,\mathrm{cov}}_q$, which is also denoted by $\partial_{\overline{\beta}_1}$. Similarly, we obtain the vector field ∂_{t_1} on $Y^{\lambda\,\mathrm{cov}}_q$.

Let U be any open subset of $Y^{\lambda\,\mathrm{cov}}_q$. Let $\mathcal{K}_{Y^{\lambda\,\mathrm{cov}}_q}(U)$ denote the space of C^∞-functions f on U such that $\partial_{\overline{\beta}_1}(f) = 0$. We obtain a sheaf of algebras $\mathcal{K}_{Y^{\lambda\,\mathrm{cov}}_q}$ on $Y^{\lambda\,\mathrm{cov}}_q$. We have the induced operator $\partial_{t_1} : \mathcal{K}_{Y^{\lambda\,\mathrm{cov}}_q} \longrightarrow \mathcal{K}_{Y^{\lambda\,\mathrm{cov}}_q}$. It is an epimorphism, and the kernel is $\mathcal{O}_{Y^{\lambda\,\mathrm{cov}}_q}$. We set $\mathcal{K}_{Y^{\lambda\,\mathrm{cov}}_q}(*H^{\lambda}_{\infty,q}) := \mathcal{K}_{Y^{\lambda\,\mathrm{cov}}_q} \otimes_{\mathcal{O}_{Y^{\lambda\,\mathrm{cov}}_q}} \mathcal{O}_{Y^{\lambda\,\mathrm{cov}}_q}(*H^{\lambda\,\mathrm{cov}}_q)$. We have the induced operator $\partial_{t_1} : \mathcal{K}_{Y^{\lambda\,\mathrm{cov}}_q}(*H^{\lambda\,\mathrm{cov}}_{\infty,q}) \longrightarrow \mathcal{K}_{Y^{\lambda\,\mathrm{cov}}_q}(*H^{\lambda}_{\infty,q})$. It is an epimorphism, and the kernel is $\mathcal{O}_{Y^{\lambda\,\mathrm{cov}}_q}(*H^{\lambda\,\mathrm{cov}}_{\infty,q})$.

Because $\mathcal{K}_{Y^{\lambda\,\mathrm{cov}}_q}$ is naturally $\mathbb{Z}$-equivariant, we obtain the sheaf of algebras $\mathcal{K}_{Y^{\lambda}_q}$ on Y^{λ}_q as the descent. It is naturally equipped with the operator $\partial_{t_1} : \mathcal{K}_{Y^{\lambda}_q} \longrightarrow \mathcal{K}_{Y^{\lambda}_q}$, which is an epimorphism, and the kernel is $\mathcal{O}_{Y^{\lambda}_q}$. Similarly, $\mathcal{K}_{Y^{\lambda\,\mathrm{cov}}_q}(*H^{\lambda\,\mathrm{cov}}_{\infty,q})$ is naturally $\mathbb{Z}$-equivariant, we obtain the sheaf of algebras $\mathcal{K}_{Y^{\lambda}_q}(*H^{\lambda}_{\infty,q})$ on Y^{λ}_q as the descent. It is naturally equipped with the operator $\partial_{t_1} : \mathcal{K}_{Y^{\lambda}_q}(*H^{\lambda}_{\infty,q}) \longrightarrow \mathcal{K}_{Y^{\lambda}_q}(*H^{\lambda}_{\infty,q})$, which is an epimorphism, and the kernel is $\mathcal{O}_{Y^{\lambda}_q}(*H^{\lambda}_{\infty,q})$. We have $\mathcal{K}_{Y^{\lambda}_q}(*H^{\lambda}_{\infty,q}) = \mathcal{O}_{Y^{\lambda}_q}(*H^{\lambda}_{\infty,q}) \otimes_{\mathcal{O}_{Y^{\lambda}_q}} \mathcal{K}_{Y^{\lambda}_q}$.

Because the complex vector field $\partial_{\overline{\beta}_1}$ on $Y_q^{\lambda\,\mathrm{cov}}$ is invariant with respect to the $\mathbb{Z}$-action, we obtain $\partial_{\overline{\beta}_1}$ on Y_q^{λ}. We may obtain $\mathcal{K}_{Y_q^\lambda}$ as the sheaf of C^∞-functions f such that $\partial_{\overline{\beta}_1} f = 0$.

4.5.2 The Induced $\mathcal{O}_{\mathcal{B}_q^\lambda}$-Modules from λ-Connections

We explain the analytic version of the construction in Sect. 3.6, which is also a ramified version of the construction in Sect. 2.8.3.4. Note that the variable x_q in this subsection is related with the variable w in Sect. 2.8.3.4 by $w = (1+|\lambda|^2)^{-1}x_q^q$.

There exists the natural map $\mathbb{R}_u \times \mathbb{P}^1_{x_q} \longrightarrow \mathbb{P}^1_{x_q}$ induced by $(u, x_q) \longmapsto x_q$. It induces the proper map $Y_q^{\lambda\,\mathrm{cov}} \longrightarrow \mathbb{P}^1_{x_q} \setminus \{0\}$. In terms of the local coordinate systems $(t_1, \beta_{1,q}^{-1})$ and x_q^{-1}, it is described as

$$(t_1, \beta_{1,q}^{-1}) \longmapsto \beta_{1,q}^{-1}(1 - 2\sqrt{-1}\lambda t_1 \beta_{1,q}^{-q})^{-1/q}.$$

It induces the map $\Psi_q^\lambda : Y_q^\lambda \longrightarrow \mathbb{P}^1_{x_q} \setminus \{0\}$. It is equal to the map Ψ_q^λ in Sect. 4.3.2 under the identification $x_q = (1+|\lambda|^2)^{1/q} w_q$. Clearly, $(\Psi_q^\lambda)^{-1}(\infty) = H^\lambda_{\infty,q}$. Note that the restriction of Ψ_q^λ to $Y_q^{\lambda *}$ is equal to Ψ_q^0 to Y_q^{0*} under the C^∞-identification $Y_q^{\lambda *} = Y_q^{0*}$. We also note that we obtain the morphism $(\Psi_q^\lambda)^* : (\Psi_q^\lambda)^{-1}\mathcal{O}_{\mathbb{P}^1_{x_q}\setminus\{0\}} \longrightarrow \mathcal{K}_{Y_q^\lambda}$ by the pull back of functions.

Let $U_{x,q}$ be any open subset of $\mathbb{P}^1_{x_q} \setminus \{0\}$. We set $\mathcal{B}_q^\lambda := (\Psi_q^\lambda)^{-1}(U_{x,q})$. Let $\mathcal{V}$ be a locally free $\mathcal{O}_{U_{x,q}}(*\infty)$-module with a λ-connection $\nabla^\lambda : \mathcal{V} \longrightarrow \mathcal{V} \otimes \Omega^1_{U_{x,q}}$ whose Poincaré rank is strictly less than q. By the inner product with $\partial_x = q^{-1}x_q^{-q+1}\partial_{x_q}$, we obtain a morphism of sheaves $\nabla_x^\lambda : \mathcal{V} \longrightarrow \mathcal{V}$. We set

$$\begin{aligned}\widetilde{\mathcal{V}}^\infty &:= (\Psi_q^\lambda)^{-1}(\mathcal{V}) \otimes_{(\Psi_q^\lambda)^{-1}\mathcal{O}_{U_{x,q}}} \mathcal{K}_{\mathcal{B}_q^\lambda} \\ &= (\Psi_q^\lambda)^{-1}(\mathcal{V}) \otimes_{(\Psi_q^\lambda)^{-1}\mathcal{O}_{U_{x,q}}(*\infty)} \mathcal{K}_{\mathcal{B}_q^\lambda}(*H^\lambda_{\infty,q}).\end{aligned} \tag{4.13}$$

There uniquely exists $\partial_{\widetilde{\mathcal{V}}^\infty, t_1} : \widetilde{\mathcal{V}}^\infty \longrightarrow \widetilde{\mathcal{V}}^\infty$ determined by the following condition for any local sections f and s of $\mathcal{K}_{\mathcal{B}_q^\lambda}$ and $\mathcal{V}$:

$$\partial_{\widetilde{\mathcal{V}}^\infty, t_1}\big(f \cdot (\Psi_q^\lambda)^{-1}(s)\big) = \partial_{t_1}(f)(\Psi_q^\lambda)^{-1}(s) + f(\Psi_q^\lambda)^{-1}(-2\sqrt{-1}\nabla_x^\lambda s). \tag{4.14}$$

Let $(\Psi_q^\lambda)^*(\mathcal{V}, \nabla^\lambda)$ denote the $\mathcal{O}_{\mathcal{B}_q^\lambda}(*H^\lambda_{\infty,q})$-module obtained as the kernel of $\partial_{\widetilde{\mathcal{V}}^\infty, t_1} : \widetilde{\mathcal{V}}^\infty \longrightarrow \widetilde{\mathcal{V}}^\infty$.

Similarly, for any lattice $\mathcal{L}$ of $\mathcal{V}$ such that $\nabla_x^\lambda \mathcal{L} \subset \mathcal{L}$, we set

$$\widetilde{\mathcal{L}}^\infty := (\Psi_q^\lambda)^{-1}(\mathcal{L}) \otimes_{(\Psi_q^\lambda)^{-1}(\mathcal{O}_{U_{x,q}})} \mathcal{K}_{\mathcal{B}_q^\lambda}.$$

It is equipped with the induced operator $\partial_{\widetilde{\mathcal{L}}^\infty,t_1} : \widetilde{\mathcal{L}}^\infty \longrightarrow \widetilde{\mathcal{L}}^\infty$. The kernel is denoted by $(\Psi_q^\lambda)^*(\mathcal{L}, \nabla^\lambda)$ which is an $\mathcal{O}_{\mathcal{B}_q^\lambda}$-module.

Let $(\mathcal{V}, \nabla^\lambda)_{|\widehat{\infty}_{x,q}}$ denote the formal completion of $(\mathcal{V}, \nabla^\lambda)$ at ∞, i.e., $\mathcal{V}_{|\widehat{\infty}_{x,q}} := \mathcal{V}_\infty \otimes_{\mathcal{O}_{U_{x,q}}} \mathbb{C}[\![x_q^{-1}]\!]$. By the construction in Sect. 3.6.3, we obtain a locally free $\mathcal{O}_{\widehat{H}^\lambda_{\infty,q}}(*H^\lambda_{\infty,q})$-module $(\Psi_q^\lambda)^*\big((\mathcal{V}, \nabla^\lambda)_{|\widehat{\infty}_{x,q}}\big)$. From the completion $\mathcal{L}_{|\widehat{\infty}_{x,q}}$ of $\mathcal{L}$ which is a ∇_x^λ-invariant lattice of $(\mathcal{V}, \nabla^\lambda)_{|\widehat{\infty}_{x,q}}$, a lattice $(\Psi_q^\lambda)^*\big((\mathcal{L}, \nabla^\lambda)_{|\widehat{\infty}_{x,q}}\big)$ of $(\Psi_q^\lambda)^*\big((\mathcal{V}, \nabla^\lambda)_{|\widehat{\infty}_{x,q}}\big)$ is obtained.

Proposition 4.5.1

- $(\Psi_q^\lambda)^*(\mathcal{V}, \nabla^\lambda)$ *is a locally free* $\mathcal{O}_{\mathcal{B}_q^\lambda}(*H^\lambda_{\infty,q})$*-module. Moreover, there exists a natural isomorphism:*

$$(\Psi_q^\lambda)^*(\mathcal{V}, \nabla^\lambda)_{|\widehat{H}^\lambda_{\infty,q}} \simeq (\Psi_q^\lambda)^*\big((\mathcal{V}, \nabla^\lambda)_{|\widehat{\infty}_{x,q}}\big). \tag{4.15}$$

- *For any* ∇_x^λ*-invariant lattice* $\mathcal{L}$ *of* $\mathcal{V}$, $(\Psi_q^\lambda)^*(\mathcal{L}, \nabla^\lambda)$ *is a lattice of* $(\Psi_q^\lambda)^*(\mathcal{V}, \nabla^\lambda)$. *Moreover, there exists a natural isomorphism:*

$$(\Psi_q^\lambda)^*(\mathcal{L}, \nabla^\lambda)_{|\widehat{H}^\lambda_{\infty,q}} \simeq (\Psi_q^\lambda)^*\big((\mathcal{L}, \nabla^\lambda)_{|\widehat{\infty}_{x,q}}\big). \tag{4.16}$$

Proof It is easy to see that $(\Psi_q^\lambda)^*(\mathcal{V}, \nabla^\lambda)_{|\mathcal{B}_q^\lambda \setminus H^\lambda_{\infty,q}}$ is a locally free $\mathcal{O}_{\mathcal{B}_q^\lambda \setminus H^\lambda_{\infty,q}}$-module. Let $\mathcal{L}$ be a ∇_x^λ-invariant lattice of $\mathcal{V}$. Take any frame $\boldsymbol{v}$ of $\mathcal{L}$. We set $r := \operatorname{rank}(\mathcal{V})$. We obtain a section A of $M_r(\mathcal{O}_{U_{x,q}})$ determined by $\nabla_x^\lambda \boldsymbol{v} = \boldsymbol{v} \cdot A$. We set $\widetilde{\boldsymbol{v}} = (\Psi_q^\lambda)^{-1}(\boldsymbol{v})$ which is frame of $\widetilde{\mathcal{V}}^\infty$. Let P be any point of $H^\lambda_{\infty,q}$. We take a mini-complex local coordinate system $(t_1, \beta_{1,q}^{-1})$ around P. We obtain the following local section of $M_r\big(\mathcal{K}_{\mathcal{B}_q^\lambda}\big)$ around P:

$$\widetilde{A} = -2\pi\sqrt{-1}A\big(\beta_{1,q}^{-1}(1 - 2\sqrt{-1}\lambda t \beta_{1,q}^{-q})^{-1/q}\big).$$

We obtain $\partial_{\widetilde{\mathcal{V}}^\infty,t_1} \widetilde{\boldsymbol{v}} = \widetilde{\boldsymbol{v}} \cdot \widetilde{A}$. There exists a local section G of $M_r\big(\mathcal{K}_{\mathcal{B}_q^\lambda}\big)$ around P such that (i) G is invertible, i.e., $G^{-1} \in M_r\big(\mathcal{K}_{\mathcal{B}_q^\lambda}\big)$, (ii) $G^{-1}\partial_t G = -\widetilde{A}$. Then, we obtain $\partial_{\widetilde{\mathcal{V}}^\infty,t_1}(\widetilde{\boldsymbol{v}} \cdot G) = 0$. Because $\widetilde{\boldsymbol{v}} \cdot G$ is a local frame of $(\Psi_q^\lambda)^*(\mathcal{V}, \nabla^\lambda)$ around P, we obtain that $(\Psi_q^\lambda)^*(\mathcal{L}, \nabla^\lambda)$ is a locally free $\mathcal{O}_{\mathcal{B}_q^\lambda}$-module. Then, the claims are clear. ∎

Let $\mathcal{P}_*\mathcal{V}$ be a filtered bundle over $\mathcal{V}$ such that each $\mathcal{P}_a\mathcal{V}$ is ∇_x^λ-invariant. The family $(\Psi_q^\lambda)^*(\mathcal{P}_a\mathcal{V}, \nabla^\lambda)$ $(a \in \mathbb{R})$ induces a filtered bundle over $(\Psi_q^\lambda)^*(\mathcal{V}, \nabla^\lambda)$, denoted by $(\Psi_q^\lambda)^*(\mathcal{P}_*\mathcal{V}, \nabla^\lambda)$. We obtain the following lemma from Proposition 3.6.15 and Proposition 4.5.1.

Proposition 4.5.2 *$(\mathcal{P}_*\mathcal{V}, \nabla^\lambda)$ is a good filtered λ-flat bundle if and only if the induced objects $(\Psi_q^\lambda)^*(\mathcal{P}_*\mathcal{V}, \nabla^\lambda)$ is a good filtered bundle.* ■

Proposition 4.5.3 *Let $(\mathcal{P}_*\mathcal{V}_i, \nabla^\lambda)$ be good filtered λ-flat bundles on $(U_{x,q}, \infty_{x,q})$ whose Poincaré rank are strictly smaller than q. There exist the following natural isomorphisms:*

$$\Psi_q^*\big((\mathcal{P}_*\mathcal{V}_1, \nabla^\lambda) \oplus (\mathcal{P}_*\mathcal{V}_2, \nabla^\lambda)\big) \simeq \Psi_q^*(\mathcal{P}_*\mathcal{V}_1, \nabla^\lambda) \oplus \Psi_q^*(\mathcal{P}_*\mathcal{V}_2, \nabla^\lambda). \tag{4.17}$$

$$\Psi_q^*\big((\mathcal{P}_*\mathcal{V}_1, \nabla^\lambda) \otimes (\mathcal{P}_*\mathcal{V}_2, \nabla^\lambda)\big) \simeq \Psi_q^*(\mathcal{P}_*\mathcal{V}_1, \nabla^\lambda) \otimes \Psi_q^*(\mathcal{P}_*\mathcal{V}_2, \nabla^\lambda). \tag{4.18}$$

$$\Psi_q^*\Big(\mathcal{H}om\big((\mathcal{P}_*\mathcal{V}_1, \nabla^\lambda), (\mathcal{P}_*\mathcal{V}_2, \nabla^\lambda)\big)\Big) \simeq \mathcal{H}om\Big(\Psi_q^*(\mathcal{P}_*\mathcal{V}_1, \nabla^\lambda), \Psi_q^*(\mathcal{P}_*\mathcal{V}_2, \nabla^\lambda)\Big). \tag{4.19}$$

In particular, for a good filtered λ-flat bundle $(\mathcal{P}_*\mathcal{V}, \nabla^\lambda)$ on $(U_{x,q}, \infty_{x,q})$ whose Poincaré rank is strictly smaller than q, we obtain

$$\Psi_q^*(\mathcal{P}_*\mathcal{V}, \nabla^\lambda)^\vee \simeq \Psi_q^*\big((\mathcal{P}_*\mathcal{V}, \nabla^\lambda)^\vee\big).$$

■

Remark 4.5.4 The construction is available in the case $\infty \notin U_{x,q}$, and it is the same as the construction in Sect. 2.8.3. ■

4.5.3 Norm Estimates for λ-Connections

Let us recall the norm estimate condition for λ-flat bundles with a Hermitian metric. Let $U_{x,q}$ denote a neighbourhood of ∞ in $\mathbb{P}^1_{x_q}$. We set $U_{x,q}^* := U_{x,q} \setminus \{\infty\}$. Let $(\mathcal{P}_*\mathcal{V}, \nabla^\lambda)$ be a good filtered λ-flat bundle on $(U_{x,q}, \infty)$. We set $\mathrm{Gr}_a^{\mathcal{P}}(\mathcal{V}) := \mathcal{P}_a\mathcal{V}/\mathcal{P}_{<a}\mathcal{V}$. There exists the residue endomorphism $\mathrm{Res}(\nabla^\lambda)$ on each $\mathrm{Gr}_b^{\mathcal{P}}(\mathcal{V})$. (See [64, §2.5.2].) Let W be the monodromy weight filtration of the nilpotent part of $\mathrm{Res}(\nabla^\lambda)$. We set $(V, \nabla^\lambda) := (\mathcal{V}, \nabla^\lambda)_{|U_{x,q}^*}$.

A frame $\boldsymbol{v}$ of $\mathcal{P}_a\mathcal{V}$ is called convenient for $(\mathcal{P}_*\mathcal{V}, \nabla^\lambda)$ if the following conditions are satisfied.

- There exists a decomposition $\boldsymbol{v} = \coprod_{a-1<b\leq a} \boldsymbol{v}_b$ such that $\boldsymbol{v}_b$ is a tuple of sections of $\mathcal{P}_b\mathcal{V}$ and induces a base of $\mathrm{Gr}_b^{\mathcal{P}}(\mathcal{V})$.
- There exists a decomposition $\boldsymbol{v}_b = \coprod_{k\in\mathbb{Z}} \boldsymbol{v}_{b,k}$ such that $\coprod_{k\leq\ell} \boldsymbol{v}_{b,k}$ induces a frame of $W_\ell \mathrm{Gr}_b^{\mathcal{P}}(\mathcal{V})$.

For a convenient frame $\boldsymbol{v}$, we define a Hermitian metric h_0 of V as follows. For $v_i \in \boldsymbol{v}_{b,k}$, we put $b(v_i) := b$ and $k(v_i) := k$, and we set

$$h_0(v_i, v_j) := \begin{cases} |x_q|^{2b(v_i)}(\log|x_q|)^{k(v_i)} & (i = j) \\ 0 & (i \neq j). \end{cases}$$

The following lemma is clear.

Lemma 4.5.5 *Let h_0' be a Hermitian metric of V obtained from another convenient frame $\boldsymbol{v}'$ as above. Then, for any relatively compact neighbourhood U' of ∞ in $U_{x,q}$, h_0 and h_0' are mutually bounded on $U' \setminus \{\infty\}$.* ■

Definition 4.5.6 Let h_V be a Hermitian metric of V. We say that (V, ∇^λ, h) satisfies the norm estimate with respect to $(\mathcal{P}_*\mathcal{V}, \nabla^\lambda)$ if h_V and h_0 are mutually bounded, where h_0 is a Hermitian metric of V obtained from a convenient frame $\boldsymbol{v}$ for $(\mathcal{P}_*\mathcal{V}, \nabla^\lambda)$. ■

4.5.4 Comparison of the Norm Estimates

Let $U_{x,q}$ be a neighbourhood of ∞ in $\mathbb{P}^1_{x_q} \setminus \{0\}$. We set $\mathcal{B}_q^\lambda := (\Psi_q^\lambda)^{-1}(U_{x,q})$. Let $(\mathcal{P}_*\mathcal{V}, \nabla^\lambda)$ be a good filtered λ-flat bundle on $(U_{x,q}, \infty)$. We obtain a good filtered bundle $(\Psi_q^\lambda)^*(\mathcal{P}_*\mathcal{V}, \nabla^\lambda)$ on $(\mathcal{B}_q^\lambda, H_{\infty,q}^\lambda)$. We set $U_{x,q}^* := U_{x,q} \setminus \{\infty\}$, $\mathcal{B}_q^{\lambda*} := (\Psi_q^\lambda)^{-1}(U_{x,q}^*)$ and $(V, \nabla^\lambda) := (\mathcal{V}, \nabla^\lambda)_{|U_{x,q}^*}$.

Proposition 4.5.7 *Let h_V be a Hermitian metric of V such that (V, ∇^λ, h_V) satisfies the norm estimate with respect to $(\mathcal{P}_*\mathcal{V}, \nabla^\lambda)$. Suppose that the Poincaré rank of $(\mathcal{V}, \nabla^\lambda)$ is strictly less than q. Then, $\Psi_q^*(\mathcal{V}, \nabla^\lambda)_{|\mathcal{B}_q^{\lambda*}}$ with $\Psi_q^{-1}(h_V)$ satisfies the norm estimate with respect to the filtered bundle $(\Psi_q^\lambda)^*(\mathcal{P}_*\mathcal{V}, \nabla^\lambda)$.*

Proof Let us study the case where there exists $\mathfrak{a} \in x_q\mathbb{C}[x_q]$ such that $\nabla^\lambda - d\mathfrak{a}\,\mathrm{id}$ is logarithmic with respect to $\mathcal{P}_*\mathcal{V}$. Note that $\deg_{x_q} \mathfrak{a} < q$. Set $r := \mathrm{rank}(\mathcal{V})$. Let $I_r \in M_r(\mathbb{C})$ denote the identity matrix. Let $\boldsymbol{v}$ be a convenient frame of $\mathcal{P}_a\mathcal{V}$. There exists a decomposition $\boldsymbol{v} = \coprod_{a-1<b\leq a}\coprod_{k\in\mathbb{Z}} \boldsymbol{v}_{b,k}$ such that (i) $\coprod_{k\in\mathbb{Z}} \boldsymbol{v}_{b,k}$ is a tuple of sections of $\mathcal{P}_b\mathcal{V}$, and induces a base of $\mathrm{Gr}_b^{\mathcal{P}}(\mathcal{V})$, (ii) $\coprod_{k\leq\ell} \boldsymbol{v}_{b,k}$ induces a base of $W_\ell\,\mathrm{Gr}_b^{\mathcal{P}}(\mathcal{V})$. For $v_i \in \boldsymbol{v}_{b,k}$, we set $b(v_i) = b$ and $k(v_i) = k$. Let $h_{V,0}$ be the Hermitian metric determined by $h_{V,0}(v_i, v_i) = |x_q|^{2b(v_i)}(\log|x_q|)^{k(v_i)}$ and $h_{V,0}(v_i, v_j) = 0$ $(i \neq j)$. We set $\widetilde{h}_0 := (\Psi_q^\lambda)^{-1}(h_{V,0})$.

Let A be the matrix valued function determined by $\nabla_x^\lambda \boldsymbol{v} = \boldsymbol{v} \cdot \big(\partial_x\mathfrak{a} I_r + A(x_q^{-1})x_q^{-q}\big)$. Then, $A(x_q^{-1})$ is holomorphic with respect to x_q^{-1}. Set $\widetilde{\boldsymbol{v}} := (\Psi_q^\lambda)^{-1}(\boldsymbol{v})$ which is a frame of $\widetilde{\mathcal{V}}^\infty$. We obtain

$$\partial_{\widetilde{\mathcal{V}}^\infty,t_1}\widetilde{\boldsymbol{v}} = \widetilde{\boldsymbol{v}}(-2\sqrt{-1})(\Psi_q^\lambda)^{-1}\Big(\partial_x\mathfrak{a} I_r + x_q^{-q}A(x_q^{-1})\Big). \tag{4.20}$$

Take $P \in H^{\lambda}_{\infty,q}$. Set $t_1^0 := \pi_q^{\lambda}(P)$. The restriction $\widetilde{\boldsymbol{v}}_{|\mathcal{B}_q^{\lambda}\langle t_1^0\rangle}$ is a holomorphic frame of $(\Psi_q^{\lambda})^*(\mathcal{V},\nabla^{\lambda})_{|\mathcal{B}_q^{\lambda}\langle t_1^0\rangle}$. (See Sect. 4.3.3 for $\mathcal{B}_q^{\lambda}\langle t_1^0\rangle$.) On a neighbourhood U_P of P, there exists the frame $\boldsymbol{u}_P$ of $(\Psi_q^{\lambda})^*(\mathcal{V},\nabla^{\lambda})$ such that $\boldsymbol{u}_{P|\mathcal{B}_q^{\lambda}\langle t_1^0\rangle} = \widetilde{\boldsymbol{v}}_{|\mathcal{B}_q^{\lambda}\langle t_1^0\rangle}$. By Proposition 3.6.17, $\boldsymbol{u}_P$ is a convenient frame of the filtered bundle $(\Psi_q^{\lambda})^*(\mathcal{P}_*\mathcal{V},\nabla^{\lambda})$. Let h_P be the Hermitian metric of $(\Psi_q^{\lambda})^*(\mathcal{V},\nabla^{\lambda})_{|U_P\setminus H^{\lambda}_{\infty,q}}$ determined by $h_P(u_i,u_i) = |x_q|^{2b(v_i)}(\log|x_q|)^{k(v_i)}$ and $h_P(u_i,u_j) = 0$ $(i\neq j)$.

Let G be the $M_r(\mathbb{C})$-valued C^{∞}-function on $U_P \setminus H^{\lambda}_{\infty,q}$ determined by $\boldsymbol{u}_P = \boldsymbol{v}_P G$. By (4.20), there exists a function $\mathfrak{c}$ such that (i) $\mathfrak{c} = O(|x_q|^{-1})$, (ii) $|G-(1+\mathfrak{c})I_r| = O(|x_q|^{-q})$. Hence, $\widetilde{h}_0$ and h_P are mutually bounded, and we obtain the claim of the proposition in this case.

Let us study the case that there exists a finite subset $\mathcal{I} \subset S(q) = \{\sum_{j=1}^{q-1}\mathfrak{a}_j x_q^j\}$ and an isomorphism

$$F : (\mathcal{P}_*\mathcal{V},\nabla^{\lambda})_{|\widehat{\infty}} \simeq \bigoplus_{\mathfrak{a}\in\mathcal{I}} (\mathcal{P}_*\mathcal{V}_{\mathfrak{a}},\nabla^{\lambda}_{\mathfrak{a}})_{|\widehat{\infty}}, \tag{4.21}$$

where $\nabla^{\lambda}_{\mathfrak{a}} - d\mathfrak{a}\,\mathrm{id}$ are logarithmic with respect to $\mathcal{P}_*\mathcal{V}_{\mathfrak{a}}$. We obtain the induced isomorphism:

$$(\Psi_q^{\lambda})^*(F) : (\Psi_q^{\lambda})^*(\mathcal{P}_*\mathcal{V},\nabla^{\lambda})_{|\widehat{H}^{\lambda}_{\infty,q}} \simeq \bigoplus_{\mathfrak{a}\in\mathcal{I}} (\Psi_q^{\lambda})^*(\mathcal{P}_*\mathcal{V}_{\mathfrak{a}},\nabla^{\lambda}_{\mathfrak{a}})_{|\widehat{H}^{\lambda}_{\infty,q}}. \tag{4.22}$$

Let $\mathcal{C}^{\infty}_{U_{x,q}}$ and $\mathcal{C}^{\infty}_{\mathcal{B}_q^{\lambda}}$ denote the sheaf of C^{∞}-functions on $U_{x,q}$ and $\mathcal{B}_q^{\lambda}$, respectively. There exists an isomorphism of sheaves

$$F_{C^{\infty}} : \mathcal{V}\otimes_{\mathcal{O}_{U_{x,q}}} \mathcal{C}^{\infty}_{U_{x,q}} \simeq \bigoplus_{\mathfrak{a}\in\mathcal{I}} \mathcal{V}_{\mathfrak{a}}\otimes_{\mathcal{O}_{U_{x,q}}} \mathcal{C}^{\infty}_{U_{x,q}}$$

which induces (4.21). It induces an isomorphism of sheaves

$$(\Psi_q^{\lambda})^*(F_{C^{\infty}}) : (\Psi_q^{\lambda})^*(\mathcal{P}_*\mathcal{V},\nabla^{\lambda})\otimes_{\mathcal{O}_{\mathcal{B}_q^{\lambda}}} \mathcal{C}^{\infty}_{\mathcal{B}_q^{\lambda}} \simeq \bigoplus_{\mathfrak{a}\in\mathcal{I}} (\Psi_q^{\lambda})^*(\mathcal{P}_*\mathcal{V}_{\mathfrak{a}},\nabla^{\lambda}_{\mathfrak{a}})\otimes_{\mathcal{O}_{\mathcal{B}_q^{\lambda}}} \mathcal{C}^{\infty}_{\mathcal{B}_q^{\lambda}},$$

which induces (4.22). It is easy to see that a Hermitian metric h_V of $\mathcal{V}_{|U^*_{x,q}}$ satisfies the norm estimate for $(\mathcal{P}_*\mathcal{V},\nabla^{\lambda})$ if and only if the induced Hermitian metric $F_{C^{\infty}*}(h_V)$ of $\bigoplus \mathcal{V}_{\mathfrak{a}|U^*_{x,q}}$ satisfies the norm estimate for $\bigoplus(\mathcal{P}_*\mathcal{V}_{\mathfrak{a}},\nabla^{\lambda}_{\mathfrak{a}})$. Similarly, it is easy to see that $(\Psi_q^{\lambda})^{-1}(h_V)$ satisfies the norm estimate for $(\Psi_q^{\lambda})^*(\mathcal{P}_*\mathcal{V},\nabla^{\lambda})$ if and only if $(\Psi_q^{\lambda})^{-1}(F_{C^{\infty}*}(h_V))$ satisfies the norm estimate for $\bigoplus(\Psi_q^{\lambda})^*(\mathcal{P}_*\mathcal{V}_{\mathfrak{a}},\nabla^{\lambda}_{\mathfrak{a}})$. Hence, we obtain the claim of the proposition in this case.

In the general case, there exists $p \in q\mathbb{Z}_{>0}$ such that $\varphi^*_{q,p}(\mathcal{P}_*\mathcal{V},\nabla)$ has a decomposition as in (4.21). For a Hermitian metric h_V of $\mathcal{V}_{|U^*_{x,q}}$, it is easy to see that h_V satisfies the norm estimate for $(\mathcal{P}_*\mathcal{V}, \nabla^\lambda)$ if and only if $\varphi_{q,p}^{-1}(h_V)$ satisfies the norm estimate for $\varphi_{q,p}^*(\mathcal{P}_*\mathcal{V}, \nabla^\lambda)$. Then, by using Lemma 4.4.6, we obtain the claim of the proposition in the general case. ■

4.5.5 Non-integrable Case

Let us explain the ramified version of the construction in Sect. 2.8.3.2, where the integrability of a λ-connection is not assumed. Let $U_{x,q}$ be an open subset of $\mathbb{P}^1_{x_q}$ such that $\infty \in U_{x,q}$.

4.5.5.1 A General Construction

Let $(V, \overline{\partial}_V)$ be a holomorphic vector bundle on $U_{x,q}$. Let $\mathbb{D}^{\lambda\,(1,0)} : C^\infty(U_{x,q}, V) \longrightarrow C^\infty(U_{x,q}, V \otimes \Omega^{1,0}_{U_{x,q}}((q+1)\infty))$ be the differential operator $\mathbb{D}^{\lambda\,(1,0)}(fs) = \lambda\partial_{U_{x,q}}(f)s + f\mathbb{D}^{\lambda\,(1,0)}(s)$ for any $f \in C^\infty(U_{x,q})$ and $s \in C^\infty(U_{x,q}, V)$. We do not assume $[\overline{\partial}_V, \mathbb{D}^{\lambda\,(1,0)}] = 0$ at this stage.

We set $\widetilde{V} = (\Psi^\lambda_q)^{-1}(V)$ on $\mathcal{B}^\lambda_q := (\Psi^\lambda_q)^{-1}(U_{x,q})$. We set $\mathcal{B}^{\lambda *}_q := \mathcal{B}^\lambda_q \setminus H^\lambda_{\infty,q}$. Let $\widetilde{V}^*$ denote the restriction of $\widetilde{V}$ to $\mathcal{B}^{\lambda *}_q$. In Sect. 2.8.1.3, we constructed an S^1_T-equivariant differential operator $\overline{\partial}_{\widetilde{V}^*} : C^\infty(\mathcal{B}^{\lambda *}_q, \widetilde{V}^*) \longrightarrow C^\infty(\mathcal{B}^{\lambda *}_q, \widetilde{V}^* \otimes \Omega^{0,1}_{\mathcal{B}^{\lambda *}_q})$ satisfying the mini-complex Leibniz rule.

Lemma 4.5.8 *The differential operator $\overline{\partial}_{\widetilde{V}^*}$ uniquely extends to a differential operator $\overline{\partial}_{\widetilde{V}} : C^\infty(\mathcal{B}^\lambda_q, \widetilde{V}) \longrightarrow C^\infty(\mathcal{B}^\lambda_q, \widetilde{V} \otimes \Omega^{0,1}_{\mathcal{B}^\lambda_q})$ satisfying the mini-complex Leibniz rule.*

Proof We just repeat the argument in the proof of Lemma 2.8.6. We may assume that $U_{x,q}$ is an open disc around ∞. We have the complex vector fields $\partial_{\overline{w}} = (1+|\lambda|^2)q^{-1}\overline{x}_q^{-q+1}\partial_{\overline{x}_q}$ and $\partial_w = (1+|\lambda|^2)q^{-1}\overline{x}_q^{-q+1}\partial_{x_q}$ on $U_{x,q}$. Let $\partial_{V,\overline{w}}$ (resp. $\mathbb{D}^\lambda_w$) denote the differential operator on V induced by $\partial_{\overline{w}}$ and $\overline{\partial}_V$ (resp. ∂_w and $\mathbb{D}^{\lambda\,(1,0)}$).

Let $\boldsymbol{v}$ be a holomorphic frame of $(V, \overline{\partial}_V)$ on $U_{x,q}$. We have $\partial_{V,\overline{w}}\boldsymbol{v} = 0$. Let A be the matrix valued C^∞-function on $U_{x,q}$ determined by $\mathbb{D}^\lambda_w \boldsymbol{v} = \boldsymbol{v}A$. The pull back $\widetilde{\boldsymbol{v}} := (\Psi^\lambda_q)^{-1}\boldsymbol{v}$ is a C^∞-frame of $\widetilde{V}$ on $\mathcal{B}^\lambda_q$.

We have the projection $\varpi^\lambda_q : Y^{\lambda\,\mathrm{cov}}_q \longrightarrow Y^\lambda_q$. For any $P \in H^\lambda_{\infty,q}$, we choose $\widetilde{P} \in H^{\lambda\,\mathrm{cov}}_{\infty,q}$ such that $\varpi^\lambda_q(\widetilde{P}) = P$. Then, $(t_1, \tau_{1,q}) = (t_1, \beta_{1,q}^{-1})$ around $\widetilde{P}$ induces a local mini-complex coordinate system on a neighbourhood U_P of P. By the formulas (2.16) and (2.17), on $U_P \setminus H^\lambda_{\infty,q}$, the operators $\partial_{\widetilde{V}^*,t_1}$ and $\partial_{\widetilde{V}^*,\overline{\tau}_{1,q}}$ are described as follows with respect to the frame $\widetilde{\boldsymbol{v}}$:

$$\partial_{\widetilde{V}^*,t_1}\widetilde{\boldsymbol{v}} = \widetilde{\boldsymbol{v}}\frac{-2\sqrt{-1}}{1+|\lambda|^2}(\Psi^\lambda)^*(A), \quad \partial_{\widetilde{V}^*,\overline{\tau}_{1,q}}\widetilde{\boldsymbol{v}} = 0.$$

Note that A is C^∞ with respect to x_q^{-1} and $\overline{x}_q^{-1}$, and that $(\Psi^\lambda)^*(x_q^{-1}) = \tau_{1,q}^q(1 - 2\sqrt{-1}\lambda t_1 \tau_{1,q}^q)^{-1}$ is C^∞ around P. Hence, $\partial_{\widetilde{V}^*,t_1}$ and $\partial_{\widetilde{V}^*,\overline{\tau}_1}$ uniquely extend to operators $\partial_{\widetilde{V},t_1}$ and $\partial_{\widetilde{V},\overline{\tau}_{1,q}}$ on $C^\infty(\mathcal{B}_q^\lambda, \widetilde{V})$, which implies the claim of the lemma. ■

Lemma 4.5.9 *The above construction induces an equivalence between the following objects.*

- *Holomorphic vector bundles $(V, \overline{\partial}_V)$ on $U_{x,q}$ equipped with a differential operator $\mathbb{D}^{\lambda\,(1,0)} : C^\infty(U_{x,q}, V) \longrightarrow C^\infty(U_{x,q}, V \otimes \Omega^{1,0}_{U_{x,q}}((q+1)\infty))$ such that $\mathbb{D}^{\lambda\,(1,0)}(fs) = \lambda\partial_{U_{x,q}}(f)s + f\mathbb{D}^{\lambda\,(1,0)}(s)$ for any $f \in C^\infty(U_{x,q})$ and $s \in C^\infty(U_{x,q}, V)$.*
- *S^1_T-equivariant vector bundles $\widetilde{V}$ on $\mathcal{B}_q^\lambda$ equipped with an S^1_T-equivariant linear differential operator $\overline{\partial}_{\widetilde{V}} : C^\infty(\mathcal{B}_q^\lambda, \widetilde{V}) \longrightarrow C^\infty(\mathcal{B}_q^\lambda, \widetilde{V} \otimes \Omega^{0,1}_{\mathcal{B}_q^\lambda})$ satisfying the mini-complex Leibniz rule.*

Proof By repeating the argument in the proof of Lemma 2.8.7, we indicate the inverse construction. There exists a C^∞-vector bundle V on $U_{x,q}$ equipped with an S^1_T-equivariant isomorphism $\widetilde{V} \simeq (\Psi_q^\lambda)^{-1}(V)$. We set $U^*_{x,q} = U_{x,q} \setminus \{\infty\}$ and $V^* := V_{|U^*_{x,q}}$. We obtain a λ-connection $\mathbb{D}^\lambda_{V^*}$ corresponding to $\overline{\partial}_{\widetilde{V}|\mathcal{B}_q^{\lambda*}}$.

Fix $t_1^0 \in S^1_T$. Note that $\mathcal{B}_q^\lambda\langle t_1^0\rangle$ is naturally equipped with a complex structure, and $(\widetilde{V}, \overline{\partial}_{\widetilde{V}})_{|\mathcal{B}_q^\lambda\langle t_1^0\rangle}$ is naturally a holomorphic vector bundle on $\mathcal{B}_q^\lambda\langle t_1^0\rangle$. (See Sect. 4.3.3 for $\mathcal{B}_q^\lambda\langle t_1^0\rangle$.) The induced morphism $\Psi^\lambda_{q,t_1^0} : \mathcal{B}_q^\lambda\langle t_1^0\rangle \longrightarrow U_{x,q}$ is holomorphic, and the image is an open neighbourhood of ∞. We have the isomorphism of C^∞-vector bundles $\widetilde{V}_{|\mathcal{B}_1^\lambda\langle t_1^0\rangle} \simeq (\Psi_q^\lambda)^{-1}(V)_{|\mathcal{B}_q^\lambda\langle t_1^0\rangle}$. Note that $(\Psi_q^\lambda)^{-1}(V)_{|\mathcal{B}_q^{\lambda*}\langle t_1^0\rangle}$ is equipped with the holomorphic structure induced by $\overline{\partial}_{V^*}$, and that $\widetilde{V}_{|\mathcal{B}_q^{\lambda*}\langle t_1^0\rangle} \simeq (\Psi_q^\lambda)^{-1}(V)_{|\mathcal{B}_q^{\lambda*}\langle t_1^0\rangle}$ is holomorphic. (See Sect. 4.4.1 for $\mathcal{B}_1^{\lambda*}\langle t_1^0\rangle$.) It implies that $\overline{\partial}_{V^*}$ uniquely extends to a holomorphic structure $\overline{\partial}_V$ of V.

Let $\boldsymbol{v}$ be a holomorphic frame of $(V, \overline{\partial}_V)$ on $U_{x,q}$. We set $\widetilde{\boldsymbol{v}} = (\Psi_q^\lambda)^{-1}(\boldsymbol{v})$. There exists a matrix-valued C^∞-function A on $U_{x,q}$ determined by

$$\partial_{\widetilde{V},t_1}\widetilde{\boldsymbol{v}} = \widetilde{\boldsymbol{v}} \cdot \frac{-2\sqrt{-1}}{1+|\lambda|^2}(\Psi_q^\lambda)^*(A).$$

Let $\mathbb{D}^\lambda_{V^*,w}$ denote the differential operator of V^* induced by $\mathbb{D}^\lambda_{V^*}$ and ∂_w. By the construction of $\mathbb{D}^\lambda_{V^*}$ (see the proof of Lemma 2.8.1), we have $\mathbb{D}^\lambda_{V^*,w}\boldsymbol{v}_{|U^*_{x,q}} = \boldsymbol{v}_{|U^*_{x,q}} \cdot A$. It implies that $\mathbb{D}^\lambda_{V^*}$ uniquely extends to a differential operator $\mathbb{D}^\lambda$ of V.

Thus, we obtain $(V, \overline{\partial}_V, \mathbb{D}^{\lambda\,(1,0)})$ from $(\widetilde{V}, \overline{\partial}_{\widetilde{V}})$. The two constructions are mutually inverse. ■

Corollary 4.5.10 *The above construction induces an equivalence between the following objects.*

- *Locally free $\mathcal{O}_{U_{x,q}}$-module $\mathcal{L}$ equipped with a meromorphic λ-connection $\mathbb{D}^{\lambda}$: $\mathcal{L} \longrightarrow \mathcal{L} \otimes \Omega^1_{U_{x,q}}((q+1)\infty)$.*
- *S^1_T-equivariant locally free $\mathcal{O}_{\mathcal{B}^{\lambda}_q}$-modules $\widetilde{\mathcal{L}}$.* ■

4.5.5.2 Comparison with the Construction in Sect. 4.5.2

Let $\mathcal{L}$ be a holomorphic vector bundle on $U_{x,q}$ equipped with a meromorphic λ-connection $\nabla^{\lambda} : \mathcal{L} \longrightarrow \mathcal{L} \otimes \Omega^1_{U_{x,q}}((q+1)\infty)$. Let $(V, \overline{\partial}_V)$ be the holomorphic vector bundle corresponding to $\mathcal{L}$. It is equipped with $\mathbb{D}^{\lambda\,(1,0)} : C^{\infty}(U_{x,q}, V) \longrightarrow C^{\infty}\big(U_{x,q}, V \otimes \Omega^{1,0}_{U_{x,q}}((q+1)\infty)\big)$ induced by ∇^{λ}. We obtain $(\widetilde{V}, \overline{\partial}_{\widetilde{V}})$ on $\mathcal{B}^{\lambda}_q$, which is mini-holomorphic by (2.18).

Lemma 4.5.11 $(\Psi^{\lambda}_q)^*(\mathcal{L}, \nabla^{\lambda})$ *is the sheaf of mini-holomorphic sections of* $(\widetilde{V}, \overline{\partial}_{\widetilde{V}})$.

Proof By the construction, it is easy to see that $\widetilde{\mathcal{L}}^{\infty}$ in (3.24) is the sheaf of C^{∞}-sections u of $\widetilde{V}$ such that $\partial_{\widetilde{V},\overline{\beta}_1} u = 0$. Then, the claim is clear. ■

4.5.5.3 Comparison with the Construction in the Formal Case

Let $(V, \overline{\partial}_V)$ be a holomorphic vector bundle with $\mathbb{D}^{\lambda\,(1,0)}$ as in Sect. 4.5.5.1. Suppose that $[\overline{\partial}_V, \mathbb{D}^{\lambda\,(1,0)}]_{|\widehat{\infty}_{x,q}} = 0$, i.e., the Taylor series of $[\overline{\partial}_V, \mathbb{D}^{\lambda\,(1,0)}]$ at $x_q^{-1} = 0$ is 0.

Let $\mathcal{L}$ denote the $\mathcal{O}_{U_{x,q}}$-module obtained as the sheaf of holomorphic sections of $(V, \overline{\partial}_V)$, and we obtain the $\mathcal{O}_{\widehat{\infty}_{x,q}}$-module $\widehat{\mathcal{L}} := \mathcal{L}_{|\widehat{\infty}_{x,q}}$. It is equipped with the meromorphic λ-connection ∇^{λ} induced by $\mathbb{D}^{\lambda\,(1,0)}$. In Sect. 3.6.3, we obtain the $\mathcal{K}_{\widehat{H}^{\lambda}_{\infty,q}}$-module

$$\widetilde{\mathcal{L}}^{\infty} = \mathcal{K}_{\widehat{H}^{\lambda}_{\infty,q}} \otimes_{(\Psi^{\lambda}_q)^{-1}\mathcal{O}_{\widehat{\infty}x,q}} (\Psi^{\lambda}_q)^{-1}(\widehat{\mathcal{L}})$$

equipped with the action of ∂_{t_1}, and the $\mathcal{O}_{\widehat{H}^{\lambda}_{\infty,q}}$-module $(\Psi^{\lambda}_q)^*(\widehat{\mathcal{L}}, \nabla^{\lambda})$ as the kernel of ∂_{t_1}. Let us explain how to obtain $(\widetilde{\mathcal{L}}^{\infty}, \partial_{t_1})$ and $\widetilde{\mathcal{L}}$ naturally from $(\widetilde{V}, \overline{\partial}_{\widetilde{V}})$ induced by $(V, \overline{\partial}_V + \mathbb{D}^{\lambda\,(1,0)})$.

Let us introduce sheaves of algebras $\mathcal{C}^{\infty}_{\widehat{H}^{\lambda\,\mathrm{cov}}_{\infty,q}}$ on $H^{\lambda\,\mathrm{cov}}_{\infty,q}$, and $\mathcal{C}^{\infty}_{\widehat{H}^{\lambda}_{\infty,q}}$ on $H^{\lambda}_{\infty,q}$. For any open subset $U \subset H^{\lambda\,\mathrm{cov}}_{\infty,q}$, we set $\mathcal{C}^{\infty}_{\widehat{H}^{\lambda\,\mathrm{cov}}_{\infty,q}}(U) = \big\{\sum_{i,j\geq 0} a_{i,j}(t_1)\beta_{1,q}^{-i}\overline{\beta}_{1,q}^{-j}\big\}$. There exist naturally defined linear differential operators ∂_{t_1} and $\partial_{\overline{\beta}_1}$ on $\mathcal{C}^{\infty}_{\widehat{H}^{\lambda\,\mathrm{cov}}_{\infty,q}}(U)$. Thus, we obtain the sheaf of algebras $\mathcal{C}^{\infty}_{\widehat{H}^{\lambda\,\mathrm{cov}}_{\infty,q}}$ equipped with the differential operators ∂_{t_1}

and $\partial_{\overline{\beta}_1}$ on $H^{\lambda\,\mathrm{cov}}_{\infty,q}$. We note that the kernel of $\partial_{\overline{\beta}_1} : \mathcal{C}^\infty_{\widehat{H}^{\lambda\,\mathrm{cov}}_{\infty,q}} \longrightarrow \mathcal{C}^\infty_{\widehat{H}^{\lambda\,\mathrm{cov}}_{\infty,q}}$ is $\mathcal{K}_{\widehat{H}^{\lambda\,\mathrm{cov}}_{\infty,q}}$. The sheaf $\mathcal{C}^\infty_{\widehat{H}^{\lambda\,\mathrm{cov}}_{\infty,q}}$ is naturally equivariant, and we obtain $\mathcal{C}^\infty_{\widehat{H}^{\lambda}_{\infty,q}}$ on $H^\lambda_{\infty,q}$ as the descent, which is equipped with ∂_{t_1} and $\partial_{\overline{\beta}_1}$. The kernel of $\partial_{\overline{\beta}_1} : \mathcal{C}^\infty_{\widehat{H}^{\lambda}_{\infty,q}} \longrightarrow \mathcal{C}^\infty_{\widehat{H}^{\lambda}_{\infty,q}}$ is $\mathcal{K}_{\widehat{H}^{\lambda}_{\infty,q}}$.

Let $\mathcal{C}^\infty_{\mathcal{B}^\lambda_q}$ be the sheaf of C^∞-functions on $\mathcal{B}^\lambda_q$. By taking the Taylor series along $H^\lambda_{\infty,q}$, we obtain the morphism of sheaves of algebras $\mathcal{C}^\infty_{\mathcal{B}^\lambda_q|H^\lambda_{\infty,q}} \longrightarrow \mathcal{C}^\infty_{\widehat{H}^{\lambda}_{\infty,q}}$. Thus, we obtain a morphism of ringed spaces $k^\lambda_{q,C^\infty} : (H^\lambda_{\infty,q}, \mathcal{C}^\infty_{\widehat{H}^{\lambda}_{\infty,q}}) \longrightarrow (\mathcal{B}^\lambda_q, \mathcal{C}^\infty_{\mathcal{B}^\lambda_q})$. For a locally free $\mathcal{C}^\infty_{\mathcal{B}^\lambda_q}$-module $\mathcal{W}$, we obtain the locally free $\mathcal{C}^\infty_{\widehat{H}^{\lambda}_{\infty,q}}$-module $(k^\lambda_{q,C^\infty})^*(\mathcal{W})$. If $\mathcal{W}$ is equipped with linear differential operators $\partial_{\mathcal{W},\kappa}$ $(\kappa = t_1, \overline{\beta}_1)$ satisfying $\partial_{\mathcal{W},\kappa}(fu) = \partial_\kappa(f)\cdot s + f\partial_{\mathcal{W},\kappa}(s)$ for any local sections f and s of $\mathcal{C}^\infty_{\mathcal{B}^\lambda_q}$ and $\mathcal{W}$, respectively, then $(k^\lambda_{q,C^\infty})^*\mathcal{W}$ is equipped with the induced linear differential operators $\partial_{(k^\lambda_{q,C^\infty})^*\mathcal{W},\kappa}$ $(\kappa = t_1, \overline{\beta}_1)$ satisfying $\partial_{(k^\lambda_{q,C^\infty})^*\mathcal{W},\kappa}(fu) = \partial_\kappa(f)\cdot s + f\partial_{(k^\lambda_{q,C^\infty})^*\mathcal{W},\kappa}(s)$ for any local sections f and s of $\mathcal{C}^\infty_{\widehat{H}^{\lambda}_{\infty,q}}$ and $(k^\lambda_{q,C^\infty})^*\mathcal{W}$, respectively.

To simplify the description, the sheaf of C^∞-sections of V is also denoted by V. We obtain the $\mathcal{C}^\infty_{\widehat{H}^{\lambda}_{\infty,q}}$-module $(k^\lambda_{q,C^\infty})^*\widetilde{V}_{C^\infty}$, which is equipped with the induced linear differential operators $\partial_{(k^\lambda_{q,C^\infty})^*\widetilde{V},t_1}$ and $\partial_{(k^\lambda_{q,C^\infty})^*\widetilde{V},\overline{\beta}_1}$ By (2.18), the two operators $\partial_{(k^\lambda_{q,C^\infty})^*\widetilde{V},t_1}$ and $\partial_{(k^\lambda_{q,C^\infty})^*\widetilde{V},\overline{\beta}_1}$ are commutative. The following lemma is obvious by the construction of $\widehat{\mathcal{L}}$.

Lemma 4.5.12 *The kernel of $\partial_{(k^\lambda_{q,C^\infty})^*\widetilde{V},\overline{\beta}_1}$ on $(k^\lambda_{q,C^\infty})^*\widetilde{V}_{C^\infty}$ is naturally isomorphic to $\widetilde{\mathcal{L}}^\infty$ in a way compatible with the actions of ∂_{t_1}. Hence, the intersection of the kernels of $\partial_{(k^\lambda_{q,C^\infty})^*\widetilde{V},\overline{\beta}_1}$ and $\partial_{(k^\lambda_{q,C^\infty})^*\widetilde{V},t_1}$ is naturally isomorphic to $(\Psi^\lambda_q)^*(\widehat{\mathcal{L}}, \nabla^\lambda)$.* ■

Chapter 5
Basic Examples of Monopoles Around Infinity

Abstract We explain some basic example of monopoles, which are models in the study of the asymptotic behaviour of monopoles. In particular, we shall later use the examples in Sects. 5.1 and 5.2.3, which are summarized in Sect. 5.2.4.

5.1 Examples of Monopoles with Pure Slope ℓ/p

Let p be a positive integer. Let $Y_p^{0\,\mathrm{cov}*}$ be as in Sect. 4.3.1. The pull back of the function w by $Y_p^{0\,\mathrm{cov}*} \longrightarrow \mathcal{M}^0$ is also denoted by w, which is equal to w_p^p. Hence, for instance, dw also means $pw_p^{p-1}dw_p$. We also use the complex vector fields $\partial_w = p^{-1}w_p^{1-p}\partial_{w_p}$ and $\partial_{\overline{w}} = p^{-1}\overline{w}_p^{1-p}\partial_{\overline{w}_p}$.

Let $\mathbb{L}_p^{\mathrm{cov}*}(\ell)$ denote the product line bundle on $Y_p^{0\,\mathrm{cov}*}$ with a global frame e. Let h be the Hermitian metric determined by $h(e,e) = 1$. We take a positive number T and an integer ℓ, and let ∇ and ϕ be the unitary connection and the Higgs field given as follows:

$$\nabla e = e\Big(\frac{\ell}{pT}\frac{t}{2}\Big(\frac{d\overline{w}}{\overline{w}} - \frac{dw}{w}\Big)\Big), \qquad \phi = \sqrt{-1}\frac{\ell}{pT}\log|w|.$$

Lemma 5.1.1 *$(\mathbb{L}_p^{\mathrm{cov}*}(\ell), h, \nabla, \phi)$ satisfies the Bogomolny equation with respect to the metric $dt\,dt + dw\,d\overline{w}$ on $Y_p^{0\,\mathrm{cov}*}$.*

Proof We obtain the following equalities:

$$\nabla\circ\nabla = \frac{\ell}{pT}\frac{dt}{2}\Big(\frac{d\overline{w}}{\overline{w}} - \frac{dw}{w}\Big), \qquad \nabla\phi = \sqrt{-1}\frac{\ell}{pT}\frac{1}{2}\Big(\frac{dw}{w} + \frac{d\overline{w}}{\overline{w}}\Big).$$

For the expression $F = F_{t\overline{w}}dt\,d\overline{w} + F_{tw}\,dt\,dw + F_{w\overline{w}}dw\,d\overline{w}$, we obtain

$$F_{t\overline{w}} = \frac{\ell}{pT}\frac{1}{2}\overline{w}^{-1}, \quad F_{tw} = -\frac{\ell}{pT}\frac{1}{2}w^{-1}, \quad F_{w\overline{w}} = 0.$$

T. Mochizuki, *Periodic Monopoles and Difference Modules*, Lecture Notes in Mathematics 2300, https://doi.org/10.1007/978-3-030-94500-8_5

We also obtain the following equalities:

$$\nabla_t\phi = 0, \quad \nabla_w\phi = \sqrt{-1}\frac{\ell}{pT}\frac{1}{2}w^{-1}, \quad \nabla_{\overline{w}}\phi = \sqrt{-1}\frac{\ell}{pT}\frac{1}{2}\overline{w}^{-1}.$$

Recall the following relation for the Hodge star operator:

$$*(dt\,dw) = -\sqrt{-1}\,dw, \quad *(dt\,d\overline{w}) = \sqrt{-1}d\overline{w}, \quad *(dw\,d\overline{w}) = -2\sqrt{-1}\,dt.$$

We have $F_{w\overline{w}} = \nabla_t\phi = 0$. We have

$$*\big(F_{tw}\,dt\,dw\big) = -\sqrt{-1}F_{tw}\,dw = \sqrt{-1}\frac{\ell}{2pT}w^{-1}\,dw = \nabla_w\phi\,dw.$$

We also have

$$*\big(F_{t\overline{w}}\,dt\,d\overline{w}\big) = \sqrt{-1}F_{t\overline{w}}\,d\overline{w} = \sqrt{-1}\frac{\ell}{2pT}\overline{w}^{-1}\,d\overline{w} = \nabla_{\overline{w}}\phi\,d\overline{w}.$$

Hence, the Bogomolny equation is satisfied. ■

5.1.1 Equivariance with Respect to the $\mathbb{Z}$-Action

Recall that $Y_p^{0\,\mathrm{cov}*}$ is equipped with the naturally induced $\mathbb{Z}$-action κ_p (see Sect. 4.3.1). In particular, $\kappa_{p,1} : Y_p^{0\,\mathrm{cov}*} \longrightarrow Y_p^{0\,\mathrm{cov}*}$ is given by $\kappa_{p,1}(t, w_p) = (t+T, w_p)$. We obtain

$$\kappa_{p,1}^{-1}(\nabla)\kappa_{p,1}^{-1}(e) = \kappa_{p,1}^{-1}(e)\left(\frac{\ell}{pT}\frac{t}{2}\left(\frac{d\overline{w}}{\overline{w}} - \frac{dw}{w}\right) + \frac{\ell}{2p}\left(\frac{d\overline{w}}{\overline{w}} - \frac{dw}{w}\right)\right),$$

$$\kappa_{p,1}^*\phi = \sqrt{-1}\frac{\ell}{pT}\log|w|.$$

We also have

$$\nabla\Big(e\cdot\big(|w_p|w_p^{-1}\big)^{\ell}\Big) = e\big(|w_p|w_p^{-1}\big)^{\ell}\cdot\left(\frac{\ell}{pT}\frac{t}{2}\left(\frac{d\overline{w}}{\overline{w}} - \frac{dw}{w}\right) + \frac{\ell}{2p}\left(\frac{d\overline{w}}{\overline{w}} - \frac{dw}{w}\right)\right).$$

By the correspondence $\kappa_{p,1}^{-1}(e) \longmapsto e\cdot(|w_p|w_p^{-1})^{\ell}$, the monopole $(\mathbb{L}_p^{\mathrm{cov}*}(\ell), h, \nabla, \phi)$ is equivariant with respect to the $\mathbb{Z}$-action κ_p. Recall that the quotient of $Y_p^{0*\mathrm{cov}}$ by the $\mathbb{Z}$-action is denoted by Y_p^{0*}. (See Sect. 4.3.1.) As the descent, we obtain a monopole $(\mathbb{L}_p^*(\ell), h, \nabla, \phi)$ on Y_p^{0*}.

5.1.2 The Underlying Mini-holomorphic Bundle at $\lambda = 0$

Let $\mathbb{L}_p^{0\,\mathrm{cov}*}(\ell)$ denote the $\mathbb{Z}$-equivariant mini-holomorphic bundle on $Y_p^{0\,\mathrm{cov}*}$ underlying the monopole $(\mathbb{L}_p^{\mathrm{cov}*}(\ell), h, \nabla, \phi)$.

Proposition 5.1.2 *There exists a mini-holomorphic frame $u^0_{p,\ell}$ of $\mathbb{L}_p^{0\,\mathrm{cov}*}(\ell)$ on $Y_p^{0\,\mathrm{cov}*}$ such that the following holds:*

- $|u^0_{p,\ell}|_h = |w|^{-\ell t/(pT)} = |w_p|^{-\ell t/T}$.
- $\kappa^*_{p,1}(u^0_{p,\ell}) = w_p^{-\ell} u^0_{p,\ell}$ *under the identification* $\kappa^*_{p,1}(e) = e(|w_p|w_p^{-1})^\ell$.

Proof We have the following equalities:

$$(\nabla_t - \sqrt{-1}\phi)e = e\,\frac{\ell}{pT}\log|w|, \qquad \nabla_{\overline{w}}e = e\,\frac{\ell}{pT}\frac{t}{2}\frac{1}{\overline{w}}.$$

We define the section $u^0_{p,\ell}$ as follows:

$$u^0_{p,\ell} := e\cdot\exp\Big(-\frac{\ell}{pT}t\log|w|\Big).$$

Then, $u^0_{p,\ell}$ has the desired properties. ■

Let $\mathbb{L}_p^{0*}(\ell)$ denote the mini-holomorphic bundle on Y_p^0 underlying the monopole $(\mathbb{L}_p^*(\ell), h, \nabla, \phi)$. Let $\mathbb{L}_p^0(\ell)$ and $\mathcal{P}_*\mathbb{L}_p^0(\ell)$ denote the $\mathcal{O}_{Y_p^0}(*H^0_{\infty,p})$-module and the family of filtered sheaves obtained from $\mathbb{L}_p^{0*}(\ell)$ with h by the procedure in Sect. 4.4.1.

Corollary 5.1.3 *$\mathbb{L}_p^0(\ell)$ is a locally free $\mathcal{O}_{Y_p^0}(*H^0_{\infty,p})$-module, and $\mathcal{P}_*\mathbb{L}_p^0(\ell)$ is a good filtered bundle over $\mathbb{L}_p^0(\ell)$. (See Definitions* 4.3.3 *and* 4.3.4.*) The norm estimate holds for $\mathcal{P}_*\mathbb{L}_p^0(\ell)$ with the metric h. Moreover, $\mathcal{P}_*\mathbb{L}_p^0(\ell)_{|\widehat{H}^0_{\infty,p}}$ is isomorphic to $\mathcal{P}_*^{(0)}\widehat{\mathbb{L}}_p^0(\ell, 1)$ in* Sect. 3.5.3. *(See* Sect. 4.3.2 *for $\widehat{H}^0_{\infty,q}$.)* ■

5.1.3 The Underlying Mini-holomorphic Bundle at General λ

Let λ be any complex number. Let $\mathbb{L}_p^{\lambda\,\mathrm{cov}*}(\ell)$ be the $\mathbb{Z}$-equivariant mini-holomorphic bundle on $Y_p^{\lambda\,\mathrm{cov}*}$ underlying $(\mathbb{L}_p^{\mathrm{cov}*}(\ell), h, \nabla, \phi)$. (See Sect. 4.3.2 for $Y_p^{\lambda\,\mathrm{cov}*}$, and recall that $Y_p^{\lambda\,\mathrm{cov}*} = Y_p^{0*\mathrm{cov}}$ as Riemannian manifolds.) We shall prove the following propositions in Sects. 5.1.4–5.1.7.

Proposition 5.1.4 *There exists a neighbourhood $\mathcal{U}_p^\lambda$ of $H_{\infty,p}^{\lambda\,\mathrm{cov}}$ in $Y_p^{\lambda\,\mathrm{cov}}$ and a mini-holomorphic frame $u_{p,\ell}^\lambda$ of $\mathbb{L}_p^{\lambda\,\mathrm{cov}\,*}(\ell)$ on a neighbourhood $\mathcal{U}_p^\lambda \setminus H_{\infty,p}^{\lambda\,\mathrm{cov}}$ such that the following holds:*

(A1) For any $P \in H_p^{\lambda\,\mathrm{cov}}$, there exist $C(P) > 1$ and a neighbourhood U_P of P in $\mathcal{U}_p^\lambda$ such that the following holds on $U_P \setminus H_p^{\lambda\,\mathrm{cov}}$:

$$C(P)^{-1}|w|^{-\ell t_1/(pT)} \leq |u_{p,\ell}^\lambda|_h \leq C(P)|w|^{-\ell t_1/(pT)}.$$

(A2) On $\big(\mathcal{U}_p^\lambda \cap \kappa_{p,1}^{-1}(\mathcal{U}_p^\lambda)\big) \setminus H_p^{\lambda\,\mathrm{cov}}$, we have the following equality:

$$\kappa_{p,1}^*(u_{p,\ell}^\lambda) = u_{p,\ell}^\lambda \cdot (1+|\lambda|^2)^{\ell/p}(\beta_1+2\sqrt{-1}\lambda T)^{-\ell/p} \times \exp\Big(\frac{\ell}{p}G(2\sqrt{-1}\lambda T/\beta_1)\Big),$$

where $G(x) = 1 - x^{-1}\log(1+x) = x/2 + O(x^2)$.

Let $\mathbb{L}_p^{\lambda *}(\ell)$ denote the mini-holomorphic bundle on $Y_p^{\lambda *}$ underlying the monopole $(\mathbb{L}_p^*(\ell), h, \nabla, \phi)$. (See Sect. 4.3.2 for $Y_p^{\lambda *}$, and recall $Y_p^{\lambda *} = Y_p^{0*}$ as Riemannian manifolds.) Let $\mathbb{L}_p^\lambda(\ell)$ and $\mathcal{P}_*\mathbb{L}_p^\lambda(\ell)$ denote the $\mathcal{O}_{Y_p^\lambda}(*H_{\infty,p}^\lambda)$-module and the family of sheaves obtained from $\mathbb{L}_p^{\lambda *}(\ell)$ with h by the procedure in Sect. 4.4.1.

Corollary 5.1.5 *$\mathbb{L}_p^\lambda(\ell)$ is a locally free $\mathcal{O}_{Y_p^\lambda}(*H_{\infty,p}^\lambda)$-module, and $\mathcal{P}_*\mathbb{L}_p^\lambda(\ell)$ is a good filtered bundle over $\mathbb{L}_p^\lambda(\ell)$. The norm estimate holds for $\mathcal{P}_*\mathbb{L}_p^\lambda(\ell)$ with the metric h. Moreover, $\mathcal{P}_*\mathbb{L}_p^\lambda(\ell)_{|\widehat{H}_{\infty,p}^\lambda}$ is isomorphic to $\mathcal{P}_*^{(0)}\widehat{\mathbb{L}}_p^\lambda(\ell, 1)$ in* Sect. 3.5.3. *(See* Remark 4.3.2 *for the comparison of $\widehat{H}_{\infty,p}^{\lambda\,\mathrm{cov}}$ and $\widehat{H}_{\infty,p}^{\mathrm{cov}}$ in* Sect. 3.4.2.*)* ■

5.1.4 Complex Structures

We set $\widetilde{Y}^{0\,\mathrm{cov}\,*} := \mathbb{R}_s \times Y^{0\,\mathrm{cov}\,*}$ as a C^∞-manifold. By the complex coordinate system $(z, w) = (s + \sqrt{-1}t, w)$, $\widetilde{Y}^{0\,\mathrm{cov}\,*}$ is a complex manifold. Let $\widetilde{Y}^{\lambda\,\mathrm{cov}\,*}$ denote the complex manifold induced by the complex coordinate system $(\xi, \eta) = (z + \lambda\overline{w}, w - \lambda\overline{z})$.

Let $\widetilde{Y}_p^{0\,\mathrm{cov}\,*} := \mathbb{R}_s \times Y_p^{0\,\mathrm{cov}\,*}$. It is naturally a covering space of $\widetilde{Y}^{0\,\mathrm{cov}\,*}$, and hence equipped with the complex structure. The complex coordinate system (z, w) of $\widetilde{Y}^{0\,\mathrm{cov}\,*}$ induces local coordinate systems of $\widetilde{Y}_p^{0\,\mathrm{cov}\,*}$. The complex vector fields ∂_z, $\partial_{\overline{z}}$, ∂_w and $\partial_{\overline{w}}$ are naturally lifted to complex vector fields on $\widetilde{Y}_p^{0\,\mathrm{cov}\,*}$, which are denoted by the same notation.

Let $\widetilde{Y}_p^{\lambda\,\mathrm{cov}*}$ be the complex manifold induced by the local coordinate system (ξ,η). The complex vector fields $\partial_{\overline{\xi}}, \partial_{\xi}, \partial_{\eta}, \partial_{\overline{\eta}}$ on $\widetilde{Y}^{\lambda\,\mathrm{cov}*}$ are naturally lifted to the complex vector fields on $\widetilde{Y}_p^{\lambda\,\mathrm{cov}*}$ which are denoted by the same notation.

By using the local coordinate system (α_1,β_1) given as in Sect. 2.7, we obtain the natural identification $\widetilde{Y}_p^{\lambda\,\mathrm{cov}*}=\mathbb{R}_{\mathrm{Re}\,\alpha_1}\times Y_p^{\lambda\,\mathrm{cov}*}$.

5.1.5 The Induced Instanton and the Underlying Holomorphic Bundle

Let $(\mathbb{L}_p^{\mathrm{cov}*}(\ell),h,\nabla,\phi)$ be the monopole on $Y_p^{0\,\mathrm{cov}*}$ in Sect. 5.1. We obtain the induced instanton $(\widetilde{E},\widetilde{h},\widetilde{\nabla})$ on $\widetilde{Y}_p^{0\,\mathrm{cov}*}$ as explained in Sect. 2.5.1. The connection is described as follows with respect to the actions of the complex vector fields:

$$\widetilde{\nabla}_{\overline{w}}e=e\cdot\Big(\frac{\ell}{pT}\frac{t}{2}\overline{w}^{-1}\Big),\quad \widetilde{\nabla}_{w}e=e\cdot\Big(-\frac{\ell}{pT}\frac{t}{2}w^{-1}\Big),$$

$$\widetilde{\nabla}_{z}e=e\cdot\Big(\frac{\sqrt{-1}}{2}\frac{\ell}{pT}\log|w|\Big),\quad \widetilde{\nabla}_{\overline{z}}e=e\cdot\Big(\frac{\sqrt{-1}}{2}\frac{\ell}{pT}\log|w|\Big).$$

Let us look at the holomorphic bundle on $\widetilde{Y}_p^{\lambda\,\mathrm{cov}*}$ underlying $(\widetilde{E},\widetilde{h},\widetilde{\nabla})$. The actions of $\widetilde{\nabla}_{\overline{\xi}}$ and $\widetilde{\nabla}_{\overline{\eta}}$ are given as follows:

$$\widetilde{\nabla}_{\overline{\xi}}e=e\cdot\frac{1}{1+|\lambda|^2}\frac{\ell}{pT}\Big(\frac{\sqrt{-1}}{2}\log|w|-\frac{\lambda t}{2}w^{-1}\Big),$$

$$\widetilde{\nabla}_{\overline{\eta}}e=e\cdot\frac{1}{1+|\lambda|^2}\frac{\ell}{pT}\Big(\frac{t}{2}\overline{w}^{-1}-\frac{1}{2}\lambda\sqrt{-1}\log|w|\Big).$$

In particular, we obtain

$$\widetilde{\nabla}_{\overline{\eta}}e=e\cdot\frac{1}{1+|\lambda|^2}\frac{\ell}{pT}\times$$
$$\Big(\frac{1}{4\sqrt{-1}}\frac{\xi-\overline{\xi}+\overline{\lambda}\eta+|\lambda|^2\xi}{\eta+\overline{\lambda}\xi}-\frac{\sqrt{-1}}{2}\lambda\log|\eta+\lambda\overline{\xi}|-\frac{\lambda}{4\sqrt{-1}}+\frac{\sqrt{-1}}{2}\lambda\log(1+|\lambda|^2)\Big).\tag{5.1}$$

5.1.6 C^∞-Frame

We put as follows:

$$\widetilde{v}^\lambda := e \exp\Big(-\frac{1}{2\sqrt{-1}}\frac{\ell}{pT}(\xi-\overline{\xi})\log|\overline{\eta}+\overline{\lambda}\xi|$$

$$+\frac{1}{1+|\lambda|^2}\frac{\ell}{pT}\Big(-\frac{\overline{\lambda}}{2\sqrt{-1}}(\eta+\lambda\overline{\xi})\log|\overline{\eta}+\overline{\lambda}\xi| - \frac{\lambda}{2\sqrt{-1}}(\overline{\eta}+\overline{\lambda}\xi)\log|\overline{\eta}+\overline{\lambda}\xi|$$

$$-\frac{\lambda\sqrt{-1}}{2}\big(1+\log(1+|\lambda|^2)\big)(\overline{\eta}+\overline{\lambda}\xi) - \frac{\overline{\lambda}\sqrt{-1}}{2}\big(1+\log(1+|\lambda|^2)\big)(\eta+\lambda\overline{\xi})\Big)\Big). \tag{5.2}$$

Then, the following holds.

- $\widetilde{\nabla}_{\overline{\eta}}\widetilde{v}^\lambda = 0$.
- $\widetilde{v}^\lambda$ is invariant with respect to the $\mathbb{R}$-action given by $s\cdot(z,w)=(z+s,w)$ in terms of the coordinate system (z,w), or equivalently $s\cdot(\xi,\eta)=(\xi+s,\eta-\lambda s)$ in terms of the coordinate system (ξ,η).
- $|\widetilde{v}^\lambda|_h = \exp\big(-\frac{1}{2\sqrt{-1}}(\xi-\overline{\xi})\frac{\ell}{pT}\log|\overline{\eta}+\overline{\lambda}\xi|\big)$.

Under the identification $\kappa^*_{p,1}(e)=|w_p|^\ell w_p^{-\ell}e$, we obtain

$$\kappa^*_{p,1}(\widetilde{v}^\lambda)=\widetilde{v}^\lambda w_p^{-\ell}(1+|\lambda|^2)^{-\ell/p}.$$

We also obtain the following equality:

$$\widetilde{\nabla}_{\overline{\xi}}\widetilde{v}^\lambda = \widetilde{v}^\lambda\frac{\ell}{pT}\Big(\frac{-1}{2\sqrt{-1}}\frac{\lambda}{\eta+\lambda\xi-\lambda(\xi-\overline{\xi})}(\xi-\overline{\xi}) - \frac{\sqrt{-1}}{2}\log(1+|\lambda|^2)\Big)$$

In terms of the local coordinate system (α_1,β_1) given in Sect. 2.7.6, we obtain the following:

$$\widetilde{\nabla}_{\overline{\beta}_1}\widetilde{v}^\lambda = 0,$$

$$\widetilde{\nabla}_{\overline{\alpha}_1}\widetilde{v}^\lambda = \widetilde{v}^\lambda\frac{\ell}{pT}\Big(\frac{-1}{2\sqrt{-1}}\frac{\lambda(\alpha_1-\overline{\alpha}_1)}{\beta_1-\lambda(\alpha_1-\overline{\alpha}_1)} - \frac{\sqrt{-1}}{2}\log(1+|\lambda|^2)\Big).$$

We set $t_1=\operatorname{Im}\alpha_1$ as in Sect. 2.7.6. Because $\widetilde{v}^\lambda$ is $\mathbb{R}_s$-invariant, we obtain

$$\widetilde{\nabla}^\lambda_{t_1}\widetilde{v}^\lambda = \frac{2}{\sqrt{-1}}\widetilde{\nabla}_{\overline{\alpha}_1}\widetilde{v}^\lambda = \widetilde{v}^\lambda\frac{\ell}{pT}\Big(\frac{\beta_1}{\beta_1-2\sqrt{-1}\lambda t_1} - 1 - \log(1+|\lambda|^2)\Big).$$

5.1.7 Proof of Proposition 5.1.4

We take a large $R > 0$. We consider the open subset $\mathcal{U}_p^\lambda(R) \subset Y_p^{\lambda\,\mathrm{cov}*}$ determined by the condition:

$$\mathcal{U}_p^\lambda(R) := \left\{ \left|\beta_1 - 2\sqrt{-1}\lambda t_1\right| < R,\ \frac{|2\lambda(t_1+T)|}{\left|\beta_1 - 2\sqrt{-1}\lambda t_1\right|} < \frac{1}{2},\ \frac{|2\lambda t_1|}{\left|\beta_1 - 2\sqrt{-1}\lambda t_1\right|} < \frac{1}{2} \right\}. \tag{5.3}$$

Note that we have the well defined function $w_p = (\beta_1 - 2\sqrt{-1}t_1)^{1/p}$ on $Y_p^{\lambda\,\mathrm{cov}*}$. Hence, we have the well defined function

$$(\beta_1 - 2\sqrt{-1}\lambda T)^{1/p} = (\beta_1 - 2\sqrt{-1}t_1)^{1/p}\Big(1 + \frac{2\sqrt{-1}\lambda(t_1+T)}{\beta_1 - 2\sqrt{-1}\lambda t_1}\Big)^{1/p}$$

on $\mathcal{U}_p^\lambda(R)$. We also have the well defined branch of the function

$$-\log\Big(1 + \frac{2\sqrt{-1}\lambda t_1}{\beta_1 - 2\sqrt{-1}\lambda t_1}\Big),$$

which we also denote by

$$\log\Big(1 - \frac{2\sqrt{-1}t_1}{\beta_1}\Big).$$

Because $\widetilde{v}^\lambda$ is $\mathbb{R}_s$-invariant, we obtain a C^∞-frame v^λ of $\mathbb{L}_p^{\lambda\,\mathrm{cov}*}(\ell)$. On $\mathcal{U}_p^\lambda(R)$, we consider the following section:

$$u_{p,\ell}^\lambda := v^\lambda \cdot \exp\Big[\frac{\ell}{pT}\Big(t_1\log(1+|\lambda|^2) + \frac{\beta_1}{2\sqrt{-1}\lambda}\log\Big(1 - \frac{2\sqrt{-1}\lambda}{\beta_1}t_1\Big) + t_1\Big)\Big].$$

Then, we have $\widetilde{\nabla}_{\overline{\beta}_1}^\lambda u_{p,\ell}^\lambda = 0$ and $\widetilde{\nabla}_{t_1}^\lambda u_{p,\ell}^\lambda = 0$. As for the norm, we have

$$\left|u_{p,\ell}^\lambda\right|_h = \exp\Big[-t_1\frac{\ell}{pT}\log|w| + \frac{\ell}{pT}\mathrm{Re}\Big(\frac{\beta_1}{2\sqrt{-1}\lambda}\log(1 - 2\sqrt{-1}\lambda t_1/\beta_1) + t_1\Big)\Big].$$

Note that

$$\frac{\beta_1}{2\sqrt{-1}\lambda}\log\Big(1 - \frac{2\sqrt{-1}\lambda}{\beta_1}t_1\Big) + t_1 = \frac{\beta_1}{2\sqrt{-1}\lambda} \times O\Big(\Big(t_1\frac{2\sqrt{-1}\lambda}{\beta_1}\Big)^2\Big).$$

Hence, we obtain **(A1)**.

Let us check **(A2)**. We have the following:

$$\Phi^*(u^{\lambda}_{p,\ell}) = u^{\lambda}_{p,\ell} w_p^{-\ell}(1+|\lambda^2|)^{-\ell/p}\times$$
$$\exp\Big(\frac{\ell}{pT}\Big((t_1+T)\log(1+|\lambda|^2)+\frac{\beta_1+2\sqrt{-1}\lambda T}{2\sqrt{-1}\lambda}\log\Big(\frac{\beta_1-2\sqrt{-1}\lambda t_1}{\beta_1+2\sqrt{-1}\lambda T}\Big)+t_1+T\Big)\Big)\times$$
$$\exp\Big(-\frac{\ell}{pT}\Big(t_1\log(1+|\lambda|^2)+\frac{\beta_1}{2\sqrt{-1}\lambda}\log\Big(\frac{\beta_1-2\sqrt{-1}\lambda t_1}{\beta_1}\Big)+t_1\Big)\Big). \tag{5.4}$$

We have

$$\frac{\ell}{pT}(t_1+T)\log(1+|\lambda|^2)-\frac{\ell}{pT}t_1\log(1+|\lambda|^2)=\frac{\ell}{p}\log(1+|\lambda|^2).$$

We also have

$$\frac{\ell}{pT}\Big(\frac{\beta_1}{2\sqrt{-1}\lambda}\log\Big(\frac{\beta_1-2\sqrt{-1}\lambda t_1}{\beta_1+2\sqrt{-1}\lambda T}\Big)+T\log\Big(\frac{\beta_1-2\sqrt{-1}\lambda t_1}{\beta+2\sqrt{-1}\lambda T}\Big)$$
$$+t_1+T-\frac{\beta_1}{2\sqrt{-1}\lambda}\log\Big(\frac{\beta_1-2\sqrt{-1}\lambda t_1}{\beta_1}\Big)-t_1\Big)$$
$$=\frac{\ell}{pT}\Big(\frac{\beta_1}{2\sqrt{-1}\lambda}\log\Big(\frac{\beta_1}{\beta_1+2\sqrt{-1}\lambda T}\Big)+T+T\log\Big(\frac{\beta_1-2\sqrt{-1}\lambda t_1}{\beta_1+2\sqrt{-1}\lambda T}\Big)\Big)$$
$$=\frac{\ell}{pT}\Big(-\frac{\beta_1}{2\sqrt{-1}\lambda}\log\Big(1+\frac{2\sqrt{-1}\lambda}{\beta_1}T\Big)+T\Big)+\frac{\ell}{p}\cdot\log\Big(\frac{\beta_1-2\sqrt{-1}\lambda t_1}{\beta_1+2\sqrt{-1}\lambda T}\Big). \tag{5.5}$$

Hence, we have

$$\Phi^*(u^{\lambda}_{p,\ell}) = u^{\lambda}_{p,\ell}(1+|\lambda|^2)^{\ell/p}(\beta_1+2\sqrt{-1}\lambda T)^{-\ell/p}\times$$
$$\exp\Big(\frac{\ell}{pT}\Big(\frac{-\beta_1}{2\sqrt{-1}\lambda}\log\Big(1+\frac{2\sqrt{-1}\lambda}{\beta_1}T\Big)+T\Big)\Big). \tag{5.6}$$

Then, the claim of the lemma follows. ■

5.2 Examples of Monopoles Induced by wild Harmonic Bundles

Let w_p be a p-th root of the variable w as in Sect. 4.3.1. We take $\mathfrak{a}(w_p) \in \mathbb{C}[w_p]$ such that $\deg_{w_p} \mathfrak{a} \le p$ and $\mathfrak{a}(0) = 0$. We obtain the harmonic bundle on $\mathbb{C}^*_{w_p}$ given as $\mathcal{L}^*(\mathfrak{a}) = \mathcal{O}_{\mathbb{C}^*_{w_p}} e_0$ with the Higgs field $\theta_0 = d\mathfrak{a}$ and the metric $h_0(e_0, e_0) = 1$. As explained in Sect. 2.6.1, there exists the monopole $(\mathcal{L}^*(\mathfrak{a}), h, \nabla, \phi)$ on Y_p^{0*} induced by $(\mathcal{L}^*(\mathfrak{a}), \theta_0, h_0)$. Let us describe the monopole explicitly.

Let $\Psi_p : Y_p^{0*} \longrightarrow \mathbb{C}^*_{w_p}$ denote the projection as in Sect. 4.3.1, i.e., the map induced by $(t, w_p) \longmapsto w_p$. We have $\mathcal{L}^*(\mathfrak{a}) = \Psi_p^{-1}\mathcal{L}^*(\mathfrak{a})$ as C^∞-bundle with the induced metric $h = \Psi_p^{-1}(h_0)$. We set $e := \Psi_p^{-1}(e_0)$. The unitary connection and the Higgs field are given as follows:

$$\nabla e = e\Big(-\sqrt{-1}\big(\partial_w \mathfrak{a} + \overline{\partial_w \mathfrak{a}}\big)\, dt\Big), \qquad \phi e = e\big(\partial_w \mathfrak{a} - \overline{\partial_w \mathfrak{a}}\big).$$

Here, $\partial_w \mathfrak{a} = p^{-1} w_p^{1-p} \partial_{w_p} \mathfrak{a}$. Note that $\partial_w \mathfrak{a}$ is of the form $\sum_{j=0}^{p-1} a_j w_p^{-j}$, i.e., it is an element of the set $S(p)$. (See Sect. 3.3.2.)

5.2.1 The Underlying Mini-holomorphic Bundle at $\lambda = 0$

Let us describe the mini-holomorphic bundle $\mathcal{L}^{0*}(\mathfrak{a})$ on Y_p^{0*} underlying the monopole $(\mathcal{L}^*(\mathfrak{a}), h, \nabla, \phi)$. Recall that $\varpi_p^0 : Y_p^{0\,\mathrm{cov}} \longrightarrow Y_p^0$ denotes the projection (see Sect. 4.3.1). The restriction $Y_p^{0\,\mathrm{cov}*} \longrightarrow Y_p^{0*}$ is also denoted by ϖ_p^0.

Proposition 5.2.1 *There exists a mini-holomorphic frame $u_\mathfrak{a}^0$ of $(\varpi_p^0)^*\mathcal{L}^{0*}(\mathfrak{a})$ on $Y_p^{0\,\mathrm{cov}*}$ satisfying the following:*

(A1) *For any $P \in H^0_{\infty,p}$, there exists a neighbourhood U_P of P in $Y_p^{0\,\mathrm{cov}}$ and a constant $C(P) \geq 1$ such that*

$$C(P)^{-1} \leq |u_\mathfrak{a}^0|_h \leq C(P).$$

(A2) $\kappa^*_{p,1} u_\mathfrak{a}^0 = u_\mathfrak{a}^0 \exp\big(2\sqrt{-1}T\partial_w \mathfrak{a}\big)$.

Proof We have $\partial_t e = e\big(-2\sqrt{-1}\partial_w \mathfrak{a}\big)$ and $\partial_{\overline{w}} e = 0$. Hence, we have the following mini-holomorphic frame of $(\varpi_p^0)^*\mathcal{L}^{0*}(\mathfrak{a})$ on $Y_p^{0\,\mathrm{cov}*}$:

$$u_\mathfrak{a}^0 := (\varpi_p^0)^{-1}(e) \cdot \exp\big(2\sqrt{-1}t\partial_w \mathfrak{a}\big).$$

It has the desired properties. ∎

5.2.2 The Associated λ-Connection and the Induced Mini-holomorphic Bundle at λ

The λ-flat bundle associated with the harmonic bundle $(\mathcal{L}^*(\mathfrak{a}), \theta_0, h_0)$ is described as follows:

$$(\overline{\partial} + \lambda\theta_0^\dagger)e_0 = e_0\,\lambda d\overline{\mathfrak{a}}, \qquad (\lambda\partial + \theta)e_0 = e_0 d\mathfrak{a}.$$

We set $v_0^\lambda = e_0 \exp(-\lambda\overline{\mathfrak{a}} + \overline{\lambda}\mathfrak{a})$ on $\mathbb{C}^*_{w_p}$. By the construction, it is holomorphic with respect to $\overline{\partial} + \lambda\theta^\dagger$. We also have

$$\mathbb{D}^\lambda v_0^\lambda = (\lambda\partial + \theta)v_0^\lambda = v_0^\lambda(1 + |\lambda|^2)d\mathfrak{a}.$$

Hence, we obtain $\mathbb{D}^\lambda_w v_0^\lambda = v_0^\lambda(1 + |\lambda|^2)\partial_w\mathfrak{a}$.

Let us describe the mini-holomorphic bundle $\mathcal{L}^{\lambda*}(\mathfrak{a})$ on the mini-complex manifold $Y_p^{\lambda*}$ underlying the monopole $(\mathcal{L}^*(\mathfrak{a}), h, \nabla, \phi)$. Recall that $\varpi_p^\lambda : Y_p^{\lambda\,\mathrm{cov}} \longrightarrow Y_p^\lambda$ denote the projection (see Sect. 4.3.2). The restriction $Y_p^{\lambda\,\mathrm{cov}*} \longrightarrow Y_p^{\lambda*}$ is also denoted by ϖ_p^λ. Let us emphasize that $Y_p^{\lambda\,\mathrm{cov}*} = Y_p^{0\,\mathrm{cov}*}$ and $Y_p^{\lambda*} = Y_p^{0*}$ as Riemannian manifolds, and that the restrictions of ϖ_p^λ and ϖ_p^0 to $Y_p^{\lambda\,\mathrm{cov}*} = Y_p^{0\,\mathrm{cov}*}$ are the same.

Proposition 5.2.2 *There exists a mini-holomorphic frame $u_\mathfrak{a}^\lambda$ of $(\varpi_p^\lambda)^{-1}\mathcal{L}^{\lambda*}(\mathfrak{a})$ such that the following holds:*

(A1) *For any $P \in H^{\lambda\,\mathrm{cov}}_{\infty,p}$, there exists a constant $C(P) \geq 1$ and a neighbourhood U_P of P in $Y_p^{\lambda\,\mathrm{cov}}$ such that the following holds on $U_P \setminus H^{\lambda\,\mathrm{cov}}_{\infty,p}$:*

$$C(P)^{-1} \leq \left|u_\mathfrak{a}^\lambda\right|_h \leq C(P).$$

(A2) *We obtain the following equality:*

$$\kappa^*_{p,1}u_\mathfrak{a}^\lambda = u_\mathfrak{a}^\lambda\cdot\exp\Big[(\lambda^{-1}+\overline{\lambda})\Big(\mathfrak{a}_{|w=(1+|\lambda|^2)^{-1}(\beta_1+2\sqrt{-1}\lambda T)}-\mathfrak{a}_{|w=(1+|\lambda|^2)^{-1}\beta_1}\Big)\Big].$$

Proof Set $v^\lambda := \Psi_p^{-1}(v_0^\lambda)$. By Corollary 2.8.4 with the formulas (2.16) and (2.17), we have the following:

$$\partial_{\overline{\beta}_1}v^\lambda = 0, \qquad \partial_{t_1}v^\lambda = v^\lambda\big(-2\sqrt{-1}\Psi_p^*(\partial_w\mathfrak{a})\big).$$

We remark the following:

$$\partial_{t_1}\Big(\mathfrak{a}_{|w=(1+|\lambda|^2)^{-1}(\beta_1-2\sqrt{-1}\lambda t_1)}\Big)=$$

$$\big(\partial_w\mathfrak{a}\big)_{|w=(1+|\lambda|^2)^{-1}(\beta_1-2\sqrt{-1}\lambda t_1)}\times\partial_{t_1}(\beta_1-2\sqrt{-1}\lambda t_1)$$

$$=\frac{-2\sqrt{-1}\lambda}{1+|\lambda|^2}\big(\partial_w\mathfrak{a}\big)_{|w=(1+|\lambda|^2)^{-1}(\beta_1-2\sqrt{-1}\lambda t_1)}. \tag{5.7}$$

Hence, we obtain the following mini-holomorphic frame $u^{\lambda}_{\mathfrak{a}}$ on $Y^{\lambda *}_p$:

$$u^{\lambda}_{\mathfrak{a}}:=(\varpi^{\lambda}_p)^{-1}(v^{\lambda})\times$$

$$\exp\Big(-(\lambda^{-1}+\overline{\lambda})\big(\mathfrak{a}_{|w=(1+|\lambda|^2)^{-1}(\beta_1-2\sqrt{-1}\lambda t_1)}-\mathfrak{a}_{|w=(1+|\lambda|^2)^{-1}\beta_1}\big)\Big). \tag{5.8}$$

We can easily check that $u^{\lambda}_{\mathfrak{a}}$ has the desired properties. ■

5.2.3 Special Case

Let us consider the case $\mathfrak{a}=\gamma w^p_p=\gamma w$. In this case, the associated unitary connection and the Higgs fields are given as follows:

$$\nabla e=e\Big(-\sqrt{-1}(\gamma+\overline{\gamma})\,dt\Big),\qquad \phi e=e\cdot(\gamma-\overline{\gamma}).$$

We obtain the mini-holomorphic bundle $\mathcal{L}^{\lambda *}(\gamma w)$ on $Y^{\lambda *}_p$ underlying the monopole $(\mathcal{L}^*(\gamma w),h,\nabla,\phi)$. Proposition 5.2.1 and Proposition 5.2.2 are restated as follows. (See also Sect. 8.2.)

Proposition 5.2.3 *There exists a mini-holomorphic frame $u^{\lambda}_{\gamma w}$ of $(\varpi^{\lambda}_p)^*\mathcal{L}^{\lambda *}(\gamma w)$ on $Y^{\lambda\,\mathrm{cov}\,*}_p$ satisfying the following:*

(A1) *We have $|u^{\lambda}_{\gamma w}|_h=\exp\big(-2\,\mathrm{Im}(\gamma)t_1\big)$. In particular, for any $P\in H^{\lambda}_{\infty,p}$, there exists a neighbourhood U_P of P in $Y^{\lambda\,\mathrm{cov}}_p$ and a constant $C(P)\geq 1$ such that $C(P)^{-1}\leq|u^{\lambda}_{\gamma w}|_h\leq C(P)$.*

(A2) $\kappa^*_{p,1}u^{\lambda}_{\gamma w}=u^{\lambda}_{\gamma w}\exp\big(2\sqrt{-1}T\gamma\big)$. ■

Let $\mathcal{L}^{\lambda}(\gamma w)$ and $\mathcal{P}_*\mathcal{L}^{\lambda}(\gamma w)$ denote the $\mathcal{O}_{Y^{\lambda}_p}(*H^{\lambda}_{\infty,p})$-module and the family of sheaves obtained from $\mathcal{L}^{\lambda *}(\gamma w)$ with h as in Sect. 4.4.1.

Corollary 5.2.4 *$\mathcal{L}^{\lambda}(\gamma w)$ is a locally free $\mathcal{O}_{Y^{\lambda}_p}(*H^{\lambda}_{\infty,p})$-module, and $\mathcal{P}_*\mathcal{L}^{\lambda}(\gamma w)$ is a good filtered bundle over $\mathcal{L}^{\lambda}(\gamma w)$ in the sense of Definition 4.3.3 and Definition 4.3.4. The norm estimate holds for $\mathcal{P}_*\mathcal{L}^{\lambda}(\gamma w)$ with the metric h.*

Moreover, $\mathcal{P}_*\mathcal{L}^{\lambda}(\gamma w)_{|\widehat{H}^{\lambda}_{\infty,p}}$ *is isomorphic to* $\mathcal{P}^{(0)}_*\widehat{\mathbb{L}}^{\lambda}_p(0,\alpha)$ *in* Sect. 3.5.3, *where* $\alpha := \exp(2\sqrt{-1}T\gamma)$. *(See Remark* 4.3.2 *for the comparison of* $\widehat{H}^{\lambda\,\mathrm{cov}}_{\infty,p}$ *and* $\widehat{H}^{\mathrm{cov}}_{\infty,p}$ *in* Sect. 3.4.2.*)* ■

5.2.4 Notation

For $\ell \in \mathbb{Z}$, let $(\mathbb{L}^*_p(\ell), h_{p,\ell}, \nabla_{p,\ell}, \phi_{p,\ell})$ be the monopole on Y^{0*}_p introduced in Sect. 5.1.1. For $\alpha \in \mathbb{C}^*$, we choose $\gamma \in \mathbb{C}$ such that $\alpha = \exp(2\sqrt{-1}T\gamma)$, and we obtain the monopole $(\mathcal{L}^*(\gamma w^p_p), h_\gamma, \nabla_\gamma, \phi_\gamma)$ as above. We obtain the following monopole:

$$(\mathbb{L}^*_p(\ell,\alpha), h_{\mathbb{L},p,\ell,\alpha}, \nabla_{\mathbb{L},p,\ell,\alpha}, \phi_{\mathbb{L},p,\ell,\alpha}) = (\mathbb{L}^*_p(\ell,0), h, \nabla, \phi) \otimes (\mathcal{L}^*(\gamma w^p_p), h_\gamma, \nabla_\gamma, \phi_\gamma). \tag{5.9}$$

For each λ, let $\mathbb{L}^{\lambda *}_p(\ell,\alpha)$ denote the mini-holomorphic bundle on $Y^{\lambda *}_p$ underlying the monopole (5.9). By the procedure in Sect. 4.4.1, it extends to an $\mathcal{O}_{Y^{\lambda}_p}(*H^{\lambda}_p)$-module $\mathbb{L}^{\lambda}_p(\ell,\alpha)$, and the filtered bundle $\mathcal{P}_*\mathbb{L}^{\lambda}_p(\ell,\alpha)$. Let $\mathbb{L}^{\lambda}_p(\ell)$ and $\mathcal{P}_*\mathbb{L}^{\lambda}_p(\ell)$ be as in Sects. 5.1.2–5.1.3. Let $\mathcal{L}^{\lambda}(\gamma w^p_p)$ and $\mathcal{P}_*\mathcal{L}^{\lambda}(\gamma w^p_p)$ be as in Sect. 5.2.3. There exist the following isomorphisms:

$$\mathbb{L}^{\lambda}_p(\ell,\alpha) \simeq \mathbb{L}^{\lambda}_p(\ell) \otimes \mathcal{L}^{\lambda}(\gamma w^p_p), \qquad \mathcal{P}_*\mathbb{L}^{\lambda}_p(\ell,\alpha) \simeq \mathcal{P}_*\mathbb{L}^{\lambda}_p(\ell) \otimes \mathcal{P}_*\mathcal{L}^{\lambda}(\gamma w^p_p).$$

By Corollary 5.1.3, Corollary 5.1.5 and Corollary 5.2.4, there exists an isomorphism

$$\mathcal{P}_*\mathbb{L}^{\lambda}_p(\ell,\alpha)_{|\widehat{H}^{\lambda}_{\infty,p}} \simeq \mathcal{P}^{(0)}_*\widehat{\mathbb{L}}^{\lambda}_p(\ell,\alpha). \tag{5.10}$$

(See Sect. 3.5.3 for $\mathcal{P}^{(0)}_*\widehat{\mathbb{L}}^{\lambda}_p(\ell,\alpha)$.) In particular, $\mathcal{P}_*\mathbb{L}^{\lambda}_p(\ell,\alpha)$ is a good filtered bundle. We also obtain the following lemma from Corollaries 5.1.3, 5.1.5 and 5.2.4.

Lemma 5.2.5 *The norm estimate holds for* $\mathcal{P}_*\mathbb{L}^{\lambda}_q(\ell,\alpha)$ *with* $h_{\mathbb{L},q,\ell,\alpha}$. ■

Recall $\varpi^{\lambda}_p : Y^{\lambda\,\mathrm{cov}}_p \longrightarrow Y^{\lambda}_p$ denote the projection. We set

$$\mathbb{L}^{\lambda\,\mathrm{cov}}_p(\ell,\alpha) := (\varpi^{\lambda}_p)^{-1}\mathbb{L}^{\lambda}_p(\ell,\alpha), \quad \mathcal{P}_*\mathbb{L}^{\lambda\,\mathrm{cov}}_p(\ell,\alpha) := (\varpi^{\lambda}_p)^{-1}\mathcal{P}_*\mathbb{L}^{\lambda}_p(\ell,\alpha).$$

We also set $\mathbb{L}^{\lambda\,\mathrm{cov}*}_p(\ell,\alpha) := (\varpi^{\lambda}_p)^{-1}(\mathbb{L}^{\lambda *}_p(\ell,\alpha))$. As the tensor product of the frames $u^{\lambda}_{p,\ell}$ in Propositions 5.1.2 and 5.1.4 and $u^{\lambda}_{\gamma w}$ in Proposition 5.2.3, we obtain the mini-holomorphic frame $u^{\lambda}_{p,\ell,\alpha}$ of $\mathbb{L}^{\lambda\,\mathrm{cov}}_p(\ell,\alpha)$ on a neighbourhood $\mathcal{U}^{\lambda}_p$ of $H^{\lambda\,\mathrm{cov}}_{\infty,p}$.

The $\mathbb{Z}$-action is described as

$$\kappa_{p,1}^{*}(u_{p,\ell,\alpha}^{\lambda}) = u_{p,\ell,\alpha}^{\lambda} \cdot \alpha(1+|\lambda|^2)^{\ell/p}(\beta_1 + 2\sqrt{-1}\lambda T)^{-\ell/p} \times \exp\Big(\frac{\ell}{p}G(2\sqrt{-1}\lambda T/\beta_1)\Big), \tag{5.11}$$

on $\kappa_{p,1}^{-1}(\mathcal{U}_p^{\lambda}) \cap \mathcal{U}_p^{\lambda}$, where $G(x) = 1 - x^{-1}\log(1+x)$. For any $P \in H_{\infty,p}^{\lambda\,\mathrm{cov}}$, and for any relatively compact neighbourhood U_P of P in $\mathcal{U}_p^{\lambda}$, there exists $C(P) > 1$ such that

$$C(P)^{-1}|w|^{-\ell t_1/(pT)} \leq |u_{p,\ell,\alpha}^{\lambda}|_{h_{\mathbb{L}^{\mathrm{cov}},p,\ell,\alpha}} \leq C(P)|w|^{-\ell t_1/(pT)}. \tag{5.12}$$

If $\lambda = 0$, we may choose $\mathcal{U}_p^0 = Y_p^{0\,\mathrm{cov}}$, and (5.11) is rewritten as $\kappa_{p,1}^{*}(u_{p,\ell,\alpha}^{0}) = \alpha w_p^{-\ell} u_{p,\ell,\alpha}$.

5.3 Examples of Monopoles Induced by Tame Harmonic Bundles (1)

Take $(a,\alpha) \in \mathbb{R}\times\mathbb{C}$. We have the harmonic bundle given as $\mathcal{L}(a,\alpha) = \mathcal{O}_{\mathbb{C}^*_{w_p}} e_0$ with the Higgs field $\theta_0 e_0 = e_0(-\alpha dw/w)$ and the metric h_0 determined by $h_0(e_0,e_0) = |w|^{2a}$. We have $\partial_w e_0 = e_0 \cdot a dw/w$ and $\theta_0^{\dagger} e_0 = e_0(-\overline{\alpha}\, d\overline{w}/\overline{w})$.

We obtain the monopole $(\mathcal{L}(a,\alpha), h, \nabla, \phi)$ on Y_p^{0*} induced by $(\mathcal{L}(a,\alpha), \theta_0, h_0)$ as in Sect. 2.6.1. We describe it explicitly.

We have $\mathcal{L}(a,\alpha) = \Psi_p^{-1}\mathcal{L}(a,\alpha)$, which is equipped with the induced metric $h = \Psi_p^{-1}(h_0)$. We set $e := \Psi_p^{-1}(e_0)$. The unitary connection ∇ and the Higgs field ϕ are given as follows:

$$\nabla e = e\big(aw^{-1}dw + \sqrt{-1}(\alpha w^{-1} + \overline{\alpha w}^{-1})\,dt\big), \qquad \phi e = e\big(-\alpha w^{-1} + \overline{\alpha w}^{-1}\big).$$

5.3.1 The Mini-holomorphic Bundle at λ = 0

Let us describe the mini-holomorphic bundle $\mathcal{L}^0(a,\alpha)$ on Y_p^{0*} underlying the monopole $(\mathcal{L}(a,\alpha), h, \nabla, \phi)$.

Proposition 5.3.1 *There exists a mini-holomorphic frame $u_{a,\alpha}^0$ of $(\varpi_p^0)^{-1}\mathcal{L}^0(a,\alpha)$ on $Y_p^{0\,\mathrm{cov}*}$ such that the following holds.*

(A1) *For any $P \in H^0_{\infty,p}$, there exists a neighbourhood U_P of P in $Y_p^{0\,\mathrm{cov}*}$ and a constant $C(P) \geq 1$ such that the following holds:*

$$C(P)^{-1}|w|^a \leq \left|u^0_{a,\alpha}\right|_h \leq C(P)|w|^a.$$

(A2) $\kappa^*_{p,1}u^0_{a,\alpha} = u^0_{a,\alpha}\exp\big(-2\sqrt{-1}\alpha T w^{-1}\big)$.

Proof We obtain $\nabla_{\overline{w}}e = 0$, $\nabla_w e = eaw^{-1}$, and $\nabla_t e = e\big(\sqrt{-1}(\alpha w^{-1} + \overline{\alpha}\overline{w}^{-1})\big)$. We also obtain

$$(\nabla_t - \sqrt{-1}\phi)e = e(2\sqrt{-1}\alpha w^{-1}).$$

Hence, we obtain the following mini-holomorphic frame of $(\varpi^0_p)^{-1}\mathcal{L}^0(a,\alpha)$:

$$u^0_{a,\alpha} := (\varpi^0_p)^{-1}(e)\exp\big(2\sqrt{-1}\alpha t w^{-1}\big).$$

It has the desired properties. ∎

5.3.2 The Mini-holomorphic Bundle at λ

The λ-flat bundle associated with the harmonic bundle $(\mathcal{L}(a,\alpha), \theta_0, h_0)$ is described as follows:

$$(\overline{\partial} + \lambda\theta_0^\dagger)e_0 = e_0\big(-\overline{\alpha}\lambda\big)d\overline{w}/\overline{w}, \qquad (\lambda\partial + \theta_0)e_0 = e_0\big(\lambda a - \alpha\big)dw/w.$$

We set $v_0^\lambda := e_0\exp\big(\overline{\alpha}\lambda\log|w|^2\big)$ on $\mathbb{C}^*_{w_p}$, which is holomorphic with respect to $\overline{\partial} + \lambda\theta_0^\dagger$. We obtain

$$\mathbb{D}^\lambda v_0^\lambda = v_0^\lambda\big(-\alpha + \lambda a + \lambda^2\overline{\alpha}\big)dw/w, \qquad \left|v_0^\lambda\right|_{h_0} = |w|^{a+2\,\mathrm{Re}(\overline{\alpha}\lambda)}.$$

Let us describe the mini-holomorphic bundle $\mathcal{L}^\lambda(a,\alpha)$ on $Y_p^{\lambda\,\mathrm{cov}*}$ underlying $(\mathcal{L}(a,\alpha), h, \nabla, \phi)$.

Proposition 5.3.2 *There exist a neighbourhood $\mathcal{U}_p^\lambda$ of $H^{\lambda\,\mathrm{cov}}_{\infty,p}$ in $Y_p^{\lambda\,\mathrm{cov}}$ and a mini-holomorphic frame $u^\lambda_{a,\alpha}$ of $(\varpi^\lambda_p)^{-1}\mathcal{L}^\lambda(a,\alpha)$ on $Y_p^{*\,\mathrm{cov}}$ such that the following holds.*

(A1) *For any $P \in H^{\lambda\,\mathrm{cov}}_{\infty,p}$, there exist a constant $C(P) \geq 1$ and a neighbourhood U_P of P in $\mathcal{U}_p^\lambda$ such that the following holds on $U_P \setminus H^{\lambda\,\mathrm{cov}}_{\infty,p}$:*

$$C(P)^{-1}|w|^{a+2\,\mathrm{Re}(\lambda\overline{\alpha})} \leq \left|u^\lambda_{a,\alpha}\right|_h \leq C(P)|w|^{a+2\,\mathrm{Re}(\lambda\overline{\alpha})}.$$

(A2) *On* $\left(\kappa_{p,1}^{-1}(\mathcal{U}_p^{\lambda})\cap\mathcal{U}_p^{\lambda}\right)\setminus H_{\infty,p}^{\lambda\,\mathrm{cov}}$, *the following holds:*

$$\kappa_{p,1}^{*}u_{a,\alpha}^{\lambda}=u_{a,\alpha}^{\lambda}\exp\Big(\big(-\alpha\lambda^{-1}+a+\lambda\overline{\alpha}\big)\log\big(\beta_1^{-1}(\beta_1+2\sqrt{-1}\lambda T)\big)\Big).$$

Proof We set $v^{\lambda}:=\Psi_p^{-1}(v_0^{\lambda})$. By Corollary 2.8.4 with the formulas (2.16) and (2.17), we obtain $\partial_{\overline{\beta}_1}v^{\lambda}=0$ and

$$\partial_{t_1}v^{\lambda}=v^{\lambda}\Big(-2\sqrt{-1}\frac{-\alpha+a\lambda+\lambda^2\overline{\alpha}}{\beta_1-2\sqrt{-1}\lambda t_1}\Big).$$

On an appropriate open neighbourhood $\mathcal{U}_p^{\lambda}$ of $H_{\infty,p}^{\lambda}$, we obtain the following mini-holomorphic frame on $\mathcal{U}_p^{\lambda}\setminus H_{\infty,p}^{\lambda\,\mathrm{cov}}$:

$$u_{a,\alpha}^{\lambda}=(\varpi_p^{\lambda})^{-1}(v^{\lambda})\exp\Big(-\lambda^{-1}\big(-\alpha+\lambda a+\lambda^2\overline{\alpha}\big)\log\big(\beta_1^{-1}(\beta_1-2\sqrt{-1}\lambda t_1)\big)\Big).$$

It has the desired properties. ∎

5.4 Examples of Monopoles Induced by Tame Harmonic Bundles (2)

5.4.1 Case of $\lambda = 0$

Let N be the $r\times r$ matrix whose (i,j)-entries are 1 $(i=j+1)$ or 0 $(i\neq j+1)$. Let $U_{w,p}^{*}:=\{w_p\in\mathbb{C}\mid |w_p^{p}|>R\}$ for some $R>1$. We set $V_N:=\bigoplus_{i=1}^{r}\mathcal{O}_{U_{w,p}^{*}}v_{0,i}$ on which we define the Higgs field θ_N by $\theta_N\boldsymbol{v}_0=\boldsymbol{v}_0N\,dw/w$. Let h_V be a harmonic metric of the Higgs bundle (V_N,θ_N) such that h_V is mutually bounded with the metric $h_{V,1}$ determined as follows:

$$h_{V,1}(v_i,v_j)=\begin{cases}(\log|w|)^{r+1-2i} & (i=j)\\ 0 & (i\neq j)\end{cases}$$

Note that such a harmonic metric h_V exists. (See [80] and [62, §6].)

We obtain the induced monopole (E_N,h,∇,ϕ) on $\Psi_p^{-1}(U_{w,p}^{*})\subset Y_p^{0*}$. Let us describe the underlying mini-holomorphic bundle $(E_N^0,\overline{\partial})$ on $\Psi_p^{-1}(U_{w,p}^{*})$.

Proposition 5.4.1 *There exist a neighbourhood* $\mathcal{U}_p^0$ *of* $H_{\infty,p}^{0\,\mathrm{cov}}$ *in* $Y_p^{0\,\mathrm{cov}}$ *and a mini-holomorphic frame* $\boldsymbol{u}_N^0=(u_{N,1}^0,\ldots,u_{N,r}^0)$ *of* $(\varpi_p^0)^{-1}(E_N^0,\overline{\partial})$ *on* $\mathcal{U}_p^0\setminus H_{\infty,p}^{0\,\mathrm{cov}}$ *satisfying the following conditions:*

(A1) *Let h_0 be the metric determined by*

$$h_0(u^0_{N,i}, u^0_{N,j}) = \begin{cases} (\log|w|)^{r+1-2i} & (i = j) \\ 0 & (i \neq j) \end{cases}$$

Then, h_0 and h are mutually bounded locally around any point of $H^{0\,\mathrm{cov}}_{\infty,p}$.

(A2) *On $(\kappa^{-1}_{p,1}(\mathcal{U}^0_p) \cap \mathcal{U}^0_p) \setminus H^{0\,\mathrm{cov}}_{\infty,p}$, we obtain $\kappa^*_{p,1}\boldsymbol{u}^0_N = \boldsymbol{u}^0_N \exp\bigl(2\sqrt{-1}NTw^{-1}\bigr)$.*

Proof Let $\boldsymbol{v}^0$ be the pull back $\Psi^{-1}_p(\boldsymbol{v}_0)$. We have $\partial_{E^0_N,\overline{w}}\boldsymbol{v}^0 = 0$ and $\partial_{E^0_N,t}\boldsymbol{v}^0 = \boldsymbol{v}^0(-2\sqrt{-1}Nw^{-1})$. We obtain the following frame:

$$\boldsymbol{u}^0_N = (\varpi^0_p)^{-1}(\boldsymbol{v}^0_N)\exp\Bigl(2\sqrt{-1}Ntw^{-1}\Bigr).$$

It has the desired property. ■

5.4.2 *Case of $\lambda \neq 0$*

We consider the λ-connection $\mathbb{D}^\lambda_N$ on V_N given by $\mathbb{D}^\lambda_N\boldsymbol{v}_0 = \boldsymbol{v}_0 N\,dw/w$. Let h_V be a harmonic metric of the λ-flat bundle $(V_N, \mathbb{D}^\lambda_N)$ such that h_V is mutually bounded with the metric $h_{V,1}$ determined as follows:

$$h_{V,1}(v_i, v_j) = \begin{cases} (\log|w|)^{r+1-2i} & (i = j) \\ 0 & (i \neq j) \end{cases}$$

Note that such a harmonic metric h_V exists (See [80] and [62, §6].) We obtain the induced monopole (E_N, h, ∇, ϕ) on $\Psi^{-1}_p(U^*_{w,p})$. Let us describe the underlying mini-holomorphic bundle $(E^\lambda_N, \overline{\partial})$ on $Y^{\lambda *}_p$.

Proposition 5.4.2 *There exist a neighbourhood $\mathcal{U}^\lambda_p$ of $H^{\lambda\,\mathrm{cov}}_{\infty,p}$ in $Y^{\lambda\,\mathrm{cov}}_p$ and a mini-holomorphic frame $\boldsymbol{u}^\lambda_N = (u^\lambda_{N,1}, \dots, u^\lambda_{N,r})$ of $(\varpi^\lambda_p)^{-1}(E^\lambda_N, \overline{\partial})$ on $\mathcal{U}^\lambda_p \setminus H^{\lambda\,\mathrm{cov}}_{\infty,p}$ with the following frame:*

(A1) *Let h_0 be the metric determined by*

$$h_0(u^\lambda_{N,i}, u^\lambda_{N,j}) = \begin{cases} (\log|w|)^{r+1-2i} & (i = j) \\ 0 & (i \neq j) \end{cases}$$

Then, h_0 and h are mutually bounded locally around any point of $H^{\lambda\,\mathrm{cov}}_{\infty,p}$.

(A2) *On $(\kappa^{-1}_{p,1}(\mathcal{U}^\lambda_p) \cap \mathcal{U}^\lambda_p) \setminus H^{\lambda\,\mathrm{cov}}_{\infty,p}$, we obtain*

$$\kappa^*_{p,1}\boldsymbol{u}^\lambda_N = \boldsymbol{u}^\lambda_N \exp\bigl(-\lambda^{-1}\log(\beta_1(\beta_1 + 2\sqrt{-1}\lambda T)^{-1})\bigr).$$

Proof Let $\boldsymbol{v}^{\lambda}$ be the pull back of $\boldsymbol{v}_0$. By the formulas (2.16) and (2.17), we have $\partial_{E_N^{\lambda},\overline{\beta}_1}\boldsymbol{v}^{\lambda} = 0$ and $\partial_{E_N^{\lambda},t_1}\boldsymbol{v}^{\lambda} = \boldsymbol{v}^{\lambda}(-2\sqrt{-1}N(\beta_1 - 2\sqrt{-1}\lambda t_1)^{-1})$. We obtain the following frame:

$$\boldsymbol{u}_N^{\lambda} = (\varpi_p^{\lambda})^{-1}(\boldsymbol{v}_N^{\lambda}) \exp\Big(- \lambda^{-1} N \log\big(\beta_1^{-1}(\beta_1 - 2\sqrt{-1}\lambda t_1)\big)\Big).$$

Then, it has the desired property. ■

Chapter 6
Asymptotic Behaviour of Periodic Monopoles Around Infinity

Abstract We shall explain that monopoles satisfying a mild asymptotic condition are close to a direct sum of tensor product of model ones in Sect. 5.

6.1 Outline of This Chapter

In Sect. 6.2, we introduce some notation for covering spaces and the decomposition of a mini-holomorphic bundle induced by the decomposition of the spectral curve.

In Sect. 6.3, first, we explain the main result of this section (Theorem 6.3.4), which is roughly the asymptotic orthogonality of the spectral decomposition in the slope level for a periodic monopole satisfying Condition 6.3.1 and Condition 6.3.2. Note that if Condition 6.3.1 is satisfied, and Condition 6.3.2 is also satisfied on an appropriate ramified covering. Then, with the aid of the estimate for asymptotic doubly periodic instantons in Sects. 10.2–10.3 developed in [65], we explain the second main result of this section (Proposition 6.3.8). As a result, with the aid of the estimates for asymptotic harmonic bundles in Sect. 10.1.1 and [65, §5.5], we may understand the asymptotic behaviour of periodic monopoles. We also note that Condition 6.3.1 is equivalent to the GCK-condition (Proposition 6.3.13).

The rest of this subsection is devoted to the proof of Theorem 6.3.4. In Sect. 6.4, we shall prove that the unitary connection asymptotically preserves an orthogonal decomposition associated with the Higgs field, under the assumption on the behaviour of the eigenvalues of the Higgs field. After the preliminary in Sects. 6.5–6.7, we shall prove that the assumption is satisfied and that the spectral decomposition is asymptotically orthogonal.

6.2 Preliminary

6.2.1 Notation

Set $U_w(R) := \{|w| > R\} \cup \{\infty\}$ in $\mathbb{P}^1_w$ for any $R > 0$. We set $U^*_w(R) := U_w(R) \setminus \{\infty\}$. We fix a q-th root w_q of w for any $q \in \mathbb{Z}_{\geq 1}$ such that $w_{qm}^m = w_q$ for any

T. Mochizuki, *Periodic Monopoles and Difference Modules*, Lecture Notes in Mathematics 2300, https://doi.org/10.1007/978-3-030-94500-8_6

$m \in \mathbb{Z}_{\geq 1}$. For any $q \in \mathbb{Z}_{\geq 1}$, let $\mathbb{P}^1_{w_q} \longrightarrow \mathbb{P}^1_w$ be the ramified covering map induced by $w_q \longmapsto w_q^q$. Let $U_{w,q}(R)$ and $U^*_{w,q}(R)$ denote the inverse image of $U_w(R)$ and $U^*_w(R)$, respectively. For any $p \in q\mathbb{Z}_{\geq 1}$, there exists the natural morphism $\varphi_{q,p} : U_{w,p}(R) \longrightarrow U_{w,q}(R)$.

Take $T > 0$. Let $\mathcal{B}^0_q(R)$ denote $(\mathbb{R}_t/T\mathbb{Z}) \times U_{w,q}(R)$ equipped with the natural mini-holomorphic structure. Similarly, we set $\mathcal{B}^{0*}_q(R) := (\mathbb{R}/T\mathbb{Z}) \times U^*_{w,q}(R) \subset \mathcal{B}^0_q(R)$, which is equipped with the metric $dt\,dt + dw\,d\overline{w}$. We also put $H^0_{\infty,q} := S^1_T \times \{\infty\} = \mathcal{B}^0_q(R) \setminus \mathcal{B}^{0*}_q(R)$.

The complex vector fields $q^{-1}w_q^{-q+1}\partial_{w_q}$ and $q^{-1}\overline{w}_q^{-q+1}\partial_{\overline{w}_q}$ on $U_{w,q}(R)$ and $\mathcal{B}^0_q(R)$ are denoted by ∂_w and $\partial_{\overline{w}}$, respectively. Similarly, the complex 1-forms $qw_q^{q-1}dw_q$ and $q\overline{w}_q^{q-1}d\overline{w}_q$ are denoted by dw and $d\overline{w}$. The function w_q^q is denoted by w.

We also set $\mathcal{B}^{0,\mathrm{cov}}_q(R) := \mathbb{R}_t \times U_{w,q}(R)$, $\mathcal{B}^{0,\mathrm{cov},*}_q(R) := \mathbb{R}_t \times U^*_{w,q}(R)$ and $H^{0,\mathrm{cov}}_q := \mathbb{R}_t \times \{\infty\}$. Let $\varpi^0_q : \mathcal{B}^{0\,\mathrm{cov}}_q(R) \longrightarrow \mathcal{B}^0_q(R)$ denote the map induced by $\mathbb{R} \longrightarrow \mathbb{R}/T\mathbb{Z}$. The restriction $\mathcal{B}^{0\,\mathrm{cov}*}_q(R) \longrightarrow \mathcal{B}^{0*}_q(R)$ is denoted by ϖ^0_q or simply ϖ_q.

Let κ be the free $\mathbb{Z}$-action on $\mathcal{B}^{0,\mathrm{cov}}_q(R)$ defined by $\kappa_n(t, w_q) = (t + nT, w_q)$ $(n \in \mathbb{Z})$. The quotient space is naturally identified with $\mathcal{B}^0_q(R)$.

By setting $(x, y) = (\mathrm{Re}(w), \mathrm{Im}(w)) = (\mathrm{Re}(w_q^q), \mathrm{Im}(w_q^q))$, we obtain the local coordinate systems (t, x, y) on $\mathcal{B}^{0*}_q(R)$. In particular, we obtain the vector fields ∂_a $(a = t, x, y)$, which are globally well defined.

Remark 6.2.1 The variables x and y are different from the variables x and y in Sects. 3–4. ■

For $R_1 > R$ and $p \in q\mathbb{Z}_{\geq 1}$, we have the naturally defined ramified coverings $\mathcal{R}_{q,p} : \mathcal{B}^0_p(R_1) \longrightarrow \mathcal{B}^0_q(R)$, and $\mathcal{R}_{q,p} : \mathcal{B}^{0\,\mathrm{cov}}_p(R_1) \longrightarrow \mathcal{B}^{0\,\mathrm{cov}}_q(R)$.

For a vector bundle E with a Hermitian metric h on $\mathcal{B}^{0*}_q(R)$ (resp. $U^*_{w,q}$) and for an $\mathrm{End}(E)$-valued differential form s on $\mathcal{B}^{0*}_q(R)$ (resp. $U^*_{w,q}$), let $|s|_h$ denote the function on $\mathcal{B}^{0*}_q(R)$ (resp. $U^*_{w,q}$) obtained as the point-wise norm of s with respect to h and the natural Riemannian metric of $\mathcal{B}^{0*}_q(R)$ (resp. $U^*_{w,q}$). We omit to denote the dependence on the metrics of the base spaces because they are fixed.

6.2.2 Decomposition of Holomorphic Bundles with an Automorphism

Let V be a locally free $\mathcal{O}_{U^*_{w,q}(R)}$-module of finite rank with an automorphism F. For each $Q \in U^*_{w,q}(R)$, let $\mathcal{S}p(F_{|Q})$ denote the set of the eigenvalues of $F_{|Q}$. Let $\mathcal{S}p(F) \subset U^*_{w,q}(R) \times \mathbb{C}^*$ denote the spectral curve given by $\mathcal{S}p(F) = \coprod_{Q \in U^*_{w,q}(R)} \mathcal{S}p(F_{|Q})$. It is a closed complex analytic curve in $U^*_{w,q}(R) \times \mathbb{C}^*$.

Lemma 6.2.2 *Suppose that the closure of $\mathcal{S}p(F)$ in $U_{w,q}(R) \times \mathbb{P}^1$ is complex analytic. Then, there exist $R_1 > R$, $p \in q\mathbb{Z}_{\geq 1}$, and a decomposition*

$$\varphi_{q,p}^{-1}(V, F) = \bigoplus_{(\ell,\alpha)\in\mathbb{Z}\times\mathbb{C}^*} (V_{p,\ell,\alpha}, F_{p,\ell,\alpha})$$

on $U_{w,p}(R_1)$, such that the following holds for some $C > 0$ and $\delta > 0$:

- *For any $Q \in U^*_{w,p}(R_1)$, any eigenvalue ρ_Q of $F_{p,\ell,\alpha|Q}$ satisfies*

$$\Big|\rho_Q \cdot \big(\alpha w_p(Q)^{-\ell}\big)^{-1} - 1\Big| \leq C|w_p(Q)|^{-\delta}.$$

Proof By replacing $U_{w,q}(R)$ with $U_{w,q}(R_2)$ for some large $R_2 > 0$, we may assume that the irreducible decomposition $\overline{\mathcal{S}p(F)} = \bigcup_{i\in\Lambda} \overline{\mathcal{S}p(F)_i}$ satisfies the following condition.

- The germs of $\overline{\mathcal{S}p(F)_i}$ at $\overline{\mathcal{S}p(F)_i} \cap (\{\infty\} \times \mathbb{P}^1)$ are also irreducible.

There exists the corresponding decomposition $(V, F) = \bigoplus_{i\in\Lambda}(V_i, F_i)$. Let $Z \longrightarrow \overline{\mathcal{S}p(F)}$ be the normalization. Let $Z = \coprod_{i\in\Lambda} Z_i$ denote the decomposition into the connected components.

The induced maps $a_i : Z_i \longrightarrow U_{w,q}(R)$ are coverings ramified at ∞. Let m_i be the ramification indexes. For each i, there exits a coordinate ζ_i of Z_i such that $a_i^* w_q = \zeta_i^{m_i}$. We set $Z_i^* = a_i^{-1}(U^*_{w,q}(R))$. The restriction $Z_i^* \longrightarrow U_{w,q}(R)$ is also denoted by a_i. Let $\upsilon_i : Z_i^* \longrightarrow \mathbb{C}^*$ be the map obtained as the composition of $Z_i^* \longrightarrow U^*_{w,q}(R) \times \mathbb{C}^* \longrightarrow \mathbb{C}^*$.

On Z_i^*, there exists an $\mathcal{O}_{Z_i^*}$-module with an automorphism $(\widetilde{V}_i, \widetilde{F}_i)$ such that $(V_i, F_i) = a_{i*}(\widetilde{V}_i, \widetilde{F})$, and that $\mathcal{S}p(\widetilde{F}_i) \subset Z_i^* \times \mathbb{C}^*$ is equal to the image of $\mathrm{id} \times \upsilon_i : Z_i^* \longrightarrow Z_i^* \times \mathbb{C}^*$. There exist an integer ℓ_i, a non-zero complex number γ_i and a holomorphic function $\mathfrak{b}_i(u)$ on a neighbourhood of $u = 0$ with $\mathfrak{b}_i(0) = 0$, such that $\upsilon_i(\zeta_i) = \gamma_i \zeta_i^{-\ell_i} \exp\big(2\sqrt{-1}T\,\mathfrak{b}_i(\zeta_i^{-1})\big)$. Then, we obtain the claim of the lemma. ■

6.2.3 *Some Basic Mini-Holomorphic Bundles on $\mathcal{B}_q^{0*}(R)$*

6.2.3.1 Basic Rank One Bundles

For $q \in \mathbb{Z}_{>0}$ and $(\ell, \alpha) \in \mathbb{Z} \times \mathbb{C}^*$, we introduced the monopole

$$(\mathbb{L}_q^*(\ell, \alpha), h_{\mathbb{L},q,\ell,\alpha}, \nabla_{\mathbb{L},q,\ell,\alpha}, \phi_{\mathbb{L},q,\ell,\alpha})$$

on Y_q^{0*} in (5.9). Let $\mathbb{L}_q^{0*}(\ell,\alpha)$ denote the mini-holomorphic bundle on Y_q^{0*} underlying the monopole. As explained in Sect. 5.2.4, it extends to an $\mathcal{O}_{Y_q^0}(*H^0_{\infty,q})$-module $\mathbb{L}_q^0(\ell,\alpha)$.

Let $\mathbb{L}_q^{0\,\mathrm{cov}}(\ell,\alpha)$ denote the pull back of $\mathbb{L}_q^0(\ell,\alpha)$ by $\varpi^0 : Y_q^{0\,\mathrm{cov}} \longrightarrow Y_q^0$. As explained in Sect. 5.2.4, it is equipped with a global mini-holomorphic frame $e_{q,\ell,\alpha}$, i.e., $\mathbb{L}_q^{0\,\mathrm{cov}}(\ell,\alpha) := \mathcal{O}_{Y_q^{0,\mathrm{cov}}}(*H^{0,\mathrm{cov}}_{\infty,q})\, e_{q,\ell,\alpha}$, for which $\mathbb{Z}$-action is described as follows:

$$\kappa_1^*(e_{q,\ell,\alpha}) = w_q^{-\ell}\alpha e_{q,\ell,\alpha}.$$

Let $\mathbb{L}_q^{0\,\mathrm{cov}\,*}(\ell,\alpha)$ denote the restriction of $\mathbb{L}_q^{0\,\mathrm{cov}}(\ell,\alpha)$ to $Y_q^{0\,\mathrm{cov}\,*}$, which is equal to the pull back of $\mathbb{L}_q^{0*}(\ell,\alpha)$. It is equipped with the metric $h_{\mathbb{L}^{\mathrm{cov}},q,\ell,\alpha}$ obtained as the pull back of $h_{\mathbb{L},q,\ell,\alpha}$, for which the following holds:

$$h_{\mathbb{L}^{\mathrm{cov}},q,\ell,\alpha}(e_{q,\ell,\alpha}, e_{q,\ell,\alpha}) = \left|\alpha w_q^{-\ell}\right|^{2t/T}.$$

The restriction of $(\mathbb{L}_q^{0*}(\ell,\alpha), h_{\mathbb{L},q,\ell,\alpha})$ to $\mathcal{B}_q^{0*}(R)$ is denoted by the same notation $(\mathbb{L}_q^{0*}(\ell,\alpha), h_{\mathbb{L},q,\ell,\alpha})$. Similarly, the restriction of $(\mathbb{L}_q^{0\,\mathrm{cov}\,*}(\ell,\alpha), h_{\mathbb{L}^{\mathrm{cov}},q,\ell,\alpha})$ to $\mathcal{B}_q^{0\,\mathrm{cov}\,*}(R)$ is also denoted by $(\mathbb{L}_q^{0\,\mathrm{cov}\,*}(\ell,\alpha), h_{\mathbb{L}^{\mathrm{cov}},q,\ell,\alpha})$.

6.2.3.2 Mini-Holomorphic Bundles Induced by Holomorphic Bundles with Automorphism

Let Ψ_q denote the projection $\mathcal{B}_q^{0*}(R) \longrightarrow U^*_{w,q}(R)$. Let V be a holomorphic vector bundle with an endomorphism g on $U^*_{w,q}$. We obtain the induced mini-holomorphic bundle $\Psi_q^*(V,g)$ on $\mathcal{B}_q^{0*}(R)$ as explained in Sect. 2.6.2.

6.2.4 Decomposition of Mini-Holomorphic Bundles

Let $\mathcal{V}$ be a locally free $\mathcal{O}_{\mathcal{B}_q^{0*}(R)}$-module of finite rank. We obtain the automorphism F of $\mathcal{V}_{|\{0\}\times U^*_{w,q}(R)}$ obtained as the monodromy along each loop $S^1_T \times \{w\}$ (see Sect. 2.6.3). We set $\mathcal{S}p(\mathcal{V}) := \mathcal{S}p(F)$.

Lemma 6.2.3 *Suppose that the closure of $\mathcal{S}p(\mathcal{V})$ in $U_{w,q}(R)\times\mathbb{P}^1$ is a complex analytic curve. Then, the following holds:*

- *There exist $R_1 > R$, $p \in q\mathbb{Z}_{\geq 1}$, a finite subset $S \subset \mathbb{Z} \times \mathbb{C}^*$, a tuple of locally free $\mathcal{O}_{U^*_{w,q}}$-module with an endomorphism $(V_{p,\ell,\alpha}, f_{p,\ell,\alpha})$ $((\ell, \alpha) \in S)$, and the following isomorphism on $\mathcal{B}^{0*}_p(R_1)$:*

$$\mathcal{R}^*_{q,p}\mathcal{V} \simeq \bigoplus_{(\ell,\alpha)\in S} \mathbb{L}^{0*}_p(\ell, \alpha) \otimes \Psi^*_p(V_{p,\ell,\alpha}, f_{p,\ell,\alpha}). \tag{6.1}$$

Proof Applying Lemma 6.2.2 to $V := \mathcal{V}_{|\{0\}\times U^*_{w,q}}$ with the automorphism F, we obtain a decomposition as in Lemma 6.2.2. On each Z^*_i, there exists the endomorphism $\widetilde{f}_i$ of $\widetilde{V}_i$ such that (i) $\exp(2\sqrt{-1}T\widetilde{f}_i) = \widetilde{F}_i$, (ii) $\mathcal{S}p(\widetilde{f}_i) \subset Z^*_i \times \mathbb{C}$ is equal to the image of $\mathfrak{b}_i : Z^*_i \longrightarrow \mathbb{C}$. Then, we obtain the claim of the lemma. ■

We identify the both sides of (6.1) by the isomorphism. By setting

$$\mathcal{V}_{p,\ell} := \bigoplus_{\alpha\in\mathbb{C}^*} \mathbb{L}^{0*}_p(\ell, \alpha) \otimes \Psi^*_p(V_{p,\ell,\alpha}, f_{p,\ell,\alpha}), \tag{6.2}$$

we obtain the following decomposition of mini-holomorphic bundles:

$$\mathcal{R}^*_{q,p}\mathcal{V} = \bigoplus_{\ell\in\mathbb{Z}} \mathcal{V}_{p,\ell}. \tag{6.3}$$

As a characterization of the decomposition (6.3), there exists $C > 1$ such that the following holds.

- For any eigenvalue ρ_Q of the monodromy of $\mathcal{V}_{p,\ell}$ along $S^1_T \times \{Q\}$, we obtain $C^{-1} < |\rho_Q / w_p(Q)^{-\ell}| < C$.

6.3 Estimate of Periodic Monopoles Around Infinity

6.3.1 Setting

We consider a monopole (E, h, ∇, ϕ) on $\mathcal{B}^{0*}_q(R)$ for some $R > 0$. Let $(E, \overline{\partial}_E)$ denote the underlying mini-holomorphic bundle on $\mathcal{B}^{0*}_q(R)$. We assume the following condition.

Condition 6.3.1

(B1) $\big|\nabla(\phi)_{|(t,w)}\big|_h \to 0$ *when* $|w| \to \infty$.

(B2) *The closure of* $\mathcal{S}p(E, \overline{\partial}_E)$ *in* $U_{w,q} \times \mathbb{P}^1$ *is a complex analytic curve.*

Moreover, we also assume the following condition.

Condition 6.3.2 *There exists a decomposition of mini-holomorphic bundles*

$$(E, \overline{\partial}_E) = \bigoplus_{(\ell,\alpha)\in\mathbb{Z}\times\mathbb{C}^*} (E_{\ell,\alpha}, \overline{\partial}_{E_{\ell,\alpha}}), \tag{6.4}$$

such that the following holds for some $C > 0$ and $\delta > 0$:

- *For any eigenvalue ρ_Q of the monodromy of $(E_{\ell,\alpha}, \overline{\partial}_{E_{\ell,\alpha}})$ along $S^1_T \times \{Q\}$, we have*

$$\bigl|\rho_Q \cdot (\alpha w_q(Q)^{-\ell})^{-1} - 1\bigr| \leq C|w_q(Q)|^{-\delta}.$$

Remark 6.3.3 Condition 6.3.2 implies **(B2)** in Condition 6.3.1. Conversely, if **(B2)** is satisfied, after the pull back by an appropriate covering map $\mathcal{R}_{q,p} : \mathcal{B}_p(R_1) \longrightarrow \mathcal{B}_q(R)$, Condition 6.3.2 is satisfied, according to Lemma 6.2.3. ■

We set $E_\ell := \bigoplus_\alpha E_{\ell,\alpha}$. We obtain the decomposition

$$(E, \overline{\partial}_E) = \bigoplus_{\ell\in\mathbb{Z}} (E_\ell, \overline{\partial}_{E_\ell}). \tag{6.5}$$

6.3.2 Asymptotic Orthogonality of the Decomposition (6.5)

We set $h^\circ_\ell := h_{|E_\ell}$. We obtain the metric $h^\circ := \bigoplus h^\circ_\ell$ on $E = \bigoplus E_\ell$. We set $I_q(w_q) := |w_q^q| \cdot \log |w_q^q|$. We shall prove the following theorem in Sects. 6.4–6.8.

Theorem 6.3.4 *Let s be the unique automorphism of E such that (i) $h^\circ = h \cdot s$, (ii) s is self-adjoint with respect to both h and h°. For any $m \in \mathbb{Z}_{\geq 0}$, there exist positive constants $A_i(m)$ $(i = 1, 2)$ such that*

$$\Bigl|\nabla_{\kappa_1} \circ \cdots \circ \nabla_{\kappa_m}(s - \mathrm{id}_E)\Bigr|_h \leq A_1(m) \exp\bigl(-A_2(m) \cdot I_q(w_q)\bigr)$$

for any $(\kappa_1, \ldots, \kappa_m) \in \{t, x, y\}^m$.

Remark 6.3.5 Theorem 6.3.4 and consequences below were originally prepared for the study of the Nahm transform of periodic monopoles [69]. ■

Let ∇° and ϕ° be the Chern connection and the Higgs field of $(E, \overline{\partial}_E, h^\circ)$ as in Sect. 2.4.2. Let $F(\nabla^\circ)$ denote the curvature of ∇°.

Corollary 6.3.6 *For any $m \in \mathbb{Z}_{\geq 0}$, there exist positive constants $A_i(m)$ $(i = 3, 4)$ such that*

$$\left|\nabla^{\circ}_{\kappa_1} \circ \cdots \circ \nabla^{\circ}_{\kappa_m}\big(\nabla - \nabla^{\circ}\big)\right|_{h^{\circ}} \leq A_3(m)\exp\big(-A_4(m)\cdot I_q(w_q)\big),$$

$$\left|\nabla^{\circ}_{\kappa_1} \circ \cdots \circ \nabla^{\circ}_{\kappa_m}(\phi - \phi^{\circ})\right|_{h^{\circ}} \leq A_3(m)\exp\big(-A_4(m)\cdot I_q(w_q)\big),$$

$$\left|\nabla^{\circ}_{\kappa_1} \circ \cdots \circ \nabla^{\circ}_{\kappa_m}\big(F(\nabla) - F(\nabla^{\circ})\big)\right|_{h^{\circ}} \leq A_3(m)\exp\big(-A_4(m)\cdot I_q(w_q)\big),$$

for any $(\kappa_1, \ldots, \kappa_m) \in \{t, x, y\}^m$. ∎

We obtain the following from Theorem 6.3.4.

Corollary 6.3.7 *For any $m \in \mathbb{Z}_{\geq 0}$, there exist positive constants $A_i(m)$ $(i = 5, 6)$ such that*

$$\left|\nabla^{\circ}_{\kappa_1} \circ \cdots \circ \nabla^{\circ}_{\kappa_m}\big(F(\nabla^{\circ}) - *\nabla^{\circ}\phi^{\circ}\big)\right|_{h^{\circ}} \leq A_5(m)\exp\big(-A_6(m)\cdot I_q(w_q)\big)$$

for any $(\kappa_1, \ldots, \kappa_m) \in \{t, x, y\}^m$. ∎

6.3.3 Eigen Decomposition in the Level 1

Let us look at the decomposition of mini-holomorphic bundles

$$(E_{\ell}, \overline{\partial}_{E_{\ell}}) = \bigoplus_{\alpha \in \mathbb{C}^*} (E_{\ell,\alpha}, \overline{\partial}_{E_{\ell,\alpha}}). \tag{6.6}$$

For each α, there exists a locally free $\mathcal{O}_{U^*_{w,q}}$-module $V_{\ell,\alpha}$ with an endomorphism $f_{\ell,\alpha}$ and an isomorphism of mini-holomorphic bundles:

$$E_{\ell,\alpha} \simeq \mathbb{L}^{0*}_q(\ell, \alpha) \otimes \Psi^*_q(V_{\ell,\alpha}, f_{\ell,\alpha}). \tag{6.7}$$

We impose that the eigenvalues of $f_{\ell,\alpha|w_q}$ goes to 0 as $|w_q| \to \infty$. Let $h'_{\ell,\alpha}$ denote the restriction of $h^{\circ}_{\ell} \otimes (h_{\mathbb{L},q,\ell,\alpha})^{-1}$ to $\Psi^*_q(V_{\ell,\alpha}, f_{\ell,\alpha})$. We have the Fourier expansion of $h'_{\ell,\alpha}$ along S^1_T:

$$h'_{\ell,\alpha} = \sum_{j \in \mathbb{Z}} \Psi_q^{-1}(h'_{V,\ell,\alpha;j})e^{2\pi\sqrt{-1}jt/T}.$$

Here, $h'_{V,\ell,\alpha,j}$ are sesqui-linear pairings $V_{\ell,\alpha} \otimes \overline{V_{\ell,\alpha}} \longrightarrow \mathbb{C}$. We set $h_{V,\ell,\alpha} := h'_{V,\ell,\alpha;0}$.

We set $h^{\sharp}_{\ell} := \bigoplus_{\alpha\in\mathbb{C}^*} h_{\mathbb{L},q,\ell,\alpha} \otimes \Psi_q^{-1}(h_{V,\ell,\alpha})$ on E_ℓ. We obtain the following as in the case of doubly periodic instantons [65].

Proposition 6.3.8 *Let s_ℓ be the unique endomorphism of E_ℓ determined by the condition $h^{\sharp}_{\ell} = h^{\circ}_{\ell} s_\ell$. For any $m \in \mathbb{Z}_{\geq 0}$, there exist $A_i(m) > 0$ $(i = 10, 11)$ such that the following holds for any $(\kappa_1, \ldots, \kappa_m) \in \{t, x, y\}^m$:*

$$\Big|\nabla^{\circ}_{\kappa_1} \circ \cdots \circ \nabla^{\circ}_{\kappa_m}\big(\mathrm{id}_{E_\ell} - s_\ell\big)\Big|_{h^{\circ}_{\ell}} \leq A_{10}(m) \cdot \exp\big(-A_{11}(m)|w_q^q|\big).$$

Proof We have only to compare $h^{-1}_{\mathbb{L},q,\ell,1} \otimes h^{\sharp}_{\ell}$ and $h^{-1}_{\mathbb{L},q,\ell,1} \otimes h^{\circ}_{\ell}$. Hence, it is enough to study the case $\ell = 0$. A similar problem was closely studied in the case of the doubly periodic instantons [65, Theorem 5.11, Theorem 5.17]. We shall explain an outline of the proof of a generalization of [65, Theorem 5.17] in Proposition 10.3.5 below. Here, we explain how the proof of Proposition 6.3.8 is reduced to Proposition 10.3.5.

We set $S^1_1 = \mathbb{R}/\mathbb{Z}$. Let $(\widetilde{E}_0, \overline{\partial}_{\widetilde{E}_0})$ be the holomorphic bundle with the metric $\widetilde{h}_0$ on $S^1_1 \times \mathcal{B}^{0*}_q(R)$ induced by the mini-holomorphic bundle $(E_0, \overline{\partial}_{E_0})$ with the metric h°_0 as explained in Sect. 2.5.2. Let $F(\widetilde{h}_0)$ denote the curvature of the Chern connection of $(\widetilde{E}_0, \overline{\partial}_{\widetilde{E}_0}, \widetilde{h}_0)$. By Corollary 6.3.7, there exists $C > 0$ such that

$$\Lambda F(\widetilde{h}_0) = O\big(\exp(-C I_q(w_q))\big).$$

By **(B1)** in Condition 6.3.1 and Corollary 6.3.7, we obtain $F(\widetilde{h}_0) \to 0$ as $|w_q| \to \infty$. Hence, $\widetilde{E}_{0|S^1_1 \times S^1_T \times \{w_q\}}$ are semistable of degree 0 if $|w_q|$ is sufficiently large, and we obtain the spectral curve $\Sigma_{\widetilde{E}_0}$ of $\widetilde{E}_0$. (See [65, §2.1].)

Let us prove that the closure of $\Sigma_{\widetilde{E}_0}$ in $U_{w,q} \times \mathcal{T}^{\vee}$ is a complex analytic curve. Let $S^1_1 \times \mathcal{B}^{0*}_q(R) \longrightarrow \mathcal{B}^{0*}_q(R)$ be the projection. Let s denote the standard local coordinates of $S^1_1 = \mathbb{R}/\mathbb{Z}$. By considering $z = s + \sqrt{-1}t$, we may regard $S^1_1 \times \mathcal{B}^{0*}_q(R)$ as the quotient of $\mathbb{C}_z \times U^*_{w,q}(R)$ by the lattice

$$\Gamma = \big\{n_1 + T\sqrt{-1}n_2 \,\big|\, n_1, n_2 \in \mathbb{Z}\big\} \subset \mathbb{C}_z.$$

The dual lattice $\Gamma^{\vee}$ is given as $\Gamma^{\vee} = \big\{T^{-1}\pi m_1 + \pi\sqrt{-1}m_2 \,\big|\, m_1, m_2 \in \mathbb{Z}\big\}$. Let $\mathcal{T}$ denote $S^1_1 \times S^1_T = \mathbb{C}_z/\Gamma$, and $\mathcal{T}^{\vee}$ denote the dual $\mathbb{C}_\zeta/\Gamma^{\vee}$. There exists the isomorphism $\mathbb{C}/2\pi\sqrt{-1}\mathbb{Z} \simeq \mathbb{C}^*$ induced by the exponential. We obtain the morphism

$$\mathbb{C}/2\pi\sqrt{-1}\mathbb{Z} \longrightarrow \mathcal{T}^{\vee} = \mathbb{C}/\Gamma^{\vee}$$

induced by $\gamma \longmapsto (2T)^{-1}\gamma$. It induces the following holomorphic map:

$$\mathcal{J} : \mathbb{C}^* \times U^*_{w,q} \longrightarrow \mathcal{T}^{\vee} \times U_{w,q}.$$

We can easily observe $\Sigma_{\widetilde{E}_0} = \mathcal{J}(\mathcal{S}p(E_0))$. Hence, the closure of $\Sigma_{\widetilde{E}_0}$ in $\mathcal{T}^{\vee} \times U_{w,q}$ is a complex analytic curve. Then, the claim of Proposition 6.3.8 follows from Proposition 10.3.5. ■

Let $\nabla^{\sharp}_{\ell}$ and $\phi^{\sharp}_{\ell}$ denote the Chern connection and the Higgs field of $(E_{\ell}, \overline{\partial}_{E_{\ell}}, h^{\sharp}_{\ell})$. We obtain a Hermitian metric $h^{\sharp} = \bigoplus h^{\sharp}_{\ell}$, a unitary connection $\nabla^{\sharp} = \bigoplus \nabla^{\sharp}_{\ell}$ and an anti-Hermitian endomorphism $\phi^{\sharp} = \bigoplus \phi^{\sharp}_{\ell}$ on E.

Corollary 6.3.9 *For any $m \in \mathbb{Z}_{\geq 0}$, there exist positive constants $A_i(m)$ ($i = 12, 13$) such that for any $(\kappa_1, \ldots, \kappa_m) \in \{t, x, y\}^m$:*

$$\Big|\nabla^{\circ}_{\kappa_1} \circ \cdots \circ \nabla^{\circ}_{\kappa_m}\big(\nabla^{\circ} - \nabla^{\sharp}\big)\Big|_{h^{\circ}} \leq A_{12}(m) \cdot \exp\big(-A_{13}(m)|w_q^q|\big).$$

$$\Big|\nabla^{\circ}_{\kappa_1} \circ \cdots \circ \nabla^{\circ}_{\kappa_m}\big(\phi^{\circ} - \phi^{\sharp}\big)\Big|_{h^{\circ}} \leq A_{12}(m) \cdot \exp\big(-A_{13}(m)|w_q^q|\big).$$

$$\Big|\nabla^{\circ}_{\kappa_1} \circ \cdots \circ \nabla^{\circ}_{\kappa_m}\big(F(\nabla^{\circ}) - F(\nabla^{\sharp})\big)\Big|_{h^{\circ}} \leq A_{12}(m) \cdot \exp\big(-A_{13}(m)|w_q^q|\big).$$

■

We obtain the following.

Corollary 6.3.10 *For any $m \in \mathbb{Z}_{\geq 0}$, we have positive constants $A_i(m)$ ($i = 14, 15$) such that the following holds for any $(\kappa_1, \ldots, \kappa_m) \in \{t, x, y\}^m$:*

$$\Big|\nabla^{\sharp}_{\kappa_1} \circ \cdots \circ \nabla^{\sharp}_{\kappa_m}\big(F(\nabla^{\sharp}) - *\nabla^{\sharp}\phi^{\sharp}\big)\Big|_{h^{\circ}} \leq A_{14}(m) \cdot \exp\big(-A_{15}(m)|w_q^q|\big).$$

■

6.3.4 Asymptotic Harmonic Bundles

We have the Higgs field

$$\theta_{\ell,\alpha} = f_{\ell,\alpha} \cdot (q w_q^{q-1} dw_q)$$

of $(V_{\ell,\alpha}, \overline{\partial}_{V_{\ell,\alpha}})$. Let $F(h_{V,\ell,\alpha})$ denote the curvature of the Chern connection $\nabla^{\ell,\alpha}$ of $(V_{\ell,\alpha}, \overline{\partial}_{V_{\ell,\alpha}}, h_{V,\ell,\alpha})$. Let $(\theta_{\ell,\alpha})^{\dagger}$ denote the adjoint of $\theta_{\ell,\alpha}$ with respect to $h_{V,\ell,\alpha}$. The following is a direct consequence of Corollary 6.3.10.

Corollary 6.3.11 *For any $m \in \mathbb{Z}_{\geq 1}$, we have $A_i(m)$ $(i = 20, 21)$ such that*

$$\Big|\nabla^{\ell,\alpha}_{\kappa_1} \circ \cdots \circ \nabla^{\ell,\alpha}_{\kappa_m}\Big(F(h_{V,\ell,\alpha}) + \big[\theta_{\ell,\alpha}, (\theta_{\ell,\alpha})^{\dagger}\big]\Big)\Big|_{h_{V,\ell,\alpha}} \leq A_{20}(m) \cdot \exp\big(-A_{21}(m)|w_q^q|\big) \tag{6.8}$$

for any $(\kappa_1, \ldots, \kappa_m) \in \{x, y\}^m$. ■

6.3.5 *Curvature*

We have the expression

$$F(\nabla) = F(\nabla)_{w,\overline{w}} dw\, d\overline{w} + F(\nabla)_{w,t} dw\, dt + F(\nabla)_{\overline{w},t} d\overline{w}\, dt.$$

Corollary 6.3.12 *The following estimates hold:*

$$\big|F(\nabla)_{w\overline{w}}\big|_h = O\Big(|w_q^q|^{-2}(\log|w_q|)^{-2}\Big),$$

$$\big|F(\nabla)_{wt}\big|_h = O(|w_q^q|^{-1}), \quad \big|F(\nabla)_{\overline{w}t}\big|_h = O(|w_q^q|^{-1}).$$

Proof Let $F(h'_{\ell,\alpha})$ be the curvature of the Chern connection of $E_{\ell,\alpha} \otimes \mathbb{L}_q^{0*}(\ell,\alpha)^{-1}$ with the metric $h'_{\ell,\alpha}$. We have the expression $F(h'_{\ell,\alpha}) = F(h'_{\ell,\alpha})_{w\overline{w}} dw\, d\overline{w} + F(h'_{\ell,\alpha})_{wt} dw\, dt + F(h'_{\ell,\alpha})_{\overline{w}t} d\overline{w}\, dt$. As in the case of doubly periodic instantons [65, Theorem 5.14], we obtain the following estimates for some $\delta > 0$ from Corollary 6.3.9, Corollary 6.3.11, and the estimates for asymptotic harmonic bundles (see [65, Theorem 5.14, §5.5]):

$$\Big|F(h'_{\ell,\alpha})_{w\overline{w}}\Big|_{h'_{\ell,\alpha}} = O\big(|w|^{-2}(\log|w|)^{-2}\big),$$

$$\Big|F(h'_{\ell,\alpha})_{wt}\Big|_{h'_{\ell,\alpha}} = O\big(|w|^{-1-\delta}\big), \quad \Big|F(h'_{\ell,\alpha})_{\overline{w}t}\Big|_{h'_{\ell,\alpha}} = O\big(|w|^{-1-\delta}\big).$$

Let $F(h_{\mathbb{L},q,\ell,\alpha})$ denote the curvature of $\mathbb{L}_q^{0*}(\ell,\alpha)$ with the metric $h_{\mathbb{L},q,\ell,\alpha}$. By an easy computation, we obtain the following:

$$F(h_{\mathbb{L},q,\ell,\alpha})_{w\overline{w}} = 0, \quad \big|F(h_{\mathbb{L},q,\ell,\alpha})_{wt}\big| = \big|F(h_{\mathbb{L},q,\ell,\alpha})_{\overline{w}t}\big| = O(|w|^{-1}).$$

Then, we obtain the claim of the lemma from Corollary 6.3.6. ■

6.3.6 *Another Equivalent Decay Condition*

Let (E, h, ∇, ϕ) be a monopole on $\mathcal{B}_q^{0*}(R)$ for some $R > 0$.

Proposition 6.3.13 *Condition* 6.3.1 *is satisfied for* (E, h, ∇, ϕ) *if and only if the following holds.*

(GCK) $|F(h)|_h \to 0$ $(|w_q| \to \infty)$ *and* $|\phi|_h = O\big(\log|w_q|\big)$.

Proof Suppose that **(GCK)** is satisfied. The condition **(B1)** follows from the Bogomolny equation and $|F(h)|_h \to 0$ $(|w_q| \to \infty)$. Let $(E, \overline{\partial}_E)$ denote the mini-holomorphic bundle on $\mathcal{B}_q^{0*}(R)$ underlying the monopole. Let $M(t, w_q)$ denote the holomorphic family of monodromy of $E_{|\{t\}\times U^*_{w,q}(R)}$. Because $|\phi|_h = O\big(\log|w_q|\big)$, we obtain the following estimate for some $N > 0$:

$$|M(t, w_q)|_h + |M(t, w_q)^{-1}|_h = O(|w_q|^N). \tag{6.9}$$

Let $P(w_q, x) := \det(x\,\mathrm{id} - M(t, w_q)) = \sum a_j(w_q)x^j$ denote the characteristic polynomials of the monodromy automorphisms. Then, $\mathcal{S}p(E, \overline{\partial}_E)$ is the zero set of $P(w_q, x)$. By (6.9), $a_j(w_q)$ are meromorphic at $w_q = \infty$. Hence, the closure of $\mathcal{S}p(E, \overline{\partial}_E)$ in $U_{w,q}(R) \times \mathbb{P}^1$ is also a complex analytic subset, i.e., **(B2)** is satisfied.

Suppose Condition 6.3.1 is satisfied. We obtain the estimate for the curvature by the Bogomolny equation. Let us study the growth order of $|\phi|_h$. By taking the pull back via an appropriate covering $\mathcal{R}_{q,p} : \mathcal{B}_p^{0*}(R_1) \longrightarrow \mathcal{B}_q^{0*}(R)$, we may assume that Condition 6.3.2 is also satisfied. The norm of the Higgs field of the monopole $\big(\mathbb{L}_q^{0*}(\ell, \alpha), h_{\mathbb{L},q,\ell,\alpha}\big)$ is $O\big(\log|w|\big)$ (see Sect. 5.1). There exists a constant $C_0 > 0$ such that any eigenvalue a of $f_{\ell,\alpha|Q}$ $(Q \in U^*_{w,q})$ satisfies $|a| < C_0$. Then, we obtain that $\big|f_{\ell,\alpha}\big|_{h_{V,\ell,\alpha}}$ are bounded according to Simpson's main estimate for asymptotic harmonic bundles. (See Proposition 10.1.2.) Hence, by the formula (2.6), the Higgs field of $\big(\Psi_q^*(V_{\ell,\alpha}, f_{\ell,\alpha}), \Psi_q^{-1}(h_{V,\ell,\alpha})\big)$ is bounded. Then, we obtain $|\phi|_h = O(\log|w|)$ from Corollary 6.3.6 and Corollary 6.3.9, i.e., **(GCK)** is satisfied. ■

6.4 Connections and Orthogonal Decompositions

6.4.1 *Statement*

We consider a monopole (E, h, ∇, ϕ) on $\mathcal{B}_q^{0*}(R_0)$ satisfying the following condition for some $R_0 > 0$ and $C_0 > 0$.

Condition 6.4.1

- $\big|\nabla(\phi)_{|(t,w_q)}\big|_h \to 0$ *when* $|w_q| \to \infty$.
- *There exist a finite subset* $\Lambda \subset \mathbb{Z}$ *and an orthogonal decomposition* $(E,\phi) = \bigoplus_{\ell\in\Lambda}(E^\bullet_\ell, \phi^\bullet_\ell)$ *such that any eigenvalue* α *of* $\phi^\bullet_{\ell|(t,w_q)}$ *satisfies*

$$\Big|\alpha - \sqrt{-1}\ell T^{-1}\log|w_q|\Big| \le C_0.$$

- *For any* $\ell_1, \ell_2 \in \Lambda$ $(\ell_1 \neq \ell_2)$, *we obtain* $|\ell_1 - \ell_2| \cdot T^{-1}\log R_0 > 10C_0$. ■

We obtain the decomposition $\nabla = \nabla^\bullet + \rho$, where $\nabla^\bullet$ is the unitary connection preserving the decomposition $E = \bigoplus E^\bullet_\ell$, and ρ is a section of

$$\bigoplus_{\ell_1\neq\ell_2} \mathrm{Hom}(E^\bullet_{\ell_1}, E^\bullet_{\ell_2}) \otimes \Omega^1.$$

The inner products of ρ and ∂_κ $(\kappa = t, x, y)$ are denoted by ρ_κ. We set $\mathrm{End}(E)^\bullet := \bigoplus \mathrm{End}(E^\bullet_\ell)$ and $\mathrm{End}(E)^\top = \bigoplus_{\ell_1\neq\ell_2}\mathrm{Hom}(E^\bullet_{\ell_1}, E^\bullet_{\ell_2})$. For any section s of $\mathrm{End}(E)\otimes\Omega^p$, let $s = s^\bullet + s^\top$ be the unique decomposition such that $s^\bullet$ and $s^\top$ are sections of $\mathrm{End}(E)^\bullet\otimes\Omega^p$ and $\mathrm{End}(E)^\top\otimes\Omega^p$, respectively. Note that $(\nabla\phi)^\top = [\rho, \phi]$.

We shall prove the following proposition in Sects. 6.4.2–6.4.4.

Proposition 6.4.2 *There exist positive constants* R_1 *and* C_i $(i = 1, 2)$ *depending only on* rank E, R_0, C_0 *and* Λ *such that* $\big|\rho\big|_h \le C_1\exp\big(-C_2 I_q(w_q)\big)$ *on* $\mathcal{B}^{0*}_q(R_1)$. *For any positive integer* k, *there exist positive constants* $C_i(k)$ $(i = 1, 2)$ *such that*

$$\big|\nabla^\bullet_{\kappa_1}\circ\cdots\circ\nabla^\bullet_{\kappa_k}\rho\big|_h \le C_1(k)\exp\big(-C_2(k)I_q(w_q)\big)$$

on $\mathcal{B}^{0*}_q(R_1)$ *for any* $(\kappa_1,\ldots,\kappa_k) \in \{t,x,y\}^k$.

We introduce a notation. Let f and g be functions on an open subset of $\mathcal{B}^{0*}_q(R)$ for some $R > R_0$. We say $f = O(g)$ if there exists a constant B depending only on rank E, C_0 and Λ such that $|f| \le B|g|$ on $\mathcal{B}^{0*}_q(R)$.

6.4.2 Preliminary

For any $\epsilon > 0$, there exists $R_{10}(\epsilon)$ such that $|\nabla\phi|_h < \epsilon$ on $\mathcal{B}^{0*}_q(R_{10}(\epsilon))$. If ϵ is sufficiently small, as remarked in Lemma 2.5.4, we obtain

$$\big|\nabla_{\kappa_1}\circ\cdots\circ\nabla_{\kappa_k}\phi\big|_h \le B(k)\epsilon$$

on $\mathcal{B}_q^{0*}(2R_{10}(\epsilon))$ for any $(\kappa_1, \ldots, \kappa_k) \in \{t, x, y\}^k$. Here, $B(k)$ are positive constants which are independent of ϵ.

Lemma 6.4.3 *We obtain* $|\rho_\kappa|_h = O\big(\big|(\nabla_\kappa\phi)^\top\big|\big)$ *for* $\kappa = t, x, y$. *We also obtain the following for any* $\kappa_1, \kappa_2 \in \{t, x, y\}$*:*

$$\big|\nabla^\bullet_{\kappa_1}\rho_{\kappa_2}\big|_h = O\Big(\epsilon\big|(\nabla_{\kappa_2}\phi)^\top\big|_h + \big|\nabla^\bullet_{\kappa_1}(\nabla_{\kappa_2}\phi)^\top\big|\Big).$$

Proof The first claim follows from $[\phi, \rho_\kappa] = -\nabla_\kappa(\phi)^\top$. We have the following equality:

$$\nabla^\bullet_{\kappa_1}\big((\nabla_{\kappa_2}\phi)^\top\big) = \nabla^\bullet_{\kappa_1}\big([\phi, \rho_{\kappa_2}]\big) = \big[\nabla^\bullet_{\kappa_1}\phi, \rho_{\kappa_2}\big] + \big[\phi, \nabla^\bullet_{\kappa_1}\rho_{\kappa_2}\big].$$

Then, the second claim follows. ■

We recall some equalities.

Lemma 6.4.4 *For* $(a, b, c) = (t, x, y), (x, y, t), (y, t, x)$, *we obtain*

$$(\nabla_x^2 + \nabla_y^2 + \nabla_t^2)(\nabla_a\phi) = 4\big[\nabla_b\phi, \nabla_c\phi\big] - \big[\phi, [\phi, \nabla_a\phi]\big].$$

Proof It is enough to consider the case $(a, b, c) = (t, x, y)$. We obtain

$$\begin{aligned}\nabla_x^2(\nabla_t\phi) &= \nabla_x(\nabla_t\nabla_x\phi + [F_{x,t}, \phi]) = \nabla_t\nabla_x^2\phi + [F_{x,t}, \nabla_x\phi] + \nabla_x[F_{x,t}, \phi]\\ &= \nabla_t\nabla_x^2\phi + 2[F_{x,t}, \nabla_x\phi] + [\nabla_x(F_{x,t}), \phi].\end{aligned} \tag{6.10}$$

Similarly, we also obtain $\nabla_y^2\nabla_t\phi = \nabla_t\nabla_y^2\phi + 2\big[F_{y,t}, \nabla_y\phi\big] + \big[\nabla_y(F_{y,t}), \phi\big]$. We have the following equality, which follows from the Bogomolny equation:

$$(\nabla_x^2 + \nabla_y^2 + \nabla_t^2)\phi = 0. \tag{6.11}$$

By using $*F = \nabla\phi$, we obtain the following:

$$\begin{aligned}&(\nabla_x^2 + \nabla_y^2 + \nabla_t^2)(\nabla_t\phi)\\ &= -2\big[\nabla_y\phi, \nabla_x\phi\big] + 2\big[\nabla_x\phi, \nabla_y\phi\big] + \big[-\nabla_x\nabla_y\phi + \nabla_y\nabla_x\phi, \phi\big]\\ &= 4\big[\nabla_x\phi, \nabla_y\phi\big] - \big[[F_{x,y}, \phi], \phi\big] = 4\big[\nabla_x\phi, \nabla_y\phi\big] + \big[[\phi, \nabla_t\phi], \phi\big]\\ &= 4\big[\nabla_x\phi, \nabla_y\phi\big] - \big[\phi, [\phi, \nabla_t\phi]\big].\end{aligned} \tag{6.12}$$

Thus, we obtain Lemma 6.4.4. ■

6.4.3 Step 1

We obtain the following equalities for any $\kappa_1, \kappa_2 \in \{t, x, y\}$:

$$\partial_{\kappa_1}^2 \big|(\nabla_{\kappa_2}\phi)^\top\big|_h^2 = 2\big|\nabla_{\kappa_1}(\nabla_{\kappa_2}\phi)^\top\big|_h^2 \\ + 2\,\mathrm{Re}\, h\big(\nabla_{\kappa_1}^2 \nabla_{\kappa_2}\phi, (\nabla_{\kappa_2}\phi)^\top\big) - 2\,\mathrm{Re}\, h\big(\nabla_{\kappa_1}^2 (\nabla_{\kappa_2}\phi)^\bullet, (\nabla_{\kappa_2}\phi)^\top\big). \tag{6.13}$$

Let us look at the last term in the right hand side of (6.13).

Lemma 6.4.5 *The following holds on $\mathcal{B}_q^{0*}(R_{10}(\epsilon))$:*

$$h\big(\nabla_{\kappa_1}^2 (\nabla_{\kappa_2}\phi)^\bullet, (\nabla_{\kappa_2}\phi)^\top\big) \\ = O\Big(\epsilon \cdot \Big(\big|(\nabla_{\kappa_1}\phi)^\top\big|_h + \big|\nabla_{\kappa_1}^\bullet (\nabla_{\kappa_1}\phi)^\top\big|_h\Big) \cdot \big|(\nabla_{\kappa_2}\phi)^\top\big|_h\Big). \tag{6.14}$$

Proof We have the following equality:

$$\Big(\nabla_{\kappa_1}^2 (\nabla_{\kappa_2}\phi)^\bullet\Big)^\top \\ = \Big(\big[\nabla_{\kappa_1}^\bullet \rho_{\kappa_1}, (\nabla_{\kappa_2}\phi)^\bullet\big] + 2\big[\rho_{\kappa_1}, \nabla_{\kappa_1}^\bullet (\nabla_{\kappa_2}\phi)^\bullet\big] + \Big[\rho_{\kappa_1}, \big[\rho_{\kappa_1}, (\nabla_{\kappa_2}\phi)^\bullet\big]\Big]\Big)^\top. \tag{6.15}$$

Note that $|(\nabla_{\kappa_2}\phi)^\bullet|_h = O(\epsilon)$. We also have $\nabla_{\kappa_1}^\bullet (\nabla_{\kappa_2}\phi)^\bullet = (\nabla_{\kappa_1}\nabla_{\kappa_2}\phi)^\bullet - [\rho_{\kappa_1}, [\rho_{\kappa_2}, \phi]]^\bullet$. Then, the claim follows from Lemma 6.4.3. ■

Let us look at the sum of the second terms in the right hand side of (6.13) for $\kappa_1 = t, x, y$.

Lemma 6.4.6 *We have the following estimate on $\mathcal{B}_q^{0*}(R_{10}(\epsilon))$:*

$$\sum_{\kappa_1=t,x,y} h\big(\nabla_{\kappa_1}^2 (\nabla_{\kappa_2}\phi), (\nabla_{\kappa_2}\phi)^\top\big) \\ = \Big|\big[\phi, (\nabla_{\kappa_2}\phi)^\top\big]\Big|_h^2 + O\Big(\epsilon \cdot \big|(\nabla\phi)^\top\big|_h \cdot \big|(\nabla_{\kappa_2}\phi)^\top\big|\Big). \tag{6.16}$$

Proof We explain the case $\kappa_2 = t$. By Lemma 6.4.4, we have the following equality:

$$\Big((\nabla_x^2 + \nabla_y^2 + \nabla_t^2)\nabla_t\phi\Big)^\top = \Big(4[\nabla_x\phi, \nabla_y\phi] - [\phi, [\phi, \nabla_t\phi]]\Big)^\top \\ = O\Big(\epsilon \cdot \Big(\big|(\nabla_x\phi)^\top\big|_h + \big|(\nabla_y\phi)^\top\big|_h\Big)\Big) - [\phi, [\phi, \nabla_t\phi]]^\top. \tag{6.17}$$

We have the following equality:

$$\begin{aligned}
&- h\big([\phi, [\phi, \nabla_t\phi]]^{\top}, (\nabla_t\phi)^{\top}\big) = -h\big([\phi, [\phi, \nabla_t\phi]], (\nabla_t\phi)^{\top}\big) \\
&= h\big([\phi, \nabla_t\phi], [\phi, (\nabla_t\phi)^{\top}]\big) = \big|[\phi, (\nabla_t\phi)^{\top}]\big|_h^2.
\end{aligned} \tag{6.18}$$

Then, we obtain the claim of the lemma. ■

Lemma 6.4.7 *There exist $R_{11} > 0$ and $C_{11} > 0$ such that the following inequality holds on $\mathcal{B}_q^{0*}(R_{11})$:*

$$-(\partial_x^2 + \partial_y^2 + \partial_t^2)\big|(\nabla\phi)^{\top}\big|_h^2 \leq -C_{11}\big|(\nabla\phi)^{\top}\big|_h^2 \cdot (\log|w|)^2.$$

Proof By the previous lemmas, we obtain the following estimate on $\mathcal{B}_q^{0*}(R_{10}(\epsilon))$:

$$\begin{aligned}
&- (\partial_x^2 + \partial_y^2 + \partial_t^2)\big|(\nabla_{\kappa_2}\phi)^{\top}\big|_h^2 \\
&= -2\big|\nabla\big((\nabla_{\kappa_2}\phi)^{\top}\big)\big|_h^2 - 2\big|[\phi, (\nabla_{\kappa_2}\phi)^{\top}]\big|_h^2 \\
&\quad + O\Big(\epsilon \cdot \big|(\nabla\phi)^{\top}\big|_h \cdot \big|(\nabla_{\kappa_2}\phi)^{\top}\big|_h\Big) \\
&\quad + O\Big(\epsilon \cdot \Big(\big|\nabla_t^{\bullet}(\nabla_t\phi)^{\top}\big|_h + \big|\nabla_x^{\bullet}(\nabla_x\phi)^{\top}\big|_h + \big|\nabla_y^{\bullet}(\nabla_y\phi)^{\top}\big|_h\Big) \cdot \big|(\nabla_{\kappa_2}\phi)^{\top}\big|_h\Big).
\end{aligned} \tag{6.19}$$

Because $|\rho|_h = O(\epsilon)$ on $\mathcal{B}_q^{0*}(R_{10}(\epsilon))$, it is equivalent to the following estimate:

$$\begin{aligned}
&- (\partial_x^2 + \partial_y^2 + \partial_t^2)\big|(\nabla_{\kappa_2}\phi)^{\top}\big|_h^2 \\
&= -2\big|\nabla\big((\nabla_{\kappa_2}\phi)^{\top}\big)\big|_h^2 - 2\big|[\phi, (\nabla_{\kappa_2}\phi)^{\top}]\big|_h^2 + O\Big(\epsilon \cdot \big|(\nabla\phi)^{\top}\big|_h \cdot \big|(\nabla_{\kappa_2}\phi)^{\top}\big|_h\Big) \\
&\quad + O\Big(\epsilon \cdot \Big(\big|\nabla_t(\nabla_t\phi)^{\top}\big|_h + \big|\nabla_x(\nabla_x\phi)^{\top}\big|_h + \big|\nabla_y(\nabla_y\phi)^{\top}\big|_h\Big) \cdot \big|(\nabla_{\kappa_2}\phi)^{\top}\big|_h\Big).
\end{aligned} \tag{6.20}$$

By taking the sum for $\kappa_2 = t, x, y$, we obtain the following estimate on $\mathcal{B}_q^{0*}(R_{10}(\epsilon))$:

$$\begin{aligned}
&- (\partial_x^2 + \partial_y^2 + \partial_t^2)\big|(\nabla\phi)^{\top}\big|_h^2 \\
&= - \sum_{\kappa_2 = t,x,y} 2\big|\nabla\big((\nabla_{\kappa_2}\phi)^{\top}\big)\big|_h^2 - 2\big|[\phi, (\nabla\phi)^{\top}]\big|_h^2 + O\Big(\epsilon \cdot \big|(\nabla\phi)^{\top}\big|_h^2\Big) \\
&\quad + O\Big(\epsilon \cdot \Big(\big|\nabla_t(\nabla_t\phi)^{\top}\big|_h + \big|\nabla_x(\nabla_x\phi)^{\top}\big|_h + \big|\nabla_y(\nabla_y\phi)^{\top}\big|_h\Big) \cdot \big|(\nabla\phi)^{\top}\big|_h\Big).
\end{aligned} \tag{6.21}$$

Note that there exists a constant $C_{10} > 0$ such that $\big|[\phi, (\nabla\phi)^{\top}]\big|_h \geq C_{10}(\log|w|) \cdot \big|(\nabla\phi)^{\top}\big|_h$ on $\mathcal{B}_q^{0*}(R_0)$. Then, we obtain the claim of the lemma. ■

Set $U^*_{w,q}(R) := \{|w_q| > R\}$. For any function f on $\mathcal{B}_q^{0*}(R)$, let $\int_{S^1_T} f$ denote the function on $U^*_{w,q}(R)$ given by $w_q \longmapsto \int_{S^1_T \times \{w_q\}} f\, dt$. By the previous lemma, we obtain the following inequality on $U^*_{w,q}(R_{11})$:

$$-(\partial_x^2 + \partial_y^2) \int_{S^1_T} \big|(\nabla\phi)^{\perp}\big|_h^2 \leq -C_{11}\big(\log|w|\big)^2 \cdot \int_{S^1_T} \big|(\nabla\phi)^{\perp}\big|_h^2. \tag{6.22}$$

Lemma 6.4.8 *There exist $R_{12} > R_{11}$ and $C_{12} > 0$ such that the following holds on $U^*_{w,q}(R_{12})$:*

$$\int_{S^1_T} \big|(\nabla\phi)^{\perp}\big|_h^2 = O\Big(\exp\big(-C_{12} I_q(w_q)\big)\Big).$$

Proof Let $w = w_q^q$. We have the following equality, which can be checked by a direct computation:

$$\frac{\partial^2}{\partial w \partial \overline{w}}\Big(\exp\big(-C|w|\log|w|\big)\Big) = \exp\big(-C|w|\log|w|\big) \cdot \frac{C^2}{4}(\log|w|+1)^2$$
$$-\exp\big(-C|w|\log|w|\big) \cdot \frac{C}{4|w|}(\log|w|+2) \tag{6.23}$$

We may assume $\log|w| \geq 1$ on $U^*_{w,q}(R_{11})$. Hence, we have the following inequality:

$$-\frac{\partial}{\partial w}\frac{\partial}{\partial \overline{w}} \exp\big(-C|w|\log|w|\big) \geq -C^2 \exp\big(-C|w|\log|w|\big)\big(\log|w|\big)^2. \tag{6.24}$$

Then, by the standard argument for Ahlfors lemma [2] using the inequalities (6.22) and (6.24), we obtain the desired estimate. (See [80] and [64]. See also the proof of Lemma 8.1.1.) ■

As a corollary, we obtain the following estimate on $\mathcal{B}_q^{0*}(R_{12})$:

$$\int_{S^1_T} |\rho_\kappa|_h^2 = O\Big(\exp\big(-C_{12} I_q(w_q)\big)\Big). \tag{6.25}$$

6.4.4 Step 2

Let F_ϕ denote the endomorphism on $\bigoplus_{\ell_1\neq\ell_2}\mathrm{Hom}(E^\bullet_{\ell_1},E^\bullet_{\ell_2})$ obtained as the inverse of the adjoint of ϕ.

Lemma 6.4.9 *For $(a,b,c)=(x,y,t),(y,t,x),(t,x,y)$, the following equalities hold:*

$$\nabla_a\rho_b-\nabla_b\rho_a-[\rho_a,\rho_b]^\top-2[\rho_a,\rho_b]^\bullet-[\rho_c,\phi]=0. \tag{6.26}$$

We also have the following equality:

$$\sum_{\kappa=x,t,y}\nabla_\kappa\rho_\kappa+F_\phi\left(\sum_{\kappa=x,y,t}\Big(2[\rho_\kappa,\nabla_\kappa\phi]-[\rho_\kappa,[\rho_\kappa,\phi]]\Big)^\top\right)=0. \tag{6.27}$$

Proof For $(a,b,c)=(x,y,t),(y,t,x),(t,x,y)$, we have the equalities

$$[\nabla_a,\nabla_b]=\nabla_c\phi.$$

We obtain the following:

$$\nabla^\bullet_a\rho_b-\nabla^\bullet_b\rho_a+[\rho_a,\rho_b]^\top-[\rho_c,\phi]=0.$$

Then, we obtain (6.26).

We have the equality $\sum_{\kappa=x,y,t}\nabla_\kappa\nabla_\kappa\phi=0$. Hence, we obtain the following:

$$\sum_\kappa[\nabla^\bullet_\kappa\rho_\kappa,\phi]+\Big(\sum_\kappa 2[\rho_\kappa,\nabla^\bullet_\kappa\phi]\Big)^\top+\Big(\sum_\kappa[\rho_\kappa,[\rho_\kappa,\phi]]\Big)^\top=0$$

Note that $\nabla^\bullet_\kappa\rho_\kappa=\nabla_\kappa\rho_\kappa$ and $[\rho_\kappa,\nabla_\kappa\phi]=[\rho_\kappa,\nabla^\bullet_\kappa\phi]+[\rho_\kappa,[\rho_\kappa,\phi]]$. Then, we obtain (6.27). ■

For any $w_{q,0}\in U^*_{w,q}(R_{12})$ and for any $r_0>0$, let $B_1(w_{q,0},r_0)$ denote the connected component of $\{w_q\in U^*_{w,q}\,\big|\,|w^q_q-w^q_{q,0}|<r_0\}$ such that $w_{q,0}\in B_1(w_{q,0},r_0)$. We set $B(w_{q,0},r_0):=S^1_T\times B_1(w_{q,0},r_0)$. By (6.25), we have $C_{13}>0$ such that the following holds if $B(w_{q,0},10)\subset\mathcal{B}^{0*}_q(2R_{12})$:

$$\|\rho_\kappa\|_{h,L^2(B(w_{q,0},10))}=O\Big(\exp\big(-C_{13}I_q(w_{q,0})\big)\Big).$$

Let us apply a standard bootstrapping argument. Let $\nu:\mathbb{R}_{\geq 0}\longrightarrow[0,1]$ be a C^∞-function such that (i) $\nu(u)=1$ $(u\leq 8)$, (ii) $\nu(u)=0$ $(u\geq 9)$, (iii) ν^α is C^∞ for any $\alpha>0$. Note that the C^∞-functions $2\partial_u\nu^{\alpha/2}$ $(\alpha>0)$ are equal to $\nu^{-\alpha/2}\partial_u\nu^\alpha$ on $\{\nu(u)\neq 0\}$. We set $\chi(w_q)=\rho(|w^q_q-w^q_{q,0}|)$ on $B_1(w_{q,0},10)$ and $\chi(w_q)=0$

for $w_q \notin B_1(w_{q,0}, 10)$. For any $\alpha > 0$, we set $\rho_\kappa^{(\alpha)} := \chi^\alpha \rho_\kappa$. We obtain

$$\nabla_a(\rho_b^{(\alpha)}) - \nabla_b(\rho_a^{(\alpha)}) = [\rho_a^{(\alpha/2)}, \rho_b^{(\alpha/2)}]^\top + 2[\rho_a^{(\alpha/2)}, \rho_b^{(\alpha/2)}]^\bullet + \chi^{-\alpha/2}(\partial_a\chi^\alpha)\rho_b^{(\alpha/2)} - \chi^{-\alpha/2}\partial_b\chi^\alpha\rho_a^{(\alpha/2)}, \tag{6.28}$$

$$\sum_\kappa \nabla_\kappa(\rho_\kappa^{(\alpha/2)}) = -F_\phi\Bigg(\sum_\kappa \Big(2\chi^{\alpha/2}[\rho_\kappa^{(\alpha/2)}, \nabla_\kappa\phi] - [\rho_\kappa^{(\alpha/2)}, [\rho_\kappa^{(\alpha/2)}, \phi]]\Big)^\top\Bigg) + \sum_\kappa(\chi^{-\alpha/2}\partial_\kappa\chi^\alpha)\rho_\kappa^{(\alpha/2)}. \tag{6.29}$$

There exists $C_{20} > 0$, which is independent of $w_{q,0}$, such that

$$\|\rho_\kappa^{(\alpha)}\|_{h,L^2(B(w_{q,0},10))} \le C_{20}\exp\big(-C_{13}I_q(w_{q,0})\big).$$

Note that the system of differential equations (6.28, 6.29) is elliptic. Moreover, because the curvature of ∇ is dominated independently from $w_{q,0}$, there exists a unitary frame on $B(w_{q,0}, 10)$ with respect to which the connection form of ∇ is dominated independently from $w_{q,0}$. Hence, for any $\alpha > 0$, there exist $C_i(1,\alpha) > 0$ ($i = 20, 21$) which are independent of $w_{q,0}$, such that

$$\|\rho_\kappa^{(\alpha)}\|_{h,L_1^2(B(w_{q,0},10))} \le C_{20}(1,\alpha)\exp\big(-C_{21}(1,\alpha)I_q(w_{q,0})\big),$$

where the Sobolev norms are considered for the connection ∇. By an induction of $k \in \mathbb{Z}_{>0}$, we can prove that there exist $C_i(k,\alpha) > 0$ ($i = 20, 21$) which are independent of $w_{q,0}$, such that

$$\|\rho_\kappa^{(\alpha)}\|_{h,L_k^2(B(w_{q,0},10))} \le C_{20}(k,\alpha)\exp\big(-C_{21}(k,\alpha)I_q(w_{q,0})\big).$$

Because $\nabla^\bullet = \nabla - \rho$, we obtain the same estimate with respect to the Sobolev norms for $\nabla^\bullet$. Thus, the proof of Proposition 6.4.2 is completed. ■

6.5 Some Lemmas from Linear Algebra

6.5.1 Eigenvalues

Let $d_{\mathbb{C}^r}$ denote the standard Euclidean distance on $\mathbb{C}^r$. Let $d_{\mathrm{Sym}^r\mathbb{C}}$ denote the distance on $\mathrm{Sym}^r\mathbb{C} = \mathbb{C}^r/\mathfrak{S}_r$ given by

$$d_{\mathrm{Sym}^r\mathbb{C}}(x,y) = \min_{\substack{x'\in\pi^{-1}(x)\\ y'\in\pi^{-1}(y)}} d_{\mathbb{C}^r}(x',y'),$$

where $\mathfrak{S}_r$ denotes the r-th symmetric group, and $\pi : \mathbb{C}^r \longrightarrow \mathrm{Sym}^r \mathbb{C}$ denotes the projection.

For any $Y \in M_r(\mathbb{C})$, let $\mathcal{S}p(Y)$ denote the set of the eigenvalues of Y, and let $\mathbb{E}_\alpha(Y)$ denote the generalized eigen space corresponding to $\alpha \in \mathcal{S}p(Y)$. Note that $\mathcal{S}p(Y)$ induces a point in $\mathrm{Sym}^r \mathbb{C}$.

Lemma 6.5.1 *There exists $C > 0$ depending only on r such that the following holds.*

- *Let $A \in M_r(\mathbb{C})$ be normal, i.e., ${}^t\overline{A}\, A = A\, {}^t\overline{A}$. For any $\epsilon > 0$, and for any $B \in M_r(\mathbb{C})$ satisfying $|B| \leq \epsilon$, we obtain $d_{\mathrm{Sym}^r \mathbb{C}}\big(\mathcal{S}p(A), \mathcal{S}p(A+B)\big) \leq C\epsilon$.*

Proof Let A, ϵ and B be as in the statement of the lemma. Let us begin with a preliminary.

Lemma 6.5.2 *If there exist $\alpha \in \mathbb{C}$ and a non-zero vector $\boldsymbol{v} \in \mathbb{C}^r$ such that $\big|A\boldsymbol{v} - \alpha\boldsymbol{v}\big| \leq \epsilon|\boldsymbol{v}|$, then there exists an eigenvalue β of A such that $|\beta - \alpha| \leq \epsilon r$.*

Proof We may assume that A is a diagonal matrix whose (i, i)-entries are γ_i. The assumption can be reworded as follows:

$$\sum_{i=1}^{r} |\gamma_i - \alpha|^2 |v_i|^2 \leq \epsilon^2 |\boldsymbol{v}|^2.$$

There exists i_0 such that $|v_{i_0}|^2 \geq r^{-1}|\boldsymbol{v}|^2$. Because

$$\frac{|\boldsymbol{v}|^2}{r^2} |\gamma_{i_0} - \alpha|^2 \leq |\gamma_{i_0} - \alpha|^2 |v_{i_0}|^2 \leq \epsilon^2 |\boldsymbol{v}|^2,$$

we obtain $|\gamma_{i_0} - \alpha| \leq \epsilon r$. ■

For any $\alpha \in \mathbb{C}$ and $\rho > 0$, we set $D(\alpha, \rho) := \big\{w \in \mathbb{C} \,\big|\, |w - \alpha| < \rho\big\}$. We set

$$D(\mathcal{S}p(A), r\epsilon) := \bigcup_{\alpha \in \mathcal{S}p(A)} D(\alpha, r\epsilon).$$

Lemma 6.5.2 implies $\mathcal{S}p(A + B) \subset B(\mathcal{S}p(A), r\epsilon)$. Let $B(\mathcal{S}p(A), r\epsilon) = \coprod_{i \in S(A)} U_i$ denote the decomposition into the connected components.

Lemma 6.5.3 *We obtain the following equality for each $i \in S(A)$:*

$$\sum_{\alpha \in U_i \cap \mathcal{S}p(A)} \dim \mathbb{E}_\alpha(A) = \sum_{\alpha \in U_i \cap \mathcal{S}p(A+B)} \dim \mathbb{E}_\alpha(A + B)$$

Proof For the continuous family of matrices $A_t := A + tB$ $(0 \leq t \leq 1)$, each $\mathcal{S}p(A_t)$ is contained in $B(\mathcal{S}p(A), r\epsilon) = \coprod_{i \in S(A)} U_i$. Because

$$\sum_{\alpha \in U_i \cap \mathcal{S}p(A_t)} \mathbb{E}_\alpha(A_t)$$

is invariant, the claim of the lemma follows. ■

Note that $\max_{\alpha,\beta \in U_i} |\alpha - \beta| < 2r^2\epsilon$. Hence, we obtain Lemma 6.5.1 from Lemma 6.5.3. ■

6.5.2 Almost Commuting Hermitian Matrix and Anti-Hermitian Matrix

Let $\mathrm{Re} : \mathbb{C} \longrightarrow \mathbb{R}$ and $\sqrt{-1}\,\mathrm{Im} : \mathbb{C} \longrightarrow \sqrt{-1}\mathbb{R}$ denote the maps defined by $\alpha = \mathrm{Re}(\alpha) + \sqrt{-1}\,\mathrm{Im}(\alpha)$. They induce the map $\mathrm{Re} : \mathrm{Sym}^r \mathbb{C} \longrightarrow \mathrm{Sym}^r \mathbb{R}$ and $\sqrt{-1}\,\mathrm{Im} : \mathrm{Sym}^r \mathbb{C} \longrightarrow \mathrm{Sym}^r(\sqrt{-1}\mathbb{R})$.

Lemma 6.5.4 *There exists a constant $C > 0$ depending only on r such that the following holds:*

- *For any $\epsilon > 0$, a Hermitian matrix $A \in M_r(\mathbb{C})$, and an anti-Hermitian matrix $B \in M_r(\mathbb{C})$ such that $\big|[A, B]\big| \leq \epsilon$, we obtain*

$$d_{\mathrm{Sym}^r \mathbb{C}}\big(\mathcal{S}p(A), \mathrm{Re}(\mathcal{S}p(A + B))\big) \leq C\epsilon^{1/2},$$

$$d_{\mathrm{Sym}^r \mathbb{C}}\big(\mathcal{S}p(B), \sqrt{-1}\,\mathrm{Im}(\mathcal{S}p(A + B))\big) \leq C\epsilon^{1/2}.$$

- *There exists a normal matrix $H \in M_r(\mathbb{C})$ such that (i) the characteristic polynomial of H is equal to the characteristic polynomial of $A + B$, (ii) $|H - (A + B)| \leq C\epsilon^{1/2}$.*

Proof Applying Lemma 6.5.5 below to $\mathcal{S}p(A)$ with $L_1 = 100$ and $\epsilon_0 = \epsilon^{1/2}$, there exist constants $C_0, C_1 > 0$ with $C_1/C_0 > 20$ and $4 \leq C_0 \leq 4(100r)^r$ and a decomposition $\mathcal{S}p(A) = \coprod_{i \in \Lambda} \mathcal{S}_i$ such that the following holds.

- For $\alpha, \beta \in \mathcal{S}_i$, we obtain $|\alpha - \beta| \leq C_0\epsilon^{1/2}$.
- For $\alpha \in \mathcal{S}_i$ and $\beta \in \mathcal{S}_j$ with $i \neq j$, we obtain $|\alpha - \beta| \geq C_1\epsilon^{1/2}$.

In the following of this proof, $C_{10}^{(i)}$ will denote positive constants depending only on r.

We may assume that A is diagonal, i.e., $A = \bigoplus_{i \in \Lambda} \Gamma_i$, such that Γ_i are diagonal whose (k, k)-entries are elements of $\mathcal{S}_i$. Because $|[A, B]| \leq \epsilon$, for the block decomposition $B = \sum_{i,j \in \Lambda} B_{ij}$, we obtain $|B_{ij}| \leq C_{10}^{(1)}\epsilon^{1/2}$ $(i \neq j)$.

We choose $a_i \in S_i$ for each $i \in \Lambda$, and we set $A' := \bigoplus_{i\in\Lambda} a_i I_{r_i}$. We also put $B' := \bigoplus B_{i,i}$. Because $[A', B'] = 0$, we obtain

$$\mathrm{Re}\,(\mathcal{S}p(A'+B')) = \mathcal{S}p(A'), \quad \sqrt{-1}\,\mathrm{Im}\,(\mathcal{S}p(A'+B')) = \mathcal{S}p(B').$$

Because $\left|(A'+B')-(A+B)\right| \leq C_{10}^{(2)}\epsilon^{1/2}$, we obtain $d_{\mathrm{Sym}^r\mathbb{C}}\big(\mathcal{S}p(A'+B'), \mathcal{S}p(A+B)\big) \leq C_{10}^{(3)}\epsilon^{1/2}$ by Lemma 6.5.1. We also obtain $d_{\mathrm{Sym}^r\mathbb{C}}\big(\mathcal{S}p(A'), \mathcal{S}p(A)\big) \leq C_{10}^{(4)}\epsilon^{1/2}$ and $d_{\mathrm{Sym}^r\mathbb{C}}\big(\mathcal{S}p(B'), \mathcal{S}p(B)\big) \leq C_{10}^{(4)}\epsilon^{1/2}$ by Lemma 6.5.1. Because $d_{\mathrm{Sym}^r\mathbb{C}}\big(\mathcal{S}p(A'+B'), \mathcal{S}p(A+B)\big) \leq C_{10}^{(3)}\epsilon^{1/2}$, and because $A'+B'$ is normal, there exists a normal matrix H such that $\mathcal{S}p(H) = \mathcal{S}p(A+B)$ and that $|H-(A'+B')| \leq C_{10}^{(5)}\epsilon^{1/2}$. Thus, we are done. ∎

6.5.3 *Decomposition of Finite Tuples in Metric Spaces (Appendix)*

Let (X, d) be a metric space. We take $(x_1, \ldots, x_n) \in X^n$. Take any $\epsilon_0 > 0$ and $L_1 > 8$.

Lemma 6.5.5 *There exist a non-negative integer $N \leq n$ and a decomposition $\{1, \ldots, n\} = \coprod_{j\in\Lambda} S_j$ with the following properties:*

- *For any $a \in S_j$ and $b \in S_k$ with $j \neq k$, we obtain $d\big(x_a, x_b\big) > (L_1-4)\big(L_1 n\big)^N \epsilon_0$.*
- *For any $a, b \in S_j$, we obtain $d(x_a, x_b) \leq 4\big(L_1 n\big)^N \epsilon_0$.*

Proof We make general preparations. Let Γ be any finite graph. Let $V(\Gamma)$ denote the set of vertexes of Γ. We obtain the decomposition of the graph $\Gamma = \coprod_{j\in C(\Gamma)} \Gamma_j$ corresponding to the decomposition of the geometric realization into the connected components. Let $|V(\Gamma_j)|$ denote the number of the vertices of $V(\Gamma_j)$, and we put $m(\Gamma) := \max\big\{|V(\Gamma_j)| \,\big|\, j \in C(\Gamma)\big\}$. For any positive number δ, a finite set I and $\boldsymbol{y} = (y_i \,|\, i \in I) \in X^I$, let $\Gamma(\boldsymbol{y}, \delta)$ denote the unique graph determined by the conditions.

- $V\big(\Gamma(\boldsymbol{y}, \delta)\big) = I$.
- $a, b \in I$ are connected by an edge if and only if $d\big(y_a, y_b\big) \leq \delta$ and $a \neq b$.

Let us construct a decomposition as in the claim of the proposition. We set $S^{(0)} := \{1, \ldots, m\}$, $\boldsymbol{x}^{(0)} := \boldsymbol{x}$, and $\Gamma^{(0)} := \Gamma(\boldsymbol{x}^{(0)}, L_1\epsilon_0)$. We shall inductively construct a decreasing sequence of subsets $S^{(0)} \supset S^{(1)} \supset \cdots \supset S^{(N)}$ and graphs $\Gamma^{(0)}, \Gamma^{(1)}, \ldots, \Gamma^{(N)}$ with $V(\Gamma^{(j)}) = S^{(j)}$ such that $m(\Gamma^{(j)}) > 1$ $(j < N)$ and $m(\Gamma^{(N)}) = 1$. Suppose that we have already constructed $(S^{(j)}, \Gamma^{(j)})$ for $j = 0, \ldots, \ell$ with $m(\Gamma^{(j)}) > 1$ $(j < \ell)$. If $m(\Gamma^{(\ell)}) = 1$, we stop here. Let us consider the case $m(\Gamma^{(\ell)}) > 1$. We have the decomposition $S^{(\ell)} = \coprod_{j\in C(\Gamma^{(\ell)})} S_j^{(\ell)}$ according to the decomposition of the graph $\Gamma^{(\ell)}$ into the connected components of the geometric realization. For each $j \in C(\Gamma^{(\ell)})$, we choose an element $a_j^{(\ell)} \in S_j^{(\ell)}$. Then, we define $S^{(\ell+1)} := \{a_j^{(\ell)} \mid j \in C(\Gamma^{(\ell)})\}$, $\boldsymbol{x}^{(\ell+1)} := (x_a \mid a \in S^{(\ell+1)})$ and

$$\Gamma^{(\ell+1)} := \Gamma\big(\boldsymbol{x}^{(\ell+1)}, L_1(L_1 n)^{\ell+1}\epsilon_0\big).$$

The inductive procedure finishes at some $\ell = N$. By the construction, there exist the maps $\pi_i : S^{(i)} \longrightarrow S^{(i+1)}$ determined by $\pi_i(a) = a_j^{(i)} \in S^{(i+1)}$ for $a \in S_j^{(i)}$. Let $\pi : S \longrightarrow S^{(N)}$ denote the induced map. For $c \in S^{(N)}$, we set $S_c := \pi^{-1}(c)$. By the construction, if a is contained in S_c, we obtain

$$d(a, c) \leq (L_1 n)^N \epsilon_0 + (L_1 n)^{N-1}\epsilon_0 + \cdots + (L_1 n)\epsilon_0 \leq 2(L_1 n)^N \epsilon_0.$$

Hence, for $a, b \in S_c$, we obtain $d(a, b) \leq 4(L_1 n)^N \epsilon_0$.

If $a_i \in S_{c_i}$ $(i = 1, 2)$ with $c_1 \neq c_2$, we obtain

$$\begin{aligned} d(a_1, a_2) \geq d(c_1, c_2) - d(a_1, c_1) - d(a_2, c_2) &\geq L_1(L_1 n)^N \epsilon_0 - 4(L_1 n)^N \epsilon_0 \\ &\geq (L_1 - 4)(L_1 n)^N \epsilon_0. \qquad (6.30) \end{aligned}$$

Hence, the decomposition $\{1, \ldots, n\} = \coprod_{c \in S^{(\ell)}} S_c$ has the desired property. ■

6.6 Vector Bundles with a Connection on a Circle (I)

6.6.1 Statement

Let E be a vector bundle on $S_T^1 := \mathbb{R}/T\mathbb{Z}$ of rank r with a Hermitian metric h, a unitary connection ∇, and a self-adjoint section ψ. Let t be the standard coordinate of $\mathbb{R}$, which induces local coordinates on S_T^1. For an End(E)-valued differential form s, let $|s|_h$ denote the norm of s with respect to h and the standard metric $dt\,dt$ of S_T^1.

We set $\widetilde{\nabla} := \nabla + \psi\, dt$. Let M denote the monodromy of the connection $\widetilde{\nabla}$ along the loop $\gamma : [0, 1] \longrightarrow S_T^1$ defined by $\gamma(s) = Ts$. For each eigenvalue α of M, we have the well defined number $T^{-1}\log|\alpha|$. The tuple of the numbers $T^{-1}\log|\alpha|$ with the multiplicity induces an element in $\mathrm{Sym}^r\,\mathbb{R}$, denoted by $T^{-1}\log|\mathcal{S}p(M)|$. For each $Q \in S_T^1$, the tuple of the eigenvalues with the multiplicity of $-\psi_{|Q} \in \mathrm{End}(E_{|Q})$ induces an element of $\mathrm{Sym}^r\,\mathbb{R}$, denoted by $\mathcal{S}p(-\psi_{|Q})$.

We shall prove the following proposition in Sects. 6.6.2–6.6.5.

Proposition 6.6.1 *There exist positive constants $\epsilon_0 > 0$ and $C > 0$ depending only on r such that the following holds for any $0 < \epsilon < \epsilon_0$*

- *If $|\nabla(\psi)|_h \leq \epsilon$, we obtain $d_{\mathrm{Sym}^r\mathbb{R}}\Big(\mathcal{S}p(-\psi_{|Q}),\ T^{-1}\log|\mathcal{S}p(M)|\Big) \leq C\epsilon^{1/2}$.*

In Sects. 6.6.2–6.6.5, we suppose $|\nabla\psi|_h \leq \epsilon$ for a positive constant $\epsilon > 0$.

6.6.2 Preliminary

Let $U \in \mathrm{GL}(E_{|0})$ denote the monodromy of the unitary connection ∇ along the loop γ. There exists $e^{\sqrt{-1}\theta_0} \in S^1$ such that (i) $0 \leq \theta_0 \leq 2\pi$, (ii) $|\theta_0 - \arg(\alpha)| > \pi/2r$ for any $\alpha \in \mathcal{S}p(U)$, where $\arg(\alpha)$ denotes any real number such that $\exp(\sqrt{-1}\arg(\alpha)) = \alpha$. There exists an unitary frame $\boldsymbol{e}$ of E such that $\nabla \boldsymbol{e} = \boldsymbol{e}\,(-T^{-1}A)\,dt$, such that the following holds.

- A is a constant diagonal anti-Hermitian matrix such that

$$\mathcal{S}p(A) \subset \left\{\sqrt{-1}a \,\middle|\, \theta_0 - 2\pi + \pi/2r < a < \theta_0 - \pi/2r\right\}.$$

Note that $U\boldsymbol{e}_{|0} = \boldsymbol{e}_{|0}\exp(A)$.

For any L^2-section ρ of $\mathrm{End}(E)$, let B_ρ be the matrix valued function determined by $\rho\boldsymbol{e} = \boldsymbol{e}\,B_\rho$. There exists the decomposition $B_\rho = B_{\rho,0} + B_{\rho,1}$, where $B_{\rho,0} = T^{-1}\int B_\rho\,dt$ and $\int B_{\rho,1}\,dt = 0$. It induces a decomposition $\rho = \rho_0 + \rho_1$. In particular, we obtain the decomposition $\psi = \psi_0 + \psi_1$. We obtain a similar decomposition of any L^2-section of E.

Lemma 6.6.2 *There exists a positive constant $C_0 > 0$ depending only on r such that $|\nabla\psi_0|_h \leq C_0\epsilon$ and $|\nabla\psi_1|_h \leq C_0\epsilon$.*

Proof Because A is constant, we obtain $\nabla(\psi_i) = (\nabla\psi)_i$ $(i = 0, 1)$. Let $B_{\nabla_t\psi}$ be the matrix expressing $\nabla_t\psi$ with respect to the frame $\boldsymbol{e}$, i.e., $\nabla_t(\psi)\boldsymbol{e} = \boldsymbol{e}\cdot B_{\nabla_t\psi}$, then $T^{-1}\int B_{\nabla_t\psi}\,dt$ expresses $\nabla_t\psi_0$ with respect to the frame $\boldsymbol{e}$. Hence, there exists $C_0' > 0$ depending only on r such that $|(\nabla_t\psi)_0|_h \leq C_0'\int|\nabla_t\psi|_h\,dt$. Then, the claim easily follows from the assumption $|\nabla\psi|_h \leq \epsilon$. ■

For a section s of $\mathrm{End}(E)$, we set

$$|s|_{L^2} := \left(\int_0^T |s|_h^2\right)^{1/2}, \quad |s|_{L^2_k} = \left(\sum_{0\leq j\leq k} |\nabla_t^j s|_{L^2}^2\right)^{1/2}.$$

Corollary 6.6.3 *There exist $C_1 > 0$, depending only on r, such that $|\psi_1|_{L^2_1} \leq C_1\epsilon$ and $\sup|\psi_1|_h \leq C_1\epsilon$.*

Proof Note that the eigenvalues $\sqrt{-1}\beta$ of $\mathrm{ad}(A)$ satisfy $|\beta| < 2\pi - (\pi/r)$. Hence, we obtain $|\psi_1|_{L^2_1} \leq C_1'\epsilon$ from the estimate of $|\nabla\psi_1|$ in the previous lemma. Because $\dim S^1_T = 1$, we obtain $\sup|\psi_1|_h \leq C_1''\epsilon$. ■

6.6.3 A Decomposition of Function Spaces

By the frame $\boldsymbol{e}$, we obtain a C^∞-isometry of vector bundles $\mathrm{End}(E) \simeq S^1_T \times M_r(\mathbb{C})$ on S^1_T. For $k \geq 0$, let $L^2_k(M_r(\mathbb{C}))$ denote the space of L^2_k-maps from S^1_T to $M_r(\mathbb{C})$. Let $L^2_k(M_r(\mathbb{C}))_1$ denote the subspace of $F \in L^2_k(M_r(\mathbb{C}))$ such that $\int_{S^1_T} F\,dt = 0$. Let $L^2_k(M_r(\mathbb{C}))_0$ denote the space of the constant maps from S^1_T to $M_r(\mathbb{C})$. We obtain the decomposition $L^2_k(M_r(\mathbb{C})) = L^2_k(M_r(\mathbb{C}))_0 \oplus L^2_k(M_r(\mathbb{C}))_1$. We obtain the corresponding decomposition

$$L^2_k\big(\mathrm{End}(E)\big) = L^2_k\big(\mathrm{End}(E)\big)_0 \oplus L^2_k\big(\mathrm{End}(E)\big)_1.$$

Namely, $L^2_k\big(\mathrm{End}(E)\big)_0$ denotes the space of sections of $\mathrm{End}(E)$, which are constant with respect to $\boldsymbol{e}$, and $L^2_k\big(\mathrm{End}(E)\big)_1$ denotes the space of L^2_k-sections of $\mathrm{End}(E)$ which are represented by matrices B with respect to $\boldsymbol{e}$ such that $\int_{S^1_T} B\,dt = 0$. There exist similar decompositions for $C^\infty(\mathrm{End}(E))$ and $L^2_k\big(\mathrm{End}(E) \otimes \Omega^1\big)$, etc.

Lemma 6.6.4 *There exist $C_2 > 0$ and $\epsilon_1 > 0$, depending only on r, with the following property.*

- *If $\epsilon \leq \epsilon_1$, we obtain $|a|_{L^2_1} \leq C_2\big|(\nabla_t + \psi_0)a\big|_{L^2}$ for any $a \in L^2_1(\mathrm{End}(E))_1$.*

Proof In this proof, $C_2^{(i)}$ denote positive constants depending only on r. Let $B \in M_r(\mathbb{C})$ be determined by $\psi_0\boldsymbol{e} = \boldsymbol{e}\,B$, which satisfies ${}^t\overline{B} = B$. By the assumption $|\nabla(\psi)| \leq \epsilon$ and Lemma 6.6.2, we obtain $\big|[A, B]\big| \leq C_2^{(1)}\epsilon$. We set $L = -T^{-1}A + B$. Because ${}^t\overline{L} = T^{-1}A + B$, we obtain $\big|[L, {}^t\overline{L}]\big| \leq C_2^{(2)}\epsilon$. By Lemma 6.5.4, there exists a decomposition $L = L_1 + L_2$ such that (i) L_1 is normal, and the characteristic polynomials of L_1 and L are the same, (ii) $|L_2| \leq C_2^{(3)}\epsilon^{1/2}$ holds. Note that L_1 is diagonalizable by a unitary matrix, and that any eigenvalue γ of $\mathrm{ad}\,L_1$ satisfies $|\operatorname{Im}(\gamma)| < 2\pi - (\pi/2r)$. Let $\nabla^{(0)}$ be determined by $\nabla^{(0)}_t\boldsymbol{e} = \boldsymbol{e}\,L_1$, and let $g^{(0)}$ be determined by $g^{(0)}\boldsymbol{e} = \boldsymbol{e}\,L_2$. We can easily check that there exists $C_2^{(4)} > 0$ depending only on r such that $|a|_{L^2_1} \leq C_2^{(4)}\big|\nabla^{(0)}_t a\big|_{L^2}$ for any $a \in L^2_1(\mathrm{End}(E))_1$. Because $|g^{(0)}| \leq C_2^{(3)}\epsilon^{1/2}$ and $\big|\nabla^{(0)}_t a\big|_{L^2} \leq \big|(\nabla_t + \psi_0)a\big|_{L^2} + \big|[g^{(0)}, a]\big|_{L^2}$, there exists $C_2^{(5)} > 0$ such that $\big|\nabla^{(0)}_t a\big|_{L^2} \leq C_2^{(5)}\big|(\nabla_t + \psi_0)a\big|_{L^2}$ if ϵ_1 is sufficiently small. Then, the claim of the lemma follows. ■

6.6.4 Gauge Transformation

Let ϵ_1 be as in Lemma 6.6.4. We assume that $\epsilon < \epsilon_1$. We take a neighbourhood $\mathcal{U}_0 \subset L^2_1(\mathrm{End}(E))_1$ of 0 such that $1 + a$ is invertible for any $a \in \mathcal{U}_0$.

Proposition 6.6.5 *There exist positive constants ϵ_2 and C_3, depending only on r, such that the following holds:*

- *If $\epsilon \leq \epsilon_2$, there exists $(a, b) \in \mathcal{U}_0 \times L^2(\mathrm{End}(E))_0$ satisfying $\big|(\nabla_t + \psi_0)a\big|_{L^2} \leq C_3\epsilon$, $|b| \leq C_3\epsilon$, and*

$$(1+a)^{-1} \circ (\nabla_t + \psi_0 + b) \circ (1+a) = \nabla_t + \psi.$$

Proof We define the map $\Psi : \mathcal{U}_0 \times L^2(\mathrm{End}(E))_0 \longrightarrow L^2(\mathrm{End}(E))$ by

$$\Psi(a, b) = (1+a)^{-1} \circ (\nabla_t + \psi_0 + b) \circ (1+a) - (\nabla_t + \psi_0).$$

The derivative $T_{(a,b)}\Psi : L_1^2(\mathrm{End}(E))_1 \oplus L^2(\mathrm{End}(E))_0 \longrightarrow L^2(\mathrm{End}(E))$ of Ψ at (a, b) is as follows:

$$\begin{aligned}
&T_{(a,b)}\Psi(u, v) \\
&\equiv (1+a+u)^{-1} \circ (\nabla_t + \psi_0 + b + v) \circ (1+a+u) \\
&- (1+a)^{-1} \circ (\nabla_t + \psi_0 + b) \circ (1+a) \\
&\equiv -(1+a)^{-1} \circ u \circ (1+a)^{-1} \circ (\nabla_t + \psi_0 + b) \circ (1+a) \\
&+ (1+a)^{-1} \circ (\nabla_t + \psi_0 + b) \circ (1+a) \circ (1+a)^{-1} \circ u \\
&+ (1+a)^{-1} \circ v \circ (1+a).
\end{aligned} \tag{6.31}$$

Let us obtain a bound of $T_{(a,b)}\Psi(u, v) - \big((\nabla_t + \psi_0)u + v\big)$. Note that $(\nabla_t + \psi_0)u = (\nabla_t + \psi_0) \circ u - u \circ (\nabla_t + \psi_0)$. We clearly have $\mathrm{Ad}(1+a)^{-1}(v) - v = O\big(|a|_{L_1^2} \cdot |v|\big)$. We set

$$\begin{aligned}
\mathfrak{A} := (1+a)^{-1} \circ b \circ (1+a) \circ (1+a)^{-1} \circ u \\
- (1+a)^{-1} \circ u \circ (1+a)^{-1} \circ b \circ (1+a),
\end{aligned} \tag{6.32}$$

$$\begin{aligned}
&\mathfrak{B} := (1+a)^{-1} \circ (\nabla_t + \psi_0) \circ (1+a) \circ (1+a)^{-1} \circ u \\
&- (1+a)^{-1} \circ u \circ (1+a)^{-1} \circ (\nabla_t + \psi_0) \circ (1+a) - (\nabla_t + \psi_0)\big((1+a)^{-1}u\big),
\end{aligned} \tag{6.33}$$

$$\mathfrak{C} := (\nabla_t + \psi_0)\big((1+a)^{-1} - 1\big)u. \tag{6.34}$$

We obtain

$$\begin{aligned}
&(1+a)^{-1} \circ \big(\nabla_t + \psi_0 + b\big) \circ (1+a) \circ (1+a)^{-1} \circ u \\
&- (1+a)^{-1} \circ u \circ (1+a)^{-1} \circ \big(\nabla_t + \psi_0 + b\big) \circ (1+a) - (\nabla_t + \psi_0)u = \mathfrak{A} + \mathfrak{B} + \mathfrak{C}.
\end{aligned} \tag{6.35}$$

We have $|\mathfrak{A}|_{L^2} = O\big(|b| \cdot |u|_{L_1^2}(1 + |a|_{L_1^2})\big)$. Because $\mathfrak{B} = \big[(1+a)^{-1} \circ (\nabla_t + \psi_0)(a), (1+a)^{-1} \circ u\big]$, we obtain

$$|\mathfrak{B}|_{L^2} = O\Big(C_3^{(0)}(1 + |a|_{L_1^2})^2 \big|(\nabla_t + \psi_0)a\big|_{L^2} |u|_{L_1^2}\Big).$$

For the term $\mathfrak{C}$, we have the following equality:

$$\begin{aligned}\mathfrak{C} &= (\nabla_t + \psi_0)\big(((1+a)^{-1} - 1)u\big)\\ &= \big((1+a)^{-1} - 1\big) \circ (\nabla_t + \psi_0)(u) + (\nabla_t + \psi_0)\big((1+a)^{-1}\big) \circ u. \end{aligned} \tag{6.36}$$

There exist $C_3^{(1)}, C_3^{(2)} > 0$ such that

$$\begin{aligned}|\mathfrak{C}|_{L^2} &\le C_3^{(1)} |a|_{L_1^2} \big|(\nabla_t + \psi_0)u\big|_{L^2} + C_3^{(1)} \big|(\nabla_t + \psi_0)\big((1+a)^{-1}\big)\big|_{L^2} |u|_{L_1^2}\\ &\le C_3^{(2)} |a|_{L_1^2} \big|(\nabla_t + \psi_0)u\big|_{L^2} + C_3^{(2)} (1 + |a|_{L_1^2}) \big|(\nabla_t + \psi_0)a\big|_{L^2} |u|_{L_1^2}.\end{aligned} \tag{6.37}$$

Therefore, there exists $C_3^{(3)} > 0$ such that the following holds:

$$\begin{aligned}&\Big|T_{(a,b)}\Psi(u,v) - \big((\nabla_t + \psi_0)u + v\big)\Big|_{L^2}\\ &\le C_3^{(3)}\Big(|a|_{L_1^2}|v| + |b|\,|u|_{L_1^2}(1 + |a|_{L_1^2}) + (1 + |a|_{L_1^2})^2\big|(\nabla_t + \psi_0)a\big|_{L^2}|u|_{L_1^2}\Big)\\ &+ C_3^{(3)}\Big(|a|_{L_1^2}\big|(\nabla_t + \psi_0)u\big|_{L^2} + (1 + |a|_{L_1^2})\big|(\nabla_t + \psi_0)(a)\big|_{L^2}|u|_{L_1^2}\Big).\end{aligned} \tag{6.38}$$

Let Γ be the composite of the following maps:

$$\begin{aligned}&\mathcal{U}_0 \times L^2(\mathrm{End}(E))_0 \xrightarrow{\Psi} L^2(\mathrm{End}(E))\\ &\qquad\qquad \xrightarrow{(\nabla_t + \psi_0)^{-1} \oplus \mathrm{id}} L_1^2(\mathrm{End}(E))_1 \oplus L^2(\mathrm{End}(E))_0.\end{aligned} \tag{6.39}$$

We regard $L_1^2(\mathrm{End}(E))_1$ as a Banach space by the norm $a \longmapsto \big|(\nabla_t + \psi_0)a\big|_{L^2}$. Hence, the operator norm of $T_{a,b}\Gamma - \mathrm{id}$ is $O\Big(\big(|(\nabla_t + \psi_0)a|_{L_1^2} + |b|\big) \cdot (1 + |a|_{L_1^2})^2\Big)$. There exists $C_4 > 0$ depending only on r such that if $\big|(\nabla_t + \psi_0)a\big|_{L^2} + |b| \le C_4$ then $|T_{a,b}\Gamma - \mathrm{id}\,| < 1/2$. As in the proof of the inverse mapping theorem (see [52]), the image of Γ contains

$$\Big\{c + d \in L_1^2(\mathrm{End}(E))_1 \oplus L^2(\mathrm{End}(E))_0 \,\Big|\, \big|(\nabla_t + \psi_0)c\big|_{L^2} + |d| \le C_4/2\Big\}.$$

It means that the image of Ψ contains $\Big\{c+d \in L^2(\mathrm{End}(E))\oplus L^2(\mathrm{End}(E))_0 \,\Big|\, |c|_{L^2}+|d| \leq C_4/2\Big\}$. Because we have $|\psi_1|_{L^2} \leq C_1\epsilon$, we obtain the claim of the proposition. ■

6.6.5 *Proof of Proposition 6.6.1*

Let us prove Proposition 6.6.1. We use the notation in Proposition 6.6.5. There exist $B, B' \in M_r(\mathbb{C})$ determined by $\psi_0 \boldsymbol{e} = \boldsymbol{e}B$ and $b\boldsymbol{e} = \boldsymbol{e}B'$. Let $\widetilde{\nabla}' := \nabla + (\psi_0 + b)\,d\theta$. We have $\widetilde{\nabla}'\boldsymbol{e} = \boldsymbol{e}\,(-T^{-1}A + B + B')\,d\theta$. Let $\mathrm{Re}\,\mathcal{S}p(T^{-1}A - B - B')$ denote the point of $\mathrm{Sym}^r\,\mathbb{R}$ induced by the real part of the eigenvalues of $T^{-1}A - B - B'$ with the multiplicity. Let M' denote the monodromy of $\widetilde{\nabla}'$, which is conjugate to $\exp\big(A - TB - TB'\big)$. Because M and M' are conjugate, we have $T^{-1}\log|\mathcal{S}p(M)| = T^{-1}\log|\mathcal{S}p(M')| = \mathrm{Re}\,\mathcal{S}p(T^{-1}A - B - B')$. Then, the claim follows from Lemma 6.5.4. ■

6.7 Vector Bundles with a Connection on a Circle (II)

6.7.1 *Additional Assumption on the Eigenvalues of the Monodromy*

We continue to use the notation in Sect. 6.6.1. We assume that $|\nabla(\psi)|_h \leq \epsilon < \epsilon_0$ as in Proposition 6.6.1. We impose the following additional assumptions on the eigenvalues of ψ.

Condition 6.7.1

(A1) *There exist positive constants C_i $(i = 10, 11)$ with $1 < C_{10}$ and $100C_{10} < C_{11}$, and an element $\mathcal{S} \in \mathrm{Sym}^r\,\mathbb{R}$ for which the following holds for any $Q \in S^1_T$:*

$$d_{\mathrm{Sym}^r\,\mathbb{C}}\big(\mathcal{S}p(-\psi_{|Q}), \mathcal{S}\big) < C_{10}/2, \qquad C_{11} < \min\big\{|\gamma_1 - \gamma_2| \,\big|\, \gamma_i \in \mathcal{S}, \gamma_1 \neq \gamma_2\big\}.$$

We also assume that ϵ_0 is so small that $d_{\mathrm{Sym}^r\,\mathbb{C}}\big(T^{-1}\log\big|\mathcal{S}p(M(\widetilde{\nabla}))\big|, \mathcal{S}\big) < C_{10}$. (See Proposition 6.6.1.)

(A2) *There exist $k_0 \geq 1$ and $C_{12} > 0$ such that $|\nabla^j(\psi)|_h \leq C_{12}\epsilon$ for any $1 \leq j \leq k_0$.* ■

We naturally regard $\mathcal{S}$ as a subset of $\mathbb{R}$. We obtain the orthogonal decomposition $(E, \psi) = \bigoplus_{\alpha\in\mathcal{S}}(E^\bullet_\alpha, \psi_\alpha)$ such that the eigenvalues γ of $-\psi_{\alpha|Q}$ satisfy $|\gamma - \alpha| < C_{10}/2$. It induces the decomposition $\nabla_t = \nabla^\bullet_t + \rho$, where $\nabla^\bullet_t$ is a unitary connection preserving the decomposition, and ρ is a section of $\bigoplus_{\alpha\neq\beta}\mathrm{Hom}(E^\bullet_\alpha, E^\bullet_\beta)$. We obtain the decomposition $\nabla_t^\bullet = \bigoplus_{\alpha\in S} \nabla_{\alpha,t}^\bullet$. In general, for any vector bundle E' and for any section s of $\mathrm{End}(E)\otimes E'$, we obtain the decomposition $s = s^\bullet + s^\top$ into the sections of $\bigoplus_{\alpha\in\mathbb{C}} \mathrm{End}(E_\alpha^\bullet)\otimes E'$ and $\bigoplus_{\alpha\neq\beta} \mathrm{Hom}(E_\alpha^\bullet, E_\beta^\bullet)\otimes E'$.

Lemma 6.7.2 *There exists $C_{13} > 0$, depending only on r, k_0 and C_i $(i = 10, 11, 12)$, such that the following holds for any $1 \leq j \leq k_0$:*

$$\big|(\nabla_t^\bullet)^j \psi\big|_h \leq C_{13}\epsilon, \quad \big|[\psi, (\nabla_t^\bullet)^{j-1}\rho]\big|_h \leq C_{13}\epsilon, \quad \big|(\nabla_t^\bullet)^{j-1}\rho\big|_h \leq C_{13}\epsilon. \tag{6.40}$$

Proof There exists the following description:

$$\nabla_t^j \psi = (\nabla_t^\bullet)^j \psi + \mathrm{ad}\left((\nabla_t^\bullet)^{j-1}\rho\right)\psi + \sum_{i\leq j-1} \mathcal{P}_{j,i}\cdot(\nabla_t^\bullet)^i(\psi).$$

Here, $\mathcal{P}_{j,i}$ are given as a sum of some compositions of $\mathrm{ad}(\nabla_t^m\rho)$ $(m < j-1)$. Note the equalities $(\nabla_t^\bullet)^j\psi = \left((\nabla_t^\bullet)^j\psi\right)^\bullet$ and $\mathrm{ad}\left((\nabla_t^\bullet)^{j-1}\rho\right)\psi = \left(\mathrm{ad}\left((\nabla_t^\bullet)^{j-1}\rho\right)\psi\right)^\top$. We also remark that an estimate for $\big|[\psi, (\nabla_t^\bullet)^{j-1}\rho]\big|_h$ implies an estimate for $\big|(\nabla_t^\bullet)^{j-1}\rho\big|_h$. Then, we obtain (6.40) by an easy induction. ■

Lemma 6.7.3 *There exists $C_{14} > 0$ and $\epsilon_{10} > 0$ depending only on r, k_0 and C_i $(i = 10, 11, 12)$, such that the following holds for any $k \leq k_0 - 1$ and for any $a \in \bigoplus_{\alpha\neq\beta} L^2_{k+1}(\mathrm{Hom}(E_\alpha^\bullet, E_\beta^\bullet))$ if $\epsilon \leq \epsilon_{10}$:*

$$|a|_{L^2_{k+1}} \leq C_{14}\big|(\nabla_t^\bullet + \psi)(a)\big|_{L^2_k}.$$

Proof We set $\Gamma_0 := \bigoplus \alpha\cdot\mathrm{id}_{E_\alpha^\bullet}$. Because Γ_0 is flat with respect to $\nabla^\bullet$, and self-adjoint with respect to the metric, we obtain $|(\nabla_t^\bullet + \Gamma_0)a|_{L^2} \geq |\nabla_t^\bullet a|_{L^2}$ and $|(\nabla_t^\bullet + \Gamma_0)a|_{L^2} \geq |a|_{L^2}$. We also obtain

$$\big|(\nabla_t^\bullet)^j(\nabla_t^\bullet + \Gamma_0)(a)\big|_{L^2} \geq \big|(\nabla_t^\bullet)^{j+1}a\big|_{L^2},$$

$$\big|(\nabla_t^\bullet)^j(\nabla_t^\bullet + \Gamma_0)(a)\big|_{L^2} \geq \big|(\nabla_t^\bullet)^j a\big|_{L^2}.$$

Then, the claim of the lemma follows. ■

6.7.2 Gauge Transformations

Let $0 \leq k \leq k_0$. We use the norm on $\bigoplus_{\alpha\neq\beta} L^2_{k+1}(\mathrm{Hom}(E_\alpha^\bullet, E_\beta^\bullet))$ given by $\big|(\nabla_t^\bullet + \psi)(a)\big|_{L^2_k}$. Let

$$\mathcal{U}_0 \subset \bigoplus_{\alpha\neq\beta} L^2_{k+1}(\mathrm{Hom}(E_\alpha^\bullet, E_\beta^\bullet))$$

be a small neighbourhood of 0 such that $1+a$ is invertible for any $a \in \mathcal{U}_0$. Let

$$\mathcal{U}_1 \subset \bigoplus_{\alpha} L_k^2(\mathrm{End}(E_\alpha^\bullet))$$

be a small neighbourhood of 0. We define the map $\Phi : \mathcal{U}_0 \times \mathcal{U}_1 \longrightarrow L_k^2\big(\mathrm{End}(E)\big)$ given by

$$\Phi(a,b) := (1+a)^{-1}(\nabla_t^\bullet + \psi)(a) + (1+a)^{-1}b(1+a).$$

Proposition 6.7.4 *There exist positive constants C_{15} and ϵ_{11} depending only on r, k_0 and C_i $(i = 10, 11, 12)$ such that the following holds:*

- *If $\epsilon < \epsilon_{11}$, there uniquely exists $(a,b) \in \mathcal{U}_0 \times \mathcal{U}_1$ satisfying the following conditions for any $k \leq k_0$:*

$$\Phi(a,b) = \rho, \quad \big|(\nabla_t^\bullet + \psi)(a)\big|_{L_k^2} \leq C_{15}|\rho|_{L_k^2}, \quad |b|_{L_k^2} \leq C_{15}|\rho|_{L_k^2}. \tag{6.41}$$

Moreover, a and b are C^∞.

Proof The derivative of Φ at (a,b) is as follows:

$$\begin{aligned} T_{(a,b)}\Phi(u,v) = (1+a)^{-1}(\nabla_t^\bullet + \psi)(u) - (1+a)^{-1}u(1+a)(\nabla_t^\bullet + \psi)(a) \\ + (1+a)^{-1}v(1+a) - (1+a)^{-1}u(1+a)^{-1}b(1+a) \\ + (1+a)^{-1}bu. \end{aligned} \tag{6.42}$$

Hence, there exists a positive constant C_{16} depending only on r, k_0 and C_i $(i = 10, 11, 12)$ such that the following holds for $(u,v) \in T_a\mathcal{U}_0 \oplus v \in T_b\mathcal{U}_1$:

$$\begin{aligned} \big|T_{a,b}\Phi(u,v) - (\nabla_t^\bullet + \psi)u - v\big|_{L_k^2} \leq C_{16}(1 + |a|_{L_{k+1}^2})^2|(\nabla_t^\bullet + \psi)a|_{L_k^2}\big|(\nabla_t^\bullet + \psi)u\big|_{L_k^2} \\ + C_{16}(1 + |a|_{L_{k+1}^2})^3|b|_{L_k^2}|u|_{L_k^2} + C_{16}(1 + |a|_{L_{k+1}^2})|a|_{L_{k+1}^2}|v|_{L_k^2}. \end{aligned} \tag{6.43}$$

Then, we obtain (a,b) satisfying (6.41) as in the case of Proposition 6.6.5.

Let us observe that (a,b) is C^∞. Note that $(\nabla_t^\bullet + \psi)(a) + ba = \big((1+a)\rho\big)^\top$ and $b = \big((1+a)\rho\big)^\bullet$. By the second equality, b is L_{k+1}^2. By the first, we obtain that a is L_{k+2}^2. By an easy inductive argument, we obtain that a and b are L_ℓ^2 for any ℓ. ∎

6.7.3 Comparison of the Decompositions

Let $\widetilde{\nabla} := \nabla + \psi\, dt$. By Proposition 6.6.1, there exists the $\widetilde{\nabla}$-invariant decomposition $E = \bigoplus_{\alpha \in \mathcal{S}} E_\alpha$ with the following property:

- For any eigenvalue γ of $M(\widetilde{\nabla})$ on E_α, we obtain $\left|T^{-1}\log|\gamma| - \alpha\right| \leq C_{10}$.

We assume that $\epsilon < \epsilon_{11}$. Let (a, b) be as in Proposition 6.7.4.

Proposition 6.7.5 *We obtain* $E^{\bullet}_{\alpha} = (1 + a)E_\alpha$.

Proof We set $\widetilde{E}_\alpha := (1 + a)^{-1}E^{\bullet}_{\alpha}$. The connection $\nabla^{\bullet}_t + \psi + b$ preserves the decomposition $E = \bigoplus E^{\bullet}_{\alpha}$. Because $\nabla_t + \psi = (1 + a)^{-1}(\nabla^{\bullet}_t + \psi + b)(1 + a)$, the connection $\nabla_t + \psi$ preserves the decomposition $\bigoplus \widetilde{E}_\alpha$. By considering the eigenvalues of the monodromy on $\widetilde{E}_\alpha$, we obtain that $\widetilde{E}_\alpha = E_\alpha$. ■

Corollary 6.7.6 *There exists a constant* C_{20} *depending only on* r, C_i $(i = 10, 11, 12)$ *such that the following holds:*

- *For any* $Q \in S^1_T$, *and for any* $u_\alpha \in E_{\alpha|Q}$ *and* $u_\beta \in E_{\beta|Q}$ *with* $\alpha \neq \beta$, *we obtain*

$$\left|h(u_\alpha, u_\beta)\right| \leq C_{20}\left|\rho\right|_{L^2} \cdot |u_\alpha|_h \cdot |u_\beta|_h.$$

Proof Note that $\sup |a|_h \leq C_{21}|a|_{L^1_k} \leq |\rho|_{L^2}$. Then, the claim follows from $h\big((1 + a_{|Q})u_\alpha, (1 + a_{|Q})u_\beta\big) = 0$. ■

We set $\operatorname{End}(E)^{\circ} := \bigoplus_{\alpha \in \mathcal{S}} \operatorname{End}(E_\alpha)$ and $\operatorname{End}(E)^{\perp} := \bigoplus_{\alpha \neq \beta} \operatorname{Hom}(E_\alpha, E_\beta)$. For any vector bundle E', and for any section g of $\operatorname{End}(E) \otimes E'$, there exists the corresponding decomposition $g = g^{\circ} + g^{\perp}$.

Corollary 6.7.7 *There exists a positive constant* $C_{21} > 0$ *such that the following holds:*

- *Let* g *be any section of* $\operatorname{End}(E)^{\bullet} \otimes \Omega^p$. *Then,* $\left|g^{\perp}\right|_{L^2_k} \leq C_{21} \cdot \left|\rho\right|_{L^2_k} \cdot \left|g\right|_{L^2_k}$.

Proof Let π_α denote the projection of E onto $E_\alpha \subset E$ with respect to the decomposition $E = \bigoplus E_\alpha$. Let $\pi^{\bullet}_{\alpha}$ denote the orthogonal projection of E onto $E^{\bullet}_{\alpha} \subset E$. We have $\pi_\alpha = (1 + a) \circ \pi^{\bullet}_{\alpha} \circ (1 + a)^{-1}$.

We have $g^{\perp} = \sum_{\alpha \neq \beta} \pi_\alpha \circ g \circ \pi_\beta$. By using $\pi^{\bullet}_{\alpha} \circ g \circ \pi^{\bullet}_{\beta} = 0$ for any $\alpha \neq \beta$, we can easily obtain the desired estimate. ■

6.8 Proof of Theorem 6.3.4

Let (E, h, ∇, ϕ) be a monopole on $\mathcal{B}^{0*}_q(R)$ satisfying Condition 6.3.1 and Condition 6.3.2 as in Sect. 6.3. There exists the decomposition (6.5). We put $S(E) := \{\ell \mid E_\ell \neq 0\}$. We set $\psi = -\sqrt{-1}\phi$.

Lemma 6.8.1 *There exist positive numbers* C_0, R_0 *and an orthogonal decomposition* $(E,\phi)=\bigoplus_{\ell\in S(E)}(E^{\bullet}_{\ell},\phi^{\bullet}_{\ell})$ *on* $\mathcal{B}^{0*}_q(R_0)$ *such that any eigenvalue* α *of* $\phi^{\bullet}_{\ell|(t,w_q)}$ *satisfies* $\left|\alpha-\sqrt{-1}\ell T^{-1}\log|w_q|\right|\leq C_0$. *In other words, Condition 6.4.1 is satisfied.*

Proof It follows from Proposition 6.6.1. ■

Proposition 6.8.2 *There exists a section* a *of* $\bigoplus_{\ell_1\neq\ell_2}\mathrm{Hom}(E^{\bullet}_{\ell_1},E^{\bullet}_{\ell_2})$ *such that the following holds:*

- $E^{\bullet}_{\ell}=(1+a)E_{\ell}$ *for any* $\ell\in S(E)$.
- *For any* $k\in\mathbb{Z}_{\geq 0}$, *there exist positive constants* $C_i(k)>0$ $(i=10,11)$ *such that*

$$\left|\nabla_{\kappa_1}\circ\cdots\circ\nabla_{\kappa_k}(a)\right|_h\leq C_{10}(k)\cdot\exp\big(-C_{11}(k)I_q(w_q)\big)$$

for any $(\kappa_1,\ldots,\kappa_k)\in\{t,x,y\}^k$*:*

Proof From the decomposition $E=\bigoplus E^{\bullet}_{\ell}$, we obtain the decomposition

$$\mathrm{End}(E)=\mathrm{End}(E)^{\bullet}\oplus\mathrm{End}(E)^{\top}.$$

For any section s of $\mathrm{End}(E)\otimes\Omega^p$, we use the decomposition $s=s^{\bullet}+s^{\top}$ into the section $s^{\bullet}$ of $\mathrm{End}(E)^{\bullet}\otimes\Omega^p$ and the section $s^{\top}$ of $\mathrm{End}(E)^{\top}\otimes\Omega^p$. We obtain the decomposition $\nabla=\nabla^{\bullet}+\rho$ as in Sect. 6.4.1. The norm of ρ is dominated by Proposition 6.4.2.

By Proposition 6.7.4, there exist $R_1>R$ and a section (a,b) of $\mathrm{End}(E)^{\top}\oplus\mathrm{End}(E)^{\bullet}$ on $\mathcal{B}^{0*}_q(R_1)$ such that

$$(1+a)^{-1}\circ(\nabla^{\bullet}_t+\psi+b)\circ(1+a)=\nabla^{\bullet}_t+\rho_t+\psi.$$

Here, for each $w_q\in U^*_{w,q}(R_1)$, the restriction $(a,b)_{|S^1_T\times\{w_q\}}$ is C^{∞}, and we obtain

$$\left|(\nabla^{\bullet}_t+\psi)a_{|S^1_T\times\{w_q\}}\right|_{L^2_k}+\left|b_{|S^1_T\times\{w_q\}}\right|_{L^2_k}\leq C(k)\left|\rho_{t|S^1_T\times\{w_q\}}\right|_{L^2_k}$$

for any k. It is easy to see that (a,b) are C^{∞}-sections of $\mathrm{End}(E)^{\top}\oplus\mathrm{End}(E)^{\bullet}$ by using the inverse function theorem.

Because $(\nabla^{\bullet}_t+\psi)(a)+ba=\big((1+a)\rho_t\big)^{\top}$ and $b=\big((1+a)\rho_t\big)^{\bullet}$, we obtain

$$(\nabla^{\bullet}_t+\psi)(a)+\big((1+a)\rho_t\big)^{\bullet}a-\big((1+a)\rho_t\big)^{\top}=0.$$

For any $\kappa=(\kappa_1,\ldots,\kappa_k)\in\{t,x,y\}^k$, we set $k:=|\kappa|$ and $\nabla^{\bullet}_{\kappa}:=\nabla^{\bullet}_{\kappa_1}\circ\cdots\circ\nabla^{\bullet}_{\kappa_k}$. By an inductive argument, we obtain equalities

$$(\nabla^{\bullet}_t+\psi+\mathcal{B}_{\kappa})\nabla^{\bullet}_{\kappa}a+\mathcal{C}_{\kappa}=0,$$

where $\mathcal{B}_\kappa$ and $\mathcal{C}_\kappa$ are constructed by some linear algebraic operations from $\nabla^{\bullet}_{\kappa_1}\psi$, $\nabla^{\bullet}_{\kappa_2}a$ and $\nabla^{\bullet}_{\kappa_3}\rho$, where $|\kappa_i| < |\kappa|$. We obtain the estimates for $\big|\nabla^{\bullet}_{\kappa}(a)_{|S^1_T\times\{w_q\}}\big|_{L^2}$ by an easy induction. Therefore, we obtain the claim of Proposition 6.8.2. ■

Let us complete the proof of Theorem 6.3.4. For any endomorphism f of E, let $f_h^{\dagger}$ denote the adjoint with respect to h. Let $\Pi^{\bullet}_{\ell}$ denote the projection $E \longrightarrow E^{\bullet}_{\ell} \subset E$ with respect to the orthogonal decomposition $E = \bigoplus E^{\bullet}_{\ell}$. Note that $(\Pi^{\bullet}_{\ell})^{\dagger}_h = \Pi^{\bullet}_{\ell}$ and $\sum_{\ell\in S(E)}(\Pi^{\bullet}_{\ell})^{\dagger}_h \circ \Pi^{\bullet}_{\ell} = \mathrm{id}_E$. Let Π_ℓ denote the projection $E \longrightarrow E_\ell$ with respect to the decomposition $E = \bigoplus_{\ell\in S(E)} E_\ell$. Note that

$$\Pi_\ell = (\mathrm{id}_E + a)^{-1} \circ \Pi^{\bullet}_{\ell} \circ (\mathrm{id}_E + a).$$

Let s be the automorphism of E determined by $h^{\circ} = h\cdot s$. Then, the following holds.

$$s = \sum_{\ell\in S(E)} (\Pi_\ell)^{\dagger}_h \circ \Pi_\ell.$$

We obtain the desired estimate for s from Proposition 6.8.2. Thus, we obtain Theorem 6.3.4. ■

Chapter 7
The Filtered Bundles Associated with Periodic Monopoles

Abstract In Sect. 7.1, we summarize some notation, which will be used in this chapter. In Sects. 7.2–7.3, we shall prove that a periodic monopole satisfying the GCK-condition induces a good filtered bundle. Moreover, the norm estimate is satisfied. In Sect. 7.4, we shall study the converse. Namely, we shall prove that if a monopole $(E, \overline{\partial}_E, h)$ is strongly adapted to a good filtered bundle, then the monopole satisfies the GCK-condition.

7.1 Notation

7.1.1 Some Spaces and Morphisms

We summarize some notation. For any $q \in \mathbb{Z}_{\geq 1}$, let w_q be a q-th root of the variable w such that $w_{q\ell}^{\ell} = w_q$ for any $q, \ell \in \mathbb{Z}_{>0}$. Let $\mathbb{P}^1_{w_q} \longrightarrow \mathbb{P}^1_w$ be the ramified covering induced by $w_q \longmapsto w_q^q$. The pull back of w is also denoted by w, i.e., $w = w_q^q$ on $\mathbb{P}^1_{w_q}$.

Let $\lambda \in \mathbb{C}$. We have the map $\Psi^\lambda : \overline{\mathcal{M}}^\lambda \longrightarrow \mathbb{C}_w$ induced by $(t_1, \beta_1) \longmapsto (1+|\lambda|^2)^{-1}(\beta_1 - 2\sqrt{-1}\lambda t_1)$. (See Sect. 2.7 for the mini-complex manifold $\overline{\mathcal{M}}^\lambda$ and mini-complex local coordinate systems (t_1, β_1).) For $q \in \mathbb{Z}_{\geq 1}$, let Y_q^λ be the fiber product of $\overline{\mathcal{M}}^\lambda$ and $\mathbb{P}^1_{w_q} \setminus \{0\}$ over $\mathbb{P}^1_{w_q}$. Let $\Psi_q^\lambda : Y_q^\lambda \longrightarrow \mathbb{P}^1_{w_q} \setminus \{0\}$ denote the induced proper map. We set $H^\lambda_{\infty,q} = (\Psi_q^\lambda)^{-1}(\infty)$ and $Y_q^{\lambda *} = Y_q^\lambda \setminus H^\lambda_{\infty,q} = (\Psi_q^\lambda)^{-1}(\mathbb{P}^1_{w_q} \setminus \{0,\infty\})$. Let $\pi_q^\lambda : Y_q^\lambda \longrightarrow S^1_T$ denote the morphism obtained as the composite $Y_q^\lambda \longrightarrow \overline{\mathcal{M}}^\lambda \longrightarrow S^1_T$.

Let $\varpi^\lambda : \overline{M}^\lambda \longrightarrow \overline{\mathcal{M}}^\lambda$ denote the projection in Sect. 2.7.8. Let $Y_q^{\lambda\,\mathrm{cov}}$ denote the fiber product of Y_q^λ and $\overline{M}^\lambda$ over $\overline{\mathcal{M}}^\lambda$. Let $\varpi_q^\lambda : Y_q^{\lambda\,\mathrm{cov}} \longrightarrow Y_q^\lambda$ denote the projection. Note that $Y_q^{\lambda\,\mathrm{cov}}$ is equipped with the naturally induced free $\mathbb{Z}$-action κ_q, and Y_q^λ is identified with the quotient space. We set $H^{\lambda\,\mathrm{cov}}_{\infty,q} := (\varpi_q^\lambda)^{-1}(H^\lambda_{\infty,q})$ and $Y_q^{\lambda\,\mathrm{cov}*} := Y^{\lambda\,\mathrm{cov}} \setminus H^{\lambda\,\mathrm{cov}}_{\infty,q}$.

T. Mochizuki, *Periodic Monopoles and Difference Modules*, Lecture Notes in Mathematics 2300, https://doi.org/10.1007/978-3-030-94500-8_7

We recall that Y_q^λ and $Y_q^{\lambda\,\mathrm{cov}}$ are naturally equipped with the mini-complex structures. (See Lemma 4.3.1.) We have the natural local mini-complex coordinate systems given by $(t_1, \beta_{1,q}^{-1})$, where $\beta_{1,q}$ denotes a q-th root of the variable β_1.

Let $\widehat{H}^\lambda_{\infty,q}$ denote the formal space obtained as the completion of Y_q^λ along $H^\lambda_{\infty,q}$. (See Sect. 4.3.2 for more details on $\widehat{H}^\lambda_{\infty,q}$.) Let $\widehat{H}^{\lambda\,\mathrm{cov}}_{\infty,q}$ denote the ringed space obtained as the completion of $Y_q^{\lambda\,\mathrm{cov}}$ along $H^{\lambda\,\mathrm{cov}}_{\infty,q}$.

From the ramified covering map $\varphi_{q,p} : \mathbb{P}^1_{w_p} \longrightarrow \mathbb{P}^1_{w_q}$ for $p \in q\mathbb{Z}_{\geq 1}$, we obtain the induced ramified covering maps $\mathcal{R}_{q,p} : Y_p^\lambda \longrightarrow Y_q^\lambda$. The induced map $H^\lambda_{\infty,p} \longrightarrow H^\lambda_{\infty,q}$ is an isomorphism, and the induced map $Y_p^{\lambda *} \longrightarrow Y_q^{\lambda *}$ is a covering map.

Note that $Y_q^{\lambda *}$ and $Y_q^{\lambda\,\mathrm{cov} *}$ are equipped with the Riemannian metric induced by the Riemannian metric of $\mathcal{M}^\lambda$, which are locally Euclidean compatible with the mini-complex structures. Because $\mathcal{M}^0 = \mathcal{M}^\lambda$ as Riemannian manifolds, we have $Y_q^{\lambda *} = Y_q^{0 *}$ and $Y_q^{\lambda\,\mathrm{cov} *} = Y_q^{0\,\mathrm{cov} *}$ as Riemannian manifolds. Because the maps $\varpi_q^\lambda : Y_q^{\lambda\,\mathrm{cov} *} \longrightarrow Y_q^{\lambda *}$ and $\varpi_q^0 : Y_q^{0 * \mathrm{cov}} \longrightarrow Y_q^{0 *}$ are the same, i.e., independent of λ, the restriction of ϖ_q^λ to $Y_q^{\lambda\,\mathrm{cov} *}$ is also denoted simply by ϖ_q. Similarly, because the maps $\Psi_q^\lambda : Y_q^{\lambda *} \longrightarrow \mathbb{P}^1_{w_q} \setminus \{0, \infty\}$ and $\Psi_q^0 : Y_q^{0 *} \longrightarrow \mathbb{P}^1_{w_q} \setminus \{0, \infty\}$ are the same, the restriction of Ψ_q^λ to $Y_q^{\lambda *}$ is also denoted simply by Ψ_q.

The composite $\Psi_q^\lambda \circ \varpi_q^\lambda : Y_q^{\lambda\,\mathrm{cov}} \longrightarrow \mathbb{P}^1_{w_q} \setminus \{0\}$ is denoted by $\Psi_q^{\lambda\,\mathrm{cov}}$. The restriction $Y_q^{\lambda\,\mathrm{cov} *} \longrightarrow \mathbb{P}^1_{w_q} \setminus \{0, \infty\}$ is independent of λ, which is also denoted by Ψ_q^{cov}.

7.1.2 *Neighbourhoods of Infinity*

Let U_w be a neighbourhood of ∞ in $\mathbb{P}^1_w$. We set $U_w^* = U_w \setminus \{\infty\}$. Let $U_{w,q}$ and $U_{w,q}^*$ denote the pull back of U_w and U_w^* by the ramified covering $\mathbb{P}^1_{w_q} \longrightarrow \mathbb{P}^1_w$, respectively. We set $\mathcal{B}_q^\lambda := (\Psi_q^\lambda)^{-1}(U_{w,q})$ and $\mathcal{B}_q^{\lambda *} := \mathcal{B}_q^\lambda \setminus H^\lambda_{\infty,q}$. The induced map $\mathcal{B}_q^\lambda \longrightarrow U_{w,q}$ is also denoted by Ψ_q^λ, and the restriction $\mathcal{B}_q^{\lambda *} \longrightarrow U_{w,q}^*$ is also denoted simply by Ψ_q.

For $t_1 \in S^1_T$, we set $\mathcal{B}_q^\lambda\langle t_1\rangle := (\pi_q^\lambda)^{-1}(t_1) \cap \mathcal{B}_q^\lambda$. Similarly, we set $H^\lambda_{\infty,q}\langle t_1\rangle = (\pi_q^\lambda)^{-1}(t_1) \cap H^\lambda_{\infty,q}$ and $\mathcal{B}_q^{\lambda *}\langle t_1\rangle = (\pi_q^\lambda)^{-1}(t_1) \cap \mathcal{B}_q^{\lambda *}$. The point $H_q^\lambda\langle t_1\rangle \in \mathcal{B}_q^\lambda\langle t_1\rangle$ is also denoted by ∞ if there is no risk of confusion. The formal completion of $\mathcal{B}_q^\lambda\langle t_1\rangle$ along $H^\lambda_{\infty,q}\langle t_1\rangle$ is denoted by $\widehat{H}^\lambda_{\infty,q}\langle t_1\rangle$.

We set $\mathcal{B}_q^{\lambda\,\mathrm{cov}} := (\varpi_q^\lambda)^{-1}(\mathcal{B}_q^\lambda)$ and $\mathcal{B}_q^{\lambda\,\mathrm{cov} *} := \mathcal{B}^{\lambda\,\mathrm{cov}} \setminus H^{\lambda\,\mathrm{cov}}_{\infty,q}$. The induced maps $\mathcal{B}_q^{\lambda\,\mathrm{cov}} \longrightarrow \mathcal{B}_q^\lambda$ and $\mathcal{B}_q^{\lambda\,\mathrm{cov}} \longrightarrow U_{w,q}$ are also denoted by ϖ_q^λ and $\Psi_q^{\lambda\,\mathrm{cov}}$, and the restrictions to $\mathcal{B}^{\lambda\,\mathrm{cov} *}$ are also denoted simply by ϖ_q and Ψ_q^{cov}.

We emphasize $\mathcal{B}_q^{\lambda *} = \mathcal{B}_q^{0 *}$ and $\mathcal{B}_q^{\lambda\,\mathrm{cov} *} = \mathcal{B}_q^{0\,\mathrm{cov} *}$ as Riemannian manifolds.

7.1.3 Norm of Differential Forms

For a vector bundle E with a Hermitian metric h on $\mathcal{B}_q^{\lambda *}$, and for an End$(E)$-valued differential form s, let $|s|_h$ denote the induced function on $\mathcal{B}_q^{\lambda *}$ obtained as the norm of s with respect to h and the natural Riemannian metric $dt\,dt + dw\,d\overline{w}$ of $\mathcal{B}_q^{\lambda *}$. Similarly, for a vector bundle E with a Hermitian metric h on $U_{w,q}^*$, and for an End(E)-valued differential form s, let $|s|_h$ denote the induced function on $U_{w,q}^*$ obtained as the norm of s with respect to h and the natural Riemannian metric $dw\,d\overline{w}$ of $U_{w,q}^*$. We omit to denote the dependence on the metrics of the base spaces, because they are fixed.

7.2 Meromorphic Prolongation

7.2.1 Statements

Let U_w be a neighbourhood of ∞ in $\mathbb{P}_w^1$. Let $q \in \mathbb{Z}_{\geq 1}$. For each $\lambda \in \mathbb{C}$, we obtain the induced neighbourhood $\mathcal{B}_q^\lambda = (\Psi_q^\lambda)^{-1}(U_{w,q})$ of $H_{\infty,q}^\lambda$ in Y_q^λ as in Sect. 7.1.2, and we set $\mathcal{B}_q^{\lambda *} := \mathcal{B}_q^\lambda \setminus H_{\infty,q}^\lambda$.

Let (E, h, ∇, ϕ) be a monopole on $\mathcal{B}_q^{0*}$ satisfying the GCK-condition. (See Proposition 6.3.13.) Fix $\lambda \in \mathbb{C}$. Let $(E^\lambda, \overline{\partial}_{E^\lambda})$ denote the mini-holomorphic bundle on $\mathcal{B}_q^{\lambda *}$ underlying the monopole (E, h, ∇, ϕ). (See Sect. 2.7.9.) We obtain the $\mathcal{O}_{\mathcal{B}_q^\lambda}(*H_{\infty,q}^\lambda)$-module $\mathcal{P}^h E^\lambda$ by the procedure in Sect. 4.4.1. We shall prove the following proposition in Sect. 7.2.2.

Proposition 7.2.1 *$\mathcal{P}^h(E^\lambda)$ is a locally free $\mathcal{O}_{\mathcal{B}_q^\lambda}(*H_{\infty,q}^\lambda)$-module.*

Let $\partial_{E^\lambda,h,\beta_1}$ be obtained from $\partial_{E^\lambda,\overline{\beta}_1}$ and h as in Sect. 2.9.3. According to Proposition 2.9.5, there exists the following relation:

$$\left[\partial_{E^\lambda,\overline{\beta}_1}, \partial_{E^\lambda,h,\beta_1}\right] = (1+|\lambda|^2)^{-1} \cdot F_{\overline{w},w}. \tag{7.1}$$

We obtain the following lemma from Corollary 2.9.7 and Corollary 6.3.12.

Lemma 7.2.2 *$(E^\lambda, \overline{\partial}_{E^\lambda}, h)_{|\mathcal{B}_q^{\lambda *}\langle t_1\rangle}$ is acceptable for any $t_1 \in S_T^1$. (See* Sect. 2.11.2 *for the acceptability condition.)* ∎

As in Sect. 2.11, for each $t_1 \in S_T^1$, we obtain the locally free $\mathcal{O}_{\mathcal{B}_q^\lambda\langle t_1\rangle}(*\infty)$-module $\mathcal{P}^h\big(E^\lambda_{|\mathcal{B}_q^{\lambda *}\langle t_1\rangle}\big)$, and the filtered bundle $\mathcal{P}_*^h\big(E^\lambda_{|\mathcal{B}_q^{\lambda *}\langle t_1\rangle}\big)$ over $\mathcal{P}^h\big(E^\lambda_{|\mathcal{B}_q^{\lambda *}\langle t_1\rangle}\big)$. We shall prove the following lemma in Sect. 7.2.2.

Lemma 7.2.3 *$\mathcal{P}^h(E^\lambda)_{|\mathcal{B}_q^\lambda\langle t_1\rangle}$ is naturally isomorphic to $\mathcal{P}^h\big(E^\lambda_{|\mathcal{B}_q^{\lambda *}\langle t_1\rangle}\big)$.*

Let P be any point of $H^{\lambda}_{\infty,q}$. We take a mini-holomorphic frame $\boldsymbol{u} = (u_1, \dots, u_r)$ of $\mathcal{P}^h(E^{\lambda})$ on a neighbourhood U_P of P. Let $H(h, \boldsymbol{u})$ be the Hermitian matrix valued function on $U_P \setminus H^{\lambda}_{\infty,q}$ determined by $H(h, \boldsymbol{u})_{ij} := h(u_i, u_j)$. We shall prove the following lemma in Sect. 7.2.2.

Lemma 7.2.4 *There exist $C_1 \geq 1$ and $N > 0$ such that $C_1^{-1}|x_q|^{-N} \leq |H(h, \boldsymbol{u})| \leq C_1|x_q|^N$.*

7.2.2 Proof

Take $P \in H^{\lambda}_{\infty,q}$. We put $t_1^0 = \pi^{\lambda}_q(P) \in S^1_T$. Let U_P be a neighbourhood of P in $\mathcal{B}^{\lambda}_q$ of the form $\{|t_1 - t_1^0| < \epsilon_1\} \times \{|\beta_{1,q}^{-1}| < \epsilon_2\}$. Let s be a holomorphic section of $\mathcal{P}^h\big(E^{\lambda}_{|\mathcal{B}^{\lambda *}_q\langle t_1\rangle}\big)$. The restriction $s_{|U_P \cap \mathcal{B}^{\lambda *}_q\langle t_1\rangle}$ uniquely extends to a mini-holomorphic section $\widetilde{s}$ of E on $U_P \setminus H^{\lambda}_{\infty,q}$.

Lemma 7.2.5 *$\widetilde{s}$ is a section of $\mathcal{P}^h(E^{\lambda})$ on U_P. More precisely, there exist $C_1 \geq 1$ and $N_1 > 0$ such that*

$$C_1^{-1}|\beta_1|^{-N_1} \cdot |s|_h(t_1^0, \beta_{1,q}^{-1}) \leq \big|\widetilde{s}\big|_h(t_1, \beta_{1,q}^{-1}) \leq C_1|\beta_1|^{N_1} \cdot |s|_h(t_1^0, \beta_{1,q}^{-1}).$$

Proof Because $\partial_{E^{\lambda},t_1} = \nabla_{t_1} - \sqrt{-1}\phi$. we obtain $\nabla_{t_1}\widetilde{s} = \sqrt{-1}\phi\widetilde{s}$. We obtain

$$\Big|\partial_{t_1}\big|\widetilde{s}\big|_h^2\Big| = \Big|2\,\mathrm{Re}\,h(\sqrt{-1}\phi\widetilde{s}, \widetilde{s})\Big| \leq C_1 \log|\beta_1 - 2\sqrt{-1}\lambda t_1| \cdot |\widetilde{s}|_h^2$$
$$\leq \big(C_2 \log|\beta_1| + C_3\big) \cdot |\widetilde{s}|_h^2. \tag{7.2}$$

Hence, we obtain the claim of the lemma. ■

By Lemma 7.2.5, the following morphism is an isomorphism:

$$H^0\Big(U_P, \mathcal{P}^h(E^{\lambda})\Big) \longrightarrow H^0\Big(U_P \cap \mathcal{B}^{\lambda}_q\langle t_1^0\rangle, \mathcal{P}^h(E^{\lambda}_{|\mathcal{B}^{\lambda *}_q\langle t_1^0\rangle})\Big). \tag{7.3}$$

Then, we obtain Lemma 7.2.3.

The following natural morphism is an isomorphism:

$$A := H^0\Big(U_P, \mathcal{O}_{\mathcal{B}^{\lambda}_q}(*H^{\lambda}_{\infty,q})\Big) \longrightarrow H^0\Big(\mathcal{B}^{\lambda}_q\langle t_1^0\rangle, \mathcal{O}_{\mathcal{B}^{\lambda}_q\langle t_1^0\rangle}(*\{\infty\})\Big).$$

As remarked in Sect. 7.2.1, $\mathcal{P}^h(E^{\lambda}_{|\mathcal{B}^{\lambda *}_q\langle t_1^0\rangle})$ is a locally free sheaf, and hence the both side of (7.3) are free A-modules. Hence, there exists a local frame of $\mathcal{P}^h(E^{\lambda})$, i.e., we obtain Proposition 7.2.1. We also obtain Lemma 7.2.4 from Lemma 7.2.5. ■

7.3 Filtered Prolongation

7.3.1 Statement

We continue to use the notation in Sect. 7.2.1. As a result of Lemma 7.2.3, the family

$$\mathcal{P}^h_*(E^\lambda) := \left(\mathcal{P}^h_*(E^\lambda_{|\mathcal{B}^{\lambda*}_q(t_1)})\,\Big|\, t_1 \in S^1_T\right)$$

is a filtered bundle over $\mathcal{P}^h(E^\lambda)$. (See Definition 4.3.3.) We shall prove the following proposition in this subsection.

Theorem 7.3.1 *The filtered bundle $\mathcal{P}^h_*(E^\lambda)$ is good. (See Definition 4.3.4.)*

7.3.2 Refined Statement

Suppose that Condition 6.3.2 is satisfied for (E, h, ∇, ϕ), i.e., there exists a decomposition of the mini-holomorphic bundle $(E^0, \overline{\partial}_{E^0})$ on $\mathcal{B}^{0*}_q$ as in (6.4):

$$(E^0, \overline{\partial}_{E^0}) = \bigoplus_{(\ell,\alpha)\in\mathbb{Z}\times\mathbb{C}^*} (E^0_{\ell,\alpha}, \overline{\partial}_{E^0_{\ell,\alpha}}). \tag{7.4}$$

(See Remark 6.3.3.) We shall explain a more precise claim in this situation.

There exist the holomorphic bundles $(V_{\ell,\alpha}, \overline{\partial}_{\ell,\alpha})$ on $U^*_{w,q}$ with a holomorphic automorphism $f_{\ell,\alpha}$, and isomorphisms of mini-holomorphic bundles as in (6.7):

$$(E^0_{\ell,\alpha}, \overline{\partial}_{E^0_{\ell,\alpha}}) \simeq \mathbb{L}^{0*}_q(\ell,\alpha) \otimes \Psi^*_q(V_{\ell,\alpha}, f_{\ell,\alpha}). \tag{7.5}$$

7.3.2.1 The Filtered Bundles Associated with the Basic Monopoles of Rank One

For $(\ell,\alpha) \in \mathbb{Z}\times\mathbb{C}^*$, there exists the monopole $(\mathbb{L}^*_q(\ell,\alpha), h_{\mathbb{L},q,\ell,\alpha}, \nabla_{\mathbb{L},q,\ell,\alpha}, \phi_{\mathbb{L},q,\ell,\alpha})$ on $Y^{\lambda*}_q = Y^{0*}_q$ as in (5.9). Let $\mathbb{L}^{\lambda*}_q(\ell,\alpha)$ denote the underlying mini-holomorphic bundle on $Y^{\lambda*}_q$. By the procedure in Sect. 4.4.1, it extends to a locally free $\mathcal{O}_{Y^\lambda_q}(*H^\lambda_q)$-module $\mathbb{L}^\lambda_q(\ell,\alpha)$, and the good filtered bundle $\mathcal{P}_*\mathbb{L}^\lambda_q(\ell,\alpha)$. There exists an isomorphism

$$\mathcal{P}_*\mathbb{L}^\lambda_q(\ell,\alpha)_{|\widehat{H}^\lambda_{\infty,p}} \simeq \mathcal{P}^{(0)}_*\widehat{\mathbb{L}}^\lambda_q(\ell,\alpha). \tag{7.6}$$

(See Sect. 3.5.3 for $\mathcal{P}^{(0)}_*\widehat{\mathbb{L}}^\lambda_q(\ell,\alpha)$.)

7.3.2.2 The Filtered Bundles Associated with the Asymptotic Harmonic Bundles

We set $\theta_{\ell,\alpha} = f_{\ell,\alpha}\,dw$. As explained in Sects. 6.3.3–6.3.4, we obtain Hermitian metrics $h_{V,\ell,\alpha}$ of $(V_{\ell,\alpha}, \overline{\partial}_{\ell,\alpha}, \theta_{\ell,\alpha})$ for which the estimate (6.8) holds. Let $\overline{\partial}_{\ell,\alpha}+\partial_{\ell,\alpha}$ be the Chern connection determined by $\overline{\partial}_{\ell,\alpha}$ and $h_{V,\ell,\alpha}$. Let $\theta^{\dagger}_{\ell,\alpha}$ denote the adjoint of $\theta_{\ell,\alpha}$. According to the estimate for asymptotic harmonic bundles in [65, §5.5] (see also Proposition 10.1.2 below), we obtain

$$\big[\overline{\partial}_{\ell,\alpha}, \partial_{\ell,\alpha}\big] = O\big(|w_q|^{-2}(\log|w_q|)^{-2} dw_q\, d\overline{w}_q\big),$$

$$\big[\theta_{\ell,\alpha}, \theta^{\dagger}_{\ell,\alpha}\big] = O\big(|w_q|^{-2}\log|w_q|^{-2} dw_q\, d\overline{w}_q\big).$$

Let $V^{\lambda}_{\ell,\alpha}$ be the holomorphic bundle $(V_{\ell,\alpha}, \overline{\partial}_{\ell,\alpha} + \lambda\theta^{\dagger}_{\ell,\alpha})$ with the metric $h_{V,\ell,\alpha}$. The Chern connection is $\overline{\partial}_{\ell,\alpha} + \lambda\theta^{\dagger}_{\ell,\alpha} + \partial_{\ell,\alpha} - \overline{\lambda}\theta_{\ell,\alpha}$, and the curvature is

$$\big[\overline{\partial}_{\ell,\alpha} + \lambda\theta^{\dagger}_{\ell,\alpha}, \partial_{\ell,\alpha} - \overline{\lambda}\theta_{\ell,\alpha}\big] = O\big(|w_q|^{-2}(\log|w_q|)^{-2} dw_q\, d\overline{w}_q\big). \tag{7.7}$$

Hence, we obtain the locally free $\mathcal{O}_{U_{w,p}}(*\infty)$-module $\mathcal{P}^{h_{V,\ell,\alpha}}V^{\lambda}_{\ell,\alpha}$ and the filtered bundle $\mathcal{P}^{h_{V,\ell,\alpha}}_{*}V^{\lambda}_{\ell,\alpha}$ over $\mathcal{P}^{h_{V,\ell,\alpha}}V^{\lambda}_{\ell,\alpha}$.

Set $\mathbb{D}^{\lambda\,(1,0)} := \lambda\partial_{\ell,\alpha} + \theta_{\ell,\alpha}$ and $\mathbb{D}^{\lambda} = \mathbb{D}^{\lambda\,(1,0)} + \overline{\partial}_{\ell,\alpha} + \lambda\theta^{\dagger}_{\ell,\alpha}$. Let $\boldsymbol{v}_{\ell,\alpha}$ be a holomorphic frame of $\mathcal{P}^{h_{V,\ell,\alpha}}_{a}V^{\lambda}_{\ell,\alpha}$. Let A be the matrix valued function on $U^{*}_{w,q}$ determined by $\mathbb{D}^{\lambda(1,0)}\boldsymbol{v}_{\ell,\alpha} = \boldsymbol{v}_{\ell,\alpha}A\,dw$. We obtain the following lemma. (See Sect. 10.1.3 below.)

Lemma 7.3.2 *A is a C^{∞}-function on $U_{w,q}$, and the Taylor series of $\overline{\partial}A$ at ∞ is 0. In particular, the Taylor series of A at ∞ is an element of $M_r(\mathbb{C}[\![w_q^{-1}]\!])$. Moreover, $A_{|\infty}$ is nilpotent.* ■

We use the variable $x_q = (1+|\lambda|^2)^{1/q}w_q$ instead of w_q. We obtain a filtered bundle $\mathcal{P}_{*}\widehat{\mathcal{V}}^{\lambda}_{\ell,\alpha} := \mathcal{P}^{h_{V,\ell,\alpha}}_{*}V^{\lambda}_{\ell,\alpha|\widehat{\infty}_{x,q}}$ on $(\widehat{\infty}_{x,q}, \infty)$. (See Sect. 3.4.2 for the ringed space $\widehat{\infty}_{x,q}$, and Definition 3.5.1 for filtered bundle in this situation.) By $\widehat{\mathbb{D}}^{\lambda}_{\ell,\alpha}\boldsymbol{v}_{|\widehat{\infty}_{x,q}} = \boldsymbol{v}_{|\widehat{\infty}_{x,q}}A_{|\widehat{\infty}_{x,q}}$, we obtain the formal λ-connection $\widehat{\mathbb{D}}^{\lambda}_{\ell,\alpha}$. It is independent of the choice of $\boldsymbol{v}$. By Proposition 10.1.6, we obtain the following lemma.

Lemma 7.3.3 *$(\mathcal{P}_{*}\widehat{\mathcal{V}}^{\lambda}_{\ell,\alpha}, \widehat{\mathbb{D}}^{\lambda}_{\ell,\alpha})$ are good filtered λ-flat bundles.* ■

Applying the procedure in Sect. 3.6.3, we obtain filtered bundles $(\Psi^{\lambda}_{q})^{*}(\mathcal{P}_{*}\widehat{\mathcal{V}}^{\lambda}_{\ell,\alpha}, \widehat{\mathbb{D}}^{\lambda}_{\ell,\alpha})$ over locally free $\mathcal{O}_{\widehat{H}^{\lambda}_{\infty,q}}(*H^{\lambda}_{\infty,q})$-modules $(\Psi^{\lambda}_{q})^{*}(\mathcal{P}\widehat{\mathcal{V}}^{\lambda}_{\ell,\alpha}, \widehat{\mathbb{D}}^{\lambda})$. According to Proposition 3.6.15, they are good filtered bundles.

7.3.2.3 Refined Statement

Theorem 7.3.4 *There exists an isomorphism*

$$\mathcal{P}^h E^{\lambda}_{|\widehat{H}^{\lambda}_{\infty,q}} \simeq \bigoplus_{(\ell,\alpha)} \widehat{\mathbb{L}}^{\lambda}_q(\ell,\alpha) \otimes (\Psi^{\lambda}_q)^*(\mathcal{P}\widehat{\mathcal{V}}^{\lambda}_{\ell,\alpha}, \widehat{\mathbb{D}}^{\lambda}_{\ell,\alpha}) \tag{7.8}$$

such that it induces an isomorphism of filtered bundles, i.e., the following holds for any $t_1 \in S^1_T$:

$$\mathcal{P}^h_* E^{\lambda}_{|\widehat{H}^{\lambda}_{\infty,q}\langle t_1\rangle} \simeq \bigoplus_{(\ell,\alpha)} \mathcal{P}^{(0)}_* \widehat{\mathbb{L}}^{\lambda}_q(\ell,\alpha)_{|\widehat{H}^{\lambda}_{\infty,q}\langle t_1\rangle} \otimes (\Psi^{\lambda}_q)^*(\mathcal{P}_*\widehat{\mathcal{V}}^{\lambda}_{\ell,\alpha}, \widehat{\mathbb{D}}^{\lambda}_{\ell,\alpha})_{|\widehat{H}^{\lambda}_{\infty,q}\langle t_1\rangle}.$$

Theorem 7.3.1 immediately follows from Theorem 7.3.4, Proposition 3.6.15 and Lemma 4.3.6. We shall prove Theorem 7.3.4 in Sects. 7.3.4–7.3.10.

7.3.3 *Norm Estimate*

Before going to the proof of Theorem 7.3.4, we state the norm estimate for monopoles satisfying the GCK-condition, which will be proved in Sect. 7.3.11.

Proposition 7.3.5 *The norm estimate holds for* $(E^{\lambda}, \overline{\partial}_{E^{\lambda}}, h)$ *with respect to the good filtered bundle* $\mathcal{P}^h_* E^{\lambda}$. *(See Definition* 4.4.5 *for the norm estimates for good filtered bundles with a Hermitian metric.)*

Remark 7.3.6 According to Proposition 7.3.5, the GCK-condition for monopoles implies the norm estimate with respect to the associated filtered bundles $\mathcal{P}^h_* E^{\lambda}$. In particular, the metric h is strongly adapted to $\mathcal{P}^h_* E^{\lambda}$. We shall study the converse in Proposition 7.4.1. ■

7.3.4 *Step 0*

We use the notation in Sect. 6.3. Note that $\mathcal{B}^{\lambda*}_q = \mathcal{B}^{0*}_q$ as Riemannian manifolds. We may assume that $U_w = \{w \in \mathbb{C} \mid |w| > R\} \cup \{\infty\}$.

Let (E, h, ∇, ϕ) be a monopole as in Sect. 7.3.2. There exist the decomposition (7.4) on $\mathcal{B}^{0*}_q$, holomorphic vector bundles with an automorphism $(V_{\ell,\alpha}, f_{\ell,\alpha})$ on $U^*_{w,q}$, and an isomorphism (7.5). As in Sect. 7.3.2.2 (see also Sect. 6.3.3), we have the Hermitian metrics $h_{V,\ell,\alpha}$ of $V_{\ell,\alpha}$, and the Hermitian metric $h^\sharp$ of E as

$$h^\sharp = \bigoplus_{\ell,\alpha} h_{\mathbb{L},q,\ell,\alpha} \otimes \Psi_q^{-1}(h_{V,\ell,\alpha}).$$

Let $\nabla^\sharp$ and $\phi^\sharp$ denote the Chern connection and the anti-Hermitian endomorphism associated with the mini-holomorphic bundle $(E^0, \overline{\partial}_{E^0})$ with $h^\sharp$ as in Sect. 6.3.3. They are compatible with the decomposition (7.4), i.e., we have the decomposition $\nabla^\sharp = \bigoplus \nabla^\sharp_{\ell,\alpha}$ and $\phi^\sharp = \bigoplus \phi^\sharp_{\ell,\alpha}$.

Let s be the automorphism of E such that (i) $h = h^\sharp s$, (ii) s is self-adjoint with respect to both h and $h^\sharp$.

Proposition 7.3.7 *For any $m \in \mathbb{Z}_{\geq 0}$, there exist positive constants $A_i(m)$ $(i = 1, 2)$ such that the following holds for any $(\kappa_1, \ldots, \kappa_m) \in \{t, w, \overline{w}\}^m$.*

$$\Big|\nabla^\sharp_{\kappa_1} \circ \cdots \circ \nabla^\sharp_{\kappa_m}(s - \mathrm{id}_E)\Big|_{h^\sharp} \leq A_1(m) \exp\big(-A_2(m) \cdot |w_q^q|\big),$$

$$\Big|\nabla^\sharp_{\kappa_1} \circ \cdots \circ \nabla^\sharp_{\kappa_m}(\nabla - \nabla^\sharp)\Big|_{h^\sharp} \leq A_1(m) \exp\big(-A_2(m) \cdot |w_q^q|\big),$$

$$\Big|\nabla^\sharp_{\kappa_1} \circ \cdots \circ \nabla^\sharp_{\kappa_m}(\phi - \phi^\sharp)\Big|_{h^\sharp} \leq A_1(m) \exp\big(-A_2(m) \cdot |w_q^q|\big),$$

$$\Big|\nabla^\sharp_{\kappa_1} \circ \cdots \circ \nabla^\sharp_{\kappa_m}(F(\nabla) - F(\nabla^\sharp))\Big|_{h^\sharp} \leq A_1(m) \exp\big(-A_2(m) \cdot |w_q^q|\big).$$

Proof It follows from Theorem 6.3.4, Corollary 6.3.6, Proposition 6.3.8 and Corollary 6.3.9. ∎

7.3.5 Step 1

We define the differential operators $\partial^\sharp_{E^\lambda,\overline{\beta}_1}$ and $\partial^\sharp_{E^\lambda,t_1}$ on $C^\infty(\mathcal{B}_q^{\lambda *}, E)$ as follows:

$$\partial^\sharp_{E^\lambda,\overline{\beta}_1} := \frac{1}{1+|\lambda|^2}\left(\frac{\lambda\sqrt{-1}}{2}\nabla^\sharp_t + \nabla^\sharp_{\overline{w}} - \frac{\lambda}{2}\phi^\sharp\right),$$

$$\partial^\sharp_{E^\lambda,t_1} := \frac{1-|\lambda|^2}{1+|\lambda|^2}\nabla^\sharp_t - \frac{2\lambda\sqrt{-1}}{1+|\lambda|^2}\nabla^\sharp_w + \frac{2\overline{\lambda}\sqrt{-1}}{1+|\lambda|^2}\nabla^\sharp_{\overline{w}} - \sqrt{-1}\phi^\sharp.$$

We set $\nu_{\overline{\beta}_1} := \partial^{\sharp}_{E^\lambda,\overline{\beta}_1} - \partial_{E^\lambda,\overline{\beta}_1}$ and $\nu_{t_1} := \partial^{\sharp}_{E^\lambda,t_1} - \partial_{E^\lambda,t_1}$. By the construction, we obtain the following equalities:

$$\nu_{\overline{\beta}_1} = \frac{1}{1+|\lambda|^2}\left(\frac{\lambda\sqrt{-1}}{2}(\nabla^{\sharp}_t - \nabla_t) + (\nabla^{\sharp}_{\overline{w}} - \nabla_{\overline{w}}) - \frac{\lambda}{2}(\phi^{\sharp} - \phi)\right),$$

$$\nu_{t_1} = \frac{1-|\lambda|^2}{1+|\lambda|^2}(\nabla^{\sharp}_t - \nabla_t) - \frac{2\lambda\sqrt{-1}}{1+|\lambda|^2}(\nabla^{\sharp}_w - \nabla_w) + \frac{2\overline{\lambda}\sqrt{-1}}{1+|\lambda|^2}(\nabla^{\sharp}_{\overline{w}} - \nabla_{\overline{w}}) - \sqrt{-1}(\phi^{\sharp} - \phi).$$

We obtain the following estimates from Proposition 7.3.7.

Lemma 7.3.8 *For any $m \in \mathbb{Z}_{\geq 0}$, there exists $\epsilon_m > 0$ such that*

$$\left(\partial^{\sharp}_{E^\lambda,\kappa_1} \circ \cdots \circ \partial^{\sharp}_{E^\lambda,\kappa_m}\right)\nu_{\overline{\beta}_1} = O\Big(\exp\big(-\epsilon_m |w_q|^q\big)\Big),$$

$$\left(\partial^{\sharp}_{E^\lambda,\kappa_1} \circ \cdots \circ \partial^{\sharp}_{E^\lambda,\kappa_m}\right)\nu_{t_1} = O\Big(\exp\big(-\epsilon_m |w_q|^q\big)\Big),$$

for any $(\kappa_1, \ldots, \kappa_m) \in \{t_1, \overline{\beta}_1\}^m$. ■

7.3.6 Step 2

Let $\mathbb{L}^{\lambda\,\mathrm{cov}}_q(\ell,\alpha)$ denote the $\mathcal{O}_{Y^{\lambda\,\mathrm{cov}}_q}(*H^{\lambda\,\mathrm{cov}}_{\infty,q})$-module obtained as the pull back of $\mathbb{L}^{\lambda}_q(\ell,\alpha)$ by the projection ϖ^{λ}_q. The restriction of $\mathbb{L}^{\lambda\,\mathrm{cov}}_q(\ell,\alpha)$ to $Y^{\lambda\,\mathrm{cov}*}_q$ is denoted by $\mathbb{L}^{\lambda\,\mathrm{cov}*}_q(\ell,\alpha)$ which is equal to the pull back of $\mathbb{L}^{\lambda*}_q(\ell,\alpha)$ by ϖ_q. It is equipped with the metric $h_{\mathbb{L}^{\mathrm{cov}},q,\ell,\alpha}$ obtained as the pull back of $h_{\mathbb{L},q,\ell,\alpha}$.

There exists the induced decomposition of C^∞-vector bundles on $\mathcal{B}^{\lambda\,\mathrm{cov}*}_q$:

$$E^{\lambda\,\mathrm{cov}} := \varpi_q^{-1}(E^\lambda) = \bigoplus_{\ell,\alpha} \mathbb{L}^{\lambda\,\mathrm{cov}*}_q(\ell,\alpha) \otimes (\Psi^{\mathrm{cov}}_q)^{-1}(V_{\ell,\alpha}). \tag{7.9}$$

The bundle $E^{\lambda\,\mathrm{cov}}$ is equipped with the operators $\partial^{\sharp}_{E^{\lambda\,\mathrm{cov}},\overline{\beta}_1}$ and $\partial^{\sharp}_{E^{\lambda\,\mathrm{cov}},t_1}$ induced by $\partial^{\sharp}_{E^\lambda,\overline{\beta}_1}$ and $\partial^{\sharp}_{E^\lambda,t_1}$, respectively. They preserve the decomposition (7.9).

Let $b \subset \mathbb{R}$. Let $\mathcal{P}^{h_{V,\ell,\alpha}} V^{\lambda}_{\ell,\alpha}$ be the filtered bundle on $(U_{w,q}, \infty)$ as in Sect. 7.3.2.2. Let $\boldsymbol{v}_{\ell,\alpha}$ be holomorphic frames of $\mathcal{P}^{h_{V,\ell,\alpha}}_b V^{\lambda}_{\ell,\alpha}$ on $U_{w,q}$. We obtain the C^∞-frame

$$\widetilde{\boldsymbol{v}}_{\ell,\alpha} = (\widetilde{v}_{\ell,\alpha,i} \mid i = 1, \ldots, \operatorname{rank} V_{\ell,\alpha}) := (\Psi^{\mathrm{cov}}_q)^{-1}(\boldsymbol{v}_{\ell,\alpha|U^*_{w,q}}) \tag{7.10}$$

of $(\Psi^{\mathrm{cov}}_q)^{-1}(V_{\ell,\alpha})$ on $\mathcal{B}^{\lambda\,\mathrm{cov}*}_q$.

As explained in Sect. 5.2.4, there exist a neighbourhood $\mathcal{U}_q^\lambda$ of $H_{\infty,q}^{\lambda\,\mathrm{cov}}$ in $\mathcal{B}_q^{\lambda\,\mathrm{cov}}$ and mini-holomorphic frames $u_{q,\ell,\alpha}^\lambda$ of $\mathbb{L}_q^{\lambda\,\mathrm{cov}}(\ell,\alpha)$ on $\mathcal{U}_q^\lambda$ for which (5.11) and (5.12) hold.

We set $\mathcal{U}_q^{\lambda *} := \mathcal{U}_q^\lambda \setminus H_{\infty,q}^\lambda$. We obtain the C^∞-frame $u_{q,\ell,\alpha}^\lambda \otimes \widetilde{\boldsymbol{v}}_{\ell,\alpha}$ of the C^∞-vector bundle $\mathbb{L}_q^{\lambda\,\mathrm{cov}\,*}(\ell,\alpha)\otimes(\Psi_q^{\mathrm{cov}})^{-1}(V_{\ell,\alpha})$ on $\mathcal{U}_q^{\lambda *}$.

We define the matrix valued C^∞-function $A_{\ell,\alpha}$ on $U_{w,q}$ by $\mathbb{D}^{\lambda(1,0)}\boldsymbol{v}_{\ell,\alpha} = \boldsymbol{v}_{\ell,\alpha}A_{\ell,\alpha}\,dw$. By Proposition 2.8.3 with the formulas (2.16) and (2.17), we obtain

$$\partial^\sharp_{E^{\lambda\,\mathrm{cov}},\overline{\beta}_1}\big(u_{q,\ell,\alpha}^\lambda \otimes \widetilde{\boldsymbol{v}}_{\ell,\alpha}\big)=0, \tag{7.11}$$

$$\partial^\sharp_{E^{\lambda\,\mathrm{cov}},t_1}\big(u_{q,\ell,\alpha}^\lambda \otimes \widetilde{\boldsymbol{v}}_{\ell,\alpha}\big) = \big(u_{q,\ell,\alpha}^\lambda \otimes \widetilde{\boldsymbol{v}}_{\ell,\alpha}\big)\frac{1}{1+|\lambda|^2}\Big(-2\sqrt{-1}\big((\Psi_q^{\mathrm{cov}})^{-1}(A_{\ell,\alpha|U_{w,q}^*})\big)\Big). \tag{7.12}$$

Let $H\big(h^\sharp, u_{\ell,\alpha}^\lambda \otimes \widetilde{\boldsymbol{v}}_{\ell,\alpha}\big)$ denote the Hermitian-matrix valued function on $\mathcal{B}_q^{\lambda\,\mathrm{cov}\,*}$ determined by

$$H\big(h^\sharp, u_{\ell,\alpha}^\lambda \otimes \widetilde{\boldsymbol{v}}_{\ell,\alpha}\big)_{i,j} := (\varpi_q^{-1}h^\sharp)\Big(u_{\ell,\alpha}^\lambda \otimes \widetilde{v}_{\ell,\alpha,i},\, u_{\ell,\alpha}^\lambda \otimes \widetilde{v}_{\ell,\alpha,j}\Big).$$

Lemma 7.3.9 *Let P be any point of $H_{\infty,q}^{\lambda\,\mathrm{cov}}$. Let U_P be a relatively compact neighbourhood of P in $\mathcal{B}_q^{\lambda\,\mathrm{cov}} = (\varpi_q^\lambda)^{-1}(\mathcal{B}_q^\lambda)$. Then, there exist $C_P \geq 1$ and $N_P > 0$ such that*

$$C_P^{-1}|\beta_1|^{-N_P} \leq \big|H\big(h^\sharp, u_{\ell,\alpha}^\lambda \otimes \widetilde{\boldsymbol{v}}_{\ell,\alpha}\big)\big| \leq C_P|\beta_1|^{N_P}.$$

Proof It follows from Lemma 2.11.17, Proposition 5.1.2, Proposition 5.1.4 and Proposition 5.2.3. ■

7.3.7 Step 3

By using the frame $u_{q,\ell,\alpha}^\lambda \otimes \widetilde{\boldsymbol{v}}_{\ell,\alpha}$, we extend the bundle $\mathbb{L}_q^{\lambda\,\mathrm{cov}\,*}(\ell,\alpha)\otimes(\Psi_q^{\mathrm{cov}})^{-1}(V_{\ell,\alpha})$ to a C^∞-bundle $\mathfrak{P}_b\big(\mathbb{L}_q^{\lambda\,\mathrm{cov}\,*}(\ell,\alpha),(V_{\ell,\alpha},\mathbb{D}_{\ell,\alpha}^\lambda)\big)$ on $\mathcal{B}_q^{\lambda\,\mathrm{cov}}$. Namely, we construct $\mathfrak{P}_b\big(\mathbb{L}_q^{\lambda\,\mathrm{cov}\,*}(\ell,\alpha),(V_{\ell,\alpha},\mathbb{D}_{\ell,\alpha}^\lambda)\big)$ from the product bundle $\mathbb{C}^{\mathrm{rank}\,V_{\ell,\alpha}}\times\mathcal{U}_q^\lambda$ and $\mathbb{L}_q^{\lambda\,\mathrm{cov}\,*}(\ell,\alpha)\otimes(\Psi_q^{\mathrm{cov}})^{-1}(V_{\ell,\alpha})$ by the gluing isomorphism

$$\Big(\mathbb{C}^{\mathrm{rank}\,V_{\ell,\alpha}}\times\mathcal{U}_q^\lambda\Big)_{|\mathcal{U}_q^{\lambda *}} \simeq \Big(\mathbb{L}_q^{\lambda\,\mathrm{cov}\,*}(\ell,\alpha)\otimes(\Psi_q^{\mathrm{cov}})^{-1}(V_{\ell,\alpha})\Big)_{|\mathcal{U}_q^{\lambda *}}$$

induced by the identification of the canonical frame and $u^{\lambda}_{q,\ell,\alpha}\otimes\widetilde{\boldsymbol{v}}_{\ell,\alpha}$. We naturally regard $u^{\lambda}_{q,\ell,\alpha}\otimes\widetilde{\boldsymbol{v}}_{\ell,\alpha}$ as a C^{∞}-frame of $\mathfrak{P}_b\big(\mathbb{L}_q^{\lambda\,\mathrm{cov}*}(\ell,\alpha),(V_{\ell,\alpha},\mathbb{D}^{\lambda}_{\ell,\alpha})\big)$ on $\mathcal{U}^{\lambda}_q$. We define

$$\mathfrak{P}_b\big(E^{\lambda\,\mathrm{cov}}\big):=\bigoplus_{\ell,\alpha}\mathfrak{P}_b\big(\mathbb{L}_q^{\lambda\,\mathrm{cov}*}(\ell,\alpha),(V_{\ell,\alpha},\mathbb{D}^{\lambda}_{\ell,\alpha})\big).$$

In this way, $E^{\lambda\,\mathrm{cov}}$ extends to a C^{∞}-vector bundle $\mathfrak{P}_b\big(E^{\lambda\,\mathrm{cov}}\big)$ on $\mathcal{B}_q^{\lambda\,\mathrm{cov}}$.

We can naturally regard $\partial^{\sharp}_{E^{\lambda\,\mathrm{cov}},\overline{\beta}_1}$ and $\partial^{\sharp}_{E^{\lambda\,\mathrm{cov}},t_1}$ as C^{∞}-differential operators on $\mathfrak{P}_b\big(E^{\lambda\,\mathrm{cov}}\big)$ by the expressions (7.11) and (7.12). We may assume that there exists a q-th root $\tau_{1,q}$ of β_1^{-1} on $\mathcal{U}^{\lambda}_q$. We set $\partial^{\sharp}_{E^{\lambda},\overline{\tau}_{1,q}}=-q\overline{\tau}_{1,q}^{-q-1}\partial^{\sharp}_{E^{\lambda},\overline{\beta}_1}$. Because

$$\partial^{\sharp}_{E^{\lambda},\overline{\tau}_{1,q}}u^{\lambda}_{q,\ell,\alpha}\otimes\widetilde{\boldsymbol{v}}_{\ell,\alpha}=0, \tag{7.13}$$

it also induces a differential operator on $\mathfrak{P}_b\big(E^{\lambda\,\mathrm{cov}}\big)$.

7.3.8 *Step 4*

We have the C^{∞}-frame $\boldsymbol{a}=\bigcup_{\ell,\alpha}\big(u^{\lambda}_{q,\ell,\alpha}\otimes\boldsymbol{v}_{\ell,\alpha}\big)$ of $\mathfrak{P}_b\big(E^{\lambda\,\mathrm{cov}}\big)$ on $\mathcal{U}^{\lambda}_q$. Let $\mathcal{A}_{\overline{\beta}_1}$ and $\mathcal{A}_{t_1}$ be the matrix valued functions on $\mathcal{U}^{\lambda*}_q$ determined by

$$\nu_{\overline{\beta}_1}\boldsymbol{a}=\boldsymbol{a}\cdot\mathcal{A}_{\overline{\beta}_1},\qquad \nu_{t_1}\boldsymbol{a}=\boldsymbol{a}\cdot\mathcal{A}_{t_1}.$$

Lemma 7.3.10 *$\mathcal{A}_{t_1}$ and $\mathcal{A}_{\overline{\beta}_1}$ are C^{∞}-functions on $\mathcal{U}^{\lambda}_q$ whose Taylor series along $H^{\lambda\,\mathrm{cov}}_{\infty,q}$ are 0.*

Proof From Lemma 7.3.8 and the expressions (7.11) (7.12) and (7.13), we obtain

$$\partial_{t_1}^{\ell_1}\partial_{\overline{\tau}_{1,p}}^{\ell_2}\mathcal{A}_{t_1}=O\Big(\exp\big(-\epsilon_{\ell_1,\ell_2}|w_q|^q\big)\Big),\quad \partial_{t_1}^{\ell_1}\partial_{\overline{\tau}_{1,p}}^{\ell_2}\mathcal{A}_{\overline{\beta}_1}=O\Big(\exp\big(-\epsilon_{\ell_1,\ell_2}|w_q|^q\big)\Big)$$

for any $(\ell_1,\ell_2)\in\mathbb{Z}^2_{\geq 0}$ for some $\epsilon_{\ell_1,\ell_2}>0$. Then, by the elliptic regularity (see Lemma 7.3.11 below), for any $(\ell_1,\ell_2,\ell_3)\in\mathbb{Z}^3_{\geq 0}$ there exists $\epsilon'_{\ell_1,\ell_2,\ell_3}>0$ such that

$$\partial_{t_1}^{\ell_1}\partial_{\overline{\tau}_{1,p}}^{\ell_2}\partial_{\tau_{1,p}}^{\ell_3}\mathcal{A}_{t_1}=O\Big(\exp\big(-\epsilon_{\ell_1,\ell_2,\ell_3}|w_q|^q\big)\Big),$$

$$\partial_{t_1}^{\ell_1}\partial_{\overline{\tau}_{1,p}}^{\ell_2}\partial_{\tau_{1,p}}^{\ell_3}\mathcal{A}_{\overline{\beta}_1}=O\Big(\exp\big(-\epsilon_{\ell_1,\ell_2,\ell_3}|w_q|^q\big)\Big).$$

Then, the claim of the lemma follows. ∎

Lemma 7.3.11 *For $r_1, r_2 > 0$, we set $B(r_1, r_2) = \{(\sigma, \zeta) \in \mathbb{R} \times \mathbb{C} \,|\, |\sigma| < r_1, |\zeta| < r_2\}$. Let f be a C^∞-function on $B(r_1, r_2)$. Suppose that there exists $\ell \in \mathbb{Z}_{\geq 0}$ and $A > 0$ such that $|\partial_\sigma^{j_1} \partial_{\overline{\zeta}}^{j_2} f| < A$ for any $j_1 + j_2 \leq \ell + 1$ on $B(r_1, r_2)$. Then, for any $0 < r_i' < r_i$ $(i = 1, 2)$, there exist A', depending only on r_i, r_i', ℓ and A such that $|\partial_\sigma^{k_1} \partial_\zeta^{k_2} \partial_{\overline{\zeta}}^{k_3} f| < A'$ on $B(r_1', r_2')$ for any $k_1 + k_2 + k_3 \leq \ell$.*

Proof Let $r_i' < r_i'' < r_i$. Let $\chi : \mathbb{R} \times \mathbb{C} \longrightarrow [0, 1]$ be a C^∞-function such that $\chi(\sigma, \zeta) = 1$ on $B(r_1'', r_2'')$ and that the support is contained in $B(r_1, r_2)$. We set $\widetilde{f} = \chi \cdot f$, which we can naturally regard as a C^∞-function on $\mathbb{R} \times \mathbb{C}$ with compact support. There exists $A_1 > 0$, depending only on r_i, r_i'' and A, such that $|\partial_\sigma^{j_1} \partial_{\overline{\zeta}}^{j_2} \widetilde{f}| < A_1$ for any $j_1 + j_2 \leq \ell + 1$ on $\mathbb{R} \times \mathbb{C}$. Choose any $p > 3$. There exists $A_2 > 0$, depending only on r_i, r_i'' and A such that the $L^p_{\ell+1}$-norm of $\widetilde{f}$ is smaller than A_2. By the Sobolev embedding theorem (for example, see [28, Theorem 7.10]), there exists $A_3 > 0$, depending only on r_i, r_i'' and A such that the C^ℓ-norm $\widetilde{f}$ is smaller than A_3. ■

7.3.9 Step 5

By Lemma 7.3.10, the mini-holomorphic structure $\overline{\partial}_{E^{\lambda\,\mathrm{cov}}}$ of $E^{\lambda\,\mathrm{cov}}$ extends to a mini-holomorphic structure on $\mathfrak{P}_b(E^{\lambda\,\mathrm{cov}})$. We obtain the locally free $\mathcal{O}_{\mathcal{B}_q^{\lambda\,\mathrm{cov}}}$-module as the associated sheaf of mini-holomorphic sections, which is also denoted by $\mathfrak{P}_b(E^{\lambda\,\mathrm{cov}})$. We obtain the following locally free $\mathcal{O}_{\mathcal{B}_q^{\lambda\,\mathrm{cov}}}(*H^{\lambda\,\mathrm{cov}}_{\infty,q})$-module:

$$\mathcal{P}' E^{\lambda\,\mathrm{cov}} := \mathfrak{P}_b(E^{\lambda\,\mathrm{cov}}) \otimes \mathcal{O}_{\mathcal{B}_q^{\lambda\,\mathrm{cov}}}(*H^{\lambda\,\mathrm{cov}}_{\infty,q}).$$

We can easily observe that it is independent of the choice of b up to canonical isomorphisms. The $\mathbb{Z}$-action on E^λ also naturally extends to a $\mathbb{Z}$-action on $\mathcal{P}' E^{\lambda\,\mathrm{cov}}$.

By the construction, there exists the naturally defined $\mathbb{Z}$-equivariant morphism

$$\mathcal{P}' E^{\lambda\,\mathrm{cov}} \longrightarrow \mathcal{P}^h E^{\lambda\,\mathrm{cov}}. \tag{7.14}$$

Lemma 7.3.12 *The morphism* (7.14) *is an isomorphism.*

Proof Both sides are locally free $\mathcal{O}_{\mathcal{B}_q^{\lambda\,\mathrm{cov}}}(*H_q^{\lambda\,\mathrm{cov}})$-modules, and the restriction of (7.14) to $\mathcal{B}_q^{\lambda\,\mathrm{cov}*}$ is an isomorphism. Then, the claim follows. ■

As a result of Lemma 7.3.12, the $\mathbb{Z}$-equivariant morphism

$$\mathcal{P}'(E^{\lambda\,\mathrm{cov}})_{|\widehat{H}^{\lambda\,\mathrm{cov}}_{\infty,q}} \longrightarrow \mathcal{P}^h(E^{\lambda\,\mathrm{cov}})_{|\widehat{H}^{\lambda\,\mathrm{cov}}_{\infty,q}} \tag{7.15}$$

is an isomorphism.

Lemma 7.3.13 *There exists a natural $\mathbb{Z}$-equivariant isomorphism*

$$\mathfrak{P}_b(E^{\lambda\,\mathrm{cov}})_{|\widehat{H}^{\lambda\,\mathrm{cov}}_{\infty,q}} \simeq \bigoplus_{(\ell,\alpha)} \Big(\big(\mathcal{O}_{\widehat{H}^{\lambda\,\mathrm{cov}}_{\infty,q}} u^{\lambda}_{q,\ell,\alpha}\big) \otimes (\varpi_q)^{-1} (\Psi^{\lambda})^{*} \big(\mathcal{P}_b \widehat{\mathcal{V}}_{\ell,\alpha}, \widehat{\mathbb{D}}^{\lambda}_{\ell,\alpha}\big)\Big). \tag{7.16}$$

As a result, there exists a natural $\mathbb{Z}$-equivariant isomorphism

$$\begin{aligned}&\mathcal{P}'(E^{\lambda\,\mathrm{cov}})_{|\widehat{H}^{\lambda\,\mathrm{cov}}_{\infty,q}} \\ &\simeq \bigoplus_{(\ell,\alpha)} \Big(\big(\mathcal{O}_{\widehat{H}^{\lambda\,\mathrm{cov}}_{\infty,q}} (*H^{\lambda\,\mathrm{cov}}_{\infty,q}) u^{\lambda}_{q,\ell,\alpha}\big) \otimes (\varpi_q)^{-1} (\Psi^{\lambda})^{*} \big(\mathcal{P} \widehat{\mathcal{V}}_{\ell,\alpha}, \widehat{\mathbb{D}}^{\lambda}_{\ell,\alpha}\big)\Big).\end{aligned} \tag{7.17}$$

Proof Let $\mathcal{C}^{\infty}_{\widehat{H}^{\lambda\,\mathrm{cov}}_{\infty,q}}$ denote the sheaf of algebras on $H^{\lambda\,\mathrm{cov}}_{\infty,q}$ as in Sect. 4.5.5.3. By taking the Taylor series along $H^{\lambda\,\mathrm{cov}}_{\infty,q}$, we obtain the morphism of the sheaf of algebras $\mathcal{C}^{\infty}_{\mathcal{B}^{\lambda}_q|H^{\lambda\,\mathrm{cov}}_{\infty,q}} \longrightarrow \mathcal{C}^{\infty}_{\widehat{H}^{\lambda\,\mathrm{cov}}_{\infty,q}}$. Hence, we obtain the morphism of the ringed spaces $k^{\lambda\,\mathrm{cov}}_{q,\mathcal{C}^{\infty}} : (H^{\lambda\,\mathrm{cov}}_{\infty,q}, \mathcal{C}^{\infty}_{\widehat{H}^{\lambda\,\mathrm{cov}}_{\infty,q}}) \longrightarrow (\mathcal{B}^{\lambda\,\mathrm{cov}}_{\infty,q}, \mathcal{C}^{\infty}_{\mathcal{B}^{\lambda\,\mathrm{cov}}_{\infty,q}})$. For any $\mathbb{Z}$-equivariant C^{∞}-bundle W on $\mathcal{B}^{\lambda\,\mathrm{cov}}_{\infty,q}$, we naturally obtain the $\mathcal{C}^{\infty}_{\widehat{H}^{\lambda\,\mathrm{cov}}_{\infty,q}}$-module $(k^{\lambda\,\mathrm{cov}}_{q,\mathcal{C}^{\infty}})^{*} W$.

By the construction of $\mathfrak{P}_b(E^{\lambda\,\mathrm{cov}})$, there exists the following C^{∞}-isomorphism:

$$\mathfrak{P}_b(E^{\lambda}) \otimes \mathcal{C}^{\infty}_{\mathcal{B}^{\lambda}_q} \simeq \bigoplus_{\ell,\alpha} \Big(\big(\mathcal{C}^{\infty}_{\mathcal{B}^{\lambda}_q} \cdot u^{\lambda}_{q,\ell,\alpha}\big) \otimes_{\mathcal{C}^{\infty}_{\mathcal{B}^{\lambda}_q}} (\Psi^{\lambda\,\mathrm{cov}}_q)^{-1} \big(\mathcal{P}_b V^{\lambda}_{\ell,\alpha}\big) \otimes \mathcal{C}^{\infty}_{\mathcal{B}^{\lambda}_q}\Big). \tag{7.18}$$

Let W_1 and W_2 denote the left hand side and the right hand side of (7.18), respectively. Both W_i are equipped with the linear differential operator $\overline{\partial}_{W_i}$: $C^{\infty}(\mathcal{B}^{\lambda\,\mathrm{cov}}_q, W_i) \longrightarrow C^{\infty}(\mathcal{B}^{\lambda\,\mathrm{cov}}_q, W_i \otimes \Omega^{0,1}_{\mathcal{B}^{\lambda\,\mathrm{cov}}_q})$ satisfying the mini-complex Leibniz rule. (See Sect. 2.8.1.3 for the mini-complex Leibniz rule.) Let ∂_{W_i,t_1} (resp. $\partial_{W,\overline{\beta}_1}$) denote the differential operators on W_i induced by $\overline{\partial}_{W_i}$ and ∂_{t_1} (resp. $\partial_{\overline{\beta}_1}$). Under the induced isomorphism of $\mathcal{C}^{\infty}_{\widehat{H}^{\lambda\,\mathrm{cov}}_{\infty,q}}$-modules

$$(k^{\lambda\,\mathrm{cov}}_{q,\mathcal{C}^{\infty}})^{*} W_1 \simeq (k^{\lambda\,\mathrm{cov}}_{q,\mathcal{C}^{\infty}})^{*} W_2, \tag{7.19}$$

we have $\partial_{(k^{\lambda\,\mathrm{cov}}_{q,\mathcal{C}^{\infty}})^{*}W_1,t_1} = \partial_{(k^{\lambda\,\mathrm{cov}}_{q,\mathcal{C}^{\infty}})^{*}W_2,t_1}$ and $\partial_{(k^{\lambda\,\mathrm{cov}}_{q,\mathcal{C}^{\infty}})^{*}W_1,\overline{\beta}_1} = \partial_{(k^{\lambda\,\mathrm{cov}}_{q,\mathcal{C}^{\infty}})^{*}W_2,\overline{\beta}_1}$. The intersection of the kernels of $\partial_{(k^{\lambda\,\mathrm{cov}}_{q,\mathcal{C}^{\infty}})^{*}W_1,t_1}$ and $\partial_{(k^{\lambda\,\mathrm{cov}}_{q,\mathcal{C}^{\infty}})^{*}W_1,\overline{\beta}_1}$ is identified with the left hand side of (7.16). By using Lemma 4.5.12, we obtain that the intersection of the kernel of $\partial_{(k^{\lambda\,\mathrm{cov}}_{q,\mathcal{C}^{\infty}})^{*}W_2,t_1}$ and $\partial_{(k^{\lambda\,\mathrm{cov}}_{q,\mathcal{C}^{\infty}})^{*}W_2,\overline{\beta}_1}$ is identified with the right hand side of (7.16). Hence, we obtain the isomorphism (7.16) from (7.19). ■

We obtain the isomorphism (7.8) from (7.15) and (7.17).

7.3.10 Step 6

We take $t_1^0 \in \mathbb{R}_{t_1}$ and $c \in \mathbb{R}$. For $\ell \in \mathbb{Z}$, we set

$$b(t_1^0, c, \ell) := c + \frac{\ell t_1^0}{T}.$$

Let $\boldsymbol{v}_{\ell,\alpha} = (v_{\ell,\alpha,i} \,|\, i = 1, \ldots, \operatorname{rank} V_{\ell,\alpha})$ be a holomorphic frame of $\mathcal{P}^{h_{V,\ell,\alpha}}_{b(t_1^0,c,\ell)} V^\lambda_{\ell,\alpha}$ such that it is compatible with the parabolic structure. (See Sect. 2.11.1.) Let $a(\ell, \alpha, i)$ denote the parabolic degree of $v_{\ell,\alpha,i}$ (see (2.54)). Recall the following (see Lemma 2.11.17).

Lemma 7.3.14 *Let $H^{\ell,\alpha}$ be the Hermitian matrix valued function on $U^*_{w,q}$ determined by*

$$H^{\ell,\alpha}_{i,j} := h_{V,\ell,\alpha}\big(v_{\ell,\alpha,i}, v_{\ell,\alpha,j}\big)|w_q|^{-a(\ell,\alpha,i)-a(\ell,\alpha,j)}.$$

Then, there exist $C_1 > 1$ and $N > 0$ such that $C_1^{-1}\Big(\log|w_q|\Big)^{-N} \leq |H^{\ell,\alpha}| \leq C_1\Big(\log|w_q|\Big)^{N}$ and $C_1^{-1}\Big(\log|w_q|\Big)^{-N} \leq |(H^{\ell,\alpha})^{-1}| \leq C_1\Big(\log|w_q|\Big)^{N}$. ■

As before, we obtain the C^∞-frame $u^\lambda_{q,\ell,\alpha} \otimes \widetilde{\boldsymbol{v}}_{\ell,\alpha}$ of

$$\mathbb{L}_q^{\lambda\,\mathrm{cov}\,*}(\ell, \alpha) \otimes (\Psi_q^{\mathrm{cov}})^{-1}(V_{\ell,\alpha})$$

on $\mathcal{U}^\lambda_q$. Let $\pi_q^{\lambda\,\mathrm{cov}} : \mathcal{B}_q^{\lambda\,\mathrm{cov}} \longrightarrow \mathbb{R}$ be the map induced by $(t_1, \beta_1) \longmapsto t_1$. Let $\mathcal{B}_q^{\lambda\,\mathrm{cov}}\langle t_1\rangle$ denote the fiber. We set $\mathcal{U}^\lambda\langle t_1\rangle := \mathcal{B}_q^{\lambda\,\mathrm{cov}}\langle t_1\rangle \cap \mathcal{U}^\lambda$ and $\mathcal{U}^{\lambda *}\langle t_1\rangle := \mathcal{B}_q^{\lambda\,\mathrm{cov}}\langle t_1\rangle \cap \mathcal{U}^{\lambda *}$. The restriction of $u^\lambda_{q,\ell,\alpha} \otimes \widetilde{\boldsymbol{v}}_{\ell,\alpha}$ to $\mathcal{U}_q^{\lambda *}\langle t_1\rangle$ induces a C^∞-frame

$$\big(s_{\ell,\alpha,i} \,\big|\, i = 1, \ldots, \operatorname{rank} V_{\ell,\alpha}\big)$$

of the restriction of $\mathbb{L}_q^{\lambda\,\mathrm{cov}\,*}(\ell, \alpha) \otimes (\Psi_q^{\mathrm{cov}})^{-1}(V_{\ell,\alpha})$ to $\mathcal{U}_q^{\lambda *}\langle t_1\rangle$. We obtain a C^∞-frame

$$\bigcup_{(\ell,\alpha)} \big(s_{\ell,\alpha,i} \,\big|\, i = 1, \ldots, \operatorname{rank} V_{\ell,\alpha}\big)$$

of $E^{\lambda\,\mathrm{cov}}_{|\mathcal{U}_q^{\lambda *}\langle t_1\rangle}$. By Proposition 7.3.7, h and $h^\sharp$ are mutually bounded. By Lemma 7.3.14, we obtain the following lemma.

Lemma 7.3.15 *Let H be the Hermitian-matrix valued function on $\mathcal{U}_q^{\lambda *}\langle t_1\rangle$ determined by*

$$H_{(\ell_1,\alpha_1,i_1),(\ell_2,\alpha_2,i_2)} := h\big(s_{\ell_1,\alpha_1,i_1}, s_{\ell_2,\alpha_2,i_2}\big)\cdot |\beta_{1,q}|^{-a(\ell_1,\alpha_1,i_1)-a(\ell_2,\alpha_2,i_2)+(\ell_1+\ell_2)t_1^0/T}. \tag{7.20}$$

Then, there exist $C_1 > 1$ and $N_1 > 0$ such that $C_1^{-1}(\log|\beta_1|)^{-N_1} < |H| < C_1(\log|\beta_1|)^{N_1}$. ■

The tuple $(s_{\ell,\alpha,i})$ induces a holomorphic frame of

$$\bigoplus_{(\ell,\alpha)}\Big(\mathcal{P}^{(0)}_{-\ell t_1/T}\mathbb{L}_q^{\lambda\,\mathrm{cov}}(\ell,\alpha)_{|\widehat{H}^{\lambda\,\mathrm{cov}}_{\infty,q}}\otimes(\varpi_q^{\lambda})^*\Psi_q^*\big(\mathcal{P}_{b(t_1^0,c,\ell)}\widehat{\mathcal{V}}^{\lambda}_{\ell,\alpha},\widehat{\mathbb{D}}^{\lambda}_{\ell,\alpha}\big)\Big)_{|\{t_1\}\times\widehat{\infty}}. \tag{7.21}$$

Note that $\partial_{E^{\lambda},\overline{\beta}_1}s_{\ell,\alpha,i} = O\big(\exp(-\epsilon|\beta_1|)\big)$ for some $\epsilon > 0$. For any large $M > 0$, there exist holomorphic sections $\overline{s}_{\ell,\alpha,i}$ of $E^{\lambda\,\mathrm{cov}}_{|\mathcal{U}_q^{\lambda *}\langle t_1\rangle}$ such that

$$|\overline{s}_{\ell,\alpha,i}-s_{\ell,\alpha,i}|_h = O\Big(|\beta_1|^{-M}\Big).$$

If M is sufficiently large, $\bigcup_{(\ell,\alpha)}\big(\overline{s}_{\ell,\alpha,i}\,\big|\,i = 1,\dots,\mathrm{rank}\,V_{\ell,\alpha}\big)$ is a holomorphic frame of $\mathcal{P}^h_c\Big(E^{\lambda\,\mathrm{cov}}_{|\mathcal{U}_q^{\lambda *}\langle t_1\rangle}\Big)$. Hence, we obtain that the completion of $\mathcal{P}^h_c\Big(E^{\lambda\,\mathrm{cov}}_{|\mathcal{U}_q^{\lambda *}\langle t_1\rangle}\Big)$ at ∞ is equal to (7.21). Therefore, the proof of Theorem 7.3.4 is completed. ■

7.3.11 Proof of Proposition 7.3.5

According to Lemma 4.4.6, it is enough to study the case where Condition 6.3.2 is satisfied for (E, h, ∇, ϕ), i.e., there exists a decomposition (7.4). We continue to use the notation in Sects. 7.3.4–7.3.10.

Lemma 7.3.16 *For each pair (ℓ,α), there exists a good filtered λ-flat bundle $(\mathcal{P}_*\mathcal{V}^{\lambda}_{\ell,\alpha},\nabla^{\lambda})$ on $(U_{x,q},\infty)$ with an isomorphism*

$$(\mathcal{P}_*\mathcal{V}^{\lambda}_{\ell,\alpha},\nabla^{\lambda})_{|\widehat{\infty}_{x,q}} \simeq (\mathcal{P}_*\widehat{\mathcal{V}}^{\lambda}_{\ell,\alpha},\widehat{\mathbb{D}}^{\lambda}_{\ell,\alpha}). \tag{7.22}$$

If $\lambda\neq 0$, we may assume that the Stokes structure of $(\mathcal{V}_{\ell,\alpha},\mathbb{D}^{\lambda}_{\ell,\alpha})$ is trivial.

Proof If $\lambda = 0$, we may choose $(\mathcal{P}_*\mathcal{V}^0_{\ell,\alpha},\nabla^0)$ as $(\mathcal{P}^{h_{V,\ell,\alpha}}_*\mathcal{V}^0_{\ell,\alpha}, f_{\ell,\alpha}dw)$. Let us consider the case $\lambda\neq 0$. For each $\mathfrak{a}\in w_{q'}\mathbb{C}[w_{q'}]$, we set $L(\mathfrak{a}) := (\mathbb{C}(\!(w_{q'}^{-1})\!), \lambda d + d\mathfrak{a})$. We obtain the $\mathbb{C}(\!(w_q)\!)$-module with the λ-connection $\varphi_{q,q'*}L(\mathfrak{a})$. According to the Hukuhara-Levelt-Turrittin theorem, for an appropriate q', there exist a finite subset $W(\ell,\alpha) \subset w'_q\mathbb{C}[w'_q]$, a tuple of $\mathbb{C}(\!(w_q)\!)$-modules with regular singular λ-connection $R_{\mathfrak{a}}$ ($\mathfrak{a} \in W(\ell,\alpha)$), and an isomorphism

$$(\widehat{\mathcal{V}}_{\ell,\alpha}, \widehat{\mathbb{D}}^{\lambda}_{\ell,\alpha}) \simeq \bigoplus_{\mathfrak{a}\in W(\ell,\alpha)} \varphi_{q,q'*}L(\mathfrak{a}) \otimes R_{\mathfrak{a}}.$$

Because $\varphi_{q',q*}L(\mathfrak{a})$ and $R_{\mathfrak{a}}$ are obtained as the formal completions of meromorphic flat bundles on $(U_{w,q},\infty)$, there exist meromorphic λ-flat bundles $(\mathcal{V}_{\ell,\alpha}, \mathbb{D}^{\lambda}_{\ell,\alpha})$ on $(U_{w,q},\infty)$ with isomorphisms $(\mathcal{V}_{\ell,\alpha}, \mathbb{D}^{\lambda}_{\ell,\alpha})_{|\widehat{\infty}_{w,q}} \simeq (\widehat{\mathcal{V}}_{\ell,\alpha}, \widehat{\mathbb{D}}^{\lambda}_{\ell,\alpha})$. Hence, we obtain good filtered λ-flat bundles $(\mathcal{P}_*\mathcal{V}_{\ell,\alpha}, \mathbb{D}^{\lambda}_{\ell,\alpha})$ over $(\mathcal{V}_{\ell,\alpha}, \mathbb{D}^{\lambda}_{\ell,\alpha})$ with isomorphisms $(\mathcal{P}_*\mathcal{V}_{\ell,\alpha}, \mathbb{D}^{\lambda}_{\ell,\alpha})_{|\widehat{\infty}_{w,q}} \simeq (\mathcal{P}_*\widehat{\mathcal{V}}_{\ell,\alpha}, \widehat{\mathbb{D}}^{\lambda}_{\ell,\alpha})$. ■

Let $\mathcal{C}^{\infty}_{U_{x,q}}$ denote the sheaf of C^{∞}-functions on $U_{x,q}$. There exists an isomorphism

$$F_{\ell,\alpha} : \mathcal{P}_*\mathcal{V}^{\lambda}_{\ell,\alpha} \otimes \mathcal{C}^{\infty}_{U_{x,q}} \simeq \mathcal{P}^{h_{V,\ell,\alpha}}_* V^{\lambda}_{\ell,\alpha} \otimes \mathcal{C}^{\infty}_{U_{x,q}}$$

which induces (7.22). By applying Proposition 10.1.9 to $(V^{\lambda}_{\ell,\alpha}, \mathbb{D}^{\lambda}_{\ell,\alpha})$ with $h_{V,\ell,\alpha}$, we obtain that the norm estimate holds for $(\mathcal{P}_*\mathcal{V}^{\lambda}_{\ell,\alpha}, \nabla^{\lambda})$ with $h_{0,\ell,\alpha} := F^*_{\ell,\alpha}(h_{V,\ell,\alpha})$.

We obtain the following filtered bundle

$$\mathcal{P}_*\mathcal{E}_0 := \bigoplus_{\ell,\alpha} \mathcal{P}_*\mathbb{L}^{\lambda}_q(\ell,\alpha) \otimes \Psi^*_q(\mathcal{P}_*\mathcal{V}^{\lambda}_{\ell,\alpha}, \nabla^{\lambda}).$$

By Proposition 4.5.7 and Lemma 5.2.5, the norm estimate holds for $\mathcal{P}_*\mathcal{E}_0$ with $h_{\mathcal{E}_0} = \bigoplus_{\ell,\alpha} h_{\mathbb{L},q,\ell,\alpha} \otimes \Psi_q^{-1}(h_{0,\ell,\alpha})$.

Let $\mathcal{C}^{\infty}_{\mathcal{B}^{\lambda}_q}$ denote the sheaf of C^{∞}-functions on $\mathcal{B}^{\lambda}_q$. By the proof of Theorem 7.3.4, the morphisms $F_{\ell,\alpha}$ induce the following isomorphism:

$$F : \mathcal{E}_0 \otimes \mathcal{C}^{\infty}_{\mathcal{B}^{\lambda}_q} \simeq \mathcal{P}^h(E^{\lambda}) \otimes \mathcal{C}^{\infty}_{\mathcal{B}^{\lambda}_q}.$$

Moreover, it induces the following isomorphism of the filtered bundles:

$$\left(\mathcal{P}_*\mathcal{E}_0\right)_{|\widehat{H}^{\lambda}_{\infty,q}} \simeq \mathcal{P}^h_*(E^{\lambda})_{|\widehat{H}^{\lambda}_{\infty,q}}.$$

Note that $h_{\mathcal{E}_0} = F^*(h^{\sharp})$, and hence $h_{\mathcal{E}_0}$ and $F^*(h)$ are mutually bounded. Then, we can check that the norm estimate holds for $(\mathcal{P}^h_* E^{\lambda}, h)$ by using the argument in Sect. 7.3.10. ■

7.4 Strong Adaptedness and the GCK-Condition

7.4.1 Statements

Let $\mathcal{B}_q^\lambda$ be a neighbourhood of $H_{\infty,q}^\lambda$ in Y_q^λ as in Sect. 7.1.2. We set $\mathcal{B}_q^{\lambda *} := \mathcal{B}_q^\lambda \setminus H_{\infty,q}^\lambda$. Let $\mathcal{P}_*\mathcal{E}$ be a good filtered bundle over a locally free $\mathcal{O}_{\mathcal{B}_q^\lambda}(*H_{\infty,q}^\lambda)$-module $\mathcal{E}$. We obtain the mini-holomorphic bundle $(E, \overline{\partial}_E)$ on $\mathcal{B}_q^{\lambda *}$ as the restriction of $\mathcal{E}$. Let h be a Hermitian metric of E such that $(E, \overline{\partial}_E, h)$ is a monopole.

Proposition 7.4.1 *Suppose that h is strongly adapted to $\mathcal{P}_*\mathcal{E}$ in the sense of Definition 4.4.7. Then, the monopole satisfies the GCK-condition. Moreover, $(E, \overline{\partial}_E, h)$ satisfies the norm estimate with respect to $\mathcal{P}_*\mathcal{E}$.*

We obtain the following corollary immediately.

Corollary 7.4.2 *Suppose that $(E, \overline{\partial}_E, h)$ satisfies the norm estimate with respect to $\mathcal{P}_*\mathcal{E}$. Then, the monopole satisfies the GCK-condition.*

Proof If $(E, \overline{\partial}_E, h)$ satisfies the norm estimate with respect to $\mathcal{P}_*\mathcal{E}$, then h is strongly adapted to $\mathcal{P}_*\mathcal{E}$. Hence, the claim follows from Proposition 7.4.1. ■

For the proof of Proposition 7.4.1, we shall use the following auxiliary proposition, which is also useful for the proof of Theorem 9.1.2.

Proposition 7.4.3 *There exists a Hermitian metric h_1 of E such that the following conditions are satisfied.*

(i) The norm estimate holds for $(\mathcal{P}_*\mathcal{E}, h_1)$.
(ii) We have the estimate $|G(h_1)|_{h_1} = O(|w|^{-2-\epsilon}) = O(|w_q|^{-q(2+\epsilon)})$ for some $\epsilon > 0$, and the convergence $|\overline{\partial}_E G(h_1)|_{h_1} \to 0$ as $|w_q| \to \infty$. (See Sect. 7.1.2 *for $|\cdot|_{h_1}$.)*
(iii) $|F(h_1)|_{h_1} \to 0$ and $|\nabla_{h_1}\phi_{h_1}|_{h_1} \to 0$ as $|w_q| \to \infty$.
(iv) $|\phi_{h_1}|_{h_1} = O\big(\log |w_q|\big)$.
(v) Let ∂_{E,h,β_1} denote the operator induced by $\partial_{E,\overline{\beta}_1}$ and h as in Sect. 2.9.3. *Then, we obtain*

$$\Big|\big[\partial_{E,\overline{\beta}_1}, \partial_{E,h_1,\beta_1}\big]\Big|_{h_1} = O\big(|w|^{-2}(\log|w|)^{-2}\big).$$

If $\operatorname{rank} E = 1$, *we may also assume that $\partial_{t_1}^{\ell_1}\partial_{\beta_1}^{\ell_2}\partial_{\overline{\beta}_1}^{\ell_3} G(h_1) = O(|w|^{-k})$ for any $(\ell_1, \ell_2, \ell_3) \in \mathbb{Z}_{\geq 0}^3$ and for any $k > 0$.*

7.4.2 Some Estimates for Tame Harmonic Bundles (1)

Let U_w be a neighbourhood of ∞ in $\mathbb{P}^1_w$. Let $U_{w,q}$ be the pull back of U_w by the ramified covering $\mathbb{P}^1_{w_q} \longrightarrow \mathbb{P}^1_w$, as in Sect. 7.1.2. We set $U^*_{w,q} = U_{w,q} \setminus \{\infty\}$. The pull back of w by the ramified covering is also denoted by w, i.e., $w = w_q^q$.

Let $(V, \overline{\partial}_V, \theta, h)$ be a tame harmonic bundle on $U^*_{w,q}$. For simplicity, we assume that there exists a neighbourhood $U'_{w,q}$ of $U_{w,q}$ in $\mathbb{P}^1_{w,q}$ such that $(V, \overline{\partial}_V, \theta, h)$ is the restriction of a harmonic bundle on $U'_{w,q} \setminus \{\infty\}$. We obtain the associated filtered Higgs bundle $(\mathcal{P}^h_* V, \theta)$ on $(U_{w,q}, \infty)$. Let $\overline{\partial}_V + \partial_V$ denote the Chern connection determined by $\overline{\partial}_V$ and h. Let $F(h)$ denote the curvature of $\overline{\partial}_V + \partial_V$. Let $\theta^\dagger$ denote the adjoint of θ with respect to h. We have the expressions $\theta = f\, dw/w$ and $\theta^\dagger = f^\dagger d\overline{w}/\overline{w}$.

Lemma 7.4.4

- *f is bounded with respect to h.*
- *We have the estimates $\overline{\partial}_V \partial_V f = O\big(|w|^{-2}(\log|w|)^{-2} dw\, d\overline{w}\big)$ and $\partial_V \overline{\partial}_V f^\dagger = O\big(|w|^{-2}(\log|w|)^{-2} dw\, d\overline{w}\big)$ with respect to h.*
- *$\partial_V f = O(|w|^{-1} dw)$ and $\overline{\partial}_V f^\dagger = O(|w|^{-1} d\overline{w})$ with respect to h.*
- *$\partial_{V,w} \partial_{V,w} f = O(|w|^{-2})$ and $\partial_{V,\overline{w}} \partial_{V,\overline{w}} f^\dagger = O(|w|^{-2})$ with respect to h, where $\partial_{V,w}$ and $\partial_{V,\overline{w}}$ denote the inner product of ∂_w and $\partial_{\overline{w}}$ with the Chern connection, respectively.*

Proof The first claim is proved in [80]. It is also proved

$$F(h) = O\big(|w|^{-2}(\log|w|)^{-2} dw\, d\overline{w}\big)$$

in [80]. Because $\overline{\partial}_V \partial_V f = (\overline{\partial}_V \partial_V + \partial_V \overline{\partial}_V) f = [F(h), f]$, we obtain $\overline{\partial}_V \partial_V f = O\big(|w|^{-2}(\log|w|)^{-2} dw\, d\overline{w}\big)$. By taking the adjoint, we obtain the estimate for $\partial_V \overline{\partial}_V f^\dagger$.

We have the holomorphic map $g_q : \mathbb{C}_a \longrightarrow \mathbb{C}^*_{w_q}$ given by $g_q(a) = e^{a/q}$, i.e., $e^a = w$. Let $\widetilde{U}$ denote the pull back of $U_{w,q}$ by g_q. There exists $L \in \mathbb{R}$ such that $\{a \mid \mathrm{Re}(a) > L\} \subset \widetilde{U}$. Let $(\widetilde{V}, \partial_{\widetilde{V}}, \widetilde{\theta}, \widetilde{h})$ denote the pull back of $(V, \overline{\partial}_V, \theta, h)$ by g_q. Let $\widetilde{f}$ and $\widetilde{f}^\dagger$ denote the pull back of f and $f^\dagger$ by g_q. We have $\widetilde{\theta} = f\, da$. For the adjoint $\widetilde{\theta}^\dagger$ of $\widetilde{\theta}$ with respect to $\widetilde{h}$, we have $\widetilde{\theta}^\dagger = f^\dagger d\overline{a}$. Let $\partial_{\widetilde{V},a}$ and $\partial_{\widetilde{V},\overline{a}}$ denote the inner product of ∂_a and $\partial_{\overline{a}}$ with the pull back $\partial_{\widetilde{V}} + \overline{\partial}_{\widetilde{V}}$ of $\partial_V + \overline{\partial}_V$ by g_q. Note that the curvature of the Chern connection $\partial_{\widetilde{V}} + \overline{\partial}_{\widetilde{V}}$ is $O\big(\mathrm{Re}(a)^{-1}\big)$.

There exists $C_0 > 0$ such that $|\widetilde{f}|_h < C_0$ on $\widetilde{U}$. There exists $C_1 > 0$ such that the following holds for any a_0 with $\mathrm{Re}(a_0) > L + 10$.

- $|\partial_{\widetilde{V},\overline{a}} \partial_{\widetilde{V},a} \widetilde{f}|_{\widetilde{h}} \leq C_1 |a_0|^{-2}$ on the disc $\{|a - a_0| < 2\}$.

By Proposition 7.4.5 below, there exists $C_2 > 0$ such that the following holds for any a_0 satisfying $\text{Re}(a_0) > L + 10$.

- $|\partial_{\widetilde{V},a}\widetilde{f}|_{\widetilde{h}} \leq C_2$ on the disc $\{|a - a_0| < 1\}$.

It implies that $\partial_V f = O(|w|^{-1}dw)$ and $\overline{\partial}_V f^\dagger = O(|w|^{-1}d\overline{w})$ as $|w| \to \infty$.

We have the relation $F(\widetilde{h}) + [\widetilde{\theta}, \widetilde{\theta}^\dagger] = 0$, i.e., $F(\widetilde{h})_{a,\overline{a}} + [\widetilde{f}, \widetilde{f}^\dagger] = 0$. We obtain $\partial_{\widetilde{V},a}F(\widetilde{h})_{a,\overline{a}} = O(1)$ and $\partial_{\widetilde{V},\overline{a}}F(\widetilde{h})_{a,\overline{a}} = O(1)$. Because $\partial_{\widetilde{V},\overline{a}}\partial_{\widetilde{V},a}\partial_{\widetilde{V},a}\widetilde{f} = F(\widetilde{h})_{\overline{a},a}\partial_{\widetilde{V},a}\widetilde{f} + \partial_{\widetilde{V},a}(F(\widetilde{h})_{\overline{a},a})\widetilde{f} = O(1)$, we obtain $\partial_{\widetilde{V},a}\partial_{\widetilde{V},a}\widetilde{f} = O(1)$ by Proposition 7.4.5 below. We also obtain $\partial_{\widetilde{V},\overline{a}}\partial_{\widetilde{V},\overline{a}}\widetilde{f}^\dagger = O(1)$. Thus, we obtain the fourth claim. ■

7.4.2.1 Appendix

For $r > 0$, we set $B(r) = \{\zeta \in \mathbb{C} \mid |\zeta| < r\}$. Let E be a vector bundle on $B(r)$ with a Hermitian metric h_E with a unitary connection ∇. Let $F(\nabla) = F \cdot d\zeta\, d\overline{\zeta}$ denote the curvature of ∇. Suppose that $|F|_h \leq C_1$ on $B(r)$ for a positive constant $C_1 > 0$.

Proposition 7.4.5 *Let s be any section of E on $B(r)$ such that $|s|_h + |\nabla_{\overline{\zeta}}\nabla_\zeta s|_h \leq C_2$ for a positive constant $C_2 > 0$. Let $0 < r' < r$. Then, there exists $C_3 > 0$, depending only on C_i $(i = 1, 2)$, r, r' and* rank(E)*, such that $|\nabla_\zeta s|_h + |\nabla_{\overline{\zeta}} s|_h \leq C_3$ on $B(r')$.*

Proof Let (x, y) denote the real coordinate system determined by $\zeta = x + \sqrt{-1}y$. Let $\boldsymbol{v}$ be a unitary C^∞-frame of E such that $\nabla_x(\boldsymbol{v}_{|y=0}) = 0$ and $\nabla_y \boldsymbol{v} = 0$. Let A_x be the matrix valued function determined by $\nabla_x \boldsymbol{v} = \boldsymbol{v} A_x$. Because $\nabla_y A_x$ represents F with respect to the frame $\boldsymbol{v}$, and because $A_{x|y=0} = 0$, there exists C_{10}, depending only on r, C_1 and rank(E) such that $|A_x| \leq C_{10}$.

Let s' be a C^∞-section of E with compact support. For any $p \geq 1$, we set $\|s'\|_{L^p(B(r))} = \left(\int_{B(r)} |s'|_h^p\right)^{1/p}$.

Lemma 7.4.6 *Suppose that $\|s'\|_{L^p(B(r))} + \|\nabla_{\overline{\zeta}} s'\|_{L^p(B(r))} \leq C_{11}$ for a constant $C_{11} > 0$.*

The case $p = 2$ *For any $q > 2$ and $0 < r_1 < r$, there exists $C_{12} > 0$, depending only on q, r, r_1, C_i $(i = 10, 11)$ and* rank(E) *such that $\|s'\|_{L^q(B(r_1))} \leq C_{12}$.*

The case $p > 2$ *For any $0 < r_1 < r$, there exists $C_{13} > 0$, depending only on p, r, r_1, C_i $(i = 10, 11)$ and* rank(E) *such that $\sup_{B(r_1)} |s'|_h \leq C_{13}$.*

Proof For the expression $s' = \sum s'_i v_i$, we obtain

$$\nabla_{\overline{\zeta}}(s') = \sum_i \partial_{\overline{\zeta}}(s'_i)v_i + \sum_{i,j} \frac{1}{2} A_{x,ji} s'_i v_j.$$

By the assumption, there exists $C_{14} > 0$, depending only on p, r, C_i $(i = 10, 11)$ and $\mathrm{rank}(E)$ such that $\|s'_i\|_{L^p(B(r))} + \|\partial_{\overline{\zeta}} s'_i\|_{L^p(B(r))} \leq C_{14}$. Then, the claim follows from the elliptic regularity and the Sobolev embedding theorem. (For example, see [1, Theorem 10.1] and [28, Theorem 7.10].) ∎

We set $r'' = (r + r')/2$. There exists a decreasing C^∞-function $\rho : \mathbb{R}_{\geq 0} \longrightarrow [0, 1]$ such that (i) $\rho(u) = 1$ $(u \leq r')$, (ii) $\rho(u) > 0$ $(u < r'')$ and $\rho(u) = 0$ $(u \geq r'')$, (iii) ρ^a is C^∞ on $\mathbb{R}_{\geq 0}$ for any $a > 0$. We note the conditions imply that the C^∞-function $(\partial_u \rho^a)\rho^{-a/2} = 2\partial_u \rho^{a/2}$ on $u < r''$ extends to a C^∞-function on $\mathbb{R}_{\geq 0}$ which are constantly 0 on $u \geq r''$. We obtain the C^∞-function $\chi : \mathbb{C} \longrightarrow [0, 1]$ by setting $\chi(\zeta) = \rho(|\zeta|)$.

We set $s'_a := \chi^{a/2} \cdot \nabla_\zeta(s)$ for $a > 0$. We have

$$\begin{aligned} \|s'_a\|^2_{L^2(B(r))} &= \int_{B(r)} \chi^a h(\nabla_\zeta s, \nabla_\zeta s) \\ &= -\int_{B(r)} \partial_\zeta(\chi^a) \cdot h(s, \nabla_\zeta s) - \int_{B(r)} \chi^a h(s, \nabla_{\overline{\zeta}} \nabla_\zeta s). \end{aligned} \tag{7.23}$$

There exists $C_{20}(a) > 0$ depending only on a, r, r', C_i $(i = 1, 2)$ and $\mathrm{rank}(E)$ such that

$$\left| \int_{B(r)} \chi^a h(s, \nabla_{\overline{\zeta}} \nabla_\zeta(s)) \right| \leq C_{20}(a).$$

There exists $C_{21}(a) > 0$ depending only on a, r, r', C_i $(i = 1, 2)$ and $\mathrm{rank}(E)$ such that the first term in the right hand side of (7.23) is dominated as follows:

$$\begin{aligned} \left| \int_{B(r)} \partial_\zeta(\chi^a) \cdot h(s, \nabla_\zeta s) \right| &\leq \left(\int_{B(r)} \left(\chi^{-a/2} \partial_\zeta(\chi^a)\right)^2 |s|_h^2 \right)^{1/2} \cdot \|s'_a\|_{L^2(B(r))} \\ &\leq C_{21}(a) \|s'_a\|_{L^2(B(r))}. \end{aligned} \tag{7.24}$$

We obtain $\|s'_a\|^2_{L^2(B(r))} \leq C_{20}(a) + C_{21}(a)\|s'_a\|_{L^2(B(r))}$. Hence, there exists $C_{22}(a) > 0$ depending only on a, r, r', C_i $(i = 1, 2)$ and $\mathrm{rank}(E)$ such that $\|s'_a\|_{L^2(B(r))} \leq C_{22}(a)$.

Note the following equality:

$$\nabla_{\overline{\zeta}}(s'_a) = (\chi^{-a/4} \partial_{\overline{\zeta}}(\chi^{a/2})) s'_{a/2} + \chi^{a/2} \nabla_{\overline{\zeta}} \nabla_\zeta(s). \tag{7.25}$$

Hence, there exists $C_{23}(a) > 0$ depending only on a, r, r', C_i $(i = 1, 2)$ and $\mathrm{rank}(E)$ such that $\|\nabla_{\overline{\zeta}}(s'_a)\|_{L^2(B(r))} \leq C_{23}(a)$.

Let $r'' < r_1 < r$. By Lemma 7.4.6, for any $p > 2$, there exists $C_{24}(a, p) > 0$ depending only on p, a, r, r', r_1, C_i $(i = 1, 2)$ and $\mathrm{rank}(E)$ such that $|s'_a|_{L^p(B(r_1))} \leq C_{24}(a, p)$. By (7.25), there exists $C_{25}(a, p) > 0$ depending only on p, a, r, r', r_1, C_i

$(i = 1, 2)$ and $\mathrm{rank}(E)$ such that $|\nabla_{\overline{\zeta}}(s'_a)|_{L^p(B(r_1))} \leq C_{25}(a, p)$. By Lemma 7.4.6, there exists $C_{26}(a, p) > 0$ depending only on p, a, r, r', r_1, C_i $(i = 1, 2)$ and $\mathrm{rank}(E)$ such that $\sup_{B(r'')} |s'_a|_h \leq C_{26}(a, p)$. Thus, we obtain the estimate for $|\nabla_\zeta(s)|_h$ on $B(r')$. Similarly, we obtain the estimate for $|\nabla_{\overline{\zeta}}(s)|_h$ on $B(r')$. Thus, the proof of Proposition 7.4.5 is completed. ■

7.4.3 Some Estimates for Tame Harmonic Bundles (2)

We continue to use the notation in Sect. 7.4.2. For any λ we obtain the holomorphic structure $\overline{\partial}_V + \lambda\theta^\dagger$ on V. Let V^λ denote the holomorphic bundle $(V, \overline{\partial}_V + \lambda\theta^\dagger)$. Let ∇ denote the Chern connection determined by $\overline{\partial}_V + \lambda\theta^\dagger$ and h, i.e., $\nabla = \overline{\partial}_V + \lambda\theta^\dagger + \partial_V - \overline{\lambda}\theta$. Let $F(\nabla)$ be the curvature of ∇. We have $F(\nabla) = -(1 + |\lambda|^2)[\theta, \theta^\dagger]$. Because V^λ with h is acceptable, we obtain the filtered bundle $\mathcal{P}^h_* V^\lambda$ on $(U_{w,q}, \infty)$.

Lemma 7.4.7 *We obtain $\nabla_w F(\nabla)_{w,\overline{w}} = O(|w|^{-3})$ and $\nabla_{\overline{w}} F(\nabla)_{w,\overline{w}} = O(|w|^{-3})$ with respect to h as $|w_q| \to \infty$.*

Proof We have the estimate $F(\nabla)_{w,\overline{w}} = -(1 + |\lambda|^2)|w|^{-2} \cdot [f, f^\dagger]$. We also have $\nabla_w(f) = \partial_{V,w} f - \overline{\lambda}[w^{-1} f, f] = O(|w|^{-1})$ and $\nabla_w(f^\dagger) = \partial_{V,w} f^\dagger - \overline{\lambda}[w^{-1} f, f^\dagger] = O(|w|^{-1})$ by Lemma 7.4.4. Then, we can easily deduce the claim of the lemma. ■

Let s be a holomorphic section of $\mathcal{P}^h_{<0} V^\lambda$. Note that $\nabla_{\overline{w}} s = 0$.

Lemma 7.4.8 *There exists $\epsilon > 0$ such that $\nabla_w s = O(|w|^{-1-\epsilon})$ and $\nabla_w^2 s = O(|w|^{-2-\epsilon})$ with respect to h as $|w_q| \to \infty$.*

Proof We consider the map $g_q : \mathbb{C}_a \longrightarrow \mathbb{C}^*_{w_q}$ given by $g_q(a) = e^{a/q}$ as in the proof of Lemma 7.4.4. Let $(\widetilde{V}, \widetilde{h}, \widetilde{\nabla})$ denote the pull back of (V, h, ∇) by g_q. Let $\widetilde{s}$ denote the pull back of s by g_q. Because $\widetilde{\nabla}_{\overline{a}} \widetilde{\nabla}_a \widetilde{s} = F(\widetilde{\nabla})_{\overline{a},a} \widetilde{s}$, there exists $\epsilon_1 > 0$ such that the following holds for any a_0 with $\mathrm{Re}(a_0) > L + 10$.

- On the disc $\{|a - a_0| < 3\}$, we obtain $|\widetilde{s}|_{\widetilde{h}} = O(e^{-\epsilon_1 |a_0|})$, $|\widetilde{\nabla}_{\overline{a}} \widetilde{\nabla}_a \widetilde{s}| = O(e^{-\epsilon_1 |a_0|})$.

We obtain $\nabla_a s = O(e^{-\epsilon_1 |a_0|})$ on $\{|a - a_0| < 2\}$ by Proposition 7.4.5 for any a_0 with $\mathrm{Re}(a_0) > L + 10$. Moreover, because $\nabla_{\overline{a}} \nabla_a \nabla_a s = 2F(\nabla)_{\overline{a},a} \nabla_a s + \nabla_a(F(\nabla)_{\overline{a},a}) s$, there exists $\epsilon_2 > 0$ with the following property:

- On discs $\{|a - a_0| < 2\}$, we obtain $|\nabla_a s| = O(e^{-\epsilon_2 |a_0|})$ and $|\nabla_{\overline{a}} \nabla_a \nabla_a s| = O(e^{-\epsilon_2 |a_0|})$.

We obtain $\nabla_a \nabla_a s = O(e^{-\epsilon_2 |a_0|})$ on $\{|a - a_0| < 1\}$ by Proposition 7.4.5, which implies the claim of the lemma. ■

7.4.4 λ-Connections

Let $(\mathcal{P}_*\mathcal{V}, \mathbb{D}^\lambda)$ be a good filtered λ-flat bundle on $(U_{w,q}, \infty)$ whose Poincaré rank is strictly smaller than q. For simplicity, we assume that the Stokes structure is trivial, i.e., for an appropriate covering $\varphi_{q,p} : U_{w,p} \longrightarrow U_{w,q}$, there exists a decomposition

$$\varphi_{q,p}^*(\mathcal{P}_*\mathcal{V}, \mathbb{D}^\lambda) = \bigoplus_{\mathfrak{a}\in w_p\mathbb{C}[w_p]} \varphi_{q,p}^*(\mathcal{P}_*\mathcal{V}_{\mathfrak{a}}, \mathbb{D}^\lambda_{\mathfrak{a}}),$$

such that $\mathbb{D}^\lambda_{\mathfrak{a}} - d\mathfrak{a}\,\mathrm{id}$ are logarithmic with respect to $\mathcal{P}_*\mathcal{V}_{\mathfrak{a}}$.

Let $(V, \mathbb{D}^\lambda)$ be a λ-flat bundle obtained as the restriction of $(\mathcal{V}, \mathbb{D}^\lambda)$ to $U^*_{w,q}$. We obtain the decomposition $\mathbb{D}^\lambda = d''_V + d'_V$ into the $(0,1)$-part and the $(1,0)$-part. For any Hermitian metric h of V, we obtain the operators δ'_h, δ''_h and $\mathbb{D}^{\lambda\star}_h := \delta'_h - \delta''_h$ as in Sect. 2.9.5.

Lemma 7.4.9 *There exists a Hermitian metric h satisfying the following conditions, where we consider the norms with respect to h.*

- *h satisfies the norm estimate for $(\mathcal{P}_*\mathcal{V}, \mathbb{D}^\lambda)$.*
- $|\Lambda[\mathbb{D}^\lambda, \mathbb{D}^{\lambda\star}_h]| = O(|w_q|^{-2q-2\epsilon})$, *where Λ is determined by $\Lambda(dw\,d\overline{w}) = -2\sqrt{-1}$.*
- $\big|[d''_{\overline{w}}, \delta'_{h,w}]\big|$, $\big|[d''_{\overline{w}}, \delta''_{h,\overline{w}}]\big|$, $\big|[d'_w, \delta'_{h,w}]\big|$ *and* $\big|[d'_w, \delta''_{h,\overline{w}}]\big|$ *go to 0 as $w_q \to \infty$. We also obtain*

$$\big|[d''_{\overline{w}}, \delta'_{h,w}]\big| = O\big(|w|^{-2}(\log|w|)^{-2}\big). \tag{7.26}$$

- $|d'_w - \lambda\delta'_w|$ *and* $|\delta''_{\overline{w}} - \overline{\lambda} d'_{\overline{w}}|$ *go to 0 as $w_q \to \infty$.*
- $\big|d''_{\overline{w}}\Lambda[\mathbb{D}^\lambda, \mathbb{D}^{\lambda\star}_h]\big|$, $\big|d'_w\Lambda[\mathbb{D}^\lambda, \mathbb{D}^{\lambda\star}_h]\big|$, $\big|\delta'_w\Lambda[\mathbb{D}^\lambda, \mathbb{D}^{\lambda\star}_h]\big|$ *and* $\big|\delta''_{\overline{w}}\Lambda[\mathbb{D}^\lambda, \mathbb{D}^{\lambda\star}_h]\big|$ *go to 0 as $w_q \to \infty$.*

If rank $V = 1$, *we can impose* $\Lambda[\mathbb{D}^\lambda, \mathbb{D}^{\lambda\star}_h] = 0$.

Proof First, let us study the case rank $V = 1$. Let v be a frame of $\mathcal{P}_a\mathcal{V}$. Let h be the Hermitian metric of V determined by $h(v, v) = |w_q|^{2a}$. Then, the Chern connection associated with (V, d''_V) with h is flat, and hence it is a harmonic metric, i.e., $[\mathbb{D}^\lambda, \mathbb{D}^{\lambda\star}] = 0$. (See [63, Lemma 2.31].) The norm estimate is satisfied for $(\mathcal{P}_*\mathcal{V}, \mathbb{D}^\lambda)$ with h. We obtain the meromorphic function $\mathfrak{a}(w_q)$ determined by $\mathbb{D}^\lambda v = v\,d\mathfrak{a}$. Because the Poincaré rank of $(\mathcal{V}, \mathbb{D}^\lambda)$ is strictly smaller than q, we obtain $|\mathfrak{a}| = O(|w_q|^{q-1})$ as $|w_q| \to \infty$. Note that $\partial_w(\mathfrak{a}) = O(|w_q|^{-1})$. We obtain

$$d''_{\overline{w}}v = 0, \quad d'_w v = \partial_w(\mathfrak{a})v, \quad \delta'_w v = (aq^{-1}w^{-1})v, \quad \delta''_{\overline{w}}v = \Big(-\overline{\partial_w\mathfrak{a}} - \overline{\lambda}a\overline{w}^{-1}\Big)v.$$

Then, the claims are easily proved.

To study the general case, we use the constructions in [62–64, 80]. We explain only an indication. By pulling back via a ramified covering, we may assume that there exists the following decomposition from the beginning:

$$(\mathcal{P}_*\mathcal{V}, \mathbb{D}^\lambda) = \bigoplus_{\mathfrak{a}\in w_q\mathbb{C}[w_q]} (\mathcal{P}_*\mathcal{V}_\mathfrak{a}, \mathbb{D}^\lambda_\mathfrak{a}).$$

Here, $\mathbb{D}^\lambda_\mathfrak{a} - d\mathfrak{a}\,\mathrm{id}$ are logarithmic with respect to $\mathcal{P}_*\mathcal{V}_\mathfrak{a}$. It is enough to construct a Hermitian metric for $(\mathcal{P}_*\mathcal{V}_\mathfrak{a}, \mathbb{D}^\lambda_\mathfrak{a})$ for each $\mathfrak{a}$. By considering the λ-connection associated with the harmonic bundle $(\mathcal{O}_{U^*_{w,q}} \cdot e, (1+|\lambda|^2)^{-1}d\mathfrak{a})$ with $h(e,e)=1$, it is enough to consider the case $\mathfrak{a}=0$, i.e., $\mathbb{D}^\lambda$ is logarithmic with respect to $\mathcal{P}_*\mathcal{V}$.

By using the model harmonic bundles in [62, §6], we can construct an endomorphism $\mathfrak{F}$ of $\mathcal{V}$ such that the following holds (see [62]):

- $\mathfrak{F}\mathcal{P}_a\mathcal{V} \subset \mathcal{P}_{<a}\mathcal{V}$ for any $a \in \mathbb{R}$.
- We set $\mathbb{D}^\lambda_1 := \mathbb{D}^\lambda - \mathfrak{F}\,dw/w$. Then, there exists a harmonic metric h of $(V, \mathbb{D}^\lambda_1)$ such that h satisfies the norm estimate for $(\mathcal{P}_*\mathcal{V}, \mathbb{D}^\lambda_1)$.

We obtain the operators δ'_h, $\delta''_{1,h}$ and $\mathbb{D}^{\lambda\star}_{1,h}$ from $\mathbb{D}^\lambda_1$ and h. Because h is a harmonic metric for $(V, \mathbb{D}^\lambda_1)$, $[\mathbb{D}^\lambda_1, \mathbb{D}^{\lambda\star}_{1,h}] = 0$ holds.

There exists the decomposition $\mathbb{D}^\lambda_1 = d''_V + d'_{1,V}$ into the $(0,1)$-part and the $(1,0)$-part. There exist the operators $\overline{\partial}_{1,V}$, $\partial_{1,V}$, θ_1 and $\theta_1^\dagger$ such that $d''_V = \overline{\partial}_{1,V} + \lambda\theta_1^\dagger$, $\delta''_{1,h} = \overline{\lambda}\overline{\partial}_{1,V} - \theta_1^\dagger$, $d'_{1,V} = \lambda\partial_{1,V} + \theta_1$, and $\delta'_h = \partial_{1,V} - \overline{\lambda}\theta_1$. We have the expressions $\theta_1 = g_1\,dw$ and $\theta_1^\dagger = g_1^\dagger\,d\overline{w}$. Let $\partial_{1,V,\overline{w}}$ denote the inner product of $\overline{\partial}_{1,V}$ and $\partial_{\overline{w}}$. Similarly, $\partial_{1,V,w}$ denote the inner product of $\partial_{1,V}$ and ∂_w. Then, we obtain $d''_{\overline{w}} = \partial_{1,V,\overline{w}} + \lambda g_1^\dagger$, $d'_{1,w} = \lambda\partial_{1,V,w} + g_1$, $\delta''_{1,h,\overline{w}} = \overline{\lambda}\partial_{1,V,\overline{w}} - g_1^\dagger$ and $\delta'_{h,w} = \partial_{1,V,w} - \overline{\lambda}g_1$. Because h is a harmonic metric of $(V, \mathbb{D}^\lambda)$, we obtain $\partial_{1,V,\overline{w}}g_1 = 0$, $\partial_{1,V,w}g_1^\dagger = 0$, and $[\partial_{1,V,\overline{w}}, \partial_{1,V,w}] + [g_1^\dagger, g_1] = 0$. We obtain the following equalities:

$$[d''_{\overline{w}}, \delta'_{h,w}] = -(1+|\lambda|^2)[g_1^\dagger, g_1], \qquad [d''_{\overline{w}}, \delta''_{1,h,\overline{w}}] = -(1+|\lambda|^2)\partial_{1,V,\overline{w}}g_1^\dagger,$$

$$[d'_{1,w}, \delta'_{h,w}] = -(1+|\lambda|^2)\partial_{1,V,w}g_1, \qquad [d'_{1,w}, \delta''_{1,h,\overline{w}}] = -(1+|\lambda|^2)[g_1, g_1^\dagger].$$

By Lemma 7.4.4, they go to 0 as $|w_q| \to \infty$. By the estimate for the curvature of wild harmonic bundles (see [64, Corollary 7.2.10]), we also obtain $[d''_{\overline{w}}, \delta'_{h,w}] = O\big(|w|^{-2}(\log|w|)^{-2}\big)$.

Note that $[d'_w, \delta'_{h,w}] = [d'_{1,w}, \delta'_{h,w}] - \delta'_{h,w}(\mathfrak{F}w^{-1})$. By Lemma 7.4.8, we obtain $\delta'_{h,w}(\mathfrak{F}w^{-1}) = O(|w|^{-2-\epsilon_1})$ for some $\epsilon_1 > 0$. Hence, $[d'_w, \delta'_{h,w}]$ goes to 0. By taking the adjoint, we also obtain that $[d''_{\overline{w}}, \delta''_{h,\overline{w}}]$ goes to 0.

Note that

$$[d'_w, \delta''_{h,\overline{w}}] = [d'_{1,w}, \delta''_{1,h,\overline{w}}] - d'_{1,w}(\mathfrak{F}^\dagger\overline{w}^{-1}) - \delta''_{1,h,\overline{w}}(\mathfrak{F}w^{-1}) - |w|^{-2}[\mathfrak{F}, \mathfrak{F}^\dagger].$$

We have $[\mathfrak{F}, \mathfrak{F}^{\dagger}] = O(|w|^{-2\epsilon})$. We have

$$\delta''_{1,h,\overline{w}}(\mathfrak{F}w^{-1}) = \overline{\lambda} d''_{\overline{w}}(\mathfrak{F})w^{-1} - (1+|\lambda|^2)[g_1^{\dagger}, \mathfrak{F}]w^{-1} = O(|w|^{-2-\epsilon}).$$

By taking the adjoint, we also obtain $d'_{1,w}(\mathfrak{F}^{\dagger}\overline{w}^{-1}) = O(|w|^{-2-\epsilon})$. Therefore, $[d'_w, \delta''_{1,\overline{w}}]$ goes to 0 as $|w_q| \to \infty$. As the adjoint, $[d''_{\overline{w}}, \delta'_{1,w}]$ goes to 0 as $|w_q| \to \infty$.

We note that $d'_{1,w} - \lambda\delta'_{h,w} = (1+|\lambda|^2)g_1 = O(|w|^{-1})$ and $\delta''_{1,h,\overline{w}} - \overline{\lambda} d''_{\overline{w}} = -(1+|\lambda|^2)g_1^{\dagger} = O(|w|^{-1})$. Because $d'_w - \lambda\delta'_{h,w} = (d'_{1,w} - \lambda\delta'_{h,w}) + \mathfrak{F}w^{-1}$ and $\delta''_{\overline{w}} - \overline{\lambda} d'_{\overline{w}} = (\delta''_{1,\overline{w}} - \overline{\lambda} d'_{\overline{w}}) - \mathfrak{F}_h^{\dagger}\overline{w}^{-1}$, we obtain the desired estimate for $d'_w - \lambda\delta'_{h,w}$ and $\delta''_{\overline{w}} - \overline{\lambda} d'_{\overline{w}}$.

Note that $\frac{\sqrt{-1}}{2}\Lambda\big[\mathbb{D}^{\lambda}, \mathbb{D}_h^{\lambda\star}\big] = [\delta'_{h,w}, d''_{\overline{w}}] + [d'_w, \delta''_{h,\overline{w}}]$ and $0 = \frac{\sqrt{-1}}{2}\Lambda\big[\mathbb{D}_1^{\lambda}, \mathbb{D}_{1,h}^{\lambda\star}\big] = [\delta'_{h,w}, d''_{\overline{w}}] + [d'_{1,w}, \delta''_{1,h,\overline{w}}]$. Hence, we obtain

$$\begin{aligned}\frac{\sqrt{-1}}{2}\Lambda\big[\mathbb{D}^{\lambda}, \mathbb{D}_h^{\lambda\star}\big] &= [d'_w, \delta''_{h,\overline{w}}] - [d'_{1,w}, \delta''_{1,h,\overline{w}}]\\ &= -d'_{1,w}\big(\mathfrak{F}^{\dagger}\overline{w}^{-1}\big) - \delta''_{1,h,\overline{w}}(\mathfrak{F}w^{-1}) - \big[\mathfrak{F}, \mathfrak{F}^{\dagger}\big]\cdot|w|^{-2}.\end{aligned} \tag{7.27}$$

By the previous consideration, we obtain $\frac{\sqrt{-1}}{2}\Lambda\big[\mathbb{D}^{\lambda}, \mathbb{D}_h^{\lambda\star}\big] = O(|w|^{-2-\epsilon})$ for some $\epsilon > 0$.

We obtain

$$\begin{aligned}\delta'_{h,w}\delta''_{1,h,\overline{w}}(\mathfrak{F}) &= \delta''_{1,h,\overline{w}}\delta'_{h,w}\mathfrak{F}\\ &= \overline{\lambda}(d''_{\overline{w}}\delta'_{h,w} - \delta'_{h,w}d''_{\overline{w}})\mathfrak{F} - (1+|\lambda|^2)g_1^{\dagger}\delta'_{h,w}(\mathfrak{F}) = O(|w|^{-2-\epsilon}).\end{aligned} \tag{7.28}$$

We obtain

$$d'_w\delta''_{1,h,\overline{w}}(\mathfrak{F}) = \big(\lambda\delta'_{h,w} + (1+|\lambda|^2)g_1 + \mathfrak{F}w^{-1}\big)\delta''_{1,h,\overline{w}}(\mathfrak{F}) = O(|w|^{-2-\epsilon}).$$

We have

$$\begin{aligned}d''_{\overline{w}}\delta''_{1,h,\overline{w}}(\mathfrak{F}) &= d''_{\overline{w}}\big(\overline{\lambda} d''_{\overline{w}} - (1+|\lambda|^2 g_1^{\dagger})\mathfrak{F}\big)\\ &= -(1+|\lambda|^2)d''_{\overline{w}}(g_1^{\dagger})\mathfrak{F} = -(1+|\lambda|^2)\partial_{V,\overline{w}}(g_1^{\dagger})\cdot\mathfrak{F} = O(|w|^{-2-\epsilon}).\end{aligned} \tag{7.29}$$

We obtain

$$\delta''_{h,\overline{w}}\delta''_{1,h,\overline{w}}(\mathfrak{F}) = \Big(\overline{\lambda} d''_{\overline{w}} - (1+|\lambda|^2)g_1^{\dagger} - \mathfrak{F}\overline{w}^{-1}\Big)\delta''_{1,h,\overline{w}}(\mathfrak{F}) = O(|w|^{-2-2\epsilon}).$$

By considering the adjoint, we obtain the estimate for $d''_{\overline{w}}d'_{1,w}(\mathfrak{F}^{\dagger})$, $d'_w d'_{1,w}(\mathfrak{F}^{\dagger})$, $\delta''_{h,\overline{w}}d'_{1,w}(\mathfrak{F}^{\dagger})$ and $\delta'_{h,w}d'_{1,w}(\mathfrak{F}^{\dagger})$. We have $d''_{\overline{w}}[\mathfrak{F},\mathfrak{F}^{\dagger}] = [\mathfrak{F}, d''_{\overline{w}}\mathfrak{F}^{\dagger}] = O(|w|^{-1-\epsilon})$. We also have

$$\delta''_{h,\overline{w}}[\mathfrak{F},\mathfrak{F}^{\dagger}] = \big(\overline{\lambda} d''_{\overline{w}} - (1+|\lambda|^2)g_1^{\dagger} - \mathfrak{F}^{\dagger}\overline{w}^{-1}\big)[\mathfrak{F},\mathfrak{F}^{\dagger}] = O(|w|^{-1-\epsilon}).$$

By taking the adjoint, we obtain the estimate for $\delta'_{h,w}[\mathfrak{F},\mathfrak{F}^{\dagger}]$ and $d'_w[\mathfrak{F},\mathfrak{F}^{\dagger}]$. Thus, we obtain the claim of the lemma. ■

7.4.5 *Proof of Proposition 7.4.3*

It is enough to consider the case where $\mathcal{E}_{|\widehat{H}^{\lambda}_{\infty,q}}$ is unramified modulo level < 1. There exist a finite subset $S \subset \mathbb{Z} \times \mathbb{C}^*$, a tuple of $\mathbb{C}((w_q^{-1}))$-modules with λ-connection $(\widehat{\mathcal{V}}_{\ell,\alpha}, \widehat{\mathbb{D}}^{\lambda}_{\ell,\alpha})$ $((\ell,\alpha) \in S)$ and an isomorphism

$$\mathcal{E}_{|\widehat{H}^{\lambda}_{\infty,q}} \simeq \bigoplus_{(\ell,\alpha)\in S} \widehat{\mathbb{L}}^{\lambda}_q(\ell,\alpha) \otimes (\Psi^{\lambda}_q)^*(\widehat{\mathcal{V}}_{\ell,\alpha}, \widehat{\mathbb{D}}^{\lambda}_{\ell,\alpha}). \tag{7.30}$$

Moreover, there exist good filtered λ-flat bundles $(\mathcal{P}_*\widehat{\mathcal{V}}_{\ell,\alpha}, \widehat{\mathbb{D}}^{\lambda}_{\ell,\alpha})$ over $(\widehat{\mathcal{V}}_{\ell,\alpha}, \widehat{\mathbb{D}}^{\lambda}_{\ell,\alpha})$ such that the following holds for any $t_1 \in S^1_T$:

$$\mathcal{P}_*(\mathcal{E}_{|\widehat{H}^{\lambda}_{\infty,q}\langle t_1\rangle}) \simeq \bigoplus \Big(\mathcal{P}^{(0)}_*\widehat{\mathbb{L}}^{\lambda}_q(\ell,\alpha)_{|\widehat{H}^{\lambda}_{\infty,q}\langle t_1\rangle}\Big) \otimes \Big(\Psi^*_q(\mathcal{P}_*\widehat{\mathcal{V}}_{\ell,\alpha}, \widehat{\mathbb{D}}^{\lambda}_{\ell,\alpha})_{|\widehat{H}^{\lambda}_{\infty,q}\langle t_1\rangle}\Big). \tag{7.31}$$

Lemma 7.4.10 *For each pair (ℓ,α), there exists a good filtered λ-flat bundle $(\mathcal{P}_*\mathcal{V}_{\ell,\alpha}, \mathbb{D}^{\lambda}_{\ell,\alpha})$ on $(U_{w,q},\infty)$ with an isomorphism*

$$\big(\mathcal{P}_*\mathcal{V}_{\ell,\alpha}, \mathbb{D}^{\lambda}_{\ell,\alpha}\big)_{|\widehat{\infty}_{w,q}} \simeq (\mathcal{P}_*\widehat{\mathcal{V}}_{\ell,\alpha}, \widehat{\mathbb{D}}^{\lambda}_{\ell,\alpha}).$$

For simplicity, if $\lambda \neq 0$, we assume that the Stokes structure of $(\mathcal{V}_{\ell,\alpha}, \mathbb{D}^{\lambda}_{\ell,\alpha})$ is trivial.

Proof If $\lambda \neq 0$, the claim is the same as Lemma 7.3.16. If $\lambda = 0$, by shrinking $U_{w,q}$, we obtain the decomposition $\mathcal{E} = \bigoplus_{(\ell,\alpha)\in S}\mathcal{E}_{\ell,\alpha}$ which induces the decomposition (7.30), i.e., $\mathcal{E}_{\ell,u|\widehat{H}^{0}_{\infty,q}} \simeq \widehat{\mathbb{L}}^0_q(\ell,\alpha) \otimes (\Psi^0_q)^*(\widehat{\mathcal{V}}_{\ell,\alpha}, \widehat{\mathbb{D}}^0_{\ell,\alpha})$. There exist meromorphic Higgs bundles $(\mathcal{V}_{\ell,\alpha}, \mathbb{D}^0_{\ell,\alpha})$ on $(U_{w,q},\infty)$ which induce $\mathcal{E}_{\ell,\alpha} \otimes \mathbb{L}^0_q(\ell,\alpha)^{-1}$. There exist isomorphisms $(\mathcal{V}_{\ell,\alpha}, \mathbb{D}^0_{\ell,\alpha})_{|\widehat{\infty}_{w,q}} \simeq (\widehat{\mathcal{V}}_{\ell,\alpha}, \mathbb{D}^0_{\ell,\alpha})$. Hence, we obtain good filtered Higgs bundles $(\mathcal{P}_*\mathcal{V}_{\ell,\alpha}, \mathbb{D}^0_{\ell,\alpha})$ over $(\mathcal{P}_{\ell,\alpha}, \mathbb{D}^0_{\ell,\alpha})$ for which $(\mathcal{P}_*\mathcal{V}_{\ell,\alpha}, \mathbb{D}^0_{\ell,\alpha})_{|\widehat{\infty}_{w,q}} \simeq (\mathcal{P}_*\widehat{\mathcal{V}}_{\ell,\alpha}, \mathbb{D}^0_{\ell,\alpha})$ holds. ■

Recall that there exists a Hermitian metric $h_{\mathbb{L},q,\ell,\alpha}$ of $\mathbb{L}_q^{\lambda*}(\ell,\alpha) = \mathbb{L}_q^{\lambda}(\ell,\alpha)_{|\mathcal{B}_q^{\lambda*}}$ such that $\left(\mathbb{L}_q^{\lambda*}(\ell,\alpha), h_{\mathbb{L},q,\ell,\alpha}\right)$ is a monopole and satisfies the norm estimate with respect to $\mathcal{P}_*\mathbb{L}_q^{\lambda}(\ell,\alpha)$. (See Sect. 5.2.4.)

Let $(V_{\ell,\alpha}, \mathbb{D}_{\ell,\alpha}^{\lambda})$ be the λ-flat bundle on $U_{w,q}^*$ obtained as the restriction of $(\mathcal{P}_*\mathcal{V}_{\ell,\alpha}, \mathbb{D}_{\ell,\alpha}^{\lambda})$. Let $h_{1,V,\ell,\alpha}$ be a Hermitian metric of $V_{\ell,\alpha}$ as in Lemma 7.4.9. Let $h_{1,\ell,\alpha}$ be the Hermitian metric of $E_{1,\ell,\alpha} := \mathbb{L}_q^{\lambda*}(\ell,\alpha) \otimes \Psi_q^*(V_{\ell,\alpha}, \mathbb{D}_{\ell,\alpha}^{\lambda})$ induced by $h_{\mathbb{L},q,\ell,\alpha}$ and $\Psi_q^{-1}(h_{1,V,\ell,\alpha})$.

Lemma 7.4.11 *The following holds, where we consider the norms with respect to $h_{1,\ell,\alpha}$.*

- *Each $h_{1,\ell,\alpha}$ satisfies the norm estimate for the filtered bundle $\mathcal{P}_*\mathbb{L}_q^{\lambda}(\ell,\alpha) \otimes \Psi_q^*(\mathcal{P}_*\mathcal{V}_{\ell,\alpha}, \mathbb{D}_{\ell,\alpha}^{\lambda})$.*
- *$\left|G(h_{1,\ell,\alpha})\right| = O\left(|w_q|^{-2q-2\epsilon}\right)$ for some $\epsilon > 0$.*
- *$\left|\nabla_{\overline{\beta}_0} G(h_{1,\ell,\alpha})\right|$, $\left|\nabla_{\beta_0} G(h_{1,\ell,\alpha})\right|$, $\left|\partial_{E_{1,\ell,\alpha},t_0} G(h_{1,\ell,\alpha})\right|$ and $\left|\partial'_{E_{1,\ell,\alpha},t_0} G(h_{1,\ell,\alpha})\right|$ go to 0 as $|w_q| \to \infty$. (See* Sect. 2.7.5 *for mini-complex coordinate systems (t_0, β_0).)*
- *Let $\nabla_{1,\ell,\alpha}$ and $\phi_{1,\ell,\alpha}$ be the Chern connection and the Higgs field associated with the mini-holomorphic bundle $(E_{1,\ell,\alpha}, \overline{\partial}_{E_{1,\ell,\alpha}})$ and the Hermitian metric $h_{1,\ell,\alpha}$. Then, $\left|\left[\nabla_{1,\ell,\alpha}, \nabla_{1,\ell,\alpha}\right]\right| \to 0$ and $\left|\nabla_{1,\ell,\alpha}(\phi_{1,\ell,\alpha})\right| \to 0$ as $|w_q| \to \infty$. We also have $|\phi_{1,h,\ell}| = O\left(\log|w|\right)$ as $|w_q| \to \infty$.*
- *Let $\partial_{E_{1,\ell,\alpha},h_{1,\ell,\alpha},\beta_1}$ be the operator induced by $\partial_{E_{1,\ell,\alpha},\overline{\beta}_1}$ and $h_{1,\ell,\alpha}$ as in* Sect. 2.9.3. *Then,*

$$\left|\left[\partial_{E_{1,\ell,\alpha,\overline{\beta}_1}}, \partial_{E_{1,\ell,\alpha},h_{1,\ell,\alpha},\beta_1}\right]\right| = O\left(|w|^{-2}(\log|w|)^{-2}\right). \tag{7.32}$$

If rank $V_{\ell,\alpha} = 1$, $G(h_{1,\ell,\alpha}) = 0$ *holds.*

Proof The claim follows from our construction of $h_{1,\ell,\alpha}$, Proposition 4.5.7, Lemma 7.4.9 and the formulas in Lemma 2.9.11, Lemma 2.9.12 and Lemma 2.9.14. See also Remark 2.4.5. ∎

We set $\mathcal{E}_1 := \bigoplus_{\ell,\alpha} \mathbb{L}_q^{\lambda}(\ell,\alpha) \otimes (\Psi_q^{\lambda})^*(\mathcal{P}V_{\ell,\alpha}, \mathbb{D}_{\ell,\alpha}^{\lambda})$. We obtain the naturally induced good filtered bundle $\mathcal{P}_*\mathcal{E}_1$ over $\mathcal{E}_1$. By the construction and the isomorphism (7.31), there exists the isomorphism $\widehat{f} : \mathcal{E}_{|\widehat{H}_{\infty,q}^{\lambda}} \simeq \mathcal{E}_{1|\widehat{H}_{\infty,q}^{\lambda}}$ which induces an isomorphism of filtered bundles $\mathcal{P}_*\mathcal{E}_{|\widehat{H}_{\infty,q}^{\lambda}} \simeq \mathcal{P}_*\mathcal{E}_{1|\widehat{H}_{\infty,q}^{\lambda}}$.

Let $C_{\mathcal{B}_q^{\lambda}}^{\infty}$ denote the sheaf of C^{∞}-functions on $\mathcal{B}_q^{\lambda}$. There exists an isomorphism

$$f_{C^{\infty}} : \mathcal{E} \otimes_{\mathcal{O}} C_{\mathcal{B}_q^{\lambda}}^{\infty} \simeq \mathcal{E}_1 \otimes_{\mathcal{O}} C_{\mathcal{B}_q^{\lambda}}^{\infty}$$

which induces $\widehat{f}$ at $\widehat{H}_{\infty,q}^{\lambda}$. Let h_1 be the Hermitian metric on E induced by the isomorphism $f_{C^{\infty}}$ and the metric $\bigoplus h_{\mathbb{L},q,\ell,\alpha} \otimes \Psi_q^{-1}(h_{1,V,\ell,\alpha})$. Then, h_1 has the desired property. ∎

7.4.6 The Strong Adaptedness and the Norm Estimate

Let $\mathcal{P}_*\mathcal{E}$ and $(E, \overline{\partial}_E)$ be as in Sect. 7.4.1. Let h be a Hermitian metric of E which is strongly adapted to $\mathcal{P}_*\mathcal{E}$, such that $(E, \overline{\partial}_E, h)$ is a monopole. Let h_1 be a Hermitian metric of E as in Proposition 7.4.3. By the assumption of the strong adaptedness of h (see Definition 4.4.7) for any $\delta > 0$, there exist $C_\delta \geq 1$ such that $C_\delta^{-1}|w_q|^{-\delta}h_1 \leq h \leq C_\delta|w_q|^\delta h_1$.

Lemma 7.4.12 *h and h_1 are mutually bounded, i.e., $(\mathcal{P}_*\mathcal{E}, h)$ satisfies the norm estimate.*

Proof Let s_1 be the automorphism of E determined by $h = h_1 s_1$. By Corollary 2.9.9, there exist positive constants C_2 and ϵ such that

$$-\Big(\partial_{E,\overline{\beta}_0}\partial_{E,\beta_0} + \frac{1}{4}\partial_{E,t_0}\partial'_{E,t_0}\Big)\log \mathrm{Tr}(s_1) \leq \frac{1}{2}|G(h_1)|_{h_1} \leq C_2|w|^{-2-\epsilon}.$$

Note that $\partial_{\beta_0}\partial_{\overline{\beta}_0} + \frac{1}{4}\partial_{t_0}\partial_{t_0} = \partial_w\partial_{\overline{w}} + \frac{1}{4}\partial_t\partial_t$. There exists $C_3 > 0$ such that

$$-\Big(\partial_{E,\overline{\beta}_0}\partial_{E,\beta_0} + \frac{1}{4}\partial_{E,t_0}\partial'_{E,t_0}\Big)\Big(\log \mathrm{Tr}(s_1) + C_3|w|^{-\epsilon}\Big) \leq 0.$$

We set $g := \log \mathrm{Tr}(s_1) + C_3|w|^{-\epsilon}$. We take R_0 such that $\{|w_q| \geq R_0\} \subset U^*_{w,q}$. There exists $C_5 > 0$ such that $g < C_5$ on $\Psi_q^{-1}(\{|w_q| = R_0\})$. For any $\rho > 0$, we consider the function

$$G_\rho := g - \big(C_5 + \rho \log|w|\big).$$

Note that (i) G_ρ is subharmonic, (ii) $G_\rho < 0$ on $\Psi_q^{-1}(\{|w_q| = R\})$, (iii) $G_\rho \to -\infty$ as $|w_q| \to \infty$. Hence, we obtain $G_\rho \leq 0$ on $\Psi_q^{-1}(\{|w_q| \geq R\})$. By taking the limit as $\rho \to 0$, we obtain

$$\mathrm{Tr}(s_1) \leq \exp\big(C_5 - C_3|w|^{-\epsilon}\big) \leq \exp(C_5).$$

i.e., $\mathrm{Tr}(s_1)$ is bounded.

Let s_2 be the automorphism of E determined by $h_1 = h \cdot s_2$. By a similar argument, we obtain that $\mathrm{Tr}(s_2)$ is bounded. We obtain that h and h_1 are mutually bounded. ■

7.4.7 Proof of Proposition 7.4.1

Let $(E, \overline{\partial}_E, h)$ be as in Proposition 7.4.1. We take a Hermitian metric h_1 as in Proposition 7.4.3. We obtain the automorphism s of E determined by $h = h_1 \cdot s$.

According to Lemma 7.4.12, h and h_1 are mutually bounded, and hence s and s^{-1} are bounded. Let dvol denote the volume form of $\mathcal{B}_q^{\lambda *}$ with the metric g.

Let $U_{w,1}$ be a relatively compact neighbourhood of ∞ in U_w. Let $\mathcal{B}_{q,1}^{\lambda}$ be the induced relatively compact neighbourhood of $H_{\infty,q}^{\lambda}$ in $\mathcal{B}_q^{\lambda}$. (See Sect. 7.1.2.) We set $\mathcal{B}_{q,1}^{\lambda *} := \mathcal{B}_{q,1}^{\lambda} \setminus H_{\infty,q}^{\lambda}$.

Lemma 7.4.13 *We obtain* $\int_{\mathcal{B}_{q,1}^{\lambda *}} \left(\left|\partial_{E,h_1,\beta_0} s\right|_{h_1}^2 + \left|\partial_{E,h_1,t_0} s\right|_{h_1}^2 \right) \mathrm{dvol} < \infty.$

Proof By Lemma 2.9.8, we have

$$-\left(\partial_{\overline{\beta}_0}\partial_{\beta_0} + \frac{1}{4}\partial_{t_0}^2\right)\operatorname{Tr}(s) = -\frac{1}{2}\operatorname{Tr}\left(sG(h_1)\right) - \left|s^{-1/2}\partial_{E,h_1,\beta_0} s\right|_{h_1}^2 - \frac{1}{4}\left|s^{-1/2}\partial'_{E,h_1,t_0} s\right|_{h_1}^2. \tag{7.33}$$

Recall that $\mathcal{B}_q^{\lambda *} = \mathcal{B}_q^{0*} = S_T^1 \times U_{w,q}^*$ as Riemannian manifolds, and that

$$-\partial_{\overline{\beta}_0}\partial_{\beta_0} - \frac{1}{4}\partial_{t_0}^2 = -\partial_{\overline{w}}\partial_w - \frac{1}{4}\partial_t^2.$$

For any function F on $\mathcal{B}_q^{0*}$, let $\int_{S_T^1} F$ denote the function on $U_{w,q}^*$ obtained as the integration of F along the fiber direction. We set

$$b_1 := \int_{S_T^1} \operatorname{Tr}(s), \quad b_2 := \frac{1}{2}\int_{S_T^1} \operatorname{Tr}(sG(h_1)),$$

$$b_3 := \int_{S_T^1} \left|s^{-1/2}\partial_{E,h_1,\beta_0} s\right|_{h_1}^2 + \frac{1}{4}\int_{S_T^1} \left|s^{-1/2}\partial'_{E,h_1,t_0} s\right|_{h_1}^2.$$

We obtain $-\partial_{\overline{w}}\partial_w b_1 = -b_2 - b_3$. Let $\tau_q := w_q^{-1}$. Then, we obtain the following:

$$-\partial_{\tau_q}\partial_{\overline{\tau}_q} b_1 = -q^2\left|\tau_q\right|^{-2(q+1)} \cdot b_2 - q^2\left|\tau_q\right|^{-2(q+1)} \cdot b_3.$$

Because $G(h_1) = O(|w|^{-2-\epsilon}) = O\left(|\tau_q|^{2q+q\epsilon}\right)$ as in Lemma 7.4.11, we obtain

$$|\tau_q|^{-2(q+1)} \cdot b_2 = O(|\tau_q|^{-2+q\epsilon}).$$

Hence, there exists a bounded function c such that $-\partial_{\tau_q}\partial_{\overline{\tau}_q} c = -q^2\left|\tau_q\right|^{-2(q+1)} \cdot b_2$. We obtain

$$-\partial_{\tau_q}\partial_{\overline{\tau}_q}(b_1 - c) = -q^2\left|\tau_q\right|^{-2(q+1)} \cdot b_3.$$

Note that $b_1 - c$ is bounded, and $b_3 \geq 0$. Hence, according to [80, Lemma 2.2], we obtain

$$\int_{U_{w,1,q}} |\tau_q|^{-2(q+1)} \cdot b_3 \, |d\tau_q \, d\overline{\tau}_q| < \infty.$$

Here, $U_{w,1,q}$ denotes the pull back of $U_{w,1}$ by the ramified covering $\mathbb{P}^1_{w_q} \longrightarrow \mathbb{P}^1_w$. It implies that $\int_{U_{w,1,q}} b_3 \, |dw \, d\overline{w}| < \infty$. Because s and s^{-1} are bounded, we obtain the claim of Lemma 7.4.13. ■

By the construction of the Higgs field, we have $\phi_h = \phi_{h_1} - \frac{\sqrt{-1}}{2} s^{-1} \partial'_{h_1,t_0} s$. (See Sect. 2.4.2.) Note that $|\phi_{h_1}|_{h_1} = O\big(\log|w|\big)$. There exists $R_0 > 0$ such that $\{|w_q^q| > R_0\} \subset U^*_{w,q,1}$. Let $B_{(w_q,t)}(2)$ be the ball with radius 2 centered at (w_q, t), where we use the distance induced by the Euclidean metric $dt\,dt + dw\,d\overline{w}$. By Lemma 7.4.13, there exists a constant $C > 0$ such that the following holds for any (w, t) with $|w| > 2R_0 + 10$:

$$\int_{B_{(w,t)}(2)} |\phi_h|_h^2 \leq C\big(\log|w|\big)^2.$$

Let (x, y) be the local real coordinate system on $\mathbb{C}_{w_q}$ defined by $w = x + \sqrt{-1}y$. By the equality (6.11), we obtain

$$-(\partial_t^2 + \partial_x^2 + \partial_y^2)|\phi|_h^2 = -2\big(|\nabla_x \phi|_h^2 + |\nabla_y \phi|_h^2 + |\nabla_t \phi|_h^2\big) \leq 0,$$

and hence we obtain $|\phi_h|_h = O(\log|w|)$ by the mean-value theorem for subharmonic functions.

For the proof of the condition $F(\nabla) \to 0$, we shall apply an argument in [66]. We indicate only an outline. We set $\widetilde{\mathcal{B}}^{\lambda*}_q := S^1 \times \mathcal{B}^{\lambda*}_q$. It is equipped with the complex structure induced by the local complex coordinate systems $(\alpha_0, \beta_0) = (s_0 + \sqrt{-1}t_0, \beta_0)$ as in Sect. 2.7.5. It is equipped with the Kähler metric $\widetilde{g} = d\alpha_0 d\overline{\alpha}_0 + d\beta_0 d\overline{\beta}_0$. Let $(\widetilde{E}, \overline{\partial}_{\widetilde{E}})$ denote the holomorphic bundle on $\widetilde{\mathcal{B}}^{\lambda*}_q$ induced by $(E, \overline{\partial}_E)$ as in Sect. 2.5.2. The bundle is equipped with the induced metrics $\widetilde{h}$ and $\widetilde{h}_1$. As recalled in Sect. 2.5.1, $(\widetilde{E}, \overline{\partial}_{\widetilde{E}}, \widetilde{h})$ is an instanton.

Lemma 7.4.14 *The following conditions are satisfied, where we consider the norms with respect to $\widetilde{h}_1$ and the Euclidean metric $\widetilde{g}$.*

- $|F(\widetilde{h}_1)| \to 0$ *as* $|w_q| \to \infty$.
- $|\Lambda F(\widetilde{h}_1)| = O(|w_q|^{-2q-\epsilon})$.
- $|\nabla_{\widetilde{h}_1} \Lambda F(\widetilde{h}_1)| \to 0$ *as* $|w_q| \to \infty$.

Proof Let p denote the projection $\widetilde{\mathcal{B}}^{\lambda*}_q \longrightarrow \mathcal{B}^{\lambda*}_q$. We have $F(\widetilde{h}_1) = p^*F(\nabla_{h_1}) + p^*\nabla_{h_1}\phi_{h_1} \cdot ds_0$. We also have $\sqrt{-1}\Lambda F(\widetilde{h}_1) = p^*G(h_1)$. Hence, we obtain the claims of the lemma from the assumption for h_1. ■

Let $\widetilde{s}$ be determined by $\widetilde{h} = \widetilde{h}_1\widetilde{s}$. Because $\widetilde{s}$ is the pull back of s, $\widetilde{s}$ and $\widetilde{s}^{-1}$ are bounded with respect to $\widetilde{h}_1$. By Lemma 7.4.13, we have $\int |\partial_{\widetilde{h}_1}\widetilde{s}|^2_{\widetilde{h}_1,\widetilde{g}} < \infty$. Let $\widetilde{s}_1$ be determined by $\widetilde{h}_1 = \widetilde{h}\cdot\widetilde{s}_1$. Because $\widetilde{s}_1^{-1}\partial_{\widetilde{h}}\widetilde{s}_1 = -\widetilde{s}\partial_{\widetilde{h}_1}\widetilde{s}$, we obtain $\int |\partial_{\widetilde{h}}\widetilde{s}_1|^2_{\widetilde{h},\widetilde{g}} < \infty$. For any $\epsilon > 0$, there exists a compact subset $K_\epsilon \subset \widetilde{\mathcal{B}}^{\lambda*}_q$ such that the following holds.

- $\int_{\widetilde{\mathcal{B}}^{\lambda*}_q\setminus K_\epsilon} |\widetilde{s}_1^{-1}\partial_{E,\widetilde{h}}\widetilde{s}_1|^2_{\widetilde{h},\widetilde{g}} < \epsilon$.
- $|\partial_{\widetilde{E},\widetilde{h}_1}\Lambda F(\widetilde{h}_1)|_{\widetilde{h},\widetilde{g}} < \epsilon$ on $\widetilde{\mathcal{B}}^{\lambda*}_q \setminus K_\epsilon$.
- $|F(\widetilde{h}_1)|_{\widetilde{h},\widetilde{g}} < \epsilon$ on $\widetilde{\mathcal{B}}^{\lambda*}_q \setminus K_\epsilon$.

As proved in [66, §2.9.1], there exist $C_i > 0$ $(i = 1, 2)$ which are independent of (K_ϵ, ϵ), such that the following holds on $\widetilde{\mathcal{B}}^{\lambda*}_q \setminus K_\epsilon$:

$$-\Big(\partial_{\beta_0}\partial_{\overline{\beta}_0} + \frac{1}{4}\partial_{t_0}^2 - C_2\epsilon\Big)|\widetilde{s}_1^{-1}\partial_{\widetilde{h}}\widetilde{s}_1|^2_{\widetilde{h},\widetilde{g}} \leq C_1\epsilon.$$

By using [28, Theorem 9.20], we obtain the following.

Lemma 7.4.15 *For any $\epsilon > 0$, there exists a compact subset $K'_\epsilon \subset \widetilde{\mathcal{B}}^{\lambda*}_q$ such that the following holds on $\widetilde{\mathcal{B}}^{\lambda*}_q \setminus K'_\epsilon$:*

$$\sup_{\widetilde{\mathcal{B}}^{\lambda*}_q\setminus K'_\epsilon} |\widetilde{s}_1^{-1}\partial_{\widetilde{h}}\widetilde{s}_1|^2_{\widetilde{h},\widetilde{g}} \leq \epsilon.$$

■

According to [79, Lemma 3.1], we have the following relation:

$$\Delta_{\widetilde{E},\widetilde{h}_1}(\widetilde{s}) = -\widetilde{s}\sqrt{-1}\Lambda F(\widetilde{h}_1) + \sqrt{-1}\Lambda\overline{\partial}(\widetilde{s})\widetilde{s}^{-1}\partial_{\widetilde{h}_1}(\widetilde{s}). \tag{7.34}$$

Hence, for any $\epsilon > 0$, there exists a compact subset $K^{(3)}_\epsilon \subset \widetilde{\mathcal{B}}^{\lambda*}_q$ such that $|\Delta_{\widetilde{E},\widetilde{h}_1}(\widetilde{s})|_{\widetilde{h}_1} \leq \epsilon$ on $\widetilde{\mathcal{B}}^{\lambda*}_q \setminus K^{(3)}_\epsilon$.

Take a large p. We obtain that the L^p_2-norm of $\widetilde{s} - \mathrm{id}$ on the disc with radius r_0 centered at $P \in \widetilde{\mathcal{B}}^{\lambda*}_q$ goes to 0 as P goes to ∞. By using (7.34), we obtain that the L^p_3-norm of $\widetilde{s}_1 - \mathrm{id}$ on the disc with radius r_0 centered at $P \in \widetilde{\mathcal{B}}^{\lambda*}_q$ goes to 0 as P goes to ∞. It implies that $|F(\widetilde{h})|_{\widetilde{h},\widetilde{g}} \to 0$ as P goes to ∞. Therefore, $|F(h)|_h \to 0$. Thus, we obtain the claim of Proposition 7.4.1. ■

7.5 Some Functoriality

Let $\mathcal{B}_q^{\lambda *}$ and $\mathcal{B}_q^{\lambda}$ be as in Sect. 7.2.1. Let $(E_i, h_i, \nabla_i, \phi_i)$ $(i = 1, 2)$ be monopoles on $\mathcal{B}_q^{\lambda *}$ satisfying GCK-condition. (See Sect. 6.2.1.) We obtain the locally free $\mathcal{O}_{\mathcal{B}_q^{\lambda}}(*H_{\infty,q}^{\lambda})$-module $\mathcal{P}E_i^{\lambda}$ and the associated good filtered bundles $\mathcal{P}_*E_i^{\lambda}$ over $\mathcal{P}E_i^{\lambda}$ (Proposition 7.2.1 and Theorem 7.3.1). Clearly, the monopoles

$$(E_1, h_1, \nabla_1, \phi_1) \oplus (E_2, h_2, \nabla_2, \phi_2),$$

$$(E_1, h_1, \nabla_1, \phi_1) \otimes (E_2, h_2, \nabla_2, \phi_2),$$

$$\mathrm{Hom}((E_1, h_1, \nabla_1, \phi_1), (E_2, h_2, \nabla_2, \phi_2))$$

also satisfy GCK-condition.

Proposition 7.5.1 *There exists the following natural isomorphisms of good filtered bundles:*

$$\mathcal{P}_*(E_1^{\lambda} \oplus E_2^{\lambda}) \simeq \mathcal{P}_*(E_1^{\lambda}) \oplus \mathcal{P}_*(E_2^{\lambda}) \tag{7.35}$$

$$\mathcal{P}_*(E_1^{\lambda} \otimes E_2^{\lambda}) \simeq \mathcal{P}_*(E_1^{\lambda}) \otimes \mathcal{P}_*(E_2^{\lambda}) \tag{7.36}$$

$$\mathcal{P}_*(\mathrm{Hom}(E_1, E_2)^{\lambda}) \simeq \mathcal{H}om\Big(\mathcal{P}_*(E_1^{\lambda}), \mathcal{P}_*(E_2^{\lambda})\Big). \tag{7.37}$$

Proof It follows from Lemma 2.11.18. ■

Chapter 8
Global Periodic Monopoles of Rank One

Abstract We shall establish an equivalence between monopoles of GCK type and good filtered bundles *in the rank one case*, as a preliminary for more general cases.

8.1 Preliminary

8.1.1 Ahlfors Type Lemma

Take $R > 0$ and $C_0 > 0$. Set $U^*_w(R) := \{w \in \mathbb{C} \,|\, |w| > R\}$. Let $g : U^*_w(R) \longrightarrow \mathbb{R}_{\geq 0}$ be a C^∞-function such that

$$-\partial_w \partial_{\overline{w}} g \leq -C_0 g.$$

We assume that $g = O(|w|^N)$ for some $N > 0$.

Lemma 8.1.1 *There exists* $\epsilon_1 > 0$*, depending only on* C_0*, such that the estimate* $g = O\big(\exp(-\epsilon_1|w|)\big)$ *holds.*

Proof There exist $\epsilon_1 > 0$ and $R_1 \geq R$ such that the following holds on $\{w \in \mathbb{C} \,|\, |w| > R_1\}$:

$$-\partial_w \partial_{\overline{w}} \exp\big(-\epsilon_1|w|\big) \geq -C_0 \exp\big(-\epsilon_1|w|\big),$$

$$-\partial_w \partial_{\overline{w}} \exp\big(\epsilon_1|w|\big) \geq -C_0 \exp\big(\epsilon_1|w|\big).$$

There exists $C_2 > 0$ such that $g < C_2 \exp\big(-\epsilon_1|w|\big)$ on $\{|w| = R_1\}$. For any $\delta > 0$, we set

$$F_\delta := C_2 \exp\big(-\epsilon_1|w|\big) + \delta \exp\big(\epsilon_1|w|\big).$$

T. Mochizuki, *Periodic Monopoles and Difference Modules*, Lecture Notes in Mathematics 2300, https://doi.org/10.1007/978-3-030-94500-8_8

We obtain $g < F_\delta$ on $\{|w| = R_1\}$ and $-\partial_w \partial_{\overline{w}}(g - F_\delta) \leq -C_0(g - F_\delta)$. We set

$$Z(\delta) := \Big\{ w \in \mathbb{C} \,\Big|\, |w| \geq R_1,\ g(w) > F_\delta(w) \Big\}.$$

Because we have $g(w) = O(|w|^N)$ around ∞ and $g < F_\delta$ on $\{|w| = R_1\}$, $Z(\delta)$ is relatively compact in $\{|w| > R_1\}$. Hence, $g - F_\delta = 0$ holds on $\partial Z(\delta)$. On $Z(\delta)$, we have

$$-\partial_w \partial_{\overline{w}}(g - F_\delta) < 0.$$

Hence, if $Z(\delta) \neq \emptyset$, we obtain $g - F_\delta \leq 0$ on $Z(\delta)$, which contradicts the construction of $Z(\delta)$. Hence, $Z(\delta) = \emptyset$ holds. Namely, we obtain $g \leq F_\delta$ on $\{|w| > R_1\}$ for any $\delta > 0$. Therefore, we obtain $g \leq C_2 \exp(-\epsilon_1 |w|)$. ■

We give a variant of the estimate. Let $g : U_w^*(R) \longrightarrow \mathbb{R}_{\geq 0}$ be a C^∞-function such that the following holds for some $c > 0$ and $N > 0$.

- $g = O(|w|^N)$.
- For any $k > 0$, there exists $b_k > 0$ such that $-\partial_w \partial_{\overline{w}} g \leq b_k |w|^{-k} + (-c + |w|^{-2})g$.

Lemma 8.1.2 *$g = O(|w|^{-k})$ for any $k > 0$.*

Proof There exists $R_0 > R$ such that $g(w) < 2^{-1}|w|^{2N}$ on $|w| \geq R_0$. We fix $k > 0$, and we take $R_k \geq R_0$ such that

$$cR_k^2 \geq \frac{b_k}{R_k^{k-2+2N}} + \frac{(k-2)^2}{4} + 1.$$

We set $e_k = R_k^{k-2+2N}$. We obtain $g(w) < e_k |w|^{-k+2}$ on $\{|w| = R_k\}$. We also have the following inequality on $\{|w| \geq R_k\}$:

$$c - |w|^{-2} \geq \Big(\frac{b_k}{e_k} + \frac{(k-2)^2}{4} \Big) |w|^{-2}.$$

For any small $\epsilon > 0$, we set $\rho_{k,\epsilon} := e_k |w|^{-k+2} + \epsilon |w|^{2N}$. For any large k, we obtain the following:

$$\begin{aligned}
-\partial_w \partial_{\overline{w}} \rho_{k,\epsilon} &= -e_k \frac{(k-2)^2}{4} |w|^{-k} - \epsilon N^2 |w|^{2N-2} \\
&= b_k |w|^{-k} - \Big(\frac{b_k}{e_k} + \frac{(k-2)^2}{4} \Big) |w|^{-2} e_k |w|^{-k+2} - \epsilon N^2 |w|^{2N-2} \\
&\geq b_k |w|^{-k} - \Big(\frac{b_k}{e_k} + \frac{(k-2)^2}{4} \Big) |w|^{-2} \rho_{k,\epsilon}. \qquad (8.1)
\end{aligned}$$

Let us consider the set $Z_\epsilon := \{w \in \mathbb{C} \mid |w| \geq R_k,\ g(w) > \rho_{k,\epsilon}(w)\}$. Because Z_ϵ is relatively compact in $\{|w| \geq R_k\}$, we have $g(w) - \rho_{k,\epsilon}(w) = 0$ on ∂Z_ϵ. On Z_ϵ, we have the following:

$$-\partial_w \partial_{\overline{w}}(g - \rho_{k,\epsilon}) \leq -\Big(\frac{b_k}{e_k} + \frac{(k-2)^2}{4}\Big)|w|^{-2}(g - \rho_{k,\epsilon}) < 0.$$

We obtain $g - \rho_{k,\epsilon} \leq 0$ on Z_ϵ, which contradicts the choice of Z_ϵ. Thus, we obtain $Z_\epsilon = \emptyset$, i.e., $g \leq \rho_{k,\epsilon}$ on $\{|w| \geq R_k\}$ for any ϵ. By taking the limit $\epsilon \to 0$, we obtain $g \leq e_k |w|^{-k+2}$ on $\{|w| \geq R_k\}$. ■

8.1.2 *Poisson Equation (1)*

Let a be a C^∞-function on $S^1_T \times \{|w| > R\}$ satisfying the following conditions.

- For any $(\ell_1, \ell_2, \ell_3) \in \mathbb{Z}^2_{\geq 0}$, we have $|\partial_t^{\ell_1} \partial_w^{\ell_2} \partial_{\overline{w}}^{\ell_3} a(t,w)| = O(|w|^{-k})$ for any k as $w \to \infty$.
- $\int_{S^1_T} a(t,w)\,dt = 0$ for any w.

Let f be an $\mathbb{R}$-valued C^∞-function on $S^1_T \times \{|w| > R\}$ such that (i) $\Delta f = a$, (ii) $\int_{S^1_T} f(t,w)\,dt = 0$ for any w, (iii) $f = O(|w|^N)$ for some $N > 0$. Here, $\Delta = -\partial_t^2 - \partial_x^2 - \partial_y^2$ for the real coordinate system (x, y) on $\mathbb{C}_w$ determined by $w = x + \sqrt{-1}y$.

Lemma 8.1.3 *For any $\boldsymbol{\ell} = (\ell_1, \ell_2, \ell_3) \in \mathbb{Z}^3_{\geq 0}$, we obtain $\big|\partial_t^{\ell_1} \partial_w^{\ell_2} \partial_{\overline{w}}^{\ell_3} f\big| = O\big(|w|^{-k}\big)$ for any k. If the support of a is compact, there exists $\epsilon_{\boldsymbol{\ell}} > 0$ such that $\big|\partial_t^{\ell_1} \partial_w^{\ell_2} \partial_{\overline{w}}^{\ell_3} f\big| = O\Big(\exp\big(-\epsilon_{\boldsymbol{\ell}}|w|\big)\Big)$.*

Proof The following holds:

$$\Delta |f|^2 = -2\Big(\big|\partial_t f\big|^2 + \big|\partial_x f\big|^2 + \big|\partial_y f\big|^2\Big) + 2af.$$

We set $g(x,y) := \int_{S^1_T} |f|^2(t,x,y)dt$. Because $\int_{S^1_T} f(t,x,y)\,dt = 0$, there exists $C > 0$ such that $\int_{S^1_T} \big|\partial_t f\big|^2\,dt \geq Cg$. We obtain

$$\Delta g \leq -Cg + 2\Big(\int_{S^1_T} |a|^2\Big)^{1/2} g^{1/2}.$$

For any $k > 0$, there exists $b_k > 0$ such that the following holds:

$$\Delta g \leq -Cg + b_k|w|^{-k} g^{1/2} \leq b_k^2 |w|^{-2k+2} + (-C + |w|^{-2})g.$$

We also have $g = O(|w|^N)$ for some $N > 0$. By Lemma 8.1.2, we obtain $g = O\big(|w|^{-k}\big)$ for any k. Then, we obtain the first claim by using Lemma 8.1.4 below. We obtain the second claim by using a similar argument with Lemma 8.1.1. ■

8.1.2.1 Appendix

For $r > 0$, let $B(r) = \big\{(x_1, x_2, x_3) \in \mathbb{R}^3 \mid \sum x_i^2 < r^2\big\}$. For any $\boldsymbol{\ell} = (\ell_1, \ell_2, \ell_3) \in \mathbb{Z}^3_{\geq 0}$, we set $|\boldsymbol{\ell}| = \sum \ell_i$ and $\partial_{\boldsymbol{x}}^{\boldsymbol{\ell}} := \partial_{x_1}^{\ell_1}\partial_{x_2}^{\ell_2}\partial_{x_3}^{\ell_3}$. For any $\ell \in \mathbb{Z}_{\geq}0$, we set $\mathcal{S}(\ell) = \big\{\boldsymbol{\ell} \in \mathbb{Z}^3_{\geq 0} \,|, |\boldsymbol{\ell}| \leq \ell\big\}$. We set $\Delta = -\sum \partial_{x_i}^2$.

Let a be a C^∞-function on $B(r)$ such that the following holds.

- For any $\ell \in \mathbb{Z}_{\geq 0}$, there exists $C_\ell > 0$ such that $|\partial_{\boldsymbol{x}}^{\boldsymbol{\ell}} a| < C_\ell$ for any $\boldsymbol{\ell} \in \mathcal{S}(\ell)$.

Let f be a C^∞-function on $B(r)$ such that (i) $\Delta f = a$ and that (ii) $\int_{B(r)} |f|^2 < A_0$ for a constant $A_0 > 0$.

Lemma 8.1.4 *Let $0 < r' < r$. For any $\ell \in \mathbb{Z}_{\geq 0}$, there exists $\widetilde{C}_{1,r',\ell} > 0$, depending only on r, r', A_0 and $C_{\ell'}$ $(\ell' \leq \ell)$ such that $|\partial_{\boldsymbol{x}}^{\boldsymbol{\ell}} f| < \widetilde{C}_{1,r',\ell}$ for any $\boldsymbol{\ell} \in \mathcal{S}(\ell)$ on $B(r')$.*

Proof Let $0 < r' < r_0 < r$. By [28, Theorem 9.20], there exists $\widetilde{C}_{1,r_0,1} > 0$, depending only on r, r_0, A_0 and C_0 such that $|f| < \widetilde{C}_{1,r_0,0}$ on $B(r_0)$. Let $r' < r_1 < r_0$. We apply [28, Theorem 4.6] to the Poisson equation $\Delta f = a$. Then, there exists $\widetilde{C}_{1,r_1,1} > 0$, depending only on r, r_i $(i = 0, 1)$, A_0 and $C_{\ell'}$ $(\ell' \leq 1)$ such that $|\partial_{x_j} f| < \widetilde{C}_{1,r_1,1}$ $(j = 1, 2, 3)$ on $B(r_1)$. Let $r' < r_2 < r_1$. We apply [28, Theorem 4.6] to the Poisson equations $\Delta(\partial_{x_i} f) = \partial_{x_i} a$ $(i = 1, 2, 3)$. Then, there exists $\widetilde{C}_{1,r_2,2} > 0$, depending only on r, r_i $(i = 0, 1, 2)$, A_0 and $C_{\ell'}$ $(\ell' \leq 2)$ such that $|\partial_{x_j}\partial_{x_i} f| < \widetilde{C}_{1,r_2,2}$ $(i, j = 1, 2, 3)$ on $B(r_2)$. Then, by an easy induction, we obtain the claim of Lemma 8.1.4. ■

8.1.3 Poisson Equation (2)

Let b be a C^∞-function on $S^1_T \times \mathbb{C}_w$ such that (i) $\int_{S^1_T\times\mathbb{C}_w} b \,\mathrm{dvol}_{S^1_T\times\mathbb{C}_w} = 0$, (ii) $|\partial_t^{\ell_1}\partial_w^{\ell_2}\partial_{\overline{w}}^{\ell_3} b| = O\big(|w|^{-k}\big)$ as $|w| \to \infty$ for any $(\ell_1, \ell_2, \ell_3) \in \mathbb{Z}^3_{\geq 0}$ and for any k.

Lemma 8.1.5 *There exists a C^∞-function f on $S^1_T \times \mathbb{C}_w$ such that $\Delta_{S^1_T\times\mathbb{C}_w} f = b$ and $|f| = O(|w|^{-1})$ as $|w| \to \infty$.*

Proof Set $c := \int_{S^1_T} b\, dt$. We obtain $\int_{\mathbb{R}^2} c\, dx\, dy = 0$, and $\partial_x^{\ell_1}\partial_y^{\ell_2} c = O(|w|^{-k})$ for any $(\ell_1, \ell_2) \in \mathbb{Z}^2_{\geq 0}$ and for any k. We have the natural compactification $\mathbb{C}_w \subset \mathbb{P}^1$. We may regard c as a C^∞-function on $\mathbb{P}^1$. There exists an $\mathbb{R}_{>0}$-valued C^∞-function A on $\mathbb{C}_w$ such that $A\Delta_{\mathbb{R}^2} = \Delta_{\mathbb{P}^1|\mathbb{C}}$ and $\mathrm{dvol}_{\mathbb{R}^2} = A\,\mathrm{dvol}_{\mathbb{P}^1|\mathbb{C}}$. We obtain $\int_{\mathbb{P}^1} Ac \, \mathrm{dvol}_{\mathbb{P}^1} = \int_{\mathbb{R}^2} c\, d \, \mathrm{dvol}_{\mathbb{R}^2} = 0$. By the standard theory of the Poisson equations on compact manifolds (for example, see [5, Theorem 4.7]), there exists γ such that $\gamma(\infty) = 0$ and $\Delta_{\mathbb{R}^2}\gamma = c$. We have $\gamma = O(|w|^{-1})$.

By considering $b - \frac{1}{T}\int_{S^1_T} b$, we may assume that $\int_{S^1_T} b = 0$ from the beginning. For any $P \in S^1_T \times \mathbb{C}_w$ and $s > 0$, let $B_P(s)$ denote the set of $Q \in S^1_T \times \mathbb{C}_w$ whose distance from P is less than s. Let $V_P(s)$ denote the volume of $B_P(s)$. Set $k_b(P,s) := V_P(s)^{-1}\int_{B_P(s)} |b|$. Because $k_b(P,s) = O(s^{-2})$ as $s \to \infty$, we obtain $\int_0^r s k_b(P,s)\, ds = O\big(\log(r+2)\big)$. We set $b_+(P) := \max\{b(P), 0\}$ and $b_-(P) := \max\{-b(P), 0\}$. According to [73, 341 page], there exist functions $f_\pm$ such that $\Delta f_\pm = b_\pm$ and $|f_\pm(w)| = O\big(\log(2+|w|)\big)$. We set $f = f_+ - f_-$. Then, $\Delta(f) = b$ and $|f| = O\big(\log(2+|w|)\big)$ are satisfied. By the elliptic regularity (for example, see [59, Theorem 3.23]), f is C^∞. By Lemma 8.1.3, we obtain $|f| = O\big(|w|^{-k}\big)$ for any $k > 0$ as $|w| \to \infty$. Thus, we obtain the claim of the lemma. ■

Lemma 8.1.6 *f induces a C^∞-function on $S^1_T \times \mathbb{P}^1_w$. In particular, the ℓ-th derivative of f is $O(|w|^{-1-\ell})$.*

Proof There exists the decomposition $f = f^\circ + f^\perp$, where f° is constant along S^1_T, and $\int_{S^1_T} f^\perp = 0$. By Lemma 8.1.3, $f^\perp$ and its higher derivatives are $O(|w|^{-k})$ for any k. Because f° is a C^∞-function on $\mathbb{P}^1_w$ such that $f^\circ(\infty) = 0$, we obtain the claim of the lemma. ■

8.1.4 Subharmonic Functions

Let f be a bounded function $S^1_T \times \mathbb{C}_w \longrightarrow \mathbb{R}_{\geq 0}$ such that $\Delta f \leq 0$ in the sense of distributions.

Lemma 8.1.7 *f is constant.*

Proof It is enough to consider the case where f is C^∞. We set $F := \int_{S^1_T} f$. We obtain $-\partial_w\partial_{\overline{w}} F \leq 0$ on $\mathbb{C}$, and F is bounded. Then, F induces a bounded subharmonic function on $\mathbb{P}^1$, and hence F is constant. (See [79, Proposition 2.2].)

There exists the decomposition $f = f_0 + f_1$ such that (i) f_0 is constant in the S^1_T-direction, (ii) $\int_{S^1_T} f_1 = 0$. Because $F = \int_{S^1_T} f = \int_{S^1_T} f_0$, we obtain that f_0 is constant.

We have the following inequality:

$$\Delta|f|^2 = -2\big|\partial_t f\big|^2 - 2\big|\partial_x f\big|^2 - 2\big|\partial_y f\big|^2 + 2\Delta(f)\cdot f \leq -\big|\partial_t f\big|^2.$$

We obtain the following inequality:

$$-4\partial_w\partial_{\overline{w}} \int_{S^1_T} |f|^2 \leq -\int_{S^1_T} \big|\partial_t f\big|^2.$$

Note that $\int_{S^1_T} |f|^2 = \int_{S^1_T} |f_0|^2 + \int_{S^1_T} |f_1|^2$, and that $\int_{S^1_T} |f_0|^2$ is constant. We also have

$$\int_{S^1_T} |\partial_t f|^2 = \int_{S^1_T} |\partial_t f_1|^2 \geq C_1 \int_{S^1_T} |f_1|^2.$$

We obtain the following:

$$-4\partial_w \partial_{\overline{w}} \int_{S^1} |f_1|^2 \leq -C_1 \int_{S^1_T} |f_1|^2.$$

Because $\int_{S^1_T} |f_1|^2$ is bounded, we obtain that $\int_{S^1_T} |f_1|^2 = O\big(\exp(-\epsilon|w|)\big)$ for some $\epsilon > 0$ by Lemma 8.1.1. Because $\int_{S^1_T} |f_1|^2$ is non-negative and subharmonic, we obtain that $\int_{S^1_T} |f_1|^2 = 0$, and hence $f_1 = 0$. ■

Corollary 8.1.8 *Let g be a bounded function on $S^1_T \times \mathbb{C}_w$ such that $\Delta g = 0$. Then, g is constant.* ■

8.2 Global Periodic Monopoles of Rank One (1)

We use the notation in Sect. 2.7. Take a complex number γ. Let us consider the product line bundle $L(\gamma)$ on $S^1_T \times \mathbb{C}_w$ with a global frame $\widetilde{v}$ and the metric $h_{L(\gamma)}$ given by $h_{L(\gamma)}(\widetilde{v}, \widetilde{v}) = 1$. We consider the unitary connection ∇ and the Higgs field ϕ given by

$$\nabla \widetilde{v} = \widetilde{v}\big(-\sqrt{-1}(\gamma + \overline{\gamma})\big)\,dt, \qquad \phi = \gamma - \overline{\gamma}.$$

Because $F(\nabla) = 0$ and $\nabla \phi = 0$, $(L(\gamma), h_{L(\gamma)}, \nabla, \phi)$ is a monopole on $S^1_T \times \mathbb{C}_w$.

Remark 8.2.1 The restriction of this monopole around infinity on a ramified covering appeared in Sect. 5.2.3. ■

By (2.21) and (2.22), the underlying mini-holomorphic bundle $L^\lambda(\gamma)$ on $\mathcal{M}^\lambda$ is described as follows:

$$\partial_{L^\lambda(\gamma),\overline{\beta}_1} \widetilde{v} = \widetilde{v} \frac{\lambda \overline{\gamma}}{1+|\lambda|^2}, \quad \partial_{L^\lambda(\gamma),t_1} \widetilde{v} = \widetilde{v} \frac{-2\sqrt{-1}(\gamma - |\lambda|^2 \overline{\gamma})}{1+|\lambda|^2}.$$

Let $L^{\lambda\,\mathrm{cov}}(\gamma)$ denote the pull back of $L^\lambda(\gamma)$ by the projection $\varpi^\lambda : M^\lambda \longrightarrow \mathcal{M}^\lambda$. On M^λ, we define

$$\widetilde{u} = \exp\Big(\frac{-\lambda\overline{\gamma}\overline{\beta}_1}{1+|\lambda|^2} + \frac{\overline{\lambda}\gamma\beta_1}{1+|\lambda|^2} + \frac{2\sqrt{-1}}{1+|\lambda|^2}(\gamma - |\lambda|^2\overline{\gamma})t_1\Big) \cdot (\varpi^\lambda)^{-1}(\widetilde{v}).$$

Then, $\widetilde{u}$ is a mini-holomorphic frame, i.e., $\partial_{L^{\lambda\,\mathrm{cov}}(\gamma),t_1}\widetilde{u} = 0$ and $\partial_{L^{\lambda\,\mathrm{cov}}(\gamma),\overline{\beta}_1}\widetilde{u} = 0$. Because

$$\big|\widetilde{u}\big|_h = \exp\big(-2\,\mathrm{Im}(\gamma)t_1\big),$$

$\widetilde{u}$ is a global frame of the locally free $\mathcal{O}_{\overline{M}^\lambda}(*H_\infty^{\lambda\,\mathrm{cov}})$-module $\mathcal{P}L^{\lambda\,\mathrm{cov}}(\gamma)$. Moreover, the good filtered bundle $\mathcal{P}_*L^{\lambda\,\mathrm{cov}}(\gamma)_{|\{t_1\}\times\mathbb{P}^1_{\beta_1}}$ is described as

$$\mathcal{P}_bL^{\lambda\,\mathrm{cov}}(\gamma)_{|\{t_1\}\times\mathbb{P}^1_{\beta_1}} = \mathcal{O}_{\mathbb{P}^1}([b])\widetilde{u}_{|\{t_1\}\times\mathbb{P}^1_{\beta_1}}.$$

Here, $[b] := \max\{n \in \mathbb{Z} \mid n \le b\}$.

The locally free $\mathcal{O}_{\overline{\mathcal{M}}^\lambda}(*H_\infty^\lambda)$-module $\mathcal{P}L^\lambda(\gamma)$ is described as the descent of $\mathcal{P}L^{\lambda\,\mathrm{cov}}(\gamma)$ by the action

$$\kappa_1^{-1}(\widetilde{u}) = \widetilde{u}\exp(2\sqrt{-1}\gamma T).$$

We also obtain the description of the good filtered bundle $\mathcal{P}_*L^\lambda(\gamma)$ as the descent.

Lemma 8.2.2 *Let h_1 be another Hermitian metric of $L^\lambda(\gamma)$ such that (i) $(L^\lambda(\gamma), h)$ is a monopole, (ii) h_1 is strongly adapted to the good filtered bundle $\mathcal{P}_*L^\lambda(\gamma)$. Then, there exists a positive constant a such that $h_1 = a \cdot h$.*

Proof According to Proposition 7.4.1, h and h_1 are mutually bounded. There exists the function $s : \mathcal{M}^\lambda \longrightarrow \mathbb{R}_{>0}$ determined by $h_1 = h \cdot s$. Note that $\Delta s = 0$ on $\mathcal{M}^\lambda$ according to Corollary 2.9.9. Hence, we obtain that s is constant by Corollary 8.1.8. ■

8.2.1 Reformulation

Let α be a non-zero complex number. Let $\mathcal{L}^{\mathrm{cov}}(\alpha)$ be the $\mathcal{O}_{\overline{M}^\lambda}(*H_\infty^{\lambda\,\mathrm{cov}})$-module $\mathcal{O}_{\overline{M}^\lambda}(*H_\infty^{\lambda\,\mathrm{cov}})\,e$. Let $\pi^\lambda : \overline{M}^\lambda \longrightarrow \mathbb{R}_{t_1}$ denote the projection, i.e., $(t_1, \beta_1) \longmapsto t_1$. We obtain the good filtered bundles $\mathcal{P}_*^{(0)}(\mathcal{L}^{\mathrm{cov}}(\alpha))$ by setting

$$\deg^{\mathcal{P}^{(0)}}(e_{|\pi^{-1}(t_1)}) = 0.$$

We consider the $\mathbb{Z}$-action by $\kappa_1^*e = \alpha\cdot e$. Then, we obtain a locally free $\mathcal{O}_{\overline{\mathcal{M}}^\lambda}(*H_\infty^\lambda)$-module $\mathcal{L}(\alpha)$ and the good filtered bundle $\mathcal{P}_*^{(0)}\mathcal{L}(\alpha)$ over $\mathcal{L}(\alpha)$.

Lemma 8.2.3 *There exists a Hermitian metric h of the mini-holomorphic bundle $\mathcal{L}(\alpha)_{|\mathcal{M}^\lambda}$ such that (i) $(\mathcal{L}(\alpha)_{|\mathcal{M}^\lambda}, h)$ is a monopole, (ii) h is strongly adapted to $\mathcal{P}_*^{(0)}\mathcal{L}(\alpha)$. Such a metric is unique up to the multiplication of positive constants.* ■

8.3 Global Periodic Monopoles of Rank One (2)

8.3.1 Construction of Mini-Holomorphic Bundles

Let $P \in \mathcal{M}^{\lambda}$. Let $\widetilde{P} = (t_1^0, \beta_1^0) \in (\varpi^{\lambda})^{-1}(P)$ such that $0 \leq t_1^0 < T$. We choose $0 < \epsilon < T - t_1^0$. We set $U :=]-\epsilon, T-\epsilon/2[\times \mathbb{P}^1$ and $H^{\lambda}_{\infty,\epsilon} :=]-\epsilon, T-\epsilon/2[\times\{\infty\}$. Let p_i $(i = 1, 2)$ denote the projections $p_1 : U \setminus \{\widetilde{P}\} \longrightarrow]-\epsilon, T-\epsilon/2[$ and $p_2 : U \setminus \{\widetilde{P}\} \longrightarrow \mathbb{P}^1$. Let $\mathcal{L}^{\mathrm{cov}}_{-\epsilon,T-\epsilon/2}(P, \ell)$ be the $\mathcal{O}_{U\setminus\{\widetilde{P}\}}(*H^{\lambda}_{\infty,\epsilon})$-module determined by the following conditions:

- $\mathcal{L}^{\mathrm{cov}}_{-\epsilon,T-\epsilon/2}(P, \ell)(*p_2^{-1}(\beta_1^0))$ is isomorphic to the pull back of $\mathcal{O}_{\mathbb{P}^1}(*\{\beta_1^0, \infty\})\, e$.
- We have

$$\mathcal{L}^{\mathrm{cov}}_{-\epsilon,T-\epsilon/2}(P, \ell)_{|p_1^{-1}(a)} = \begin{cases} \mathcal{O}_{\mathbb{P}^1}(*\infty)e & (-\epsilon < a < t_1^0), \\ \mathcal{O}_{\mathbb{P}^1}(-\ell\beta_1^0)(*\infty)e & (t_1^0 < a < T - \epsilon/2), \end{cases}$$

under the above isomorphism.

Let $\kappa_1 :]-\epsilon, -\epsilon/2[\longrightarrow]T-\epsilon, T-\epsilon/2[$ be the isomorphism given by $\kappa_1(t_1, \beta_1) = (t_1 + T, \beta_1 + 2\sqrt{-1}\lambda T)$. We have the isomorphism

$$\kappa_1^*\Big(\mathcal{L}^{\mathrm{cov}}_{-\epsilon,T-\epsilon/2}(P, \ell)_{|]T-\epsilon,T-\epsilon/2[\times\mathbb{P}^1}\Big) \simeq \mathcal{L}^{\mathrm{cov}}_{-\epsilon,T-\epsilon/2}(P, \ell)_{|]-\epsilon,-\epsilon/2[\times\mathbb{P}^1}$$

given by $\kappa_1^*\Big((\beta_1 - \beta_1^0)^{\ell} e\Big) \longmapsto e$, or equivalently $\kappa_1^*(e) \longmapsto (\beta_1 - \beta_1^0 + 2\sqrt{-1}\lambda T)^{-\ell} e$. Hence, by gluing, we obtain an $\mathcal{O}_{\overline{\mathcal{M}}^{\lambda}\setminus\{P\}}(*H^{\lambda}_{\infty})$-module denoted by $\mathcal{L}(P, \ell)$.

Let $\mathcal{L}^{\mathrm{cov}}(P, \ell)$ denote the pull back of $\mathcal{L}(P, \ell)$ by ϖ^{λ}. We shall describe $\mathcal{L}^{\mathrm{cov}}(P, \ell)$ explicitly. Let $\widetilde{P}_n = (t_1^{(n)}, \beta_1^{(n)})$ denote the point of $(\varpi^{\lambda})^{-1}(P)$ such that $nT \leq t_1^{(n)} < (n+1)T$. We set

$$H(\widetilde{P}_n, +) := \big\{(t_1, \beta_1^{(n)}) \,\big|\, t_1 > t_1^{(n)}\big\}, \qquad H(\widetilde{P}_n, -) := \big\{(t_1, \beta_1^{(n)}) \,\big|\, t_1 < t_1^{(n)}\big\}.$$

Then, we have the following isomorphism:

$$\mathcal{L}^{\mathrm{cov}}(P, \ell) \simeq \mathcal{O}_{\overline{\mathcal{M}}^{\lambda}\setminus(\varpi^{\lambda})^{-1}(P)}\Big(\sum_{n<0} \ell H(\widetilde{P}_n, -) - \sum_{n\geq 0} \ell H(\widetilde{P}_n, +)\Big).$$

Let us describe the isomorphism $\kappa_1^*\mathcal{L}^{\mathrm{cov}}(P, \ell) \simeq \mathcal{L}^{\mathrm{cov}}(P, \ell)$. If $n \geq 0$, the isomorphism

$$\kappa_1^*\Big(\mathcal{L}^{\mathrm{cov}}(P, \ell)_{|]t_1^0+nT,t_1^0+(n+1)T[\times\mathbb{P}^1}\Big) \simeq \mathcal{L}^{\mathrm{cov}}(P, \ell)_{|]t_1^0+(n-1)T,t_1^0+nT[\times\mathbb{P}^1}$$

is given by

$$\kappa_1^*\Big(\prod_{m=0}^{n}(\beta_1-(\beta_1^0+2\sqrt{-1}\lambda mT)^{\ell})e\Big)\longmapsto\prod_{m=0}^{n-1}(\beta_1-(\beta_1^0+2\sqrt{-1}\lambda mT)^{\ell})e.$$

If $n<0$, the isomorphism

$$\kappa_1^*\Big(\mathcal{L}^{\mathrm{cov}}(P,\ell)_{|]t_1^0+nT,t_1^0+(n+1)T[\times\mathbb{P}^1}\Big)\simeq\mathcal{L}^{\mathrm{cov}}(P,\ell)_{|]t_1^0+(n-1)T,t_1^0+nT[\times\mathbb{P}^1}$$

is given by

$$\kappa_1^*\Big(\prod_{m=0}^{-n-1}(\beta_1-(\beta_1^0-2\sqrt{-1}\lambda mT)^{-\ell})e\Big)\longmapsto\prod_{m=0}^{-n}(\beta_1-(\beta_1^0-2\sqrt{-1}\lambda mT)^{-\ell})e.$$

8.3.2 Good Filtered Bundles

Let e^{t_1} denote the restriction of e to $p_1^{-1}(t_1)$. We obtain the filtered bundle $\mathcal{P}_*^{(a)}\mathcal{L}^{\mathrm{cov}}(P,\ell)$ over $\mathcal{L}^{\mathrm{cov}}(P,\ell)$ defined by the following condition:

$$\deg^{\mathcal{P}^{(a)}}(e^{t_1})=-\ell\frac{t_1-t_1^0}{T}+a.$$

Because $\mathcal{P}_*\mathcal{L}^{\mathrm{cov}}(P,\ell)$ is $\mathbb{Z}$-equivariant, we obtain the filtered bundle $\mathcal{P}_*\mathcal{L}(P,\ell)$ over $\mathcal{L}(P,\ell)$.

Lemma 8.3.1 *We have* $\deg\big(\mathcal{P}_*^{(a)}\mathcal{L}(P,\ell)\big)=-(a+\ell/2)$.

Proof Indeed,

$$\begin{aligned}T\deg\big(\mathcal{P}_*^{(a)}\mathcal{L}(P,\ell)\big)&=\int_0^T\deg\big(\mathcal{P}_*^{(a)}\mathcal{L}(P,\ell)_{|p_1^{-1}(t_1)}\big)\,dt_1\\&=\int_0^{t_1^0}\Big(\ell\frac{t_1-t_1^0}{T}-a\Big)\,dt_1+\int_{t_1^0}^{T}\Big(\ell\frac{t_1-t_1^0}{T}-\ell-a\Big)\,dt_1\\&=\int_0^T\ell\frac{t_1}{T}\,dt_1-aT-\int_0^T\ell\frac{t_1^0}{T}\,dt_1-\ell(T-t_1^0)=-aT-\frac{\ell T}{2}.\end{aligned}\tag{8.2}$$

Thus, we obtain the desired equality. ∎

8.3.3 Monopoles

Let $\mathcal{L}^*(P,\ell)$ denote the mini-holomorphic bundle on $\mathcal{M}^\lambda \setminus \{P\}$ obtained as the restriction of $\mathcal{L}(P,\ell)$.

Proposition 8.3.2 *There exists a Hermitian metric h of $\mathcal{L}^*(P,\ell)$ such that the following holds.*

- *$(\mathcal{L}^*(P,\ell),h)$ is a monopole of Dirac type on $\mathcal{M}^\lambda \setminus \{P\}$.*
- *h is strongly adapted to $\mathcal{P}_*^{(-\ell/2)}\mathcal{L}(P,\ell)$ in the sense of Definition 4.4.7.*

Such h is unique up to the multiplication of positive constants.

Proof Let $\Psi^\lambda : \overline{\mathcal{M}}^\lambda \longrightarrow \mathbb{P}^1_w$ be given by $\Psi^\lambda(t_1,\beta_1) = (1+|\lambda|^2)^{-1}(\beta_1 - 2\sqrt{-1}\lambda t_1)$. Let $U_w := \{|w| > R\} \cup \{\infty\}$. Set $U := (\Psi^\lambda)^{-1}(U_w)$.

By using Proposition 7.4.3 in the rank one case, we can construct a Hermitian metric h_1 of $\mathcal{L}(P,\ell)$ such that (i) $(\mathcal{L}(P,\ell),h_1)$ satisfies the norm estimate with respect to $\mathcal{P}_*^{(-\ell/2)}\mathcal{L}(P,\ell)$, (ii) $\partial_{t_1}^{\ell_1}\partial_{\beta_1}^{\ell_2}\partial_{\overline{\beta}_1}^{\ell_3}G(h_1) = O\big(|w|^{-k}\big)$ for any $(\ell_1,\ell_2,\ell_3) \in \mathbb{Z}^3_{\geq 0}$ and for any k as $|w| \to \infty$, (iii) $G(h_1) = 0$ around P. By Corollary 2.9.10, we have $G(h_1e^\varphi) - G(h_1) = 2^{-1}\Delta\varphi$. We also have $\int G(h_1) = 2\pi T \deg(\mathcal{P}_*^{(-\ell/2)}\mathcal{L}) = 0$. We can take a bounded C^∞-function φ such that $-2^{-1}\Delta\varphi = -G(h_1)$ by Lemma 8.1.5. Then, $h = h_1e^\varphi$ has the desired property.

Suppose that h' is another metric satisfying the conditions. We obtain the C^∞ function φ on $\mathcal{M}^\lambda \setminus P$ such that $h' = he^\varphi$, which is bounded. We have $\Delta\varphi = 0$ on $\mathcal{M}^\lambda \setminus P$, i.e., φ is a harmonic function on $\mathcal{M}^\lambda \setminus P$. Recall that isolated singularities of bounded harmonic functions are removable. (Fore example, see [6, Theorem 2.3].) Then, we obtain that φ is a constant by Corollary 8.1.8. ■

8.4 Global Periodic Monopoles of Rank One (3)

Let Z be a finite set. Let $\mathcal{L}$ be a locally free $\mathcal{O}_{\overline{\mathcal{M}}^\lambda \setminus Z}(*H^\lambda_\infty)$-module of rank one with Dirac type singularity at Z. For each $Q \in \overline{\mathcal{M}}^\lambda \setminus Z$, let $\mathcal{L}_Q$ denote the stalk of the sheaf $\mathcal{L}$ at Q. For each $P \in Z$, let (τ,ζ) be a mini-complex coordinate system around P such that $(\tau(P),\zeta(P)) = (0,0)$. For a small positive number, we set $P_+ = (\epsilon,0)$ and $P_- = (-\epsilon,0)$. By the scattering map, we obtain the isomorphism $\mathcal{L}_{P_-}(*P_-) \simeq \mathcal{L}_{P_+}(*P_+)$ by which we identify $\mathcal{L}_{P_\pm}(*P_\pm)$. We obtain the integer

$$\ell(P) := \text{length}\Big(\mathcal{L}_{P_-}/(\mathcal{L}_{P_-}\cap\mathcal{L}_{P_+})\Big) - \text{length}\Big(\mathcal{L}_{P_+}/(\mathcal{L}_{P_-}\cap\mathcal{L}_{P_+})\Big).$$

Then, there exists a non-zero complex number α and an isomorphism

$$\mathcal{L} \simeq \mathcal{L}(\alpha) \otimes \bigotimes_{P\in Z} \mathcal{L}(P,\ell(P)).$$

There exists the unique good filtered bundle $\mathcal{P}_*\mathcal{L}$ over $\mathcal{L}$ such that $\deg(\mathcal{P}_*\mathcal{L}) = 0$, for which

$$\mathcal{P}_*\mathcal{L} \simeq \mathcal{P}_*^{(0)}\mathcal{L}(\alpha) \otimes \bigotimes_{P \in Z} \mathcal{P}_*^{(-\ell(P)/2)}\mathcal{L}(P, \ell(P)).$$

We obtain the following.

Proposition 8.4.1 *There exists a Hermitian metric h of $\mathcal{L}_{|\mathcal{M}^\lambda \setminus Z}$ such that (i) $(\mathcal{L}_{|\mathcal{M}^\lambda \setminus Z}, h)$ is a monopole of Dirac type, (ii) h is strongly adapted to $\mathcal{P}_*\mathcal{L}$. Such a metric is unique up to the multiplication of positive constants.* ■

Chapter 9
Global Periodic Monopoles and Filtered Difference Modules

Abstract We shall prove the main theorem (Theorem 9.1.2) of this monograph. In Sect. 9.3, we shall prove that for a given periodic monopole of GCK-type, the associated good filtered bundle is polystable of degree 0. In Sect. 9.4, we shall prove the existence of a monopole of GCK type which induces a given polystable good filtered bundle with degree 0. Then, Theorem 9.1.2 follows. As one of the consequences of Theorem 9.1.2, the classification of singular monopoles of GCK-type is reduced to the classification of polystable parabolic difference modules of degree 0. In Sect. 9.5, we explain smooth parabolic 0-difference modules of rank 2 are equivalent to filtered torsion-free sheaves of rank one on spectral curves, which is a variant of equivalences between Higgs bundles and sheaves on the spectral curves due to Hitchin and Beauville-Narasimhan-Ramanan, and we revisit the classification of SU(2)-monopoles of GCK-type without Dirac type singularity studied by Harland.

9.1 Statements

Let Z be a finite subset in $\mathcal{M}$. (See Sect. 2.7 for $\mathcal{M}$.) Let g denote the Euclidean metric $dt\,dt + dw\,d\overline{w}$ of $\mathcal{M}$.

Definition 9.1.1 A monopole (E, h, ∇, ϕ) on $\mathcal{M} \setminus Z$ is called of GCK type if the following holds.

- $|F(\nabla)|_h \to 0$ and $|\phi|_h = O(\log|w|)$ as $w \to \infty$.
- Each point $P \in Z$ is a Dirac type singularity of (E, h, ∇, ϕ). ■

Let (E, h, ∇, ϕ) be a monopole on $\mathcal{M} \setminus Z$ of GCK-type. Let λ be any complex number. Let $(E^{\lambda}, \overline{\partial}_{E^{\lambda}})$ be the underlying mini-holomorphic bundle on the mini-complex manifold $\mathcal{M}^{\lambda} \setminus Z$. It is prolonged to the associated good filtered bundle $\mathcal{P}^h_* E^{\lambda}$ on $(\overline{\mathcal{M}}^{\lambda}; Z, H^{\lambda}_{\infty})$ as explained in Proposition 7.2.1 and Theorem 7.3.1. This construction is compatible with direct sum, tensor product, and inner homomorphism as in Sect. 7.5. We shall prove the following theorem in Sects. 9.2–9.4.

T. Mochizuki, *Periodic Monopoles and Difference Modules*, Lecture Notes in Mathematics 2300, https://doi.org/10.1007/978-3-030-94500-8_9

Theorem 9.1.2 *The above procedure induces a bijection between the equivalence classes of the following objects:*

- *Monopoles of GCK-type* (E, h, ∇, ϕ) *on* $\mathcal{M} \setminus Z$.
- *Polystable good filtered bundles of Dirac type* $\mathcal{P}_* E^\lambda$ *on* $(\overline{\mathcal{M}}^\lambda; Z, H^\lambda_\infty)$ *with* $\deg(\mathcal{P}_* E^\lambda) = 0$. *(See Definition* 4.2.11 *for the polystablity condition.)*

We obtain the following corollary, which is an analogue of the Corlette-Simpson correspondence between stable good filtered flat bundles and stable good filtered Higgs bundles.

Corollary 9.1.3 *The above procedure induces the bijective correspondence of the equivalence classes of the following objects through monopoles of GCK-type on* $\mathcal{M} \setminus Z$.

- *Polystable good filtered bundles of Dirac type* $\mathcal{P}_* \mathcal{E}^0$ *on* $(\overline{\mathcal{M}}^0; Z, H^0_\infty)$ *with* $\deg(\mathcal{P}_* \mathcal{E}^0) = 0$.
- *Polystable good filtered bundles of Dirac type* $\mathcal{P}_* \mathcal{E}^\lambda$ *on* $(\overline{\mathcal{M}}^\lambda; Z, H^\lambda_\infty)$ *with* $\deg(\mathcal{P}_* \mathcal{E}^\lambda) = 0$. ■

As for the comparison with parabolic difference modules, together with (4.7), Proposition 4.2.16 and Proposition 4.2.17, we obtain the following.

Corollary 9.1.4 *There exists the natural bijection between the isomorphism classes of the following objects.*

- *Monopoles of GCK-type* (E, h, ∇, ϕ) *on* $\mathcal{M} \setminus Z$.
- *Polystable parabolic* $2\sqrt{-1}\lambda T$*-difference modules of degree* 0*:*

$$(\boldsymbol{V}, V, m_Z, (\boldsymbol{\tau}_{Z,x}, \boldsymbol{L}_{Z,x})_{x \in \mathbb{C}}, \mathcal{P}_*(\boldsymbol{V}_{|\widehat{\infty}})).$$

Here, Z *and* $(m_Z, \boldsymbol{\tau}_{Z,x})$ *are related as in* (2.47), (2.48) *and* (2.49). ■

9.2 Preliminary

9.2.1 Ambient Good Filtered Bundles with Appropriate Metric

Let Z be a finite subset in $\mathcal{M}^\lambda$. Let $\mathcal{P}_* \mathcal{E}^\lambda$ be a good filtered bundle with Dirac type singularity on $(\overline{\mathcal{M}}^\lambda; Z, H^\lambda_\infty)$. Let $(E, \overline{\partial}_E)$ denote the mini-holomorphic bundle with Dirac type singularity on $\mathcal{M}^\lambda \setminus Z$ obtained as the restriction of $\mathcal{P}\mathcal{E}^\lambda$.

Let h_1 be a Hermitian metric of E strongly adapted to $\mathcal{P}_* \mathcal{E}^\lambda$ such that the following holds.

Condition 9.2.1

(A1) *Around* H^λ_∞, *we have* $|G(h_1)|_{h_1} = O(|w|^{-\epsilon-2})$ *for some* $\epsilon > 0$, *and* $(E, \overline{\partial}_E, h_1)$ *satisfies the norm estimate with respect to* $\mathcal{P}_* \mathcal{E}$ *in the sense of*

Definition 4.4.5. *Moreover, we have*

$$\left|\left[\partial_{E,\overline{\beta}_1},\partial_{E,h_1,\beta_1}\right]\right|_{h_1}=O\left(|w|^{-2}(\log|w|)^{-2}\right).\tag{9.1}$$

(See Proposition 7.4.3.*)*

(A2) *Around each point of Z, $(E,\overline{\partial}_E,h_1)$ is a monopole with Dirac type singularity. In particular, it induces a C^∞-metric of the Kronheimer resolution of E. (See* Sect. 2.3.6 *for the Kronheimer resolution.)* ■

9.2.2 Degree of Filtered Subbundles

Let $\mathcal{P}_*\mathcal{E}_1\subset\mathcal{P}_*\mathcal{E}$ be a filtered subbundle on $(\overline{\mathcal{M}}^\lambda;Z,H^\lambda_\infty)$. Let $(E_1,\overline{\partial}_{E_1})$ be the mini-holomorphic bundle of Dirac type on $(\mathcal{M}^\lambda,Z)$. Let h_{1,E_1} denote the metric of E_1 induced by h_1. By the Chern-Weil formula in Lemma 2.9.4, the analytic degree $\deg(E_1,h_{1,E_1})\in\mathbb{R}\cup\{-\infty\}$ makes sense.

Proposition 9.2.2 *$2\pi T\deg(\mathcal{P}_*\mathcal{E}_1)=\deg(E_1,h_{1,E_1})$ holds.*

Proof We take a metric h_{0,E_1} of E_1 which satisfies Condition 9.2.1 for $\mathcal{P}_*\mathcal{E}_1$. Because $G(h_{0,E_1})=O(|w|^{-2-\epsilon})$ $(\epsilon>0)$ with respect to h_{0,E_1} around H^λ_∞, and because $G(h_{0,E_1})=0$ around each point of Z, $G(h_{0,E_1})$ is L^1 with respect to h_{0,E_1} and the natural volume form of $\mathcal{M}^\lambda$. Let ∇_0 and ϕ_0 be the Chern connection and the Higgs field associated with $(E_1,\overline{\partial}_{E_1})$ with h_{0,E_1}. Because $(E_1,\overline{\partial}_{E_1},h_{0,E_1})$ is a monopole with Dirac type singularity around each point P of Z, we have $(\nabla_0\phi_0)_{|Q}=O\left(d(Q,P)^{-2}\right)$ around P, and hence $\nabla_0\phi_0$ is L^1 around P, with respect to h_{0,E_1} and the Riemannian metric of $\mathcal{M}^\lambda$. Let ∂_{E_1,β_1} denote the operator induced by $\partial_{E_1,\overline{\beta}_1}$ and h_{0,E_1} as in Sect. 2.9.3. Because $\left|[\partial_{E_1,\beta_1},\partial_{E_1,\overline{\beta}_1}]\right|_{h_{0,E_1}}=O\left(|w|^{-2}(\log|w|)^{-2}\right)$ around H^λ_∞, $\left|[\partial_{E_1,\beta_1},\partial_{E_1,\overline{\beta}_1}]\right|_{h_{0,E_1}}$ is L^1 around H^λ_∞. Hence, we may apply the formula (2.32), and we obtain the following equality:

$$\int\operatorname{Tr}G(h_{0,E_1})\,\mathrm{dvol}=\int_0^T2\pi\operatorname{par-deg}\left(\mathcal{P}_*\mathcal{E}_{1|\overline{\mathcal{M}}^\lambda\langle t_1\rangle\setminus Z}\right)dt_1=2\pi T\deg(\mathcal{P}_*\mathcal{E}_1).$$

(See Sect. 4.2 for $\overline{\mathcal{M}}^\lambda\langle t_1\rangle$.)

It remains to prove the following equality:

$$\int\operatorname{Tr}G(h_{1,E_1})\,\mathrm{dvol}=\int\operatorname{Tr}G(h_{0,E_1})\,\mathrm{dvol}.\tag{9.2}$$

By considering $\det E_1\subset\bigwedge^{\operatorname{rank}E_1}E$, it is enough to consider the case $\operatorname{rank}E_1=1$. According to Proposition 8.4.1, there exists a Hermitian metric h'_{E_1} of E_1 such that (i) $(E_1,\overline{\partial}_{E_1},h'_{E_1})$ is a monopole of GCK-type, (ii) the meromorphic extension $\mathcal{P}^{h'_{E_1}}E_1$ is equal to $\mathcal{P}\mathcal{E}_1$. We obtain $\deg(\mathcal{P}^{h'_{E_1}}_* E_1) = 0$. By considering $(E, \overline{\partial}_E, h_1) \otimes (E_1, \overline{\partial}_{E_1}, h'_{E_1})^{-1}$ and $\mathcal{P}_*\mathcal{E} \otimes (\mathcal{P}^{h'_{E_1}}_* E_1)^{\vee}$, we may reduce the issue to the case where $\mathcal{P}\mathcal{E}_1$ is isomorphic to $\mathcal{O}_{\overline{\mathcal{M}}^\lambda}(*H^\lambda_\infty)$. There exists the section f of $\mathcal{P}\mathcal{E}_1$ corresponding to $1 \in \mathcal{O}_{\overline{\mathcal{M}}^\lambda}(*H^\lambda_\infty)$. Let $a \in \mathbb{R}$ be determined by $a := \inf\{b \in \mathbb{R} \mid f \in \mathcal{P}_b\mathcal{E}_1\}$. According to the following lemma, we may and will assume $a = 0$.

Lemma 9.2.3 *It is enough to study the case $a = 0$.*

Proof Let $\psi : \mathbb{C}_w \longrightarrow \mathbb{R}_{>0}$ be a C^∞-function such that $\psi(w) = |w|^{-2a}$ on $|w| > R$ for large R, and that ψ is constant around Z. The metric of E_1 induced by $\psi \cdot h_1$ is equal to $\psi \cdot h_{1,E_1}$.

Let $\mathcal{P}'_*\mathcal{E}$ be the filtered bundle over $\mathcal{E}$ determined by

$$\mathcal{P}'_c\mathcal{E}_{|\overline{\mathcal{M}}^\lambda\langle t_1\rangle\setminus Z} := \mathcal{P}_{c+a}\mathcal{E}_{|\overline{\mathcal{M}}^\lambda\langle t_1\rangle\setminus Z}.$$

Let $\mathcal{P}'_*\mathcal{E}_1$ be the filtered bundle over $\mathcal{E}_1$ induced by $\mathcal{P}'_*\mathcal{E}$. Then, $\psi \cdot h_{0,E_1}$ is a Hermitian metric of E_1 which satisfies Condition 9.2.1 for $\mathcal{P}'_*\mathcal{E}_1$.

Note that the support of $\Delta \log \psi$ is compact. According to Corollary 2.9.10, the equality $\int \operatorname{Tr} G(\psi \cdot h_{0,E_1})\, \mathrm{dvol} = \int \operatorname{Tr} G(\psi \cdot h_{1,E_1})\, \mathrm{dvol}$ implies

$$\int \operatorname{Tr} G(h_{0,E_1})\, \mathrm{dvol} = \int \operatorname{Tr} G(h_{1,E_1})\, \mathrm{dvol}.$$

Thus, Lemma 9.2.3 is proved. ■

Lemma 9.2.4 *Let $\mathcal{B}^\lambda$ be a neighbourhood of H^λ_∞ in $\overline{\mathcal{M}}^\lambda$. Let E be a mini-holomorphic bundle on $\mathcal{B}^{\lambda *} := \mathcal{B}^\lambda \setminus H^\lambda_\infty$ with a metric h such that $G(h)$ is L^1. Let f be a mini-holomorphic section of E such that*

$$C_1^{-1} \leq |f|_h (\log|w|)^{-k} \leq C_1$$

for some $C_1 > 1$ and $k \in \mathbb{R}$. Then, $\big|\nabla_{\beta_0} f\big|_h \cdot |f|_h^{-1}$ and $\big|(\nabla_{t_0} + \sqrt{-1}\,\mathrm{ad}\,\phi) f\big|_h \cdot |f|_h^{-1}$ are L^2, where ∇ and ϕ denote the Chern connection and the Higgs field associated with $(E, \overline{\partial}_E, h)$.

Proof It is enough to prove that

$$\big|\nabla_{\beta_0} f\big|_h (\log|w|)^{-k}, \quad \big|(\nabla_{t_0} + \sqrt{-1}\,\mathrm{ad}\,\phi) f\big|_h (\log|w|)^{-k}$$

are L^2. Because f is mini-holomorphic, we have $\nabla_{\overline{\beta}_0} f = 0$ and $(\nabla_{t_0} - \sqrt{-1}\,\mathrm{ad}\,\phi) f = 0$. We may assume that $\mathcal{B}^{\lambda *} = (\Psi^\lambda)^{-1}(U^*_w(R))$, where $U^*_w(R) := \{w \in \mathbb{C} \mid |w| > R\}$.

Let $\rho : \mathbb{R} \longrightarrow \{0 \leq a \leq 1\} \subset \mathbb{R}_{\geq 0}$ be a C^∞-function such that, (i) $\rho(t) = 0$ $(t \geq 1)$, (ii) $\rho(t) = 1$ $(t \leq 1/2)$, (iii) $\rho(t)^{1/2}$ and $\partial_t \rho(t)/\rho(t)^{1/2}$ induce C^∞-functions.

For any large positive integer N, by setting $\chi_N(w) := \rho\big(N^{-1} \log |w|\big)$, we obtain C^∞-functions $\chi_N : U_w^*(R) \longrightarrow \mathbb{R}_{\geq 0}$ such that $\chi_N(w) = 0$ $(|w| \geq e^N)$ and $\chi_N(w) = 1$ $(|w| \leq e^{N/2})$. Let $\mu : U_w^*(R) \longrightarrow \mathbb{R}_{\geq 0}$ be a C^∞-function such that $\mu(w) = 1 - \rho\big(\log(|w|/R)\big)$. We set $\widetilde{\chi}_N := \mu \cdot \chi_N$. We have

$$\partial_w \widetilde{\chi}_N(w) = \partial_w \mu(w) \chi_N(w) + \mu(w) \rho'(N^{-1} \log |w|^2) N^{-1} w^{-1}.$$

By the assumption on ρ, $\partial_w \widetilde{\chi}_N(w)/\widetilde{\chi}_N(w)^{1/2}$ naturally give C^∞-functions on $U_w^*(R)$, and there exists $C_2 > 0$, which is independent of N, such that the following holds:

$$\left|\partial_w \widetilde{\chi}_N(w)/\widetilde{\chi}_N(w)^{1/2}\right| \leq C_2 |w|^{-1} (\log |w|^2)^{-1}.$$

Because $\partial_{\beta_0} w = (1 + |\lambda|^2)^{-1}$ and $\partial_{\beta_0} \overline{w} = -\overline{\lambda}^2 (1 + |\lambda|^2)^{-1}$, there exists $C_3 > 0$, which is independent of N, such that the following holds:

$$\left|\partial_{\beta_0}\big(\widetilde{\chi}_N(w)\big)/\widetilde{\chi}_N(w)^{1/2}\right| \leq C_3 |w|^{-1} (\log |w|^2)^{-1}.$$

We consider the following integral:

$$\begin{aligned}
&\int_{\mathcal{B}^{\lambda*}} \widetilde{\chi}_N(w) \cdot h(\nabla_{\beta_0} f, \nabla_{\beta_0} f)(\log |w|^2)^{-2k} \,\mathrm{dvol} \\
&= -\int_{\mathcal{B}^{\lambda*}} \partial_{\beta_0}\big(\widetilde{\chi}_N(w)\big) \cdot h(f, \nabla_{\beta_0} f)(\log |w|^2)^{-2k} \,\mathrm{dvol} \\
&\quad - \int_{\mathcal{B}^{\lambda*}} \widetilde{\chi}_N(w) \cdot h(f, \nabla_{\overline{\beta}_0} \nabla_{\beta_0} f)(\log |w|^2)^{-2k} \,\mathrm{dvol} \\
&\quad + \int_{\mathcal{B}^{\lambda*}} \widetilde{\chi}_N(w) \cdot h(f, \nabla_{\beta_0} f) \cdot (-2k)(\log |w|^2)^{-2k-1} (w^{-1} \partial_{\beta_0} w + \overline{w}^{-1} \partial_{\beta_0} \overline{w}) \,\mathrm{dvol}
\end{aligned} \tag{9.3}$$

We have the following inequality:

$$\begin{aligned}
&\left|\partial_{\beta_0} \widetilde{\chi}_N \cdot h(f, \nabla_{\beta_0} f)(\log |w|^2)^{-2k}\right| \\
&\leq \Big(C_3 C_1 |w|^{-1} (\log |w|^2)^{-1}\Big) \cdot \Big(\widetilde{\chi}_N^{1/2}(w) \cdot \big|\nabla_{\beta_0} f\big|_h (\log |w|^2)^{-k}\Big).
\end{aligned} \tag{9.4}$$

We also have the following inequality:

$$\Big|\widetilde{\chi}_N \cdot h(f, \nabla_{\beta_0} f) \cdot (\log |w|^2)^{-2k-1}(w^{-1}\partial_{\beta_0} w + \overline{w}^{-1}\partial_{\beta_0}\overline{w})\Big| \\ \leq 2\Big(C_1\widetilde{\chi}_N^{1/2} \cdot |w|^{-1}(\log |w|^2)^{-1}\Big) \cdot \Big(\widetilde{\chi}_N^{1/2}|\nabla_{\beta_0} f|_h(\log |w|^2)^{-k}\Big). \tag{9.5}$$

Note that $\nabla_{\overline{\beta}_0}\nabla_{\beta_0} f = (\nabla_{\overline{\beta}_0}\nabla_{\beta_0} - \nabla_{\beta_0}\nabla_{\overline{\beta}_0})f = -[F_{\beta_0,\overline{\beta}_0}(h), f]$. There exists $C_4 > 0$ which are independent of N, such that the following holds:

$$\int_{\mathcal{B}^{\lambda*}} \widetilde{\chi}_N \cdot |\nabla_{\beta_0} f|_h^2(\log |w|^2)^{-2k}\,\mathrm{dvol} \\ \leq C_4 + C_4\Big(\int_{\mathcal{B}^{\lambda*}} \widetilde{\chi}_N \cdot |\nabla_{\beta_0} f|_h^2(\log |w|^2)^{-2k}\,\mathrm{dvol}\Big)^{1/2} \\ + \int_{\mathcal{B}^{\lambda*}} \widetilde{\chi}_N \cdot h\big(f, [F_{\beta_0,\overline{\beta}_0}, f]\big)(\log |w|^2)^{-2k}\,\mathrm{dvol}. \tag{9.6}$$

Similarly, there exists $C_5 > 0$ which is independent of N, such that the following holds:

$$\int_{\mathcal{B}^{\lambda*}} \widetilde{\chi}_N \cdot |(\nabla_{t_0} + \sqrt{-1}\,\mathrm{ad}\,\phi)f|_h^2(\log |w|^2)^{-2k}\,\mathrm{dvol} \\ \leq C_5 + C_5\Big(\int_{\mathcal{B}^{\lambda*}} \widetilde{\chi}_N \cdot |(\nabla_{t_0} + \sqrt{-1}\,\mathrm{ad}\,\phi)f|_h^2(\log |w|^2)^{-2k}\,\mathrm{dvol}\Big)^{1/2} \\ + \int_{\mathcal{B}^{\lambda*}} \widetilde{\chi}_N \cdot h\big(f, [-2\sqrt{-1}\nabla_{t_0}\phi, f]\big)(\log |w|^2)^{-2k}\,\mathrm{dvol}. \tag{9.7}$$

Because $G(h)$ is L^1, there exists a constant $C_6 > 0$, which is independent of N, such that the following holds:

$$\int_{\mathcal{B}^{\lambda*}} \widetilde{\chi}_N \cdot h\big(f, [F_{\beta_0,\overline{\beta}_0}, f]\big)(\log |w|^2)^{-2k}\,\mathrm{dvol} \\ + \frac{1}{4}\int_{\mathcal{B}^{\lambda*}} \widetilde{\chi}_N \cdot h\big(f, [-2\sqrt{-1}\nabla_{t_0}\phi, f]\big)(\log |w|^2)^{-2k}\,\mathrm{dvol} \leq C_6. \tag{9.8}$$

We put

$$A_N := \int_{\mathcal{B}^{\lambda*}} \widetilde{\chi}_N \cdot |\nabla_{\beta_0} f|_h^2(\log |w|^2)^{-2k}\,\mathrm{dvol} \\ + \frac{1}{4}\int_{\mathcal{B}^{\lambda*}} \widetilde{\chi}_N \cdot |(\nabla_{t_0} + \sqrt{-1}\,\mathrm{ad}\,\phi)f|_h^2(\log |w|^2)^{-2k}\,\mathrm{dvol}. \tag{9.9}$$

By (9.6) and (9.7), there exists $C_7 > 0$, which are independent of N, such that the following holds:

$$A_N \leq C_7 + C_7 A_N^{1/2}.$$

Hence, there exists $C_8 > 0$ such that $A_N \leq C_8$ for any large N. By taking $N \to \infty$, we obtain the claim of Lemma 9.2.4. ■

Let h_{2,E_1} be a Hermitian metric of E_1 such that the following holds.

- There exists a neighbourhood N_1 of Z such that $h_{2,E_1} = h_{0,E_1}$ on $\mathcal{M}^\lambda \setminus N_1$.
- There exists a neighbourhood N_2 of Z contained in N_1 such that $h_{2,E_1} = h_{1,E_1}$ on $N_2 \setminus Z$.

We obtain the function s determined by $h_{1,E_1} = h_{2,E_1} \cdot s$. According to Corollary 2.9.10, we have the relation $G(h_{1,E_1}) - G(h_{2,E_1}) = 2^{-1}\Delta \log s$. The support of $\log s$ is contained in $\mathcal{M}^\lambda \setminus N_2$. By using Lemma 9.2.4, we obtain $\int \Delta \log s = 0$. Hence, we obtain $\int G(h_{1,E_1}) = \int G(h_{2,E_1})$.

To compare $\int G(h_{0,E_1})$ and $\int G(h_{2,E_1})$, it is enough to compare the integrals over a neighbourhood for each $P \in Z$. We apply a variant of the argument in [54]. For simplicity of the description, we consider the case $P = (0,0)$ for the coordinate system (t_0, β_0). Let U denote the connected component of N_1 which contains P. We may regard U as an open subset of $\mathbb{R} \times \mathbb{C}$. We take the Kronheimer resolution as in Sect. 2.9.6. We have the map $\varphi : \mathbb{C}^2 \longrightarrow \mathbb{R} \times \mathbb{C}$ in (2.2). Set $\widetilde{U} = \varphi^{-1}(U)$. We have the holomorphic vector bundle $\widetilde{E}$ and $\widetilde{E}_1$ on $\widetilde{U}$, induced by E and E_1, respectively. We may regard $\widetilde{E}_1$ as a saturated subsheaf of $\widetilde{E}$. Note that $\widetilde{E}/\widetilde{E}_1$ is not necessarily locally free. The metric $\widetilde{h}_0 := \varphi^{-1}(h_{0,E_1})$ induces a C^∞-metric of $\widetilde{E}_1$. The metric $\widetilde{h}_2 := \varphi^{-1}(h_{2,E_1})$ may have singularity.

We take a projective morphism $\psi : \widetilde{U}' \longrightarrow \widetilde{U}$ such that (i) $D := \psi^{-1}(0,0)$ is simple normal crossing, (ii) $\widetilde{U}' \setminus D \simeq \widetilde{U} \setminus \{(0,0)\}$, (iii) the saturation $(\psi^*\widetilde{E}_1)^\sim$ of $\psi^*\widetilde{E}_1$ in $\psi^*\widetilde{E}$ is a subbundle, i.e., $\psi^*\widetilde{E}/(\psi^*\widetilde{E}_1)^\sim$ is locally free. We have the Hermitian metric $\widetilde{h}_0' := \psi^*(\widetilde{h}_0)$ of $\psi^*\widetilde{E}_1$. The metric $\psi^*(\widetilde{h}_2)$ induces a C^∞-metric $\widetilde{h}_2'$ of $(\psi^*\widetilde{E}_1)^\sim$. Let s be the function on $\widetilde{U}' \setminus D$ determined by $\widetilde{h}_0' = \widetilde{h}_2' \cdot s$. There exists a neighbourhood N' of D such that $s = 1$ on $\widetilde{U}' \setminus N'$. Note that $\log s$ is an L^2-function on $\widetilde{U}'$, and we obtain the following equality of (1, 1)-currents on $\widetilde{U}'$:

$$\overline{\partial}\partial \log s = F(\widetilde{h}_0') - F(\widetilde{h}_2') + \sum a_i [D_i].$$

Here, D_i denote the irreducible components of D, $[D_i]$ denote the (1, 1)-current obtained as the integrations over D_i, and a_i are constants. Because $\psi_*[D_i] = 0$, we obtain $\int_{\widetilde{U}} \Lambda F(\widetilde{h}_0) = \int_{\widetilde{U}} \Lambda F(\widetilde{h}_2)$. By Lemma 2.9.15, we obtain the desired equality (9.2). ■

9.2.3 Analytic Degree of Subbundles

Let $E_2 \subset E$ be a mini-holomorphic subbundle. Let h_{1,E_2} denote the metric of E_2 induced by h_1. By the Chern-Weil formula in Lemma 2.9.4, $\deg(E_2, h_{1,E_2}) \in \mathbb{R} \cup \{-\infty\}$ makes sense.

Proposition 9.2.5 *If* $\deg(E_2, h_{1,E_2}) \neq -\infty$, *then there exists a good filtered subbundle* $\mathcal{P}_*\mathcal{E}_2 \subset \mathcal{P}_*\mathcal{E}$ *such that* $\mathcal{E}_{2|\mathcal{M}^\lambda \setminus Z} = E_2$. *Moreover, we obtain* $\deg(E_2, h_{1,E_2}) = 2\pi T \deg(\mathcal{P}_*\mathcal{E}_2)$.

Proof We set $\mathcal{M}^\lambda\langle t_1\rangle := \overline{\mathcal{M}}^\lambda\langle t_1\rangle \cap \mathcal{M}^\lambda$ for any $t_1 \in S^1_T$. By (9.1) and [79, Lemma 10.6], $E_{2|\mathcal{M}^\lambda\langle t_1\rangle\setminus Z}$ extend to locally free $\mathcal{O}_{\overline{\mathcal{M}}^\lambda\langle t_1\rangle\setminus Z}(*\infty)$-submodules of $\mathcal{P}\mathcal{E}_{|\overline{\mathcal{M}}^\lambda\langle t_1\rangle\setminus Z}$.

Let P be any point of H^λ_∞. Let U be a neighbourhood of P in $\overline{\mathcal{M}}^\lambda$. On U, we use a local mini-complex coordinate system (t_1, β_1^{-1}). On $\widetilde{U} := \mathbb{R}_{s_1} \times U$, we use the complex coordinate system $(\alpha_1, \beta_1^{-1}) = (s_1 + \sqrt{-1}t_1, \beta_1^{-1})$ as in Sect. 2.7.6. We set $D := \mathbb{R}_{s_1} \times (U \cap H^\lambda_\infty)$. As in Sect. 2.5.2, we obtain the locally free $\mathcal{O}_{\widetilde{U}}(*D)$-module $\widetilde{\mathcal{P}\mathcal{E}}$ induced by $\mathcal{P}\mathcal{E}$. We also have the holomorphic vector subbundle $\widetilde{E}_2$ of $\widetilde{\mathcal{P}\mathcal{E}}_{|\widetilde{U}\setminus D}$ induced by E_2. Let $p : \widetilde{U} \longrightarrow D$ be the projection given by $p(\alpha_1, \beta_1^{-1}) = \alpha_1$. By the above consideration, $\widetilde{E}_{2|p^{-1}(\alpha_1)\setminus D}$ extends to $\mathcal{O}_{p^{-1}(\alpha_1)}(*\infty)$-submodule of $\widetilde{\mathcal{P}\mathcal{E}}_{|p^{-1}(\alpha_1)}$. By [84, Theorem 4.5], $\widetilde{E}_2$ extends to $\mathcal{P}_{\widetilde{U}}(*D)$-submodule $\widetilde{\mathcal{P}\mathcal{E}_2}$ of $\widetilde{\mathcal{P}\mathcal{E}}$. By the construction, $\widetilde{\mathcal{P}\mathcal{E}_2}$ is naturally $\mathbb{R}$-equivariant, and hence $E_{2|U\setminus H^\lambda_\infty}$ extends to a locally free $\mathcal{O}_U\big(*(H^\lambda_\infty \cap U)\big)$-submodule of $\mathcal{P}\mathcal{E}_{|U}$. Therefore, E_2 extends to a locally free $\mathcal{O}_{\overline{\mathcal{M}}^\lambda\setminus Z}(*H^\lambda_\infty)$-module $\mathcal{P}\mathcal{E}_2$. We obtain the good filtered bundle $\mathcal{P}_*\mathcal{E}_2$ over $\mathcal{P}\mathcal{E}_2$ as in Sect. 2.3.8 and Sect. 4.2.1. The claim for the degree follows from the previous proposition. ■

As a consequence, we obtain the following.

Corollary 9.2.6 $\mathcal{P}_*\mathcal{E}$ *is stable if and only if* (E, h_1) *is analytically stable.* ■

9.3 Good Filtered Bundles Associated with Monopoles of GCK-Type

Let Z be a finite subset of $\mathcal{M}^\lambda$. Let $(E, \overline{\partial}_E, h)$ be a monopole on $\mathcal{M}^\lambda \setminus Z$ of GCK-type. Let $\mathcal{P}^h_* E$ be the associated filtered bundle of Dirac type on $(\overline{\mathcal{M}}^\lambda; Z, H^\lambda_\infty)$.

Proposition 9.3.1 *The good filtered bundle* $\mathcal{P}^h_* E$ *is polystable with* $\deg(\mathcal{P}^h_* E) = 0$. *Moreover, the monopole* $(E, \overline{\partial}_E, h)$ *is irreducible, if and only if* $\mathcal{P}^h_* E$ *is stable.*

Proof By Corollary 6.3.12, the relation (7.1), and Proposition 7.3.5, $(E, \overline{\partial}_E, h)$ satisfies the condition in Sect. 9.2.1. Applying Proposition 9.2.2 to $\mathcal{P}^h_* E$, we obtain

$2\pi T \deg(\mathcal{P}_*^h E) = \deg(E, \overline{\partial}_E, h) = 0$. Let $\mathcal{P}_*\mathcal{E}_1$ be a good filtered subbundle of $\mathcal{P}_*^h E$. Let $(E_1, \overline{\partial}_{E_1})$ be the restriction of $\mathcal{E}_1$ to $\mathcal{M}^\lambda \setminus Z$. By Proposition 9.2.2, we obtain $2\pi T \deg(\mathcal{P}_*\mathcal{E}_1) = \deg(E_1, h_{E_1}) \leq 0$. Moreover, if $\deg(\mathcal{P}_*\mathcal{E}_1) = 0$, the Chern-Weil formula (2.24) implies that the orthogonal projection onto E_1 is flat with respect to the Chern connection and the Higgs fields, and the orthogonal decomposition $E = E_1 \oplus E_1^\perp$ is mini-holomorphic. Hence, we obtain the decomposition $\mathcal{P}_*^h E = \mathcal{P}_*^{h_{E_1}} E_1 \oplus \mathcal{P}_*^{h_{E_1^\perp}} E_1^\perp$. We also obtain that E_1 and $E_1^\perp$ with the induced metrics are monopoles. Then, by an easy induction, we can prove that $\mathcal{P}_*^h E$ is polystable. ■

Proposition 9.3.2 *Let h' be another metric of E such that (i) $(E, \overline{\partial}_E, h')$ is a monopole, (ii) any points of Z are Dirac type singularity, (iii) h' is strongly adapted to $\mathcal{P}_*E$ in the sense of Definition* 4.4.7. *Then, the following holds.*

- *There exists a mini-holomorphic decomposition $(E, \overline{\partial}_E) = \bigoplus_{i=1}^m (E_i, \overline{\partial}_{E_i})$, which is orthogonal with respect to both h and h'.*
- *There exists positive numbers a_i $(i = 1, \dots, m)$ such that $h_{E_i} = a_i h'_{E_i}$.*

Proof By Proposition 7.3.5, h and h' are mutually bounded. Hence, we obtain the claim from [66, Proposition 1.3]. ■

9.4 Construction of Monopoles

Let Z be a finite subset. Let $\mathcal{P}_*\mathcal{E}$ be a stable good filtered bundle of Dirac type on $(\overline{\mathcal{M}}^\lambda; Z, H_\infty^\lambda)$ with $\deg(\mathcal{P}_*\mathcal{E}) = 0$. Set $E := \mathcal{P}_a\mathcal{E}_{|\mathcal{M}^\lambda \setminus Z}$.

Proposition 9.4.1 *There exists a Hermitian metric h such that (i) $(E, \overline{\partial}_E, h)$ is a monopole of GCK-type, (ii) $(E, \overline{\partial}_E, h)$ satisfies the norm estimate with respect to $\mathcal{P}_*\mathcal{E}$.*

Proof By using Propositions 7.4.3 and 8.4.1, we can construct a metric h_0 of E satisfying Condition 9.2.1 and the following conditions.

- Let ∇_{h_0} and ϕ_{h_0} be the Chern connection and the Higgs field associated with $(E, \overline{\partial}_E, h_0)$. Then, we have the decay $|F(h_0)|_{h_0} \to 0$ and the estimate $|\phi_{h_0}|_{h_0} = O(\log |w|)$ when $|w| \to \infty$.
- $\det(E, \overline{\partial}_E, h_0)$ is a monopole.

Set $A^\lambda := S^1 \times \mathcal{M}^\lambda$ and $\overline{A}^\lambda := S^1 \times \overline{\mathcal{M}}^\lambda$. We have the complex structure on A^λ given by the local complex coordinate system $(\alpha_0, \beta_0) = (s_0 + \sqrt{-1}t_0, \beta_0)$ in Sect. 2.7.5. Let $p : A^\lambda \longrightarrow \mathcal{M}^\lambda$ be the projection. Let $(\widetilde{E}, \overline{\partial}_{\widetilde{E}})$ be the holomorphic bundle on $A^\lambda \setminus p^{-1}(Z)$ induced by $(E, \overline{\partial}_E)$ as in Sect. 2.5.2. It is equipped with the induced metric $\widetilde{h}_0$. We have the natural S^1-action on A^λ, for which $(\widetilde{E}, \overline{\partial}_{\widetilde{E}}, \widetilde{h}_0)$ is equivariant. We obtain that the S^1-equivariant bundle $(\widetilde{E}, \overline{\partial}_{\widetilde{E}})$ with the metric $\widetilde{h}_0$ is analytically stable with respect to the S^1-action in the sense of [66]. Applying the Kobayashi-Hitchin correspondence for analytically stable bundles in [66], we obtain a metric h such that the following holds.

- $(E, \overline{\partial}_E, h)$ is a monopole.
- h and h_0 are mutually bounded. In particular, $(E, \overline{\partial}_E, h)$ satisfies the norm estimate with respect to $\mathcal{P}_*\mathcal{E}$.
- $\det(h) = \det(h_0)$.

By Proposition 7.4.1 and Proposition 2.4.7, we obtain that $(E, \overline{\partial}_E, h)$ is of GCK-type. ■

Theorem 9.1.2 follows from Proposition 9.3.1, Proposition 9.3.2 and Proposition 9.4.1. ■

9.5 Smooth Parabolic 0-Difference Modules of Rank 2

In [30], by using the Nahm transform and the Kobayashi-Hitchin correspondence between wild harmonic bundles and filtered Higgs bundles on $\mathbb{P}^1$, Harland classified irreducible SU(2)-monopoles (E, h, ∇, ϕ) on $\mathcal{M}$ of GCK-type without Dirac type singularity in terms of torsion-free sheaves of rank one with parabolic structure on the spectral curves. We revisit his result directly by using Corollary 9.1.4 with $\lambda = 0$.

9.5.1 Smooth Parabolic 0-Difference Modules

Definition 9.5.1 A parabolic difference module

$$(\boldsymbol{V}, V, m, (\boldsymbol{\tau}_x, \boldsymbol{L}_x)_{x\in\mathbb{C}}, \mathcal{P}_*(\boldsymbol{V}_{|\widehat{\infty}}))$$

is called smooth if $m(x) = 0$ for any $x \in \mathbb{C}$. ■

Let $(\boldsymbol{V}, V, m, (\boldsymbol{\tau}_x, \boldsymbol{L}_x)_{x\in\mathbb{C}}, \mathcal{P}_*(\boldsymbol{V}_{|\widehat{\infty}}))$ be a smooth parabolic 0-difference module. Then, the data m and $(\boldsymbol{\tau}_x, \boldsymbol{L}_x)_{x\in\mathbb{C}}$ are trivial. Because $\lambda = 0$, the difference operator of $\boldsymbol{V}$ is a $\mathbb{C}(\beta_1)$-linear automorphism F. Indeed, it is a $\mathbb{C}[\beta_1]$-automorphism of V. By setting $V_{|\widehat{\infty}} := \mathbb{C}((\beta_1^{-1})) \otimes_{\mathbb{C}[\beta_1]} V$, we obtain a natural isomorphism $V_{|\widehat{\infty}} \simeq \boldsymbol{V}_{|\widehat{\infty}}$. Hence, instead of $(\boldsymbol{V}, V, m, (\boldsymbol{\tau}_x, \boldsymbol{L}_x)_{x\in\mathbb{C}}, \mathcal{P}_*(\boldsymbol{V}_{|\widehat{\infty}}))$, we prefer to consider $(V, F, \mathcal{P}_*(V_{|\widehat{\infty}}))$.

Lemma 9.5.2 *For a smooth parabolic* 0*-difference module* $(V, F, \mathcal{P}_*(V_{|\widehat{\infty}}))$, $\mathrm{tr}(F)$ *is a polynomial of* β_1, *and* $\det(F)$ *is a non-zero constant.*

Proof The first claim is obvious. Because $\det(F)$ is a $\mathbb{C}[\beta_1]$-automorphism of $\det(V) \simeq \mathbb{C}[\beta_1]$, it is a non-zero constant. ■

For a smooth parabolic 0-difference module $(V, F, \mathcal{P}_*(V_{|\widehat{\infty}}))$, let $\mathcal{F}_V$ denote the $\mathcal{O}_{\mathbb{P}^1}(*\infty)$-module induced by V. Let $\mathcal{P}_*\mathcal{F}_V$ denote the filtered bundle over $\mathcal{F}_V$ induced by $\mathcal{P}_*(V_{|\widehat{\infty}})$.

Lemma 9.5.3 *We have* $\deg(V, F, \mathcal{P}_*(V_{|\widehat{\infty}})) = \deg(\mathcal{P}_*\mathcal{F}_V)$.

Proof Recall the formula (1.9). There is no contribution of $\deg(L_{i,x}, L_{i-1,x})$. It is easy to see that $\sum r(\omega) \cdot \omega = 0$ because $\det(F)$ is a non-zero constant. Then, we obtain $\deg(V, F, \mathcal{P}_*(V_{|\widehat{\infty}})) = \deg(\mathcal{P}_*\mathcal{F}_V)$. ■

As a consequence of Corollary 9.1.4, we obtain the following.

Proposition 9.5.4 *There exists the natural bijection between the isomorphism classes of the following objects.*

- *Monopoles of GCK-type (E, h, ∇, ϕ) defined on $\mathcal{M}$ without Dirac type singularity.*
- *Polystable smooth parabolic 0-difference modules $(V, F, \mathcal{P}_*(V_{|\widehat{\infty}}))$ such that* $\deg(\mathcal{P}_*\mathcal{F}_V) = 0$.

Under this equivalence, (E, h, ∇, ϕ) is an $\mathrm{SU}(n)$*-monopole if and only if* $\det(F) = 1$ *and* $\mathrm{rank}(V) = n$ *hold for the corresponding* $(V, F, \mathcal{P}_*(V_{|\widehat{\infty}}))$. ■

Let us look at smooth polystable parabolic 0-difference modules $(V, F, \mathcal{P}_*(V_{|\widehat{\infty}}))$ of degree 0 with $\mathrm{rank}(V) \leq 2$.

9.5.2 Rank One Case

For any $\gamma \in \mathbb{C}^*$, we set $N(\gamma) = \mathbb{C}[\beta_1]$. It is equipped with the $\mathbb{C}[\beta_1]$-automorphism determined by the multiplication of γ. Under the natural identification $N(\gamma)_{|\widehat{\infty}} = \mathbb{C}((\beta_1^{-1})) \otimes_{\mathbb{C}[\beta_1]} N(\gamma) \simeq \mathbb{C}((\beta_1^{-1}))$, let $\mathcal{P}_*(N(\gamma)_{|\widehat{\infty}})$ denote the filtered bundle over $N(\gamma)_{|\widehat{\infty}}$ defined by $\mathcal{P}_a(N(\gamma)_{|\widehat{\infty}}) = \beta_1^{[a]}\mathbb{C}[\![\beta_1^{-1}]\!]$, where $[a] := \max\{n \in \mathbb{Z} \mid n \leq a\}$. Let $\mathcal{P}_*N(\gamma)$ denote the parabolic 0-difference module of rank 1. Note that $\deg(\mathcal{P}_*N(\gamma)) = 0$. The following lemma is clear.

Lemma 9.5.5 *Any smooth parabolic difference module of degree* 0 *of rank* 1 *is isomorphic to $\mathcal{P}_*N(\gamma)$ for some γ,* ■

9.5.3 Filtered Torsion-Free Sheaves of Rank One on an Integral Scheme

Let $Q \in \mathbb{C}[\beta_1]$ be a non-constant polynomial, i.e., $\deg(Q) > 0$. Let $\gamma \in \mathbb{C}^*$. Let $\Sigma^\circ(Q, \gamma) \subset \mathbb{C}_x^* \times \mathbb{C}_{\beta_1}$ denote the scheme defined by $x^2 - Q(\beta_1)x + \gamma$.

Lemma 9.5.6 *If $Q(\beta_1)$ is not constant, then the polynomial $x^2 - Q(\beta_1)x + \gamma \in \mathbb{C}[\beta_1][x]$ is irreducible. As a result, $\Sigma^\circ(Q, \gamma)$ is integral.*

Proof As remarked in [30], it follows from Eisenstein's criterion. We can also obtain the claim by the observation that the discriminant $Q(\beta_1)^2 - 4\gamma$ of $x^2 - Q(\beta_1)x + \gamma$ is not a square of a polynomial. ■

We set $\boldsymbol{P} = \mathbb{P}\big(\mathcal{O}_{\mathbb{P}_{\beta_1}} \oplus \mathcal{O}_{\mathbb{P}^1_{\beta_1}}(-k\{\infty\})\big)$, where $k = \deg(Q)$. We regard $Q(\beta_1)$ and γ as sections of $\mathcal{O}_{\mathbb{P}^1_{\beta_1}}(k\{\infty\})$ and $\mathcal{O}_{\mathbb{P}^1_{\beta_1}}(2k\{\infty\})$, respectively. We obtain the subscheme $\Sigma(Q, \gamma) \subset \boldsymbol{P}$ defined by $x^2 - Qx + \gamma$. (See [7, §3].) By naturally regarding $\mathbb{C}^*_x \times \mathbb{C}_{\beta_1}$ as an open subset of $\boldsymbol{P}$, we may regard $\Sigma(Q, \gamma)$ as the closure of $\Sigma^\circ(Q, \gamma)$ in $\boldsymbol{P}$. Let $p_{\Sigma(Q,\gamma)} : \Sigma(Q, \gamma) \longrightarrow \mathbb{P}^1_{\beta_1}$ denote the projection. We set $\Sigma_\infty(Q, \gamma) := p^{-1}_{\Sigma(Q,\gamma)}(\infty)$. In the following, $\Sigma(Q, \gamma)$ and $\Sigma_\infty(Q, \gamma)$ are also denoted by Σ and Σ_∞, respectively, if there is no risk of confusion.

Lemma 9.5.7 *$\Sigma_\infty(Q, \gamma)$ consists of two distinct points, and they are smooth point of $\Sigma(Q, \gamma)$. As a result, $\Sigma(Q, \gamma)$ is integral. On a neighbourhood of each point of $\Sigma_\infty(Q, \gamma)$, $p_{\Sigma(Q,\gamma)}$ is etale.*

Proof Let $(1 - 4\gamma Q^{-2})^{1/2} \in \mathbb{C}[\![y^{-1}]\!]$ denote the convergent power series such that (i) the constant term is 1, (ii) $\big((1 - 4\gamma Q^{-2})^{1/2}\big)^2 = 1 - 4\gamma Q^{-2}$. One of the roots of $x^2 - Qx + 1$ is

$$\mathfrak{a}_1 = \frac{Q + Q(1 - 4\gamma Q^{-2})^{1/2}}{2} \in y^{\deg(Q)}\mathbb{C}[\![y^{-1}]\!].$$

The other is $\mathfrak{a}_2 = \gamma\mathfrak{a}_1^{-1} \in y^{-\deg(Q)}\mathbb{C}[\![y^{-1}]\!]$. Note $\lim_{y^{-1}\to 0} y^{-\deg(Q)}\mathfrak{a}_1 \neq 0$ and $\lim_{y^{-1}\to 0} y^{\deg(Q)}\mathfrak{a}_2 \neq 0$. Then, the claim of the lemma is obvious. ■

9.5.3.1 Degree of Torsion-Free Sheaves on $\Sigma(Q, \gamma)$

Let us recall the degree of coherent sheaves on the integral scheme $\Sigma(Q, \gamma)$. (See [40] for a more precise explanation in the more general situation.) Let $\mathfrak{P}$ be any ample line bundle on $\Sigma(Q, \gamma)$. For any coherent sheaf $\mathcal{E}$ on $\Sigma(Q, \gamma)$, there exist rational numbers $\alpha_i(\mathcal{E})$ $(i = 0, 1)$ such that $\chi(\mathcal{E} \otimes \mathfrak{P}^m) = m\alpha_1(\mathcal{E}) + \alpha_0(\mathcal{E})$ for any $m \in \mathbb{Z}$, where $\chi(\mathcal{E} \otimes \mathfrak{P}^m)$ denotes the Euler number of $\mathcal{E} \otimes \mathfrak{P}^m$ (see [40, §1.2]). According to [40, Definition 1.2.2, Definition 1.2.11], we set

$$\operatorname{rank}(\mathcal{E}) := \frac{\alpha_1(\mathcal{E})}{\alpha_1(\mathcal{O}_{\Sigma(Q,\gamma)})}, \qquad \deg(\mathcal{E}) := \alpha_0(\mathcal{E}) - \operatorname{rank}(\mathcal{E})\alpha_0(\mathcal{O}_{\Sigma(Q,\gamma)}).$$

Because $\Sigma(Q, \gamma)$ is integral, there exists a Zariski open dense connected subset U of $\Sigma(Q, \gamma)$ such that $\mathcal{E}_{|U}$ is a locally free $\mathcal{O}_U$-module, and $\operatorname{rank}(\mathcal{E})$ is equal to the rank of $\mathcal{E}_{|U}$. Because $\dim \Sigma(Q, \gamma) = 1$, it is easy to see that $\deg(\mathcal{E})$ is independent of the choice of an ample line bundle $\mathfrak{P}$ though $\alpha_i(\mathcal{E})$ depend on $\mathfrak{P}$. If $\Sigma(Q,\gamma)$ is smooth, we have $\deg(\mathcal{E}) = \int_{\Sigma(Q,\gamma)} c_1(\mathcal{E})$ for the first Chern class $c_1(\mathcal{E})$ in the ordinary sense.

Lemma 9.5.8 *Let $\mathcal{E}$ be a torsion-free sheaf of rank* 1 *on $\Sigma(Q,\gamma)$. Then, we have* $\deg(p_{\Sigma(Q,\gamma)*}\mathcal{E}) = \deg(\mathcal{E}) - \deg(Q)$.

Proof We consider the ample line bundles $\mathcal{O}_{\mathbb{P}^1_{\beta_1}}(1)$ on $\mathbb{P}^1_{\beta_1}$ and $\mathfrak{P} := p_\Sigma^*(\mathcal{O}_{\mathbb{P}^1_\beta}(1))$ on Σ. We set $\mathcal{F} = p_{\Sigma*}\mathcal{E}$, which is a locally free $\mathcal{O}_{\mathbb{P}^1_{\beta_1}}$-module. Because $\chi(\mathcal{F} \otimes \mathcal{O}(m)) = \chi(\mathcal{E} \otimes \mathfrak{P}^m)$, we have $\alpha_i(\mathcal{F}) = \alpha_i(\mathcal{E})$. Hence, we obtain

$$\deg(\mathcal{F}) = \alpha_0(\mathcal{F}) - \operatorname{rank}(\mathcal{F})\alpha_0(\mathcal{O}_{\mathbb{P}^1}) = \alpha_0(\mathcal{E}) - 2\alpha_0(\mathcal{O}_{\mathbb{P}^1}).$$

As explained in [7], we have $p_{\Sigma*}(\mathcal{O}_\Sigma) \simeq \mathcal{O}_{\mathbb{P}^1} \oplus \mathcal{O}_{\mathbb{P}^1}(-k)$, where $k = \deg(Q)$. Hence, we obtain

$$\chi(\mathcal{O}_\Sigma \otimes \mathfrak{P}^m) = \chi(\mathcal{O}_{\mathbb{P}^1}(m)) + \chi(\mathcal{O}_{\mathbb{P}^1}(m-k)).$$

It implies that $\alpha_0(\mathcal{O}_\Sigma) = 2\alpha_0(\mathcal{O}_{\mathbb{P}^1}) - k$. Hence, we obtain

$$\deg(\mathcal{E}) = \alpha_0(\mathcal{E}) - \alpha_0(\mathcal{O}_\Sigma) = \alpha_0(\mathcal{E}) - 2\alpha_0(\mathcal{O}_{\mathbb{P}^1}) + k = \deg(\mathcal{F}) + k.$$

Thus, we are done. ■

9.5.3.2 Filtered Torsion-Free Sheaf of Rank One on $(\Sigma(Q,\gamma), \Sigma_\infty(Q,\gamma))$

Let $\mathcal{L}$ be a torsion-free $\mathcal{O}_{\Sigma(Q,\gamma)}(*\Sigma_\infty(Q,\gamma))$-module of rank 1.

Definition 9.5.9 A filtered sheaf $\mathcal{P}_*\mathcal{L}$ over $\mathcal{L}$ is a tuple of $\mathcal{O}_{\Sigma(Q,\gamma)}$-submodules $\mathcal{P}_{\boldsymbol{b}}(\mathcal{L}) \subset \mathcal{L}$ ($\boldsymbol{b} = (b_P \mid P \in \Sigma_\infty(Q,\gamma)) \in \mathbb{R}^{\Sigma_\infty(Q,\gamma)}$) such that the following holds.

- For $P \in \Sigma_\infty(Q,\gamma)$, let U_P be a neighbourhood of P in $\Sigma(Q,\gamma)$ such that (i) $U_P \cap \Sigma_\infty(Q,\gamma) = \{P\}$, (ii) U_P is smooth. Then, $\mathcal{P}_{\boldsymbol{b}}(\mathcal{L})_{|U_P}$ depends only on b_P, which we denote by $\mathcal{P}_{b_P}(\mathcal{L}_{|U_P})$. Moreover, $\mathcal{P}_*(\mathcal{L}_{|U_P})$ is a filtered bundle over $\mathcal{L}_{|U_P}$. (See Sect. 2.11.3.1.)

Such $\mathcal{P}_*\mathcal{L}$ is called a filtered torsion-free sheaf of rank one on $(\Sigma(Q,\gamma), \Sigma_\infty(Q,\gamma))$. ■

Let $\mathcal{P}_*\mathcal{L}$ be a filtered torsion-free sheaf of rank one on $(\Sigma(Q,\gamma), \Sigma_\infty(Q,\gamma))$. For $P \in \Sigma_\infty(Q,\gamma)$, let $\mathcal{L}_P$ denote the stalk of $\mathcal{L}$ at P. We have the induced filtration

$\mathcal{P}_*(\mathcal{L}_P)$. We obtain the $\mathbb{C}$-vector spaces $\mathrm{Gr}^{\mathcal{P}}_c(\mathcal{L}_P) = \mathcal{P}_c(\mathcal{L}_P)/\sum_{d<c}\mathcal{P}_d(\mathcal{L}_P)$. We set

$$\deg(\mathcal{P}_*\mathcal{L}) = \deg(\mathcal{P}_{\boldsymbol{b}}\mathcal{L}) - \sum_{P\in\Sigma_\infty}\sum_{b_P-1<c\le b_P} c\dim \mathrm{Gr}^{\mathcal{P}}_c(\mathcal{L}_P).$$

It is independent of the choice of $\boldsymbol{b}$.

9.5.3.3 The Induced Smooth Parabolic Difference Modules of Rank 2

In this subsection, $(\Sigma(Q,\gamma), \Sigma_\infty(Q,\gamma))$ is denoted by (Σ, Σ_∞). Let $\mathcal{P}_*\mathcal{L}$ be a filtered torsion-free sheaf of rank one on (Σ, Σ_∞). Let us construct a parabolic 0-difference module $\Upsilon(\mathcal{P}_*\mathcal{L})$.

We obtain a locally free $\mathcal{O}_{\mathbb{P}^1_{\beta_1}}(*\infty)$-module $p_{\Sigma*}(\mathcal{L})$ of rank 2. There exists the following isomorphism of the stalks:

$$p_{\Sigma*}(\mathcal{L})_\infty = \bigoplus_{P\in\Sigma_\infty}\mathcal{L}_P.$$

We define the filtration $\mathcal{P}_*\big(p_{\Sigma*}(\mathcal{L})_\infty\big)$ by

$$\mathcal{P}_a\big(p_{\Sigma*}(\mathcal{L})_\infty\big) = \bigoplus_{P\in\Sigma_\infty}\mathcal{P}_a(\mathcal{L}_P).$$

Let $\mathcal{P}_a\big(p_{\Sigma*}\mathcal{L}\big) \subset p_{\Sigma*}(\mathcal{L})$ be determined by $\mathcal{P}_a\big(p_{\Sigma*}\mathcal{L}\big)_\infty = \mathcal{P}_a\big(p_{\Sigma*}(\mathcal{L})_\infty\big)$. Thus, we obtain a filtered bundle $\mathcal{P}_*\big(p_{\Sigma*}(\mathcal{L})\big)$ over $p_{\Sigma*}(\mathcal{L})$.

Lemma 9.5.10 *We have* $\deg(\mathcal{P}_*(p_{\Sigma*}\mathcal{L})) = \deg(\mathcal{P}_*\mathcal{L}) - \deg(Q)$.

Proof By the construction, we have $\mathcal{P}_a(p_{\Sigma*}\mathcal{L}) = p_{\Sigma*}(\mathcal{P}_{a,a}\mathcal{L})$ for any $a \in \mathbb{R}$. Then, the claim follows from Lemma 9.5.8 and the definitions of $\deg(\mathcal{P}_*(p_{\Sigma*}\mathcal{L}))$ and $\deg(\mathcal{P}_*\mathcal{L})$. ■

As explained in [7, §3], there exists a natural morphism $F : \mathcal{P}_a(p_{\Sigma*}(\mathcal{L})) \longrightarrow \mathcal{P}_a(p_{\Sigma*}(\mathcal{L})) \otimes \mathcal{O}_{\mathbb{P}^1}(\deg(Q))$ for any $a \in \mathbb{R}$. In particular, we obtain an automorphism F of $p_{\Sigma*}(\mathcal{L})$. It induces the automorphism F of the $\mathbb{C}(\!(\beta_1^{-1})\!)$-vector space $p_{\Sigma*}(\mathcal{L})_{|\widehat{\infty}}$. It is easy to see that the decomposition

$$p_{\Sigma*}(\mathcal{L})_{|\widehat{\infty}} = \bigoplus_{P\in\Sigma_\infty}\mathcal{L}_{|\widehat{P}} \tag{9.10}$$

is the eigen decomposition of F.

We set $V := H^0(\mathbb{P}^1, p_{\Sigma*}(\mathcal{L}))$. Because $V_{|\widehat{\infty}} \simeq p_{\Sigma*}(\mathcal{L})_{|\widehat{\infty}}$, we obtain a filtered bundle $\mathcal{P}_*(V_{|\widehat{\infty}})$ over $V_{|\widehat{\infty}}$. Because the filtration is compatible with the eigen decomposition (9.10) of F, $\mathcal{P}_*(V_{|\widehat{\infty}})$ is a good parabolic structure of (V, F) at ∞. Thus, we obtain a smooth parabolic difference module $\Upsilon(\mathcal{P}_*\mathcal{L}) = (V, F, \mathcal{P}_*(V_{|\widehat{\infty}}))$. If $\deg(\mathcal{P}_*\mathcal{L}) = \deg(Q)$, then we obtain $\deg(\Upsilon(\mathcal{P}_*\mathcal{L})) = 0$. The following lemma is obvious by the construction.

Lemma 9.5.11 *Let $\mathcal{P}_*\mathcal{L}_i$ $(i = 1, 2)$ be filtered torsion-free sheaf of rank one on (Σ, Σ_∞). Isomorphisms of smooth parabolic difference modules $\Upsilon(\mathcal{P}_*\mathcal{L}_1) \simeq \Upsilon(\mathcal{P}_*\mathcal{L}_2)$ bijectively correspond to isomorphisms of filtered torsion-free sheaves $\mathcal{P}_*\mathcal{L}_1 \simeq \mathcal{P}_*\mathcal{L}_2$.* ■

We obtain the following lemma because Σ is irreducible.

Lemma 9.5.12 *For any filtered torsion-free sheaf of rank one $\mathcal{P}_*\mathcal{L}$, the induced smooth parabolic 0-difference module $\Upsilon(\mathcal{P}_*\mathcal{L})$ is stable.* ■

9.5.4 Description of Parabolic Difference Modules of Rank 2

Let $Q \in \mathbb{C}[\beta_1]$ and $\gamma \in \mathbb{C}^*$. Let $\mathfrak{S}(Q, \gamma, c)$ denote the set of the isomorphism classes of smooth parabolic difference module $(V, F, \mathcal{P}_*(V_{|\widehat{\infty}}))$ such that $\mathrm{rank}(V) = 2$ and $\deg(\mathcal{P}_*\mathcal{F}_V) = c$. Let $\mathfrak{S}^s(Q, \gamma, c) \subset \mathfrak{S}(Q, \gamma, c)$ be the subset of the isomorphism classes of stable objects. Let $\mathfrak{S}^{ps}(Q, \gamma, c) \subset \mathfrak{S}(Q, \gamma, c)$ be the subset of the isomorphism classes of polystable objects.

Parabolic Torsion Free Sheaves and Stable Parabolic Difference Modules

Suppose that Q is non-constant. Let $\mathfrak{L}(Q, \gamma, c)$ denote the set of the isomorphism classes of torsion-free sheaf of rank one $\mathcal{P}_*\mathcal{L}$ on $(\Sigma(Q, \gamma), \Sigma_\infty(Q, \gamma))$ such that $\deg(\mathcal{P}_*\mathcal{L}) = c$. In §9.5.3.3, we constructed $\Upsilon : \mathfrak{L}(Q, \gamma, c + \deg(Q)) \longrightarrow \mathfrak{S}(Q, \gamma, c)$. According to Lemma 9.5.12, the image of Υ is contained in $\mathfrak{S}^s(Q, \gamma, c)$. According to Lemma 9.5.11, Υ is injective. The following proposition is a variant of [7, Proposition 3.6]. The idea goes back to [34].

Proposition 9.5.13 *We have $\mathfrak{S}^s(Q, \gamma, c) = \mathfrak{S}(Q, \gamma, c)$, and Υ is a bijection.*

Proof Because $\Sigma(Q, \gamma)$ is irreducible, it is easy to see $\mathfrak{S}^s(Q, \gamma, c) = \mathfrak{S}(Q, \gamma)$.

Let $(V, F, \mathcal{P}_*(V_{|\widehat{\infty}}))$ be a smooth parabolic 0-difference module. According to [7, Proposition 3.6], the $\mathcal{O}_{\mathbb{P}^1_{\beta_1}}(*\infty)$-module $\mathcal{F}_V$ equipped with an automorphism F corresponds to a torsion-free $\mathcal{O}_{\Sigma(Q,\gamma)}(*\Sigma_\infty(Q, \gamma))$-module $\mathcal{L}$ of rank 1, where $\mathcal{L}$ and $\mathcal{F}_V$ are related as $p_{\Sigma*}(\mathcal{L}) = \mathcal{F}_V$. Note that there exists an eigen decomposition $V_{|\widehat{\infty}} = \mathbb{E}_{\mathfrak{a}_1} \oplus \mathbb{E}_{\mathfrak{a}_2}$, which is identified with the following isomorphism induced by $p_{\Sigma*}(\mathcal{L}) = \mathcal{F}_V$:

$$V_{|\widehat{\infty}} = \bigoplus_{P \in \Sigma_\infty(Q,\gamma)} \mathcal{L}_{|\widehat{P}}. \tag{9.11}$$

Lemma 9.5.14 *A filtered bundle $\mathcal{P}_*(V_{|\widehat{\infty}})$ over $V_{|\widehat{\infty}}$ is a good parabolic structure of the 0-difference module (V, F) if and only if $\mathcal{P}_*(V_{|\widehat{\infty}})$ is induced by filtered bundles $\mathcal{P}_*(\mathcal{L}_{|\widehat{P}})$ over $\mathcal{L}_{|\widehat{P}}$ under the isomorphism (9.11).*

Proof Suppose that $\mathcal{P}_*(V_{|\widehat{\infty}})$ is a good parabolic structure at infinity. Because good filtration is compatible with the slope decomposition (see Lemma 3.3.17), we have the following for any $a \in \mathbb{R}$:

$$\mathcal{P}_a(V_{|\widehat{\infty}}) = \big(\mathcal{P}_a(V_{|\widehat{\infty}}) \cap \mathbb{E}_{\mathfrak{a}_1}\big) \oplus \big(\mathcal{P}_a(V_{|\widehat{\infty}}) \cap \mathbb{E}_{\mathfrak{a}_2}\big).$$

It means that $\mathcal{P}_*(V_{|\widehat{\infty}})$ is induced by the filtrations $\mathcal{L}_{|\widehat{P}}$ $(P \in \Sigma_\infty(Q, \gamma))$. We can check the converse similarly. Indeed, we have already observed it in Sect. 9.5.3.3. ■

Let $\mathcal{P}_*(\mathcal{L}_{|\widehat{P}})$ $(P \in \Sigma_\infty(Q, \gamma))$ be filtered bundles over $\mathcal{L}_{|\widehat{P}}$ corresponding to $\mathcal{P}_*(V_{|\widehat{\infty}})$. For $\boldsymbol{b} = (b_P) \in \mathbb{R}^{\Sigma_\infty(Q,\gamma)}$, there exist the torsion-free $\mathcal{O}_{\Sigma(Q,\gamma)}$-submodule $\mathcal{P}_{\boldsymbol{b}}\mathcal{L} \subset \mathcal{L}$ such that $\mathcal{P}_{\boldsymbol{b}}(\mathcal{L})_{|\widehat{P}} = \mathcal{P}_{b_P}(\mathcal{L}_{|\widehat{P}})$. Thus, we obtain a filtered torsion-free sheaf of rank one $\mathcal{P}_*\mathcal{L}$. By the construction, we have $\Upsilon(\mathcal{P}_*\mathcal{L}) = (V, F, \mathcal{P}_*(V_{|\widehat{\infty}}))$. Hence, we obtain that Υ is surjective. Because Υ is injective, we obtain Proposition 9.5.13. ■

Exceptional Case

Suppose that Q is constant.

Proposition 9.5.15 *$\mathfrak{S}^s(Q, \gamma, 0)$ is empty, and $\mathfrak{S}^{ps}(Q, \gamma, 0)$ consists of the isomorphism classes of $\mathcal{P}_*N(\alpha_1) \oplus \mathcal{P}_*N(\alpha_2)$, where (α_1, α_2) is determined by $x^2 - Qx + \gamma = (x - \alpha_1)(x - \alpha_2)$. (See* Sect. 9.5.2 *for $\mathcal{P}_*N(\alpha)$.)*

Proof Let $(V, F, \mathcal{P}_*(V_{|\widehat{\infty}})) \in \mathfrak{S}^{ps}(Q, \gamma, 0)$. If $\alpha_1 \neq \alpha_2$, there exists a decomposition $(V, F) = (V_1, F_1) \oplus (V_2, F_2)$ such that F_i are the multiplication of α_i. By Lemma 3.3.18, the decomposition is compatible with the parabolic structure at infinity, i.e., $\mathcal{P}_*(V_{|\infty}) = \mathcal{P}_*(V_{1|\infty}) \oplus \mathcal{P}_*(V_{2|\infty})$. Because $(V, F, \mathcal{P}_*(V_{|\widehat{\infty}}))$ is assumed to be polystable of degree 0, we obtain that $\deg(V_i, F_i, \mathcal{P}_*(V_{i|\widehat{\infty}})) = 0$. Thus, we obtain the claim of the lemma in the case $\alpha_1 \neq \alpha_2$.

Let us study the case $\alpha_1 = \alpha_2 =: \alpha$. Note that F preserves $\mathcal{P}_*(V_{|\widehat{\infty}})$. Hence, $F - \alpha\,\mathrm{id}$ also preserves $\mathcal{P}_*(V_{|\widehat{\infty}})$. Suppose that $F - \alpha\,\mathrm{id} \neq 0$. Because rank $V = 2$, we have $(F - \alpha\,\mathrm{id})^2 = 0$. Hence, there exists a free $\mathbb{C}[\beta_1]$-submodule $V_0 \subset V$ such that (i) $\mathrm{Im}(F - \alpha_0) \subset V_0 = \mathrm{Ker}(F - \alpha\,\mathrm{id})$, (ii) $V_1 = V/V_0$ is also a free $\mathbb{C}[\beta_1]$-module. We obtain the induced parabolic 0-difference modules $(V_i, F_i, \mathcal{P}_*(V_{i|\widehat{\infty}}))$. Because $(V, F, \mathcal{P}_*(V_{|\widehat{\infty}}))$ is polystable of degree 0, we obtain

$$\mu(V_0, F_0, \mathcal{P}_*(V_{0|\widehat{\infty}})) \leq \mu(V, F, \mathcal{P}_*(V_{|\widehat{\infty}})) = 0 \leq \mu(V_1, F_1, \mathcal{P}_*(V_{1|\widehat{\infty}})).$$

Because $F - \alpha\,\mathrm{id}$ induces a non-trivial morphism

$$(V_1, F_1, \mathcal{P}_*(V_{1|\widehat{\infty}})) \longrightarrow (V_0, F_0, \mathcal{P}_*(V_{0|\widehat{\infty}})),$$

and because rank $V_i = 1$, we obtain $\mu(V_1, F_1, \mathcal{P}_*(V_{1|\widehat{\infty}})) \leq \mu(V_0, F_0, \mathcal{P}_*(V_{0|\widehat{\infty}}))$. Thus, we obtain $\mu(V_0, F_0, \mathcal{P}_*(V_{0|\widehat{\infty}})) = \mu(V_1, F_1, \mathcal{P}_*(V_{1|\widehat{\infty}})) = 0$. Because $(V, F, \mathcal{P}_*(V_{|\widehat{\infty}}))$ is assumed to be polystable, we obtain

$$(V, F, \mathcal{P}_*(V_{|\widehat{\infty}})) \simeq (V_0, F_0, \mathcal{P}_*(V_{0|\widehat{\infty}})) \oplus (V_1, F_1, \mathcal{P}_*(V_{1|\widehat{\infty}})),$$

which implies $F - \alpha\,\mathrm{id} = 0$. But, it contradicts with the assumption. Thus, we obtain $F - \alpha\,\mathrm{id} = 0$. Then, the filtered bundle $\mathcal{P}_*\mathcal{F}_V$ induced by $\mathcal{F}_V$ and $\mathcal{P}_*(V_{|\widehat{\infty}})$ is a polystable bundle of degree 0 on $(\mathbb{P}^1, \infty)$ in the ordinary sense. It is well known such a filtered bundle is isomorphic to a direct sum of obvious filtered bundles of rank one on $(\mathbb{P}^1, \infty)$. ■

Chapter 10
Asymptotic Harmonic Bundles and Asymptotic Doubly Periodic Instantons (Appendix)

Abstract In Sect. 10.1, we shall explain a formal λ-connection induced by a Higgs bundle with a Hermitian metric which asymptotically satisfies the Hitchin equation. In Sects. 10.2–10.3, the estimates for doubly-periodic instantons are generalized to the context of asymptotically doubly-periodic instantons.

10.1 Formal λ-Connections Associated with Asymptotic Harmonic Bundles

10.1.1 Asymptotic Harmonic Bundles

Let U be a neighbourhood of 0 in $\mathbb{C}$. Let $(V, \overline{\partial}_V, \theta)$ be a Higgs bundle on $U \setminus \{0\}$. We assume that $(V, \overline{\partial}_V, \theta)$ is unramifiedly good wild, i.e., after shrinking U, there exists a finite subset $\mathcal{I} \subset z^{-1}\mathbb{C}[z^{-1}]$ and a decomposition

$$(V, \overline{\partial}_V, \theta) = \bigoplus_{\mathfrak{a}\in\mathcal{I}} (V_{\mathfrak{a}}, \overline{\partial}_{V_{\mathfrak{a}}}, \theta_{\mathfrak{a}}) \tag{10.1}$$

such that the following holds for each $\mathfrak{a} \in \mathcal{I}$.

- Let $f_{\mathfrak{a}}$ be the endomorphism of $V_{\mathfrak{a}}$ determined by $\theta_{\mathfrak{a}} - d\mathfrak{a}\,\mathrm{id}_{V_{\mathfrak{a}}} = f_{\mathfrak{a}} dz/z$. Let $\det(t\,\mathrm{id}_{V_{\mathfrak{a}}} - f_{\mathfrak{a}}) = \sum a_{\mathfrak{a},j}(z)t^j$ be the characteristic polynomial of $f_{\mathfrak{a}}$. Then, $a_{\mathfrak{a},j}$ are holomorphic on U.

For any $\mathfrak{a} \in z^{-1}\mathbb{C}[z^{-1}]$, set $\mathrm{ord}(\mathfrak{a}) := -\deg_{z^{-1}} \mathfrak{a}$. Let p be a positive integer such that

$$-p \leq \min\{\mathrm{ord}(\mathfrak{a} - \mathfrak{b}) \mid \mathfrak{a}, \mathfrak{b} \in \mathcal{I}, \mathfrak{a} \neq \mathfrak{b}\}. \tag{10.2}$$

(Note that "p" is denoted as "$-p$" in [65, §5.5].)

Let h be a Hermitian metric of V. Let $F(h)$ denote the curvature of the Chern connection ∇ of $(V, \overline{\partial}_V, h)$. Let $\theta^\dagger = f^\dagger(d\overline{z}/\overline{z})$ denote the adjoint of θ with respect to h. We assume the following condition.

T. Mochizuki, *Periodic Monopoles and Difference Modules*, Lecture Notes in Mathematics 2300, https://doi.org/10.1007/978-3-030-94500-8_10

- Let $P(s_1, s_2, s_3, s_4)$ be a polynomial of non-commutative variables s_i $(i = 1, 2, 3, 4)$. Then, there exists $\epsilon(P) > 0$ such that the following estimate holds:

$$P\Big(\mathrm{ad}(f), \mathrm{ad}(f^\dagger), \nabla_z, \nabla_{\overline{z}}\Big)\big(F(h) + [\theta, \theta^\dagger]\big) = O\Big(\exp\big(-\epsilon(P)|z|^{-p}\big)\Big)\, dz\, d\overline{z}. \tag{10.3}$$

Remark 10.1.1 The above condition is not the same as that in [65, §5.5]. Although a stronger assumption $-p < \min\{\mathrm{ord}(\mathfrak{a} - \mathfrak{b}) \mid \mathfrak{a}, \mathfrak{b} \in \mathcal{I}, \mathfrak{a} \neq \mathfrak{b}\}$ is imposed in [65], the condition (10.2) is enough for our purpose. While we considered only the estimate for $F(h) + [\theta, \theta^\dagger]$ in [65], we impose the decay condition for the higher derivatives in this paper, which will be used in Proposition 10.1.4. ■

By shrinking U, we may assume that there exists the refined decomposition

$$(V_{\mathfrak{a}}, \overline{\partial}_{V_{\mathfrak{a}}}, \theta_{\mathfrak{a}}) = \bigoplus_{\alpha \in \mathbb{C}} (V_{\mathfrak{a},\alpha}, \overline{\partial}_{V_{\mathfrak{a},\alpha}}, \theta_{\mathfrak{a},\alpha}) \tag{10.4}$$

such that the following holds.

- Let $f_{\mathfrak{a},\alpha}$ be the endomorphism of $V_{\mathfrak{a},\alpha}$ determined by $\theta_{\mathfrak{a},\alpha} - (d\mathfrak{a} + \alpha dz/z)\,\mathrm{id}_{V_{\mathfrak{a},\alpha}} = f_{\mathfrak{a},\alpha} dz/z$. We consider the characteristic polynomial $\det(t\,\mathrm{id}_{V_{\mathfrak{a},\alpha}} - f_{\mathfrak{a},\alpha}) = t^{\mathrm{rank}\, V_{\mathfrak{a},\alpha}} + \sum_{j=0}^{\mathrm{rank}\, V_{\mathfrak{a},\alpha}-1} a_{\mathfrak{a},\alpha,j}(z) t^j$. Then, $a_{\mathfrak{a},\alpha,,j}$ are holomorphic on U, and $a_{\mathfrak{a},\alpha,j}(0) = 0$ for any $0 \leq j \leq \mathrm{rank}\, V_{\mathfrak{a},\alpha} - 1$.

10.1.2 Simpson's Main Estimate

The following proposition is [65, Proposition 5.18], which is a variant of Simpson's main estimate in [64, 79].

Proposition 10.1.2

- *If* $\mathfrak{a} \neq \mathfrak{b}$, *there exists* $\epsilon > 0$ *such that* $V_{\mathfrak{a},\alpha}$ *and* $V_{\mathfrak{b},\beta}$ *are* $O\big(\exp(-\epsilon|z|^{\mathrm{ord}(\mathfrak{a}-\mathfrak{b})})\big)$-*asymptotically orthogonal, i.e., there exists* $C_1 > 0$ *such that the following holds for any* $u \in V_{\mathfrak{a},\alpha|Q}$ *and* $v \in V_{\mathfrak{b},\beta|Q}$*:*

$$\big|h(u, v)\big| \leq C_1 \exp\big(-\epsilon|z(Q)|^{\mathrm{ord}(\mathfrak{a}-\mathfrak{b})}\big) \cdot |u|_h \cdot |v|_h.$$

- *If* $\alpha \neq \beta$, *there exists* $\epsilon > 0$ *such that* $V_{\mathfrak{a},\alpha}$ *and* $V_{\mathfrak{a},\beta}$ *are* $O(|z|^{\epsilon})$-*asymptotically orthogonal.*
- $\theta_{\mathfrak{a},\alpha} - (d\mathfrak{a} + \alpha\, dz/z)\,\mathrm{id}_{E_{\mathfrak{a},\alpha}}$ *is bounded with respect to h and the Poincaré metric* g_{P}.

Proof Because the condition is slightly changed, we repeat the argument in the proof of [65, Proposition 5.18] with minor modifications. By considering the tensor product with a harmonic bundle of a rank one, we may assume $-p \leq$

$\min\{\operatorname{ord}(\mathfrak{a}) \mid \mathfrak{a} \in \mathcal{I}\}$. For any $\ell \le p$, we define the map $\eta_\ell : z^{-1}\mathbb{C}[z^{-1}] \longrightarrow z^{-\ell}\mathbb{C}[z^{-1}]$ by $\eta_\ell\big(\sum \mathfrak{a}_j z^j\big) = \sum_{j\le -\ell} \mathfrak{a}_j z^j$. Let $\mathcal{I}_\ell$ denote the image of $\mathcal{I}$. For each $\mathfrak{b} \in \mathcal{I}_\ell$, we set $V_{\mathfrak{b}}^{(\ell)} := \bigoplus_{\eta_\ell(\mathfrak{a})=\mathfrak{b}} \bigoplus_{\alpha\in\mathbb{C}} V_{\mathfrak{a},\alpha}$. Let $\pi_{\mathfrak{b}}^{(\ell)}$ denote the projection of V onto $V_{\mathfrak{b}}^{(\ell)}$ with respect to the decomposition $V = \bigoplus V_{\mathfrak{b}}^{(\ell)}$.

We take total orders $\le'$ on $\mathcal{I}_\ell$ for any ℓ such that the induced maps $\mathcal{I}_{\ell_1} \longrightarrow \mathcal{I}_{\ell_2}$ are order-preserving for each $\ell_1 \le \ell_2$. Let $V_{\mathfrak{b}}^{\prime(\ell)}$ be the orthogonal complement of $\bigoplus_{\mathfrak{c}<'\mathfrak{b}} V_{\mathfrak{c}}^{(\ell)}$ in $\bigoplus_{\mathfrak{c}\le'\mathfrak{b}} V_{\mathfrak{c}}^{(\ell)}$. Let $\pi_{\mathfrak{b}}^{\prime(\ell)}$ be the orthogonal projection onto $V_{\mathfrak{b}}^{\prime(\ell)}$.

We put $\zeta_\ell := \eta_\ell - \eta_{\ell+1}$. Let f be the endomorphism of V determined by $\theta = f\,dz$. We put $f^{(\ell)} := f - \sum_{\mathfrak{a}} \partial_z\big(\eta_{\ell+1}(\mathfrak{a})\big)\pi_{\mathfrak{a}}$, $\mu^{(\ell)} := f^{(\ell)} - \sum_{\mathfrak{a}} \partial_z\zeta_\ell(\mathfrak{a})\pi'_{\mathfrak{a}}$ and $\mathcal{R}_{\mathfrak{b}}^{(\ell)} := \pi_{\mathfrak{b}}^{(\ell)} - \pi_{\mathfrak{b}}^{\prime(\ell)}$. We consider the following claims.

(P_ℓ) $|f^{(\ell')}|_h = O(|z|^{-\ell'-1})$ for any $\ell' \ge \ell$.
(Q_ℓ) $|\mu^{(\ell')}|_h = O(|z|^{-\ell'})$ for any $\ell' \ge \ell$.
(R_ℓ) There exists $C > 0$ such that $|\mathcal{R}_{\mathfrak{b}}^{(\ell')}|_h = O\big(\exp(-C|z|^{-\ell'})\big)$ for any $\ell' \ge \ell$ and for any $\mathfrak{b} \in \mathcal{I}_{\ell'}$.

The asymptotic orthogonality of $V_{\mathfrak{a},\alpha}$ and $V_{\mathfrak{b},\beta}$ ($\mathfrak{a} \ne \mathfrak{b}$) follows from ($R_1$).

We have the expression $\theta^\dagger = f^\dagger d\bar{z}$. We set $\Delta := -\partial_z\partial_{\bar{z}}$. If a holomorphic section s of $\operatorname{End}(E)$ satisfies $[s, f] = 0$, we obtain the following inequality for some $C_0 > 0$ and $\epsilon_0 > 0$, which follows from (10.3) with $P = 1$:

$$\Delta \log |s|_h^2 \le -\frac{|[f^\dagger, s]|_h^2}{|s|_h^2} + C_0 \exp(-\epsilon_0|z|^{-p}). \tag{10.5}$$

Let $f^{(\ell)\dagger}$ denote the adjoint of $f^{(\ell)}$ with respect to h. Suppose that the claims $P_{\ell+1}$, $Q_{\ell+1}$ and $R_{\ell+1}$ hold. Because $[f - f^{(\ell)}, f^{(\ell)}] = 0$, we obtain $\big[(f - f^{(\ell)})^\dagger, f^{(\ell)}\big] = O\big(\exp(-\epsilon|z|^{-\ell-1})\big)$ by $R_{\ell+1}$. By applying (10.5) to $f^{(\ell)}$, we obtain the following for some $C_1 > 0$ as in (99) of [64]:

$$\Delta \log |f^{(\ell)}|_h^2 \le -\frac{|[f^{(\ell)\dagger}, f^{(\ell)}]|_h^2}{|f^{(\ell)}|_h^2} + C_1.$$

Then, by the same argument as that in §7.3.2–§7.3.3 of [64], we obtain P_ℓ and Q_ℓ. We put

$$k_{\mathfrak{b}}^{(\ell)} := \log\big(|\pi_{\mathfrak{b}}^{(\ell)}|_h^2 / |\pi_{\mathfrak{b}}^{\prime(\ell)}|_h^2\big) = \log\big(1 + |\mathcal{R}_{\mathfrak{b}}^{(\ell)}|_h^2 / |\pi_{\mathfrak{b}}^{\prime(\ell)}|_h^2\big).$$

By applying (10.5) to $\pi_{\mathfrak{b}}^{(\ell)}$, we obtain

$$\Delta \log k_{\mathfrak{b}}^{(\ell)} \le -\frac{|[f^\dagger, \pi_{\mathfrak{b}}^{(\ell)}]|_h^2}{|\pi_{\mathfrak{b}}^{(\ell)}|_h^2} + C_0 \exp\big(-\epsilon_0|z|^{-p}\big).$$

For any $A > 0$, there exists $C_2 > 0$ and $r_2 > 0$ such that the following holds for any $|z| < r_2$:

$$\begin{aligned}\Delta \exp(-A|z|^{-\ell}) &\geq -\exp(-A|z|^{-\ell})\Big(\frac{\ell^2}{4}A^2|z|^{-2(\ell+1)}\Big)\\ &\geq -\exp(-A|z|^{-\ell})\frac{\ell^2}{4}A^2C_2|z|^{-2(\ell+1)} + C_0\exp(-\epsilon_0|z|^{-p}).\end{aligned} \tag{10.6}$$

Hence, we obtain R_ℓ by using the argument in §7.3.4 of [64]. Similarly, we obtain the asymptotic orthogonality of $V_{\mathfrak{a},\alpha}$ and $V_{\mathfrak{a},\beta}$ ($\alpha \neq \beta$), and the boundedness of $\theta_{\mathfrak{a},\alpha} - (d\mathfrak{a} + \alpha dz/z)\,\mathrm{id}_{V_{\mathfrak{a},\alpha}}$ by using the argument in §7.3.5–§7.3.7 of [64] with (10.5). ■

By applying the argument in [64, §7.2.5], we obtain the following corollary.

Corollary 10.1.3 *$(V, \overline{\partial}_V, h)$ is acceptable, i.e., the curvature $F(h)$ is bounded with respect to h and $g_{\mathbf{p}}$.* ■

10.1.3 The Associated Filtered Bundles and Formal λ-Connections

Let $\lambda \in \mathbb{C}$. We obtain the holomorphic bundle $V^\lambda = (V, \overline{\partial}_V + \lambda\theta^\dagger)$ on $U \setminus \{0\}$. As a consequence of Proposition 10.1.2 and (10.3) with $P = 1$, we obtain the following estimate with respect to h:

$$\big[\overline{\partial}_V + \lambda\theta^\dagger, \partial_V - \overline{\lambda}\theta\big] = [\overline{\partial}_V, \partial_V] - |\lambda|^2[\theta, \theta^\dagger] = O\Big(|z|^{-2}(-\log|z|)^{-2}\Big)\, dz\, d\overline{z}.$$

Namely, $(V^\lambda, \overline{\partial}_V + \lambda\theta^\dagger, h)$ is acceptable. Hence, we obtain a filtered bundle $\mathcal{P}^h_* V^\lambda$ on $(U, 0)$ as in Proposition 2.11.15.

We set $\mathbb{D}^\lambda := \overline{\partial}_V + \lambda\theta^\dagger + \lambda\partial_V + \theta$. Let $\boldsymbol{v}$ be a holomorphic frame of $\mathcal{P}^h_a V^\lambda$ which is compatible with the parabolic structure, i.e., there exists a decomposition $\boldsymbol{v} = \coprod_{a-1<b\leq a} \boldsymbol{v}_b$ such that (i) $\boldsymbol{v}_b$ is a tuple of holomorphic sections of $\mathcal{P}_b V^\lambda$, (ii) $\boldsymbol{v}_b$ induces a frame of $\mathcal{P}^h_b V^\lambda \big/ \sum_{c<b} \mathcal{P}^h_c V^\lambda$. Let A be the matrix valued function determined by $\mathbb{D}^\lambda \boldsymbol{v} = \boldsymbol{v} A\, dz$.

Proposition 10.1.4 *There exists $N \in \mathbb{Z}_{>0}$ such that $z^N A$ is C^∞, and that the following holds:*

- *For any pair $(\ell_1, \ell_2) \in \mathbb{Z}_{\geq 1} \times \mathbb{Z}_{\geq 0}$, there exists $\epsilon(\ell_1, \ell_2) > 0$ such that $\partial_{\overline{z}}^{\ell_1}\partial_z^{\ell_2}(z^N A) = O\big(\exp(-\epsilon(\ell_1, \ell_2)|z|^{-p})\big)$.*

As a result, the Taylor series of $z^N A$ at 0 is of the form $\sum_{j=0}^{\infty} B_j z^j\, dz$ ($B_j \in M_{\operatorname{rank} V}(\mathbb{C})$).

Proof Let C be the matrix valued function on $U \setminus \{0\}$ determined by $(\partial_V - \overline{\lambda}\theta)\boldsymbol{v} = \boldsymbol{v}C\,dz$. Let F_C be the endomorphism determined by $F_C\boldsymbol{v} = \boldsymbol{v}\,C$. By the estimate for the connection form of acceptable bundles [64, Lemma 21.9.3], we obtain $|F_C|_h = O\big(|z|^{-1}(-\log|z|)^{N_0}\big)$ for some N_0. According to Proposition 10.1.2, we have $|\theta|_h = O(|z|^{-N_1})$ for some N_1. Because $\lambda\partial_V + \theta - \lambda(\partial_V - \overline{\lambda}\theta) = (1+|\lambda|^2)\theta$, we obtain $A = O\big(|z|^{-N}\big)$ for some $N \in \mathbb{Z}$. Hence, $z^N A$ is bounded.

Set $G := \big[\overline{\partial}_V + \lambda\theta^\dagger, \lambda\partial_V + \theta\big] = \lambda\big([\overline{\partial}_V, \partial_V] + [\theta, \theta^\dagger]\big)$. We have $G\boldsymbol{v} = \boldsymbol{v}\overline{\partial}A$. For each $\ell \in \mathbb{Z}_{\geq 0}$, there exists $\epsilon_\ell > 0$ such that $(\partial_{\overline{z}} + \lambda\theta^\dagger_{\overline{z}})^\ell G = O\big(\exp(-\epsilon_\ell|z|^{-p})\big)$, which implies $\partial_{\overline{z}}^\ell(\overline{\partial}A) = O\big(\exp(-\epsilon_\ell|z|^{-p})\big)$. We obtain the claim of the proposition from the following lemma.

Lemma 10.1.5 *Let f be a C^∞-function on $U \setminus \{0\}$. Suppose that for any $\ell \in \mathbb{Z}_{\geq 0}$ there exists $\epsilon_\ell > 0$ such that $|\partial_{\overline{z}}^\ell f| = O\big(\exp(-\epsilon_\ell|z|^{-p})\big)$. Then, for any $(\ell_1, \ell_2) \in \mathbb{Z}_{\geq 0} \times \mathbb{Z}_{\geq 0}$, we obtain $\big|\partial_z^{\ell_1}\partial_{\overline{z}}^{\ell_2} f\big| = O\big(\exp(-\epsilon_{\ell_1,\ell_2}|z|^{-p})\big)$ for some $\epsilon_{\ell_1,\ell_2} > 0$.*

Proof We may assume that $U = \{|z| < 1\}$. Let $\Psi : \{\zeta \in \mathbb{C} \mid \mathrm{Re}(\zeta) < 0\} \longrightarrow U \setminus \{0\}$ be the covering map defined by $\Psi(\zeta) = \exp(\zeta)$. Set $f_1 := \Psi^*(f)$. By the condition, for any $\ell \in \mathbb{Z}_{\geq 0}$, there exists $\epsilon_{1,\ell} > 0$ such that $\big|\partial_{\overline{\zeta}}^\ell f_1\big| = O\big(\exp(-\epsilon_{1,\ell}|\Psi(\zeta)|^{-p})\big)$.

Take $\zeta_0 \in \mathbb{C}$ with $\mathrm{Re}(\zeta_0) < -10$. Let $B(\zeta_0, r)$ be the disc with radius r with center ζ_0. There exist $C_\ell > 0$ and $\epsilon_{2,\ell} > 0$, which are independent of ζ_0, such that the following holds on $B(\zeta_0, 2r)$:

$$|\partial_{\overline{\zeta}}^{\ell+i} f| \leq C_\ell \exp\big(-\epsilon_{2,\ell}|\Psi(\zeta_0)|^{-p}\big) \quad (i = 0, 1, 2).$$

For any $q > 1$, there exists $C_{q,\ell} > 0$, independently from ζ_0, such that the L_1^q-norm of $\partial_{\overline{\zeta}}^{\ell+1} f$ on $B(\zeta_0, 7r/4)$ is dominated by $C_{q,\ell}\exp\big(-\epsilon_{2,\ell}|\Phi(\zeta_0)|^{-p}\big)$. Then, there exists $C'_{q,\ell} > 0$, independently from ζ_0, such that the L^q-norm of $\partial_\zeta\partial_{\overline{\zeta}}^{\ell+1} f$ on $B(\zeta_0, 5r/3)$ is dominated by $C'_{q,\ell}\exp\big(-\epsilon_{2,\ell}|\Phi(\zeta_0)|^{-p}\big)$. Then, there exists $C''_\ell > 0$, independently from ζ_0, such that the sup norm of $\partial_\zeta\partial_{\overline{\zeta}}^\ell f$ on $B(\zeta_0, 3r/2)$ is dominated by $C''_\ell\exp\big(-\epsilon_{2,\ell}|\Phi(\zeta_0)|^{-p}\big)$. By an easy induction, we obtain the claim of Lemma 10.1.5. The proof of Proposition 10.1.4 is also completed. ∎

10.1.4 Formal Good Filtered λ-Flat Bundles

Let $\widehat{\mathcal{V}}^\lambda$ denote the free $\mathbb{C}((z))$-module obtained as the formal completion of $\mathcal{P}^h V_0^\lambda$, i.e., $\widehat{\mathcal{V}}^\lambda := \mathbb{C}[[z]] \otimes_{\mathcal{O}_{U,0}} \mathcal{P}^h V_0^\lambda$. As a result of Proposition 10.1.4, $\mathbb{D}^\lambda$ induces a λ-connection $\widehat{\mathbb{D}}^\lambda$ of $\widehat{\mathcal{V}}^\lambda$. As the formal completion of $\mathcal{P}_*^h V^\lambda$, we obtain a filtered bundle $\mathcal{P}_*\widehat{\mathcal{V}}^\lambda$ over $\widehat{\mathcal{V}}^\lambda$.

Proposition 10.1.6 *$(\mathcal{P}_*\widehat{\mathcal{V}}^\lambda, \widehat{\mathbb{D}}^\lambda)$ is a unramifiedly good filtered λ-flat bundle. More precisely, there exists the decomposition*

$$(\mathcal{P}_*\widehat{\mathcal{V}}^\lambda, \widehat{\mathbb{D}}^\lambda) = \bigoplus_{\mathfrak{a}\in\mathcal{I}} (\mathcal{P}_*\widehat{\mathcal{V}}_\mathfrak{a}, \widehat{\mathbb{D}}^\lambda_\mathfrak{a}), \tag{10.7}$$

where $\widehat{\mathbb{D}}^\lambda_\mathfrak{a} - (1+|\lambda|^2)d\mathfrak{a}$ are logarithmic, and $\mathcal{I}$ is the index set of the decomposition (10.1).

Proof Let $\pi_\mathfrak{a}$ be the projection onto $V_\mathfrak{a}$ with respect to the decomposition (10.1). According to Proposition 10.1.2, we obtain the following estimate:

$$\big[\overline{\partial}_V + \lambda\theta^\dagger, \pi_\mathfrak{a}\big] = O\big(\exp(-\epsilon|z|^{-p})\big).$$

Hence, for any large $N > 0$, by using the argument in [64, Lemma 7.4.7], we can construct holomorphic endomorphisms $p_\mathfrak{a}$ ($\mathfrak{a} \in \mathcal{I}$) of $\mathcal{P}^h_* V^\lambda$ satisfying the following conditions:

$$p_\mathfrak{a} - \pi_\mathfrak{a} = O(|z|^N), \quad [p_\mathfrak{a}, p_\mathfrak{b}] = 0, \quad p_\mathfrak{a}\circ p_\mathfrak{a} = p_\mathfrak{a}, \quad \sum p_\mathfrak{a} = \mathrm{id}. \tag{10.8}$$

Lemma 10.1.7 *$\widehat{\mathbb{D}}^\lambda - \sum_{\mathfrak{a}\in\mathcal{I}}(1+|\lambda|^2)d\mathfrak{a}\cdot p_\mathfrak{a}$ is logarithmic with respect to $\mathcal{P}_*\widehat{\mathcal{V}}^\lambda$.*

Proof We consider

$$\mathbb{D}^\lambda_0 := \mathbb{D}^\lambda - \sum(1+|\lambda|^2)d\mathfrak{a}\cdot p_\mathfrak{a} = \overline{\partial}_V + \lambda\theta^\dagger + \lambda(\partial_V - \overline{\lambda}\theta) + (1+|\lambda|^2)\Big(\theta - \sum d\mathfrak{a}\cdot p_\mathfrak{a}\Big).$$

Note that $\big|\theta - \sum d\mathfrak{a}\cdot p_\mathfrak{a}\big|_h = O\big(|z|^{-1}dz\big)$. Hence, we obtain that $(\mathbb{D}^\lambda_0)_{|\widehat{0}}$ is logarithmic. ■

The claim of Proposition 10.1.6 follows from Lemma 10.1.7 and [64, Proposition 2.3.6]. ■

Lemma 10.1.8 *For each $\mathfrak{a}$, there exists a good filtered λ-flat bundle $(\mathcal{P}_*\mathcal{V}_\mathfrak{a}, \mathbb{D}^\lambda_\mathfrak{a})$ on $(U, 0)$ with an isomorphism $(\mathcal{P}_*\mathcal{V}_\mathfrak{a}, \mathbb{D}^\lambda_\mathfrak{a}) \otimes_{\mathcal{O}_U} \mathbb{C}[\![z]\!] \simeq (\mathcal{P}_*\widehat{\mathcal{V}}_\mathfrak{a}, \mathbb{D}^\lambda_\mathfrak{a})$.*

Proof The claim is obvious in the case $\lambda = 0$. Let us study the case $\lambda \neq 0$. It is enough to study the case $\mathfrak{a} = 0$. Therefore, we may assume that $\mathcal{I} = \{0\}$ from the beginning, i.e., $\widehat{\mathbb{D}}^\lambda$ is logarithmic with respect to $\mathcal{P}_*\widehat{\mathcal{V}}$.

Let $\mathcal{S}p_{-1<b\leq 0}(\mathrm{Res}(\widehat{\mathbb{D}}^\lambda))$ denote the eigenvalues of the endomorphism $\mathrm{Res}(\widehat{\mathbb{D}}^\lambda)$ on $\bigoplus_{-1<b\leq 0}\mathrm{Gr}^\mathcal{P}_b(\widehat{\mathcal{V}})$. Let N be a large integer such that

$$N > 100\max\Big\{\big|\lambda^{-1}(\alpha-\beta)\big|\;\Big|\;\alpha,\beta\in\mathcal{S}p_{-1<b\leq 0}(\widehat{\mathbb{D}}^\lambda)\Big\}.$$

We set $r := \operatorname{rank} \widehat{\mathcal{V}}$. Let $\boldsymbol{v}$ be a frame of $\mathcal{P}_0\widehat{\mathcal{V}}$ compatible with a parabolic structure. We obtain $A \in M_r(\mathbb{C}[\![z]\!])$ determined by $\widehat{\mathbb{D}}^\lambda \boldsymbol{v} = \boldsymbol{v} \cdot A\,dz/z$. For the expansion $A = \sum_{j=0}^\infty A_j z^j$, the set of the eigenvalues of A_0 is equal to $\mathcal{S}p_{-1<b\leq 0}(\operatorname{Res}(\widehat{\mathbb{D}}^\lambda))$.

We inductively construct $G_i \in M_r(\mathbb{C})$ $(i = N, N+1, \ldots, M)$ such that the following condition is satisfied.

- Let $I_r \in M_r(\mathbb{C})$ denote the identity matrix. We set $\boldsymbol{v}^{(M)} := \boldsymbol{v} \cdot (I_r + G_N z^N)(I_r + G_{N+1} z^{N+1}) \cdots (I_r + G_m z^M)$. Let $A^{(M)} \in M_r(\mathbb{C}[\![z]\!])$ be determined by $\widehat{\mathbb{D}}^\lambda \boldsymbol{v}^{(M)} = \boldsymbol{v}^{(M)} \cdot A^{(M)} dz/z$. Then, for the expansion $A^{(M)} = \sum_{j=0}^\infty A_j^{(M)} z^j$, we obtain $A_i^{(M)} = A_i$ $(i < N)$ and $A_i^{(M)} = 0$ $(N \leq i \leq M)$.

Suppose that we have already constructed G_i $(i = N, \ldots, M-1)$. There uniquely exists $G_M \in M_r(\mathbb{C})$ such that $MG_M + [A_0^{(M-1)}, G_M] + A_M^{(M-1)} = 0$. Then, it is easy to check that the condition is satisfied for M.

As the limit of $\boldsymbol{v}^{(M)}$ with respect to the z-adic topology, we obtain a frame $\boldsymbol{v}^{(\infty)}$ of $\mathcal{P}_0\widehat{\mathcal{V}}$. Let $A^{(\infty)} \in M_r(\mathbb{C}[\![z]\!])$ be the matrix determined by $\widehat{\mathbb{D}}^\lambda \boldsymbol{v}^{(\infty)} = \boldsymbol{v}^{(\infty)} A^{(\infty)} dz/z$. By the construction, $A^{(\infty)} \in M_r(\mathbb{C}[z])$. Thus, we obtain the claim of the lemma. ■

10.1.5 Residues and KMS-Structure

For any $a \in \mathbb{R}$, we set $\mathcal{P}_{<a}\widehat{\mathcal{V}}^\lambda := \sum_{b<a} \mathcal{P}_b\widehat{\mathcal{V}}^\lambda$ and $\operatorname{Gr}_a^{\mathcal{P}}(\widehat{\mathcal{V}}^\lambda) := \mathcal{P}_a\widehat{\mathcal{V}}^\lambda / \mathcal{P}_{<a}\widehat{\mathcal{V}}^\lambda$. We also obtain $\mathcal{P}_{<a}\widehat{\mathcal{V}}^\lambda_{\mathfrak{a}}$ and $\operatorname{Gr}_a^{\mathcal{P}}(\widehat{\mathcal{V}}^\lambda_{\mathfrak{a}})$ from $(\mathcal{P}_*\widehat{\mathcal{V}}^\lambda_{\mathfrak{a}}, \widehat{\mathbb{D}}^\lambda_{\mathfrak{a}})$. There exists the natural decomposition $\operatorname{Gr}_a^{\mathcal{P}}(\widehat{\mathcal{V}}^\lambda) = \bigoplus_{\mathfrak{a}\in\mathcal{I}} \operatorname{Gr}_a^{\mathcal{P}}(\widehat{\mathcal{V}}^\lambda_{\mathfrak{a}})$.

We obtain the residue endomorphisms $\operatorname{Res}(\widehat{\mathbb{D}}^\lambda_{\mathfrak{a}})$ of $\operatorname{Gr}_a^{\mathcal{P}}(\widehat{\mathcal{V}}^\lambda_{\mathfrak{a}})$ for any $a \in \mathbb{R}$. (See [64, §2.5.2].) There exists the generalized eigen decomposition

$$\operatorname{Gr}_a^{\mathcal{P}}(\widehat{\mathcal{V}}^\lambda_{\mathfrak{a}}) = \bigoplus_{\alpha\in\mathbb{C}} \mathbb{E}_\alpha \operatorname{Gr}_a^{\mathcal{P}}(\widehat{\mathcal{V}}^\lambda_{\mathfrak{a}}),$$

where $\mathbb{E}_\alpha \operatorname{Gr}_a^{\mathcal{P}}(\widehat{\mathcal{V}}^\lambda_{\mathfrak{a}})$ denote the generalized eigen spaces according to α. Let $\mathcal{KMS}(\mathcal{P}_*\widehat{\mathcal{V}}^\lambda_{\mathfrak{a}})$ denote the set of (a, α) such that $\mathbb{E}_\alpha \operatorname{Gr}_a^{\mathcal{P}}(\widehat{\mathcal{V}}^\lambda_{\mathfrak{a}}) \neq 0$. We put $\mathfrak{m}^\lambda(a, \alpha, \mathfrak{a}) := \dim \mathbb{E}_\alpha \operatorname{Gr}_a^{\mathcal{P}}(\widehat{\mathcal{V}}^\lambda_{\mathfrak{a}})$. Let $N^\lambda_{a,\alpha,\mathfrak{a}}$ denote the nilpotent part of $\operatorname{Res}(\widehat{\mathbb{D}}^\lambda_{\mathfrak{a}})$ on $\mathbb{E}_\alpha \operatorname{Gr}_a^{\mathcal{P}}(\widehat{\mathcal{V}}^\lambda_{\mathfrak{a}}) \neq 0$. We obtain the monodromy weight filtration W of $N^\lambda_{a,\alpha,\mathfrak{a}}$.

10.1.6 Norm Estimate

Let $\boldsymbol{v}$ be a frame of $\mathcal{P}_a\mathcal{V}^\lambda$ which is compatible with the parabolic filtration and the weight filtration, i.e., there exists a decomposition $\boldsymbol{v} = \coprod_{a-1<b\leq a} \coprod_{k\in\mathbb{Z}} \boldsymbol{v}_{b,k}$ such

that (i) $\coprod_{k\in\mathbb{Z}} \boldsymbol{v}_{b,k}$ is a tuple of sections of $\mathcal{P}_a\mathcal{V}^\lambda$, (ii) $\coprod_{k\leq \ell} \boldsymbol{v}_{b,k}$ induces a frame of $W_\ell \operatorname{Gr}^{\mathcal{P}}(\widehat{\mathcal{V}}^\lambda)$. For $v_i \in \boldsymbol{v}_{b,k}$, we set $b(v_i) := b$ and $k(v_i) := k$.

Let h_0 be a Hermitian metric of V for which

$$h_0(v_i, v_j) := \begin{cases} |z|^{-2b(v_i)}(-\log|z|^2)^{k(v_i)} & (i = j), \\ 0 & \text{(otherwise)}. \end{cases}$$

We can prove the following proposition by using the same argument as the proof of [64, Proposition 8.1.1], which originally goes back to [80].

Proposition 10.1.9 *h_0 and h are mutually bounded.*

Proof We explain only an outline. We decompose $\mathbb{D}^\lambda = d'' + d' = (\overline{\partial}_V + \lambda\theta) + (\lambda\partial_V + \theta^\dagger)$. Note that $\big|[d'', d']\big|_h = O\big(\exp(-\epsilon|z|^{-p})\big)$. For any large N, there exists a C^∞-section r_N of $\operatorname{End}(V^\lambda)$ such that (i) $|r_N|_h + |d' r_N|_h + |d'' r_N|_h = O\big(|z|^{2N}\big)$, (ii) $[d'', d' - r_N\, dz] = 0$. We set $\mathbb{D}_1^\lambda := \mathbb{D}^\lambda - r_N\, dz$. Let $\mathbb{D}_1^\lambda = d_1'' + d_1'$ be the decomposition into the $(0,1)$-part and the $(1,0)$-part. We obtain the differential operators $\delta'_{1,h}$ and $\delta''_{1,h}$ as in Sect. 2.9.5. We set $\mathbb{D}^{\lambda\star}_{1,h} = \delta'_{1,h} - \delta''_{1,h}$. Then, we obtain $\big|[\mathbb{D}_1^\lambda, \mathbb{D}^{\lambda\star}_{1,h}]\big|_h = O\big(|z|^N\big)$. If N is sufficiently large, $(\mathcal{P}^h_* V^\lambda, \mathbb{D}_1^\lambda)$ is a good filtered λ-flat bundle by [64, Proposition 2.3.6]. The formal completion at 0 has the decomposition

$$(\mathcal{P}_*(\widehat{\mathcal{V}}^\lambda), \widehat{\mathbb{D}}_1^\lambda) = \bigoplus_{\mathfrak{a}\in\mathcal{I}} (\mathcal{P}_*\widehat{\mathcal{V}}_1^\lambda, \widehat{\mathbb{D}}^\lambda_{1,\mathfrak{a}}) \tag{10.9}$$

such that $\widehat{\mathbb{D}}^\lambda_{1,\mathfrak{a}} - d\mathfrak{a}\,\mathrm{id}$ are logarithmic. Note that the decompositions (10.7) and (10.9) are not necessarily the same. But, because $(\widehat{\mathbb{D}}^\lambda - \widehat{\mathbb{D}}_1^\lambda)\mathcal{P}_b\widehat{\mathcal{V}}^\lambda \subset \mathcal{P}_{b-N}\widehat{\mathcal{V}}^\lambda$ for any $b \in \mathbb{R}$, we obtain $\operatorname{Res}(\widehat{\mathbb{D}}^\lambda)$ and $\operatorname{Res}(\widehat{\mathbb{D}}_1^\lambda)$ and $\operatorname{Gr}^{\mathcal{P}}_a(\widehat{\mathcal{V}}^\lambda_{\mathfrak{a}}) = \operatorname{Gr}^{\mathcal{P}}_a(\widehat{\mathcal{V}}^\lambda_{1,\mathfrak{a}})$.

As in the proof of [64, Proposition 2.3.6], we can construct a Hermitian metric h_1 of V^λ such that (i) h_1 is mutually bounded with h_0, (ii) $\big[\mathbb{D}_1^\lambda, \mathbb{D}^{\lambda\star}_{1,h_1}\big] = O(|z|^{-2+\epsilon})dz\,d\overline{z}$ for some $\epsilon > 0$. Then, by [80, Corollary 4.3] (see also [60, §4.3] for the case of general λ), we obtain that h and h_1 are mutually bounded. ∎

10.1.7 Comparison of KMS-Structures

We define the map $\mathfrak{k}(\lambda) : \mathbb{R} \times \mathbb{C} \longrightarrow \mathbb{R} \times \mathbb{C}$ by

$$\mathfrak{k}(\lambda, a, \alpha) := \big(a + 2\operatorname{Re}(\lambda\overline{\alpha}), \alpha - a\lambda - \overline{\alpha}\lambda^2\big).$$

We can prove the following proposition by using the argument in [80, §7] (see also [64, Proposition 8.2.1]).

Proposition 10.1.10 *The map $\mathfrak{k}(\lambda)$ induces a bijection*

$$\mathcal{KMS}(\mathcal{P}_*\widehat{\mathcal{V}}^0_{\mathfrak{a}}) \simeq \mathcal{KMS}(\mathcal{P}_*\widehat{\mathcal{V}}^\lambda_{\mathfrak{a}}).$$

It preserves the multiplicities, i.e., $\mathfrak{m}^0(a, \alpha, \mathfrak{a}) = \mathfrak{m}^\lambda\big(\mathfrak{k}(\lambda, a, \alpha), \mathfrak{a}\big)$. *The conjugacy classes of* $N^0_{a,\alpha,\mathfrak{a}}$ *and* $N^\lambda_{\mathfrak{k}(\lambda,a,\alpha),\mathfrak{a}}$ *are the same.*

Proof We explain only an outline. Let $(\mathcal{P}^h_* V^\lambda, \mathbb{D}^\lambda_1)$ denote the good filtered λ-flat bundle as in the proof of Proposition 10.1.9.

Let $\pi_{\mathfrak{a},\alpha}$ denote the projection onto $V_{\mathfrak{a},\alpha}$ with respect to the decomposition (10.4). Let $p_{\mathfrak{a}}$ be as in the proof of Proposition 10.1.6. As in [64, Lemma 8.2.4], we can construct endomorphisms $p_{\mathfrak{a},\alpha}$ of $\mathcal{P}^h_* V^\lambda$ ($\mathfrak{a} \in \mathcal{I}$) such that

$$\big|p_{\mathfrak{a},\alpha} - \pi_{\mathfrak{a},\alpha}\big|_h = O\big(|z|^\epsilon\big), \quad p^2_{\mathfrak{a},\alpha} = p_{\mathfrak{a},\alpha},$$

$$p_{\mathfrak{a},\alpha} \circ p_{\mathfrak{b},\beta} = 0 \ \ ((\mathfrak{a},\alpha) \neq (\mathfrak{b},\beta)), \quad \sum_\alpha p_{\mathfrak{a},\alpha} = p_{\mathfrak{a}}.$$

Here, ϵ denotes a positive number. We set $\Phi := \sum_{\mathfrak{a},\alpha}(d\mathfrak{a} + \alpha dz/z) \cdot p_{\mathfrak{a},\alpha}$. The following lemma is an analogue of [64, Lemma 8.2.5].

Lemma 10.1.11 *Let v be a holomorphic section of V^λ such that*

$$|v|_h \sim |z|^{-b}(-\log|z|)^k$$

for $b \in \mathbb{R}$ and $k \in \mathbb{Z}$. Then, we obtain

$$\big|\mathbb{D}^\lambda_1 - (1 + |\lambda|^2)\Phi + \lambda \cdot a \cdot v \cdot dz/z\big|_h = O\big(|z|^{-b}(-\log|z|)^{k-1}\big)\,|dz/z|.$$

Proof It is enough to study the case $b = 0$. We obtain

$$\mathbb{D}^\lambda_1 v - (1 + |\lambda|^2)\Phi(v) = \lambda\delta'_{1,h}v + (1 + |\lambda|^2)(\theta - \Phi)v - r_N v.$$

Because $|(\theta - \Phi)v| = O\big((-\log|z|)^{k-1}\big)$ (see Proposition 10.1.2) and $|r_N v| = O(|z|^{N-1})$, it is enough to prove that

$$\int \big|\delta'_{1,h}v\big|^2_h(-\log|z|)^{1-\delta-2k}\big|dz\,d\overline{z}\big| < \infty$$

for any $\delta > 0$. We can prove it by using the argument in [80, Page 761–762]. ■

We obtain the decomposition $\mathcal{P}^h_* V^\lambda = \bigoplus \mathrm{Im}_{\mathfrak{a},\alpha}\,\mathcal{P}_*\big(p_{\mathfrak{a},\alpha}\big)$. Let

$$\boldsymbol{v}_{\mathfrak{a},\alpha} = \coprod_{-1<a\leq 0}\coprod_{k\in\mathbb{Z}} \boldsymbol{v}_{\mathfrak{a},\alpha,a,k}$$

be a frame of $\mathcal{P}^h_0(\mathrm{Im}\, p_{\mathfrak{a},\alpha})$ compatible with the parabolic structure and the weight filtration. (See Sect. 10.1.6.) We obtain the induced frame $\boldsymbol{v} = (v_1, \ldots, v_r) = \coprod_{\mathfrak{a},\alpha} \boldsymbol{v}_{\mathfrak{a},\alpha}$ of $\mathcal{P}^h_0 V^\lambda$. For each $v_i \in \boldsymbol{v}_{\mathfrak{a},\alpha,a,k}$, we set $\mathfrak{a}(v_i) = \mathfrak{a}$, $\alpha(v_i) = \alpha$, $a(v_i) = a$, and $k(v_i) = k$.

We obtain $A \in M_r(\mathbb{C})$ by $\mathrm{Res}(\mathbb{D}^\lambda_1)\boldsymbol{v}_{|0} = \boldsymbol{v}_{|0} A$. We obtain the following lemma by Proposition 10.1.9, Lemma 10.1.11 and the construction of $\boldsymbol{v}$.

Lemma 10.1.12 *$A_{i,j} = 0$ holds unless $\mathfrak{a}(v_i) = \mathfrak{a}(v_j)$, $\alpha(v_i) = \alpha(v_j)$, and $(a(v_i), k(v_i)) \leq'' (a(v_j), k(v_j))$, where $\leq''$ denotes the lexicographic order on $\mathbb{R} \times \mathbb{Z}$. Moreover, $A_{i,i} = (1 + |\lambda|^2)\alpha(v_i) - \lambda a(v_i)$ holds.* ■

As in [64, §8.2.5], there exists a wild harmonic bundle

$$(\widetilde{V}, \overline{\partial}_{\widetilde{V}}, \widetilde{\theta}, \widetilde{h}) = \bigoplus (\widetilde{V}_{\mathfrak{a},\alpha}, \overline{\partial}_{\widetilde{V}_{\mathfrak{a},\alpha}}, \widetilde{\theta}_{\mathfrak{a},\alpha}, \widetilde{h}_{\mathfrak{a},\alpha}) \tag{10.10}$$

on $U \setminus \{0\}$ with an isomorphism of filtered bundles $\Psi : \mathcal{P}^h_* \widetilde{V}^0 \simeq \mathcal{P}^h_* V^0$ such that $(\theta\Psi - \Psi\widetilde{\theta})\mathcal{P}^h_a \widetilde{V}^0 \subset \mathcal{P}^h_{<a} V^0$ for any $a \in \mathbb{R}$. We may assume that the decompositions (10.4) and (10.10) are the same under the isomorphism Ψ. By the norm estimate (Proposition 10.1.9), Ψ and Ψ^{-1} are bounded. We identify the bundles V and $\widetilde{V}$ via Ψ as C^∞-bundles. The metrics h and $\widetilde{h}$ are mutually bounded. By Proposition 10.1.2 and the construction, we obtain $|\theta - \widetilde{\theta}|_h = O\big(|z|^\epsilon\big)\, dz/|z|$ for some $\epsilon > 0$. We also obtain $|\theta^\dagger_h - \widetilde{\theta}^\dagger_{\widetilde{h}}|_h = O\big((-\log|z|)^{-1}\big)\, d\overline{z}/|z|$ by Proposition 10.1.2.

Lemma 10.1.13 *Let M be the C^∞-endomorphism of V determined by $\theta^\dagger_h - \widetilde{\theta}^\dagger_{\widetilde{h}} = M\, d\overline{z}/\overline{z}$. Let U' be any relatively compact neighbourhood of 0 in $U \cap \{|z| < 1\}$. Then, we obtain*

$$\int_{U'} |M|^2_h \cdot (-\log|z|) \cdot |dz\, d\overline{z}| < \infty. \tag{10.11}$$

Proof To clarify the argument, we regard M as a homomorphism $\widetilde{V} \longrightarrow V$ obtained as $\theta \circ \Psi - \Psi \circ \widetilde{\theta} = M\, dz/z$, which is a section of the bundle $\mathrm{Hom}(\widetilde{V}, V)$. Note that $\mathrm{Hom}(\widetilde{V}, V)$ with the induced Higgs field $\widetilde{\theta}^{(1)}$ and the induced Hermitian metric $\widetilde{h}^{(1)}$ satisfies (10.3). By the proof of Weitzenböck formula in [80, Lemma 4.1, P743], there exist $C_i > 0$ $(i = 1, 2)$ and $\epsilon_1 > 0$ such that

$$-\partial_z \partial_{\overline{z}} |\Psi|^2_{\widetilde{h}^{(1)}} \leq -C_1 |M|^2_{\widetilde{h}^{(1)}} + C_2 |z|^{2\epsilon_1 - 2}.$$

We obtain $-\partial_z \partial_{\overline{z}}\big(|\Psi|^2_{\widetilde{h}^{(1)}} + C_2 \epsilon_1^{-2} |z|^{\epsilon_1}\big) \leq -C_1 |M|^2_{\widetilde{h}^{(1)}}$. Because $|\Psi|^2_{\widetilde{h}^{(1)}} + C_2 \epsilon_1^{-2} |z|^{\epsilon_1}$ is bounded, we obtain (10.11) by the argument in [80, Lemma 7.7] (see also Lemma 10.1.15 below). ■

From $(\widetilde{V}, \overline{\partial}_{\widetilde{V}}, \widetilde{\theta}, \widetilde{h})$, we obtain a good filtered λ-flat bundle $(\mathcal{P}^h_* \widetilde{V}^\lambda, \widetilde{\mathbb{D}}^\lambda) = \bigoplus_{\mathfrak{a} \in \mathcal{I}} (\mathcal{P}^h_* \widetilde{V}^\lambda_{\mathfrak{a},\alpha}, \widetilde{\mathbb{D}}^\lambda_{\mathfrak{a},\alpha})$. Let $\widetilde{p}_{\mathfrak{a},\alpha}$ denote the projection of $\mathcal{P}^h_* \widetilde{V}^\lambda$ onto $\mathcal{P}^h_* \widetilde{V}^\lambda_{\mathfrak{a},\alpha}$ with respect to the decomposition. Note that $\widetilde{p}_{\mathfrak{a},\alpha} = \pi_{\mathfrak{a},\alpha}$ under the isomorphism Ψ. Hence, we obtain $\widetilde{p}_{\mathfrak{a},\alpha} - p_{\mathfrak{a},\alpha} = O(|z|^{\epsilon})$.

Let $\widetilde{\boldsymbol{v}}_{\mathfrak{a},\alpha} = \coprod_{-1<a\leq 0}\coprod_{k\in\mathbb{Z}} \widetilde{\boldsymbol{v}}_{\mathfrak{a},\alpha,a,k}$ be a frame of $\mathcal{P}_0\widetilde{V}^{\lambda}_{\mathfrak{a},\alpha}$ compatible with the parabolic structure and the weight filtration. We obtain a frame $\widetilde{\boldsymbol{v}} = \coprod_{\mathfrak{a},\alpha}\widetilde{\boldsymbol{v}}_{\mathfrak{a},\alpha}$ of $\mathcal{P}_0\widetilde{V}^{\lambda}$. For $\widetilde{v}_i \in \widetilde{\boldsymbol{v}}_{\mathfrak{a},\alpha,a,k}$, we set $\mathfrak{a}(\widetilde{v}_i) = \mathfrak{a}$, $\alpha(\widetilde{v}_i) = \alpha$, $a(\widetilde{v}_i) = a$, and $k(\widetilde{v}_i) = k$. Let $I = (I_{j,i})$ be determined by $\widetilde{v}_i = \sum I_{j,i}v_j$. We put $\mathcal{B}_{j,i} = I_{j,i}|z|^{a(\widetilde{v}_i)-a(v_j)-(k(\widetilde{v}_i)-k(v_j))/2}$. By using the argument in [80] (see [64, Proposition 8.2.2] for a summary) with Lemma 10.1.13 and a complement for the wild case in [64, Lemma 8.2.8], we can prove the following.

- $\mathcal{B}_{j,i} = O(|z|^{\epsilon})$ unless $(\mathfrak{a}(\widetilde{v}_i),\alpha(\widetilde{v}_i)) = (\mathfrak{a}(v_j),\alpha(v_j))$.
- $|\mathcal{B}_{j,i}| = O\big((-\log|z|)^{-1}\big)$ unless $a(\widetilde{v}_i) = a(v_j)$.
- We obtain

$$\int |\mathcal{B}_{j,i}|_h^2 \frac{|dz\,d\overline{z}|}{|z|^2(-\log|z|)^2\log(-\log|z|)} < \infty$$

unless $k(\widetilde{v}_i) = k(v_j)$.

Then, we obtain $\dim \mathrm{Gr}^W_k \mathbb{E}_{\beta}\mathrm{Gr}^{\mathcal{P}}_a(\widehat{\mathcal{V}}^{\lambda}_{\mathfrak{a}}) = \dim \mathrm{Gr}^W_k \mathbb{E}_{\beta}\mathrm{Gr}^{\mathcal{P}}_a(\mathcal{P}\widetilde{\mathcal{V}}^{\lambda}_{\mathfrak{a}})$ for any $\mathfrak{a} \in \mathcal{I}$, $-1 < a \leq 0$, $\beta \in \mathbb{C}$ and $k \in \mathbb{Z}$, by applying the argument in the proof of [80, Proposition 7.6, Theorem 7] with Lemma 10.1.12. Because the claim of Proposition 10.1.10 holds for $(\widetilde{V}, \overline{\partial}_{\widetilde{V}}, \widetilde{\theta}, \widetilde{h})$, we obtain the claim of Proposition 10.1.10 for $(V, \overline{\partial}_V, \theta, h)$. ■

10.1.8 Appendix

Let ϕ be a bounded $\mathbb{R}_{\geq 0}$-valued C^{∞}-function on $\mathbb{C}^*$ such that there exists $R_0 > 0$ such that $\phi(w) = 0$ for $|w| > R_0$. We may naturally regard $\phi(w)$ as an L^1-function on $\mathbb{C}$, and we obtain the distribution $\partial_w\partial_{\overline{w}}\phi$ on $\mathbb{C}$.

Let $\rho : \mathbb{C} \longrightarrow \mathbb{R}_{\geq 0}$ be a C^{∞}-function such that (i) $0 \leq \rho(w) \leq 1$ for any w, (ii) $\rho(w) = 1$ for $|w| \leq 1$, (iii) $\rho(w) = 0$ for $|w| \geq 2$.

For any $z \in \mathbb{C}^*$, we take $0 < \epsilon < |z|/100$, and set $\chi_{z,\epsilon}(w) := \rho\big((z-w)/\epsilon\big)$. The following integral is well defined.

$$F_{\phi}(z) := \frac{2}{\pi}\int_{\mathbb{C}} \partial_w\partial_{\overline{w}}\phi(w)\cdot\Big(\chi_{z,\epsilon}(w)\log|z-w|\Big) + \frac{2}{\pi}\int_{\mathbb{C}} \phi(w)\cdot\partial_w\partial_{\overline{w}}\Big(\big(1-\chi_{z,\epsilon}(w)\big)\cdot\log|z-w|\Big). \tag{10.12}$$

It is easy to check that $F(z)$ is independent of the choice of ϵ.

Lemma 10.1.14 $F_{\phi}(z) = \phi(z)$.

Proof For z and ϵ as above, we obtain

$$\begin{aligned}
&\frac{2}{\pi}\int_{\mathbb{C}} \partial_w\partial_{\overline{w}}\big((1-\chi_{z,\epsilon}(w))\phi(w)\big)\cdot\Big(\chi_{z,\epsilon}(w)\log|z-w|\Big)\\
&\quad+\frac{2}{\pi}\int_{\mathbb{C}}(1-\chi_{z,\epsilon}(w))\phi(w)\cdot\partial_w\partial_{\overline{w}}\Big(\big(1-\chi_{z,\epsilon}(w)\big)\cdot\log|z-w|\Big)\\
&\quad=\frac{2}{\pi}\int_{|z-w|\geq\epsilon/4}\big(1-\chi_{z,\epsilon}(w)\big)\cdot\phi(w)\cdot\partial_w\partial_{\overline{w}}\log|z-w|=0.
\end{aligned}\tag{10.13}$$

Hence, we obtain $F_\phi(z)=F_{\chi_{z,\epsilon}\phi}(z)$. Note that $\chi_{z,\epsilon}\phi$ is naturally a test function on $\mathbb{C}$. Because $\frac{1}{2\pi}\log|w|$ is the Green function for the operator $4\partial_w\partial_{\overline{w}}$, we obtain $F_{\chi_{z,\epsilon}\phi}(z)=(\chi_{z,\epsilon}\phi)(z)=\phi(z)$. ■

Let U be a neighbourhood of 0 in $\mathbb{C}$. We set $U^*:=U\setminus\{0\}$. Let $f_i: U^*\longrightarrow\mathbb{R}_{\geq 0}$ $(i=1,2)$ be C^∞-functions such that (i) f_1 is bounded, (ii) $f_2\leq\partial_w\partial_{\overline{w}}f_1$ on U^*. The following lemma is contained in [80, Lemma 7.7].

Lemma 10.1.15 *For any relatively compact neighbourhood U' of 0 in U, we obtain $\int_{U'}f_2\cdot(-\log|w|)<\infty$.*

Proof We may assume $\big\{|w|\leq 10\big\}\subset U$. It is enough to consider the case $U'=\{|w|<1/10\}$. We may naturally regard $\widetilde{f_1}:=\rho(w/5)\cdot f_1$ as a C^∞-function on $\mathbb{C}^*$. The integral $F_{\widetilde{f_1}}$ in (10.12) is well defined for $z\in\mathbb{C}^*$. By Lemma 10.1.15, we obtain $F_{\widetilde{f_1}}(z)=\widetilde{f_1}(z)$.

Suppose $|z|<1/4$. We obtain

$$\begin{aligned}
F_{\widetilde{f_1}}(z)=&\frac{2}{\pi}\int_{\mathbb{C}}\partial_w\partial_{\overline{w}}\widetilde{f_1}(w)\cdot\Big(\chi_{z,\epsilon}(w)\log|z-w|\Big)\\
&+\frac{2}{\pi}\int_{\mathbb{C}}\widetilde{f_1}(w)\cdot\partial_w\partial_{\overline{w}}\Big(\rho(3(z-w))\cdot\big(1-\chi_{z,\epsilon}(w)\big)\cdot\log|z-w|\Big)\\
&+\frac{2}{\pi}\int_{\mathbb{C}}\widetilde{f_1}(w)\cdot\partial_w\partial_{\overline{w}}\Big(\big(1-\rho(3(z-w))\big)\cdot\log|z-w|\Big).
\end{aligned}\tag{10.14}$$

It is rewritten as follows.

$$\begin{aligned}
&\frac{2}{\pi}\int_{\mathbb{C}}\partial_w\partial_{\overline{w}}\widetilde{f_1}(w)\cdot\Big(\chi_{z,\epsilon}(w)\log|z-w|\Big)\\
&\quad+\frac{2}{\pi}\int_{\mathbb{C}}\partial_w\partial_{\overline{w}}\widetilde{f_1}(w)\cdot\Big(\rho(3(z-w))\cdot\big(1-\chi_{z,\epsilon}(w)\big)\cdot\log|z-w|\Big)\\
&\quad+\frac{2}{\pi}\int_{\mathbb{C}}\widetilde{f_1}(w)\cdot\partial_w\partial_{\overline{w}}\Big(\big(1-\rho(3(z-w))\big)\cdot\log|z-w|\Big).
\end{aligned}\tag{10.15}$$

Here, $\partial_w \partial_{\overline{w}} \widetilde{f}_1$ in the second integral is the current. According to [80, Lemma 2.2], $f_2 \leq \partial_w \partial_{\overline{w}} f_1$ weakly holds on U. Note that $\bigcup_{|z|<1/4} \{|w-z| < 2/3\} \subset \{|w| < 1\}$. Hence, we obtain the following inequality:

$$\frac{2}{\pi} \int_{|z-w|\leq 1/3} f_2(w) \cdot \Big(-\rho(3(z-w)) \log|z-w| \Big) \leq$$

$$-F_{\widetilde{f}_1}(z) + \frac{2}{\pi} \int_{\mathbb{C}} \widetilde{f}_1(w) \cdot \partial_w \partial_{\overline{w}} \Big(\big(1-\rho(3(z-w))\big) \cdot \log|z-w| \Big). \tag{10.16}$$

Note that $-F_{\widetilde{f}_1}(z) = -f_1(z)$ is bounded as $|z| \to 0$. We also note that $\{|w| < 1/10\} \subset \{|z-w| < 1/3\}$ as $|z| \to 0$. Hence, there exists $C > 0$ such that the following holds for any $|z| < 1/100$:

$$\int_{U'} f_2(w) \cdot \big(-\log|w-z| \big) \leq C.$$

By Fatou's lemma, we obtain $\int_{U'} f_2(w)(-\log|w|) < \infty$. ■

10.2 Family of Vector Bundles on Torus with Small Curvature

10.2.1 Preliminary

Let $T_0 := \mathbb{R}^2/\mathbb{Z}^2$. Let (x_1, x_2) denote the local coordinate systems of T_0 induced by the standard coordinate system of $\mathbb{R}^2$. Set $U := \big\{(\xi_1, \ldots, \xi_{n-2}) \,\big|\, |\xi_i| \leq 1\big\} \subset \mathbb{R}^{n-2}$. For any non-negative integer k, we set

$$S_1(k) := \big\{(m_1, m_2) \in \mathbb{Z}^2_{\geq 0} \,\big|\, m_1 + m_2 = k\big\},$$

$$S_2(k) := \big\{(m_1, \ldots, m_{n-2}) \in \mathbb{Z}^{n-2}_{\geq 0} \,\big|\, \sum m_i = k\big\}.$$

We set $S(k_1, k_2) := S(k_1) \times S(k_2)$. We put $\partial_{\boldsymbol{x}}^{\boldsymbol{m}} := \prod \partial_{x_i}^{m_i}$ and $\partial_{\boldsymbol{\xi}}^{\boldsymbol{m}} := \prod \partial_{\xi_i}^{m_i}$. We put $N(k_i) := |S(k_i)|$ and $N(k_1, k_2) := N(k_1) \cdot N(k_2)$.

In general, for a vector space V and for $f \subset C^{\infty}(T_0 \times U, V)$, we set

$$D_{\boldsymbol{x}}^{k_1} D_{\boldsymbol{\xi}}^{k_2}(f) := \big(\partial_{\boldsymbol{x}}^{\boldsymbol{m}_1} \partial_{\boldsymbol{\xi}}^{\boldsymbol{m}_2} f \mid (\boldsymbol{m}_1, \boldsymbol{m}_2) \in S(k_1, k_2)\big) \in C^{\infty}(T_0 \times U, V^{N(k_1,k_2)}).$$

We formally set $D^0 f := f$.

Let E be a topologically trivial C^{∞} vector bundle $T_0 \times U$ with a Hermitian metric h and a unitary connection ∇. We set $r := \operatorname{rank} E$. Let $F(\nabla)$ denote the curvature of ∇. Fix a positive integer M. Let ϵ denote a small positive number. We suppose the following condition.

Condition 10.2.1 $\left|\nabla_{x_1}^{m_1}\nabla_{x_2}^{m_2}\nabla_{\xi_1}^{\ell_1}\cdots\nabla_{\xi_{n-2}}^{\ell_{n-2}}F(\nabla)\right| \leq \epsilon$ *if* $\sum m_i \leq M$ *and* $\sum \ell_i \leq M$. ■

10.2.2 Partially Almost Holomorphic Frames

Let $\tau \in \mathbb{C}$ with $\mathrm{Im}(\tau) > 0$. We set $T := \mathbb{C}/(\mathbb{Z}+\tau\mathbb{Z})$. Let z denote the standard coordinate of $\mathbb{C}$, which induces local coordinates on T. We identify the C^∞-manifolds $T_0 := \mathbb{R}^2/\mathbb{Z}^2$ and T by the diffeomorphism induced by $(x, y) \longmapsto z = x + \tau y$.

For any frame $\boldsymbol{w}$ of E, let $A_z^{\boldsymbol{w}}$ and $A_{\bar{z}}^{\boldsymbol{w}}$ be the matrix valued functions determined by $\nabla_{\bar{z}}\boldsymbol{w} = \boldsymbol{w}A_{\bar{z}}^{\boldsymbol{w}}$ and $\nabla_z\boldsymbol{w} = \boldsymbol{w}A_z^{\boldsymbol{w}}$. Similarly, let $A_{\xi_i}^{\boldsymbol{w}}$ be determined by $\nabla_{\xi_i}\boldsymbol{w} = \boldsymbol{w}A_{\xi_i}^{\boldsymbol{w}}$, and we set ${}^2A^{\boldsymbol{w}} := \sum A_{\xi_i}^{\boldsymbol{w}}d\xi_i$. Let $H(h, \boldsymbol{w})$ denote the matrix valued function whose (i, j)-th entries are $h(w_i, w_j)$.

Take $p \geq 2$. We fix a norm $\|\cdot\|_{L_k^p(T)}$ on the Banach space of L_k^p-functions on T. For a function f on $T \times U$, let $\|f\|_{L_k^p}$ denote the function $U \longrightarrow \mathbb{R}_{\geq 0}$ determined by $\|f\|_{L_k^p}(\boldsymbol{\xi}) = \|f_{|T\times\{\boldsymbol{\xi}\}}\|_{L_k^p(T)}$. The following lemma is a reword of [65, Proposition 4.7].

Lemma 10.2.2 *If* $\epsilon > 0$ *is sufficiently small, there exist* $\Gamma \in M_r(\mathbb{C})$ *and a frame* $\boldsymbol{u}$ *of* E *on* $T \times U$ *satisfying the following conditions for a positive constant* C_1 *depending only on* n, r, M *and* p.

- $\left|[\Gamma, {}^t\overline{\Gamma}]\right| \leq C_1\epsilon$.
- $A_{\bar{z}}^{\boldsymbol{u}}$ *is constant along the* T*-direction, and* $\left|D_{\boldsymbol{x}}^k(A_{\bar{z}}^{\boldsymbol{u}} - \Gamma)\right| \leq C_1\epsilon$ *for* $0 \leq k \leq M$.
- $\left\|D_{\boldsymbol{\xi}}^k(A_z^{\boldsymbol{u}} + {}^t\overline{\Gamma})\right\|_{L_M^p} \leq C_1\epsilon$ *for any* $0 \leq k \leq M$.
- $\left\|D_{\boldsymbol{\xi}}^k({}^2A^{\boldsymbol{u}})\right\|_{L_M^p} \leq C_1\epsilon$ *for* $0 \leq k \leq M$.

Moreover, $\|D_{\boldsymbol{\xi}}^k(H(h,\boldsymbol{u}) - I_r)\|_{L_{M+1}^p} \leq C_1\epsilon$ *for* $0 \leq k \leq M$, *where* $I_r \in M_r(\mathbb{C})$ *denotes the identity matrix.* ■

10.2.3 Spectra

For each $\boldsymbol{\xi} \in U_2$, we regard $E_{\boldsymbol{\xi}} := E_{|T\times\{\boldsymbol{\xi}\}}$ with $\nabla_{\bar{z}}$ as a holomorphic vector bundle on T. If ϵ is sufficiently small, $E_{\boldsymbol{\xi}}$ are semistable of degree 0 for any $\boldsymbol{\xi} \in U_2$. Let $T^\vee$ denote the dual of T. Let $\mathrm{RFM} : D^b(T) \longrightarrow D^b(T^\vee)$ denote the Fourier-Mukai transform. Because $E_{\boldsymbol{\xi}}$ is semistable of degree 0, $\mathrm{RFM}(E_{\boldsymbol{\xi}})$ is a torsion $\mathcal{O}_{T^\vee}$-module. The support and the length of $\mathrm{RFM}(E_{\boldsymbol{\xi}})$ determines a point in $\mathrm{Sym}^r(T^\vee)$, which is denoted by $[\mathcal{S}p(E_{\boldsymbol{\xi}})]$.

The eigenvalues of Γ determines the point $[\mathcal{S}p(\Gamma)] \in \mathrm{Sym}^r(\mathbb{C})$. The quotient map $\Phi : \mathbb{C} \longrightarrow T^\vee$ induces $\mathrm{Sym}^r\,\mathbb{C} \longrightarrow \mathrm{Sym}^r\,T^\vee$, which is also denoted by Φ.

Let $d_{\mathrm{Sym}^r T^\vee}$ denote the distance on $T^\vee$ induced by a C^∞-Riemannian metric of $\mathrm{Sym}^r T^\vee$. The following is proved in [65, Corollary 4.10].

Corollary 10.2.3 *Let Γ be as in Lemma 10.2.2. There exist $\epsilon_0 > 0$ and $C_0 > 0$ depending only on r such that the following holds if $\epsilon \leq \epsilon_0$:*

$$d_{\mathrm{Sym}^r(T^\vee)}\big([\mathcal{S}p(E_{\xi})], \Phi([\mathcal{S}p(\Gamma)])\big) \leq C_0\epsilon.$$

■

10.2.4 Additional Assumption on Spectra

Let ϵ_0 be as in Corollary 10.2.3. We assume that $\epsilon < \epsilon_0$. Moreover, we assume the following.

Condition 10.2.4 *There exist $0 < \rho$, $0 < \delta < 1/10$, $\zeta_0 \in \mathbb{C}$, and a finite subset*

$$Z \subset \left\{\zeta \in \mathbb{C} \,\middle|\, 0 \leq \mathrm{Im}(\zeta - \zeta_0) \leq (1-\delta)\pi,\ 0 \leq \mathrm{Im}\left(\overline{\tau}(\zeta - \zeta_0)\right) \leq (1-\delta)\pi\right\}$$

such that the following holds.

- *For any distinct points $\nu_1, \nu_2 \in Z$, we have $d_{\mathbb{C}}(\nu_1, \nu_2) > 100r^2\rho$, where $d_{\mathbb{C}}$ denotes the Euclidean distance on $\mathbb{C}$.*
- *For any $\kappa \in \mathcal{S}p(E_{\xi})$, there exists $\nu \in Z$ such that $d_{T^\vee}(\Phi(\nu), \kappa) < \rho$, where $d_{T^\vee}$ denotes the natural distance on $T^\vee$.*

We also assume that ϵ is sufficiently smaller than ρ^2. ■

We obtain the C^∞-decomposition $E = \bigoplus_{\nu \in Z} E_\nu$ such that (i) $\nabla_{\overline{z}}$ preserves the decomposition, (ii) any $\kappa \in \mathcal{S}p(E_{\nu|T\times\{\xi\}})$ satisfies $d_{T^\vee}(\Phi(\nu), \kappa) < \rho$. The following lemma is implicitly used in [65, §4.3, §5.2].

Lemma 10.2.5 *If ϵ is sufficiently small, there exist $\Gamma \in M_r(\mathbb{C})$ and a frame $\boldsymbol{u}$ of E such that the following holds.*

- *Γ and $\boldsymbol{u}$ satisfy the conditions in Lemma 10.2.2.*
- *There exists a decomposition $\boldsymbol{u} = \bigcup_{\nu \in Z} \boldsymbol{u}_\nu$ such that $\boldsymbol{u}_\nu = (u_{\nu,1}, \ldots, u_{\nu,\mathrm{rank}\, E_\nu})$ are frames of the subbundles E_ν.*
- *We may impose the following for any $\boldsymbol{\xi} \in U_2$:*

$$\frac{1}{|T|}\int_{T\times\{\xi\}} h(u_{\nu,i}, u_{\nu,j}) = \begin{cases} 1 \ (i = j), \\ 0 \ (\text{otherwise}). \end{cases} \tag{10.17}$$

Proof In the proof of this lemma, C_i denote positive constants depending only on n, r, M, p and ρ. We take Γ and $\boldsymbol{u}$ as in Lemma 10.2.2. There exists a unitary matrix $B^{(0)}$ such that the following holds.

- $\Gamma' := (B^{(0)})^{-1}\Gamma B^{(0)}$ is an upper triangular matrix.
- There exist a total ordering $\{\nu_1, \dots, \nu_m\}$ on Z and numbers r_i $(i = 1, \dots, m)$ with $\sum r_i = r$ such that (j, j)-th entries of Γ' are contained in the 2ρ-ball centered at ν_i, where $\sum_{k \le i-1} r_k + 1 \le j \le \sum_{k \le i} r_k$.

Set $\Lambda_i := \{j \in \mathbb{Z} \mid \sum_{k \le i-1} r_k + 1 \le j \le \sum_{k \le i} r_k\}$. If $j_1 \in \Lambda_1$, $j_2 \in \Lambda_2$ with $\Lambda_1 \neq \Lambda_2$, the (j_1, j_2)-entries of Γ' is smaller than $C_3\epsilon$ because $\big|[\Gamma, {}^t\overline{\Gamma}]\big| \le C_1\epsilon$. Let Γ'_i denote the r_i-square matrix given by the (j_1, j_2)-entries $(j_1, j_2 \in \Lambda_i)$ of Γ'. Then, there exists an upper triangular matrix $B^{(1)} \in M_r(\mathbb{C})$ such that the following holds.

- $B^{(1)}_{j_1, j_2} = 0$ if j_1 and j_2 are contained in the same Λ_i.
- $|B^{(1)}| \le C_4\epsilon$.
- $(I_r + B^{(1)})^{-1}\Gamma'(I_r + B^{(1)}) = \bigoplus \Gamma'_i$.

Then, $\bigoplus \Gamma'_i$ and $\boldsymbol{u}' := \boldsymbol{u} \cdot B^{(0)}(I + B^{(1)})$ satisfy the conditions in Lemma 10.2.2, and $\boldsymbol{u}'$ has a decomposition with the desired property. Hence, we may assume the existence of the decomposition $\boldsymbol{u} = \bigcup \boldsymbol{u}_\nu$ from the beginning.

Set $r_\nu := \operatorname{rank} E_\nu$. Let $H(h, \boldsymbol{u}_\nu)$ be the r_ν-th square Hermitian matrices valued function determined by

$$H(h, \boldsymbol{u}_\nu)_{k,\ell} := h(u_{\nu,k}, u_{\nu,\ell}).$$

We obtain $\big\| D^k_\xi\big(H(h, \boldsymbol{u}_\nu)_{k,\ell} - I_{r_\nu}\big)\big\|_{L^k_p} \le C_5\epsilon$ for $0 \le k \le M$, where $I_{r_\nu} \in M_{r_\nu}(\mathbb{C})$ denotes the identity matrix.

Let $H_{1,\nu}$ be the function from U to the space of r_ν-th positive definite Hermitian matrices determined by $\big(\overline{H_{1,\nu}}\big)^{-2} = \int_T H(h, \boldsymbol{u}_\nu)$. Then, we have $\big\| D^k_\xi\big(H_{1,\nu} - I_{r_\nu}\big)\big\|_{L^p_k} \le C_6\epsilon$ for $0 \le k \le M$. Then, Γ and $\bigcup_{\nu \in Z}\big(\boldsymbol{u}_\nu \cdot H_{1,\nu}\big)$ satisfies (10.17). ■

There exists the decomposition $\nabla_{\overline{z}} = \nabla_{\overline{z},0} + f$ such that (i) $(E_\xi, \nabla_{\overline{z},0})$ is holomorphically isomorphic to the product vector bundle $\mathbb{C}^r \times T$ on T for any $\xi \in U$, (ii) $\nabla_{\overline{z},0} f = 0$, (iii) $\mathcal{S}p(f_{|T \times \xi})$ is contained in the ρ-ball of Z. Let $\mathcal{V}_\xi$ denote the space of holomorphic sections of $(E_\xi, \nabla_{\overline{z},0})$. Then, $\mathcal{V}_\xi$ $(\xi \in U)$ naturally give a C^∞-bundle $\mathcal{V}$ on U, and $\boldsymbol{u}$ induces a frame of $\mathcal{V}$. Let $p : T \times U \longrightarrow U$ denote the projection. Then, there exists a naturally defined C^∞-isomorphism $E \simeq p^*(\mathcal{V})$ on $T \times U$. We obtain a decomposition $\mathcal{V} = \bigoplus \mathcal{V}_\nu$, which is compatible with the above isomorphism.

10.2.5 Spaces of Functions

Let $C^M_\xi L^p_{M,x}$ denote the space of C^M-functions $U \longrightarrow L^p_M(T)$. Let $C^M_\xi L^p_{M,x}(E)$ denote the space of sections $f = \sum f_i u_i$ of E such that $f_i \in C^M_\xi L^p_{M,x}$, where $\boldsymbol{u}$ is as in Lemma 10.2.5. The space is independent of the choice of $\boldsymbol{u}$. Because $E = p^*\mathcal{V}$, we naturally regard $C^M(U, \mathcal{V})$ as a subspace of $C^M(T \times U, E)$. There also exists the naturally defined morphism $C^M_\xi L^p_{M,x}(E) \longrightarrow C^M(U, \mathcal{V})$ induced by the integral along the fibers. The kernel is denoted by $C^M_\xi L^p_{M,x}(E)_0$. Similar spaces are defined for $\mathrm{End}(E_\nu)$ and $\mathrm{Hom}(E_\nu, E_\mu)$. We set

$$C^M_\xi L^p_{M,x}(\mathrm{End}(E))^\circ := \bigoplus_{\nu \in Z} C^M(U, \mathrm{End}(\mathcal{V}_\nu)), \tag{10.18}$$

$$C^M_\xi L^p_{M,x}(\mathrm{End}(E))^\perp := \bigoplus_{\nu} C^M_\xi L^p_{M,x}(\mathrm{End}(E_\nu))_0 \oplus \bigoplus_{\nu \neq \mu} C^M_\xi L^p_{M,x}(\mathrm{Hom}(E_\nu, E_\mu)). \tag{10.19}$$

We obtain the decomposition

$$C^M_\xi L^p_{M,x}(\mathrm{End}(E)) = C^M_\xi L^p_{M,x}(\mathrm{End}(E))^\circ \oplus C^M_\xi L^p_{M,x}(\mathrm{End}(E))^\perp.$$

For any $s \in C^M_\xi L^p_{M,x}(\mathrm{End}(E))$, we obtain the corresponding decomposition $s = s^\circ + s^\perp$.

10.2.6 Some Estimates

Let $\widetilde{g} := \overline{H(h, \boldsymbol{u})}$. We obtain $\left\| D^k_\xi(\widetilde{g} - I_r) \right\|_{L^p_k} \leq C\epsilon$ for $0 \leq k \leq M$, where C is a positive constant depending only on n, r, M, p and ρ. Let $F_{z\bar{z}}$ denote the $dz\,d\bar{z}$-component of $F(\nabla)$. We obtain the decomposition $F_{z\bar{z}} = F^\circ_{z\bar{z}} + F^\perp_{z\bar{z}}$ as in Sect. 10.2.5. The following lemma is proved by the argument in the proof of [65, Lemma 4.12, Lemma 4.15].

Lemma 10.2.6 *There exist $C_{10} > 0$ and $\epsilon_{10} > 0$, depending only on r, M, n, p and ρ, such that if $\epsilon < \epsilon_{10}$, we obtain*

$$\left\| D^k_\xi(\widetilde{g} - I) \right\|_{L^p_{M+2}} \leq C_{10} \sum_{j=0}^{k} \left\| D^j_\xi F^\perp_{z\bar{z}} \right\|_{L^p_M}.$$

In particular, we have $\sup|\widetilde{g}-I| \leq C_{11}\|F^{\perp}_{z\overline{z}}\|_{L^2}$ *for a positive constant* C_{11} *depending only on n, r and* ρ. ■

Remark 10.2.7 The frame $\boldsymbol{u}$ in [65, §4.3] should be chosen as in Lemma 10.2.5. ■

10.3 Estimates for Asymptotic Doubly Periodic Instantons

10.3.1 Setting

Let L be a lattice in $\mathbb{C}$. Let $T := \mathbb{C}/L$. The standard coordinate z of $\mathbb{C}$ induces local coordinates on T. For any $R > 0$, let $U^*_w(R) := \{w \in \mathbb{C} \,\big|\, |w| \geq R\}$ and $U_w(R) := U^*_w(R) \cup \{\infty\}$ in $\mathbb{P}^1_w$. Take $q \in \mathbb{Z}_{\geq 1}$. Let $\mathbb{P}^1_{w_q} \longrightarrow \mathbb{P}^1_w$ be the ramified covering given by $w_q \longmapsto w_q^q$. Let $U_{w,q}(R)$ and $U^*_{w,q}(R)$ denote the pull back of $U_w(R)$ and $U^*_w(R)$, respectively. We use the metric $dz\,d\overline{z} + dw\,d\overline{w} = dz\,d\overline{z} + q^2|w_q|^{2(q-1)}dw_q\,d\overline{w}_w$ on $T \times U^*_{w,q}(R)$.

Take $R_0 > 0$. Let $(E, \overline{\partial}_E)$ be a holomorphic vector bundle on $T \times U^*_{w,q}(R_0)$ with a Hermitian metric h. Let $F(h)$ denote the curvature of the Chern connection ∇_h determined by h and $\overline{\partial}_E$. For any $\mathrm{End}(E)$-valued form s, let $|s|_h$ denote the function obtained as the norm of s with respect to h and the Euclidean metric.

Condition 10.3.1 *Let $\mathcal{A}$ denote the set of polynomials of non-commutative variables t_1, t_2, t_3, t_4. For any $P \in \mathcal{A}$, there exist $C(P) > 0$ and $\epsilon(P) > 0$ such that*

$$\Big|P(\nabla_z, \nabla_{\overline{z}}, \nabla_w, \nabla_{\overline{w}})\big(\Lambda F(h)\big)\Big|_h \leq C(P)\exp\big(-\epsilon(P) I_q(w_q)\big). \tag{10.20}$$

Here, $I_q(w_q) := |w_q^q|(\log|w_q|)^{1/2}$. Namely, $(E, \overline{\partial}_E, h)$ asymptotically satisfies the condition of instantons. We also assume $|F(h)|_h \to 0$ as $|w_q| \to \infty$. ■

The following lemma is standard due to [87, 88].

Lemma 10.3.2 *For any $P \in \mathcal{A}$, we obtain $\big|P(\nabla_z, \nabla_{\overline{z}}, \nabla_w, \nabla_{\overline{w}})F(h)\big|_h \to 0$ as $|w_q| \to \infty$.*

Proof For $Q \in T \times U^*_{w,q}$ and $\rho > 0$, we set

$$B_Q(\rho) = \big\{Q' \in T \times U^*_{w,q} \,\big|\, d_{T\times U^*_{w,q}}(Q, Q') < \rho\big\},$$

where $d_{T\times U^*_{w,q}}$ denotes the distance induced by $dz\,d\overline{z} + dw\,d\overline{w}$. Let $w(Q)$ denote the image of Q via the natural map $T \times U^*_{w,q} \longrightarrow U^*_w$. Let p be any large real number. For any unitary frame $\boldsymbol{u}$ of $E_{|B_Q(2)}$, let $A^{\boldsymbol{u}}$ denote the Chern connection form with respect to $\boldsymbol{u}$. Similarly, let $F^{\boldsymbol{u}}$ denote the curvature form representing $F(h)$ with respect to $\boldsymbol{u}$. We have the elliptic system of partial differential equations

$$\Lambda(dA^{\boldsymbol{u}} + A^{\boldsymbol{u}} \wedge A^{\boldsymbol{u}}) = \Lambda F^{\boldsymbol{u}}, \quad d^*A^{\boldsymbol{u}} = 0.$$

For any $\delta > 0$ and $\ell \in \mathbb{Z}_{\geq 0}$, there exists $R(\delta, \ell) > R$ such that the following holds $(\ell_1, \ell_2, \ell_3, \ell_4) \in \mathbb{Z}^4_{\geq 0}$ with $\sum \ell_i \leq \ell$ if $|w(Q)| > R(\delta, \ell)$:

$$\left|\nabla_z^{\ell_1} \nabla_{\overline{z}}^{\ell_2} \nabla_w^{\ell_3} \nabla_{\overline{w}}^{\ell_4} (\Lambda F^{\boldsymbol{u}})\right| \leq \delta. \tag{10.21}$$

We also assume the following if $|w(Q)| > R(\delta, \ell)$.

$$|F^{\boldsymbol{u}}| \leq \delta. \tag{10.22}$$

If δ is sufficiently small, and if $|w(Q)| > R(\delta, \ell)$ which implies (10.22), according to [87, 88], there exists a unitary frame $\boldsymbol{u}$ of $E_{|B_Q(2)}$ such that (i) $d^* A^{\boldsymbol{u}} = 0$, (ii) $\|A^{\boldsymbol{u}}_{|B_Q(2)}\|_{L_1^p} \leq C_1 \|F(h)_{|B_Q(2)}\|_{L^p}$ for a positive constant C_1 which is independent of Q. Let $1 < r_1 < r_2 < 2$. Let $\kappa : \mathbb{R}_{\geq 0} \longrightarrow [0, 1]$ be a C^∞-function such that (i) $\kappa(u) = 1$ $(u \leq r_1)$, (ii) $\kappa(u) = 0$ $(u \geq r_2)$, (iii) κ^a is C^∞ for any $a > 0$. We set $\chi(Q') := \kappa(d(Q, Q'))$ and $A^{\boldsymbol{u}}_a := \chi^a A^{\boldsymbol{u}}_{|B_Q(2)}$. There exists $C_1'(a) > 0$, which is independent of Q, such that $\|A^{\boldsymbol{u}}_a\|_{L_1^p} \leq C_1'(a)\delta$. We have

$$\Lambda d(A^{\boldsymbol{u}}_a) = \Lambda\Big(\big(\chi^{-a/2} d(\chi^a)\big) A^{\boldsymbol{u}}_{a/2} - A^{\boldsymbol{u}}_{a/2} \wedge A^{\boldsymbol{u}}_{a/2}\Big) + \chi^a \Lambda F^{\boldsymbol{u}}, \tag{10.23}$$

$$d^*(A^{\boldsymbol{u}}_a) = \mathcal{P}\big(\chi^{-a/2} d\chi^a, A^{\boldsymbol{u}}_{a/2}\big), \tag{10.24}$$

where $\mathcal{P}(\chi^{-a/2} d\chi^a, A^{\boldsymbol{u}}_{a/2})$ is bilinear with respect to $\chi^{-a/2} d\chi^a$ and $A^{\boldsymbol{u}}_{a/2}$. We apply [1, Theorem 17.2] to the system (10.23, 10.24). Then, there exists $C_2(a) > 0$, which is independent of Q, such that $|A^{\boldsymbol{u}}_a|_{L_2^p} \leq C_2(a)\delta$. By an induction, we can prove that there exist $C_j(a) > 0$ $(j = 2, \ldots, \ell + 1)$, which are independent of Q, such that $|A^{\boldsymbol{u}}_a|_{L_j^p} \leq C_j(a)\delta$. Thus, we obtain Lemma 10.3.2. ■

By replacing R_0 with a larger number, we may assume that $(E, \overline{\partial}_E)_{|T \times \{w_q\}}$ are semistable of degree 0 for any $w_q \in U^*_{w,q}(R_0)$ from the beginning. We obtain the spectral curve $\Sigma_E \subset T^\vee \times U^*_{w,q}(R_0)$. We assume the following.

Condition 10.3.3 *The closure $\overline{\Sigma}_E$ of Σ_E in $T^\vee \times U_{w,q}(R_0)$ is a complex analytic curve. We set $\mathcal{S}p_\infty(E) := \overline{\Sigma}_E \cap \big(T^\vee \times \{\infty\}\big)$.* ■

10.3.2 Decomposition

We fix $\widetilde{\mathcal{S}p}_\infty(E) \subset \mathbb{C}$ such that the projection $\mathbb{C} \longrightarrow T^\vee$ induces a bijection $\widetilde{\mathcal{S}p}_\infty(E) \simeq \mathcal{S}p_\infty(E)$. Let $\Psi : T \times U^*_{w,q}(R_0) \longrightarrow U^*_{w,q}(R_0)$ denote the projection. We obtain the following lemma (see [65, §2.1]).

Lemma 10.3.4 *There exists a holomorphic vector bundle $\mathcal{V}$ with an endomorphism f on $U^*_{w,q}(R_0)$ such that the following holds.*

- *For any $\rho > 0$, there exists $R_1 \geq R_0$ such that if $|w_q^q| \geq R_1$ and if α is an eigenvalue of $f_{|w_q}$, then there exists $\beta \in \widetilde{Sp}_\infty(E)$ such that $|\beta - \alpha| \leq \rho$.*
- *Let $(\Psi^{-1}(\mathcal{V}), \overline{\partial}_{\Psi^{-1}(\mathcal{V})})$ denote the holomorphic vector bundle on $T \times U^*_{w,q}(R_0)$ obtained as the pull back of $\mathcal{V}$. As a twist, we obtain the holomorphic vector bundle $\Psi^*(\mathcal{V}, f) = \big(\Psi^{-1}(\mathcal{V}), \overline{\partial}_{\Psi^{-1}(\mathcal{V})} + f\,d\overline{z}\big)$. Then, there exists an isomorphism $\Psi^*(\mathcal{V}, f) \simeq E$.* ■

There exists the decomposition $(\mathcal{V}, f) = \bigoplus_{\beta \in \widetilde{Sp}_\infty(E)} (\mathcal{V}_\beta, f_\beta)$, where the eigenvalues of $f_{\beta|w_q}$ are convergent to β as $|w_q| \to \infty$. Under the identification $E = \Psi^*(\mathcal{V}, f)$, we obtain the decomposition $E = \bigoplus_{\alpha \in \widetilde{Sp}_\infty(E)} \Psi^*(\mathcal{V}_\alpha, f_\alpha)$.

10.3.3 Estimates

Let h_α be the restriction of h to $\Psi^*(\mathcal{V}_\alpha, f_\alpha)$. We obtain the Fourier expansion of h_α along the T-direction. Let $h_{\mathcal{V},\alpha}$ be the invariant part of h_α. In other words, for any sections u_1, u_2 of $\mathcal{V}_\alpha$, we set

$$h_{\mathcal{V},\alpha}(u_1, u_2) := |T|^{-1} \int_T h\big(\Psi^{-1}(u_1), \Psi^{-1}(u_2)\big).$$

We set $h^\circ := \bigoplus_{\alpha \in \widetilde{Sp}_\infty(E)} \Psi^{-1}(h_{\mathcal{V},\alpha})$. Let b be the automorphism of E determined by $h = h^\circ \cdot b$. The following proposition is essentially proved in [65, §5].

Proposition 10.3.5 *For any $P \in \mathcal{A}$, there exist $C(P) > 0$ and $\epsilon(P) > 0$ such that*

$$P(\nabla_z, \nabla_{\overline{z}}, \nabla_w, \nabla_{\overline{w}})(b - \mathrm{id}) \leq C(P) \exp\big(-\epsilon(P)|w|\big).$$

Proof We shall explain an outline of the proof. We set $E_\alpha := \Psi^*(\mathcal{V}_\alpha, f_\alpha)$. We obtain the decomposition $\mathrm{End}(E) = \bigoplus \mathrm{Hom}(E_\alpha, E_\beta)$.

Let U be any open subset in $U^*_w(R_0)$. We have the map $\int_T : C^\infty(T \times U, \mathrm{End}(E_\alpha)) \longrightarrow C^\infty(U, \mathrm{End}(\mathcal{V}_\alpha))$ induced by the integration along the fibers. Let $C^\infty(T \times U, \mathrm{End}(E_\alpha))_0$ denote the kernel of $\int_T$. We also have the natural inclusion $C^\infty(U, \mathrm{End}(\mathcal{V}_\alpha)) \longrightarrow C^\infty(T \times U, \mathrm{End}(E_\alpha))$ induced by the pull back. Thus, we obtain the decomposition $C^\infty(T \times U, \mathrm{End}(E_\alpha)) = C^\infty(U, \mathrm{End}(\mathcal{V}_\alpha)) \oplus C^\infty(T \times U, \mathrm{End}(E_\alpha))_0$. We set

$$C^\infty(T \times U, \mathrm{End}(E))^\circ := \bigoplus_{\alpha \in \widetilde{Sp}_\infty(E)} C^\infty(U, \mathrm{End}(\mathcal{V}_\alpha)),$$

$$C^\infty(T \times U, \mathrm{End}(E))^\perp := \bigoplus_{\alpha \in \widetilde{Sp}_\infty(E)} C^\infty(T \times U, \mathrm{End}(E_\alpha))_0 \oplus \bigoplus_{\alpha \neq \beta} C^\infty(T \times U, \mathrm{Hom}(E_\alpha, E_\beta)).$$

We obtain the decomposition $C^\infty(T \times U, \mathrm{End}(E)) = C^\infty(T \times U, \mathrm{End}(E))^\circ \oplus C^\infty(T \times U, \mathrm{End}(E))^\perp$. Any section $s \in C^\infty(T \times U, \mathrm{End}(E))$ has the corresponding decomposition $s = s^\circ + s^\perp$. We also obtain the function $\|s\|$ on U determined by $\|s\|(w_q)^2 = |T|^{-1} \int_{T \times \{w_q\}} |s|^2$. Note that there exists a constant $C > 0$ such that the following holds if $s = s^\perp$:

$$\|\nabla_{\overline{z}} s\| \geq C\|s\|. \tag{10.25}$$

Let us consider the case $U = \{w_q \mid |w_q^q - w_{q,0}^q| \leq 1\}$ for some $w_{q,0} \in U_{w,q}$. We have the expression $F(h) = F_{z\overline{z}} dz\, d\overline{z} + F_{z\overline{w}} dz\, d\overline{w} + F_{w\overline{z}} dw\, d\overline{z} + F_{w\overline{w}} dw\, d\overline{w}$. For any $\epsilon > 0$, if $|w_{q,0}|$ is sufficiently large, we obtain $|F(h)| \leq \epsilon$ on $T \times U$. If ϵ is sufficiently small, there exists a frame $\boldsymbol{u}$ as in Lemma 10.2.5. Note that $\boldsymbol{u}$ is a unitary frame with respect to the metric h°, and that $\widetilde{g}$ in Sect. 10.2.6 corresponds to the restriction of b to U. Then, we apply the argument for local estimates explained in [65, §5.2] by replacing the condition $\Lambda F(h) = 0$ with (10.20). We use the notation $g_1 = O(g_2)$ if we have a positive constant C, which is independent of $w_{q,0}$, such that $g_1 \leq C g_2$. In (10.26–10.29) below, C_1 will denote a positive constant which is independent of $w_{q,0}$. We obtain the following by the argument in the proof of [65, Proposition 5.5] and the condition (10.20):

$$\begin{aligned} &-\partial_w \partial_{\overline{w}} \|F_{z\overline{z}}^\perp\|^2 \leq -\|\nabla_{\overline{z}} F_{z\overline{z}}^\perp\|^2 - \|\nabla_{\overline{z}} F_{z\overline{z}}^\perp\|^2 - \|\nabla_{\overline{w}} F_{z\overline{z}}^\perp\|^2 - \|\nabla_{\overline{w}} F_{z\overline{z}}^\perp\|^2 \\ &\quad + O\Big(\epsilon \|F_{z\overline{z}}^\perp\|^2 + \epsilon \|F_{z\overline{z}}^\perp\| \|F_{w\overline{z}}^\perp\| + \epsilon \|\nabla_{\overline{w}} F_{w\overline{z}}^\perp\| \|F_{z\overline{z}}\| + \epsilon \|\nabla_{\overline{z}} F_{z\overline{z}}^\perp\| \|F_{z\overline{z}}^\perp\|\Big) \\ &+ O\Big(\epsilon \|\nabla_{\overline{w}} F_{z\overline{z}}^\perp\| \|F_{z\overline{z}}^\perp\| + \epsilon \|F_{w\overline{z}}^\perp\|^2 + \epsilon \|F_{w\overline{z}}^\perp\| \|\nabla_{\overline{w}} F_{z\overline{z}}^\perp\|\Big) + O\Big(\exp\big(-C_1 I_q(w_{q,0})\big)\Big). \end{aligned} \tag{10.26}$$

We obtain the following by using the condition (10.20) and the argument in the proof of [65, Proposition 5.6]:

$$\begin{aligned} &-\partial_w \partial_{\overline{w}} \|F_{z\overline{w}}^\perp\|^2 \leq -\|\nabla_{\overline{z}} F_{z\overline{w}}^\perp\|^2 - \|\nabla_z F_{z\overline{w}}^\perp\|^2 - \|\nabla_{\overline{w}} F_{z\overline{w}}^\perp\|^2 - \|\nabla_w F_{z\overline{w}}^\perp\|^2 \\ &+ O\Big(\epsilon \|F_{z\overline{w}}^\perp\|^2 \|F_{z\overline{z}}^\perp\| + \epsilon \|\nabla_{\overline{w}} F_{w\overline{z}}^\perp\| \|F_{z\overline{w}}^\perp\| + \epsilon \|F_{z\overline{w}}^\perp\| \|F_{w\overline{z}}^\perp\| + \epsilon \|\nabla_{\overline{w}} F_{z\overline{z}}^\perp\| \|F_{z\overline{w}}^\perp\|\Big) \\ &\quad + O\Big(\epsilon \|\nabla_{\overline{z}} F_{z\overline{w}}^\perp\| \|F_{z\overline{z}}^\perp\| + \epsilon \|F_{z\overline{w}}^\perp\|^2\Big) + O\Big(\exp(-C_1 I_q(w_{q,0}))\Big). \end{aligned} \tag{10.27}$$

By a similar argument, we also have the following:

$$\begin{aligned} &-\partial_w \partial_{\overline{w}} \|F_{w\overline{z}}^\perp\|^2 \leq -\|\nabla_z F_{w\overline{z}}^\perp\|^2 - \|\nabla_{\overline{z}} F_{w\overline{z}}^\perp\|^2 - \|\nabla_w F_{w\overline{z}}^\perp\|^2 - \|\nabla_{\overline{w}} F_{w\overline{z}}^\perp\|^2 \\ &+ O\Big(\epsilon \|F_{w\overline{z}}^\perp\| \|F_{z\overline{z}}^\perp\| + \epsilon \|\nabla_{\overline{w}} F_{w\overline{z}}^\perp\| \|F_{w\overline{z}}^\perp\| + \epsilon \|F_{w\overline{z}}^\perp\|^2 + \epsilon \|\nabla_{\overline{w}} F_{z\overline{z}}^\perp\| \|F_{w\overline{z}}^\perp\|\Big) \\ &\quad + O\Big(\epsilon \|\nabla_{\overline{z}} F_{w\overline{z}}^\perp\| \|F_{z\overline{z}}\| + \epsilon \|F_{w\overline{z}}^\perp\| \|F_{w\overline{w}}\|\Big) + O\Big(\exp\big(-C_1 I_q(w_{q,0})\big)\Big). \end{aligned} \tag{10.28}$$

By the condition (10.20), we have

$$-\partial_w\partial_{\overline{w}}\|F^{\perp}_{w\overline{w}}\|^2 = -\partial_w\partial_{\overline{w}}\|F^{\perp}_{z\overline{z}}\|^2 + O\Big(\exp\big(-C_1 I_q(w_{q,0})\big)\Big). \tag{10.29}$$

From local estimates (10.26–10.29) with (10.25), we obtain the following inequality on $U^*_w(R_2)$ for some $R_2 > R_0$:

$$-\partial_w\partial_{\overline{w}}\Big(\|F^{\perp}_{z\overline{z}}\|^2 + \|F^{\perp}_{w\overline{w}}\|^2 + \|F^{\perp}_{z\overline{w}}\|^2 + \|F^{\perp}_{w\overline{z}}\|^2\Big) \leq \\ -C_2\Big(\|F^{\perp}_{z\overline{z}}\|^2 + \|F^{\perp}_{w\overline{w}}\|^2 + \|F^{\perp}_{z\overline{w}}\|^2 + \|F^{\perp}_{w\overline{z}}\|^2\Big) + C_3\exp(-C_1 I_q(w_q)). \tag{10.30}$$

Here, C_i $(i = 2, 3)$ are positive constants. By a standard argument of Ahlfors lemma (see [65, Lemma 5.2]), there exists $C_4 > 0$ such that

$$\|F^{\perp}_{z\overline{z}}\|^2 + \|F^{\perp}_{w\overline{w}}\|^2 + \|F^{\perp}_{z\overline{w}}\|^2 + \|F^{\perp}_{w\overline{z}}\|^2 = O\Big(\exp\big(-C_4|w_q^q|\big)\Big).$$

Let $F(h)^{\perp} := F^{\perp}_{z\overline{z}}dz\,d\overline{z} + F^{\perp}_{z\overline{w}}dz\,d\overline{w} + F^{\perp}_{w\overline{z}}dw\,d\overline{z} + F^{\perp}_{w\overline{w}}dw\,d\overline{w}$. By using a standard boot-strapping argument as in the proof of [65, Proposition 5.8], we obtain the following.

- For any $P \in \mathcal{A}$, there exist $C(P) > 0$ and $\epsilon(P) > 0$ such that

$$\big|P(\nabla_z, \nabla_{\overline{z}}, \nabla_w, \nabla_{\overline{w}})F(h)^{\perp}\big| \leq C(P)\exp\big(-\epsilon(P)|w|\big).$$

By Lemma 10.2.6, we obtain the desired estimate for $b - \mathrm{id}$. ∎

Corollary 10.3.6 *The Higgs bundles $(\mathcal{V}_\alpha, f_\alpha dw)$ with the metric h_α satisfy the condition in* Sect. 10.1.1. ∎

References

1. S. Agmon, A. Douglis, L Nirenberg, Estimates near the boundary for solutions of elliptic partial differential equations satisfying general boundary conditions. II. Commun. Pure Appl. Math. **17**, 35–92 (1964)
2. L.V. Ahlfors, An extension of Schwarz's lemma. Trans. Am. Math. Soc. **43**, 359–364 (1938)
3. M.F. Atiyah, Magnetic monopoles in hyperbolic spaces, in *Vector Bundles on Algebraic Varieties (Bombay, 1984)*, vol. 11. Tata Institute of Fundamental Research Studies in Mathematics (1987), pp. 1–33
4. M.F. Atiyah, N. Hitchin, *The Geometry and Dynamics of Magnetic Monopoles*. M. B. Porter Lectures (Princeton University Press, Princeton, 1988)
5. T. Aubin, *Nonlinear Analysis on Manifolds. Monge-AmpÃĺre Equations*. Grundlehren der Mathematischen Wissenschaften, vol. 252 (Springer, New York, 1982)
6. S. Axler, P. Bourdon, W. Ramey, *Harmonic Function Theory*, 2nd edn. Graduate Texts in Mathematics, vol. 137 (Springer, New York, 2001), xii+259
7. A. Beauville, M.S. Narasimhan, S. Ramanan, Spectral curves and the generalised theta divisor. J. Reine Angew. Math. **398**, 169–179 (1989)
8. O. Biquard, *Fibrés de Higgs et connexions intégrables: le cas logarithmique (diviseur lisse)*. Ann. Sci. École Norm. Sup. **30**, 41–96 (1997)
9. O. Biquard, P. Boalch, Wild non-abelian Hodge theory on curves. Compos. Math. **140**, 179–204 (2004)
10. O. Biquard, M. Jardim, Asymptotic behaviour and the moduli space of doubly-periodic instantons. J. Eur. Math. Soc. **3**, 335–375 (2001)
11. B. Charbonneau, J. Hurtubise, Singular Hermitian-Einstein monopoles on the product of a circle and a Riemann surface. Int. Math. Res. Not. 175–216 (2011)
12. G. Chen, A. Fahim, Formal reduction of linear difference systems. Pacific J. Math. **182**, 37–54 (1998)
13. S.A. Cherkis, A. Kapustin, Nahm transform for periodic monopoles and $\mathcal{N} = 2$ super Yang-Mills theory. Comm. Math. Phys. **218**, 333–371 (2001)
14. S.A. Cherkis, A. Kapustin, Periodic monopoles with singularities and $\mathcal{N} = 2$ super-QCD. Comm. Math. Phys. **234**, 1–35 (2003)
15. K. Corlette, Flat *G*-bundles with canonical metrics. J. Differ. Geom. **28**, 361–382 (1988)
16. M. Cornalba, P. Griffiths, Analytic cycles and vector bundles on noncompact algebraic varieties. Invent. Math. **28**, 1–106 (1975)
17. K. Diederich, T. Ohsawa, Harmonic mappings and disc bundles over compact KÃd'hler manifolds. Publ. Res. Inst. Math. Sci. **2**(1), 819–833 (1985)

T. Mochizuki, *Periodic Monopoles and Difference Modules*, Lecture Notes in Mathematics 2300, https://doi.org/10.1007/978-3-030-94500-8

18. S.K. Donaldson, Instantons and geometric invariant theory. Comm. Math. Phys. **93**, 453–460 (1984)
19. S.K. Donaldson, Nahm's equations and the classification of monopoles. Comm. Math. Phys. **96**, 387–407 (1984)
20. S.K. Donaldson, Anti self-dual Yang-Mills connections over complex algebraic surfaces and stable vector bundles. Proc. London Math. Soc. (3) **50**, 1–26 (1985)
21. S.K. Donaldson, Infinite determinants, stable bundles and curvature. Duke Math. J. **54**, 231–247 (1987)
22. S.K. Donaldson, Twisted harmonic maps and the self-duality equations. Proc. London Math. Soc. **55**, 127–131 (1987)
23. A. Duval, Lemmes de Hensel et factorisation formelle pour les opérateurs aux différences. Funkcial. Ekvac. **26**, 349–368 (1983)
24. C. Elliott, V. Pestun, Multiplicative Hitchin systems and supersymmetric gauge theory. Selecta Math. (N.S.) **25**, 82 (2019). Paper No. 64
25. L. Foscolo, Deformation theory of periodic monopoles (with singularities). Comm. Math. Phys. **341**, 351–390 (2016)
26. L. Foscolo, A gluing construction for periodic monopoles. Int. Math. Res. Not. **2017**(24), 7504–7550 (2017). arXiv:1411.6951
27. R. García López, On the Mellin transform of a D-module. arXiv:1804.09776
28. D. Gilbarg, N.S. Trudinger, *Elliptic Partial Differential Equations of Second Order*. Reprint of the 1998 edition. Classics in Mathematics (Springer, Berlin, 2001), xiv+517 pp.
29. A. Graham-Squire, Calculation of local formal Mellin transforms. Pacific J. Math. **283**, 115–137 (2016)
30. D. Harland, Parabolic Higgs bundles and cyclic monopole chains. arXiv:2012.01083
31. D. Harland, R.S. Ward, Dynamics of periodic monopoles. Phys. Lett. B **675**, 262–266 (2009)
32. N. Hitchin, Monopoles and geodesics. Comm. Math. Phys. **83**, 579–602 (1982)
33. N. Hitchin, Construction of monopoles. Comm. Math. Phys. **89**, 145–190 (1983)
34. N. Hitchin, The self-duality equations on a Riemann surface. Proc. Lond. Math. Soc. (3) **55**, 59–126 (1987)
35. N. Hitchin, *A Note on Vanishing Theorems*. Geometry and Analysis on Manifolds. Progress in Mathematics, vol. 308 (Birkhäuser/Springer, Cham, 2015), pp. 373–382
36. J. Hurtubise, Monopoles and rational maps: a note on a theorem of Donaldson'. Comm. Math. Phys. **100**, 191–196 (1985)
37. J. Hurtubise, The classification of monopoles for the classical groups. Comm. Math. Phys. **120**, 613–641 (1989)
38. J. Hurtubise, M.K. Murray, On the construction of monopoles for the classical groups. Comm. Math. Phys. **122**, 35–89 (1989)
39. J. Hurtubise, M.K. Murray, Monopoles and their spectral data. Comm. Math. Phys. **133**, 487–508 (1990)
40. D. Huybrechts, M. Lehn, *The Geometry of Moduli Spaces of Sheaves*, 2nd edn. Cambridge Mathematical Library (Cambridge University Press, Cambridge, 2010), xviii+325 pp.
41. A. Jaffe, C. Taubes, *Vortices and Monopoles*. Structure of Static Gauge Theories, Progress in Physics, vol. 2 (Birkhäuser, Boston, 1980)
42. S. Jarvis, Euclidean monopoles and rational maps. Proc. Lond. Math. Soc. **77**, 170–192 (1998)
43. S. Jarvis, Construction of Euclidean monopoles. Proc. Lond. Math. Soc. (3) **77**, 193–214 (1998)
44. A. Kapustin, E. Witten, Electric-magnetic duality and the geometric Langlands program. Commun. Number Theory Phys. **1**, 1–236 (2007)
45. S. Kobayashi, First Chern class and holomorphic tensor fields. Nagoya Math. J. **77**, 5–11 (1980)
46. S. Kobayashi, Curvature and stability of vector bundles. Proc. Japan Acad. Ser. A Math. Sci. **58**, 158–162 (1982)
47. S. Kobayashi (noted by I. Enoki), *Differential Geometry of Holomorphic Vector Bundles (in Japanese)*. Seminary Note in the University of Tokyo, vol. 41 (1982)

48. S. Kobayashi, *Differential Geometry of Complex Vector Bundles*. Publications of the Mathematical Society of Japan, vol. 15 (Princeton University Press, Princeton, 1987), xii+305 pp.
49. M. Kontsevich, Y. Soibelman, Riemann-Hilbert correspondence in dimension one, Fukaya categories and periodic monopoles. Preprint
50. T. Kotake, T. Ochiai (eds.) Non-linear problems in geometry, in *Proceedings of the Sixth International Symposium, Division of Mathematics*. The Taniguchi Foundation (1979)
51. P.B. Kronheimer, *Monopoles and Taub-NUT Metrics*. Master's thesis, Oxford (1986)
52. S. Lang, *Real Analysis*, 2nd edn. (Addison-Wesley Publishing Company, Advanced Book Program, Reading, 1983)
53. J.M. Lee, *Introduction to Smooth Manifolds*. Graduate Texts in Mathematics, vol 218 (Springer, New York, 2003)
54. J. Li, M.S. Narasimhan, Hermitian-Einstein metrics on parabolic stable bundles. Acta Math. Sin. (Engl. Ser.) **15**, 93–114 (1999)
55. M. Lübke, Chernklassen von Hermite-Einstein-Vektorbündeln. Math. Ann. **260**, , 133–141 (1982)
56. M. Lübke, Stability of Einstein-Hermitian vector bundles. Manuscripta Math. **42**, 245–257 (1983)
57. R. Maldonado, Periodic monopoles from spectral curves. J. High Energy Phys. (2), 099 (2013). Front matter +31 pp.
58. R. Maldonado, R.S. Ward, Dynamics of monopole walls. Phys. Lett. B **734**, 328–332 (2014)
59. S. Mizohata, *The Theory of Partial Differential Equations*. Translated from the Japanese by Katsumi Miyahara (Cambridge University Press, New York, 1973)
60. T. Mochizuki, Asymptotic behaviour of tame nilpotent harmonic bundles with trivial parabolic structure. J. Differ. Geom. **62**, 351–559 (2002)
61. T. Mochizuki, Kobayashi-Hitchin correspondence for tame harmonic bundles and an application. Astérisque **309**, viii+117 (2006)
62. T. Mochizuki, Asymptotic behaviour of tame harmonic bundles and an application to pure twistor D-modules I, II. Mem. AMS **185** (2007)
63. T. Mochizuki, Kobayashi-Hitchin correspondence for tame harmonic bundles. II. Geom. Topol. **13**, 359–455 (2009)
64. T. Mochizuki, Wild harmonic bundles and wild pure twistor D-modules. Astérisque, vol. 340. Société Mathématique de France, Paris (2011)
65. T. Mochizuki, Asymptotic behaviour and the Nahm transform of doubly periodic instantons with square integrable curvature. Geom. Topol. **18**, 2823–2949 (2014)
66. T. Mochizuki, Kobayashi-Hitchin correspondence for analytically stable bundles. Trans. Amer. Math. Soc. **373**, 551–596 (2020). arXiv:1712.08978
67. T. Mochizuki, Triply periodic monopoles and difference modules on elliptic curves. SIGMA **16**, 048 (2020), 23 p. arXiv:1903.03264
68. T. Mochizuki, Good wild harmonic bundles and good filtered Higgs bundles. SIGMA Symmetry Integrability Geom. Methods Appl. **17**, 66 (2021). Paper No. 068
69. T. Mochizuki, Notes on peirodic monopoles and Nahm transforms. In preparation
70. T. Mochizuki, Doubly periodic monopoles and q-difference modules. arXiv:1902.03551(version 1)
71. T. Mochizuki, M. Yoshino, Some characterizations of Dirac type singularity of monopoles. Comm. Math. Phys. **356**, 613–625 (2017). https://doi.org/10.1007/s00220-017-2981-z
72. M.S. Narasimhan, C.S. Seshadri, Stable and unitary vector bundles on a compact Riemann surface. Ann. of Math. (2) **82**, 540–567 (1965)
73. L. Ni, Y. Shi, L.-F. Tam, Poisson equation, Poincaré-Lelong equation and curvature decay on complete Kähler manifolds. J. Differ. Geom. **57**, 339–388 (2001)
74. P. Norbury, Magnetic monopoles on manifolds with boundary. Trans. Amer. Math. Soc. **363**, 1287–1309 (2011)
75. M. Pauly, Monopole moduli spaces for compact 3-manifolds. Math. Ann. **311**, 125–46 (1998)
76. C. Praagman, The formal classification of linear difference operators. Nederl. Akad. Wetensch. Indag. Math. **45**, 249–261 (1983)

77. J.-P. Ramis, J. Sauloy, C. Zhang, Local analytic classification of q-difference equations. Astérisque **355** (2013)
78. C. Sabbah, Harmonic metrics and connections with irregular singularities. Ann. Inst. Fourier (Grenoble) **49**, 1265–1291 (1999)
79. C.T. Simpson, Constructing variations of Hodge structure using Yang-Mills theory and applications to uniformization. J. Amer. Math. Soc. **1**, 867–918 (1988)
80. C.T. Simpson, Harmonic bundles on noncompact curves. J. Amer. Math. Soc. **3**, 713–770 (1990)
81. C.T. Simpson, Higgs bundles and local systems. Publ. I.H.E.S. **75**, 5–95 (1992)
82. C. Simpson, The Hodge filtration on nonabelian cohomology, in *Proceedings of Symposia in Pure Mathematics*, vol. 62, Part 2 (American Mathematical Society, Providence, 1997), pp. 217–281
83. C.T. Simpson, Mixed twistor structures. math.AG/9705006
84. Y.T. Siu, *Techniques of Extension of Analytic Objects*. Lecture Notes in Pure and Applied Mathematics (Marcel Dekker, Inc., New York, 1974)
85. Y. Takayama, Nahm's equations, quiver varieties and parabolic sheaves. Publ. Res. Inst. Math. Sci. **52**, 1–41 (2016)
86. H.L. Turrittin, The formal theory of systems of irregular homogeneous linear difference and differential equations. Bol. Soc. Mat. Mexicana (2) **5**, 255–264 (1960)
87. K. Uhlenbeck, Removable singularities in Yang-Mills fields. Comm. Math. Phys. **83**, 11–29 (1982)
88. K. Uhlenbeck, Connections with L^p bounds on curvature. Comm. Math. Phys. **83**, 31–42 (1982)
89. K. Uhlenbeck, S.-T. Yau, On the existence of Hermitian-Yang-Mills connections in stable vector bundles. Comm. Pure Appl. Math. **39**, No. S, suppl. (1986), S257–S293
90. M. van der Put, M. Reversat, Galois theory of q-difference equations. Ann. Fac. Sci. Toulouse Math. (6) **16**, 665–718 (2007)

Index

T. Mochizuki, *Periodic Monopoles and Difference Modules*, Lecture Notes in Mathematics 2300, https://doi.org/10.1007/978-3-030-94500-8

T

U

V

W

LECTURE NOTES IN MATHEMATICS Springer

Editorial Policy

1. Lecture Notes aim to report new developments in all areas of mathematics and their applications – quickly, informally and at a high level. Mathematical texts analysing new developments in modelling and numerical simulation are welcome.

   Manuscripts should be reasonably self-contained and rounded off. Thus they may, and often will, present not only results of the author but also related work by other people. They may be based on specialised lecture courses. Furthermore, the manuscripts should provide sufficient motivation, examples and applications. This clearly distinguishes Lecture Notes from journal articles or technical reports which normally are very concise. Articles intended for a journal but too long to be accepted by most journals, usually do not have this "lecture notes" character. For similar reasons it is unusual for doctoral theses to be accepted for the Lecture Notes series, though habilitation theses may be appropriate.

2. Besides monographs, multi-author manuscripts resulting from SUMMER SCHOOLS or similar INTENSIVE COURSES are welcome, provided their objective was held to present an active mathematical topic to an audience at the beginning or intermediate graduate level (a list of participants should be provided).

   The resulting manuscript should not be just a collection of course notes, but should require advance planning and coordination among the main lecturers. The subject matter should dictate the structure of the book. This structure should be motivated and explained in a scientific introduction, and the notation, references, index and formulation of results should be, if possible, unified by the editors. Each contribution should have an abstract and an introduction referring to the other contributions. In other words, more preparatory work must go into a multi-authored volume than simply assembling a disparate collection of papers, communicated at the event.

3. Manuscripts should be submitted either online at www.editorialmanager.com/lnm to Springer's mathematics editorial in Heidelberg, or electronically to one of the series editors. Authors should be aware that incomplete or insufficiently close-to-final manuscripts almost always result in longer refereeing times and nevertheless unclear referees' recommendations, making further refereeing of a final draft necessary. The strict minimum amount of material that will be considered should include a detailed outline describing the planned contents of each chapter, a bibliography and several sample chapters. Parallel submission of a manuscript to another publisher while under consideration for LNM is not acceptable and can lead to rejection.

4. In general, **monographs** will be sent out to at least 2 external referees for evaluation.

   A final decision to publish can be made only on the basis of the complete manuscript, however a refereeing process leading to a preliminary decision can be based on a pre-final or incomplete manuscript.

   Volume Editors of **multi-author works** are expected to arrange for the refereeing, to the usual scientific standards, of the individual contributions. If the resulting reports can be forwarded to the LNM Editorial Board, this is very helpful. If no reports are forwarded or if other questions remain unclear in respect of homogeneity etc, the series editors may wish to consult external referees for an overall evaluation of the volume.

5. Manuscripts should in general be submitted in English. Final manuscripts should contain at least 100 pages of mathematical text and should always include

   – a table of contents;
   – an informative introduction, with adequate motivation and perhaps some historical remarks: it should be accessible to a reader not intimately familiar with the topic treated;
   – a subject index: as a rule this is genuinely helpful for the reader.
   – For evaluation purposes, manuscripts should be submitted as pdf files.

6. Careful preparation of the manuscripts will help keep production time short besides ensuring satisfactory appearance of the finished book in print and online. After acceptance of the manuscript authors will be asked to prepare the final LaTeX source files (see LaTeX templates online: https://www.springer.com/gb/authors-editors/book-authors-editors/manuscriptpreparation/5636) plus the corresponding pdf- or zipped ps-file. The LaTeX source files are essential for producing the full-text online version of the book, see http://link.springer.com/bookseries/304 for the existing online volumes of LNM). The technical production of a Lecture Notes volume takes approximately 12 weeks. Additional instructions, if necessary, are available on request from lnm@springer.com.

7. Authors receive a total of 30 free copies of their volume and free access to their book on SpringerLink, but no royalties. They are entitled to a discount of 33.3 % on the price of Springer books purchased for their personal use, if ordering directly from Springer.

8. Commitment to publish is made by a *Publishing Agreement*; contributing authors of multiauthor books are requested to sign a *Consent to Publish form*. Springer-Verlag registers the copyright for each volume. Authors are free to reuse material contained in their LNM volumes in later publications: a brief written (or e-mail) request for formal permission is sufficient.

Addresses:

Professor Jean-Michel Morel, CMLA, École Normale Supérieure de Cachan, France
E-mail: moreljeanmichel@gmail.com

Professor Bernard Teissier, Equipe Géométrie et Dynamique,
Institut de Mathématiques de Jussieu – Paris Rive Gauche, Paris, France
E-mail: bernard.teissier@imj-prg.fr

Springer: Ute McCrory, Mathematics, Heidelberg, Germany,
E-mail: lnm@springer.com

United States
r Publisher Services